A SURVEY OF COMBINATORIAL THEORY

A Survey of Combinatorial Theory

Edited by

JAGDISH N. SRIVASTAVA

Colorado State University
Fort Collins, Colo., U.S.A.

with the cooperation of

FRANK HARARY

University of Michigan
Ann Arbor, Mich., U.S.A.

C. R. RAO

Indian Statistical Institute
Calcutta, India

G.-C. ROTA

Massachusetts Institute of Technology
Cambridge, Mass., U.S.A.

S. S. SHRIKHANDE

University of Bombay
Bombay, India

1973

NORTH-HOLLAND PUBLISHING COMPANY – AMSTERDAM · LONDON
AMERICAN ELSEVIER PUBLISHING COMPANY, INC. – NEW YORK

Library of Congress Catalog Card Number: 72–88578
North-Holland ISBN: 0 7204 22620
American Elsevier ISBN: 0 444 10425 9

Publishers:
NORTH-HOLLAND PUBLISHING COMPANY – AMSTERDAM
NORTH-HOLLAND PUBLISHING COMPANY, LTD. – LONDON

Sole Distributors for the U.S.A. and Canada:
AMERICAN ELSEVIER PUBLISHING COMPANY, INC.
52 VANDERBILT AVENUE
NEW YORK, N.Y. 10017

PRINTED IN SCOTLAND

Professor R. C. Bose

Dedicated to Professor R. C. Bose
on the happy occasion of
his seventieth birthday

Preface

An International Symposium on Combinatorial Mathematics and its Applications was held at Colorado State University (CSU), Fort Collins, Colorado on September 9–11, 1971. Part of the present volume contains the proceedings of the Symposium. Apart from the Symposium proceedings, this volume also contains articles by many other distinguished persons. The volume is dedicated in honor of Professor R. C. Bose, on the occasion of his seventieth birthday (June 29, 1971). We wish him many happy returns.

Although every reader would probably be familiar with at least a few aspects of Professor Bose's work, there are probably only a few people who are familiar with all the various aspects of his work in statistics and mathematics. It may therefore be useful to summarize some of his major activities. Professor Bose started out in 1926 in the field of geometry, particularly non-Euclidean and differential geometry. In the early thirties, he was moved by Professor Mahalanobis to the newly developing subject of statistics. Between 1933 and 1939, Professor Bose published both in mathematics and statistics. Several important results in multivariate statistical analysis were obtained, particularly on the D^2 statistic. From about 1938 onwards, Professor Bose got interested in the statistical design of experiments, particularly combinatorial aspects of the same. Important achievements in this area were made, like the work on the balanced incomplete block designs, partially balanced incomplete block designs, orthogonal Latin squares, factorial designs, and linear estimation. Work in all those fields continued after Professor Bose's arrival in the United States in 1949. In 1959, came (joint with Shrikhande), the famous disproof of Euler's conjecture, a problem which had confronted mathematicians for 177 years. About this time, Professor Bose became interested in Coding Theory, in which area important developments (like the now well-known BCH Codes) were made. During the past decade, Professor Bose has continued working in coding, design, and graph theory, and also in finite geometries which (as he showed) connects these areas.

The above is only the barest mention of his work, as is evident from his list of publications. Moreover, new advances are still to be expected from him for he is as active as ever.

It might not be out of place to summarize briefly the activities in connection with the preparation of this "Bose Volume". The idea started with Srivastava, particularly in a conversation with P. K. Sen in April 1969. Later that year, various people were contacted, including Shrikhande and Rao. With the offer of the latter two to cooperate, the project went off the ground in

January 1970. It was decided, for the sake of homogeneity, to limit the volume to Combinatorial Mathematics, and to invite Rota and Harary to join the Editorial Board.

It turned out that just about that time, the idea of having a Symposium in honor of Professor Bose had occurred to Harary. This was a happy coincidence. The Symposium idea was taken up by Srivastava, who discussed the matter with Colonel William Trott of the Air Force Office of Scientific Research (AFOSR). Great credit is due to Trott for his role in arranging the financing of this Symposium by the AFOSR.

The Organizing Committee for the Symposium was established, which consisted of the present Editorial Board members plus Trott who represented the Air Force. Though Srivastava and Harary were formally the Organizer and the Co-Organizer for the Symposium, close contact was maintained between the committee members by various means, including their visits to Fort Collins.

The success of the Symposium is due not only to the Organizing Committee, but also to many other people and groups, and our thanks go to all of them. Firstly, we wish to thank Dr. J. P. Jordan, Associate Dean of Natural Sciences, CSU, who helped this project in many ways. Our thanks are also due to Colonel N. P. Callas, who took over as Project Director after Colonel Trott left the AFOSR. We would also like to thank Mr. James McIver, Director, Department of Conferences, Colorado State University, for the excellent arrangements made by them. Our thanks are also due to Mr. James Brown, Office of Contracts and Grants for the many pains he took to see things run smoothly. We would also like to acknowledge the excellent work done by the secretaries of the CSU Statistics and Mathematics Departments, particularly Susan Degering, Cathy Bishop and Barbara Fosberg.

Thanks are due to many of the contributors to the volume for reading the papers of other contributors. In connection with the reading of papers, we appreciate the cooperation we received from Drs. Bruen, Levinger, Manvel, Orvedahl, and Williams, all of CSU. We would like to express our gratitude to President A. R. Chamberlain of CSU, Colonel P. Daily of the AFOSR, and Dr. J. P. Jordan for their addresses during the Inauguration of the Symposium. Also, we are grateful to the various speakers at the Symposium and the other contributors to the Bose Volume for their valuable articles, and the Chairmen for the various Symposium sessions. Last, but not the least, our sincere thanks go to all the participants in the Symposium for it was their participation which truly made it a success.

F. HARARY, *Co-organizer*
C. R. RAO
G. C. ROTA
S. S. SHRIKHANDE
J. N. SRIVASTAVA, *Organizer*

Contents

List of Publications by R. C. Bose

1. (With S. Mukhopadhyaya) General theorem of cointimacy of symmetries of a hyperbolic triad, *Bull. Cal. Math. Soc.* **17** (1926), 39–54.
2. New methods in Euclidean geometry of four dimensions, *Bull. Cal. Math. Soc.* **17** (1926), 105–140.
3. (With S. Mukhopadhyaya) Triadic equations in hyperbolic geometry, *Bull. Cal. Math. Soc.* **18** (1927), 99–110.
4. The theory of associated figures in hyperbolic geometry, *Bull. Cal. Math. Soc.* **19** (1928), 101–116.
5. Theorems in the synthetic geometry of the circle on the hyperbolic plane, *Tohoku Math. J.* (Japan) **34** (1931), 42–50.
6. On a new derivation of the fundamental formulas of hyperbolic geometry, *Tohoku Math. J.* (Japan) **34** (1931), 291–294.
7. Synthetic relations between any three elements of a right-angled triangle on the hyperbolic plane, *J. Ind. Math. Soc.* **19** (1931), 126–129.
8. Generalizations of Roeser's correspondence between certain types of polyhydra in non-Euclidean space, *Math. Phys. J.* **3** (1932), 43–51.
9. (With W. Blaschke) Quadrilateral 4-webs of curves in a plane, *Math. Phys. J.* **3** (1932), 99–101.
10. Correspondence between a tetrahedron and a special type of heptahedron in hyperbolic space, *Math. Phys. J.* **3** (1932), 133–137.
11. On the number of circles of curvature perfectly enclosing or perfectly enclosed by a closed convex oval, *Math. Z.* (Leipzig) **35** (1932), 16–24.
12. Functional equations satisfied by the fundamental functions of hyperbolic geometry, and their application to the geometry of the circle, *Math. Phys. J.* **4** (1933), 37–41.
13. On the application of hyperspace geometry to the theory of multiple correlation, *Sankhya*, **1** 338–342.
14. A note on the convex oval, *Bull. Cal. Math. Soc.* **26** (1935), 55–60.
15. A theorem on the non-Euclidean triangle, *Bull. Cal. Math. Soc.* **26** (1935), 69–72.
16. (With S. N. Roy) Some properties of the convex oval with reference to its perimeter centroid, *Bull. Cal. Math. Soc.* **26** (1935), 79–86.
17. (With S. N. Roy) A note on the area centroid of a closed convex oval, *Bull. Cal. Math. Soc.* **26** (1935), 111–118.
18. (With S. N. Roy) On the four centroids of a closed convex surface, *Bull. Cal. Math. Soc.* **26** (1935), 119–147.
19. (With S. N. Roy) On the evaluation of the probability integral of the D^2 statistic, *Sci. Cult.* (Calcutta) **1** (1935), 436–437.
20. On the exact distribution and moment-coefficients of the D^2-statistic, *Sankhya*, **2** (1936), 143–154.
21. Theory of skew rectangular pentagons of hyperbolic space. Derivation of the set of associated pentagons, *Bull. Cal. Math. Soc.* **28** (1936), 159–186.

22. A note on the distribution of the differences in mean values of samples drawn from two multivariate normally distributed populations and the definition of the D^2-statistic, *Sankhya* **2** (1936) 379–384.

23. Two theorems on the convex oval, *J. Ind. Math. Soc.* (New Series) **2** (1936), 13–15.

24. A theorem on equiangular convex polygons circumscribing a convex curve, *J. Ind. Math. Soc.* (New Series) **2** (1936), 96–98.

25. Analogue of a theorem of Blaschke, *J. Ind. Math. Soc.* (New Series) **2** 105–106.

26. (With P. C. Mahalanobis and S. N. Roy) Normalization of variates and the use of rectangular coordinates in the theory of sampling distributions, *Sankhya* **3** (1936), 1–40.

27. On a criterion for the existence of a cyclic point, *Tohoku Math. J.* (Japan) **43** (1936), 84–88.

28. A note on the osculating circles of a plane curve, *Bull. Cal. Math. Soc.* **29** (1937), 29–32.

29. On the distribution of means of samples drawn from a Bessel functions population, *Sankhya* **3** (1938), 262–266.

30. On the application of the properties of Galois fields to the problem of construction of hyper Graeco-Latin squares, *Sankhya* **3** (1938), 323–339.

31. (With S. N. Roy) Distribution of the studentized D^2-statistic, *Sankhya* **4** (1938), 19–38.

32. (With K. R. Nair) Partially balanced incomplete block designs, *Sankhya* **4** (1939), 337–373.

33. (With K. Kishen) On partially balanced Youden squares, *Sci. Cult.* (Calcutta) **4** (1939), 136–137.

34. On the construction of balanced incomplete block designs, *Ann. Eugen.* (London) **9** (1939), 358–398.

35. (With S. N. Roy) The use and distribution of the studentized D^2-statistic when the variances and covariances are based on k samples, *Sankhya* **4** (1940), 535–542.

36. (With K. Kishen) On the problem of confounding in the general symmetrical factorial design, *Sankhya* **5** (1940), 21–36.

37. (With K. R. Nair) On complete sets of Latin squares, *Sankhya* **5** (1941), 361–382.

38. Some new series of balanced incomplete block designs, *Bull. Cal. Math. Soc.* **34** (1942), 17–31.

39. An affine analogue of Singer's theorem, *J. Ind. Math. Soc.* (New Series) **6** (1942), 1–15.

40. A note on two combinatorial problems having applications in the theory of design of experiments, *Sci. Cult.* (Calcutta) **8** (1942), 192–193.

41. A note on the resolvability of balanced incomplete block designs, *Sankhya* **6** (1942), 105–110.

42. A note on two series of balanced incomplete block designs, *Bull. Cal. Math. Soc.* **35** (1943), 129–130.

43. The fundamental theorem of linear estimation, *Proc. Indian Sci. Congr.* (1944), 4–5.

44. (With S. Chowla and C. R. Rao) On the integral order (mod p) of quadratics x^2+ax+b, with applications to the construction of minimum functions for $GF(p^2)$, and to some number theory results, *Bull. Cal. Math. Soc.* **36** (1944), 153–174.

45. (With S. Chowla) On the construction of affine difference sets, *Bull. Cal. Math. Soc.* **37** (1945), 107–112.
46. Mathematical theory of the Symmetrical factorial design, *Sankhya* **8** (1947), 107–166.
47. On a resolvable series of balanced incomplete block designs, *Sankhya* **8** (1947), 249–256.
48. Recent work on "Incomplete Block Designs" in India, *Biometrics* **3** (1947), 176–178.
49. The design of experiments. Presidential Address, Section of Statistics, *Proc. 34th Indian Sci. Congr.* (1947).
50. A note on Fisher's inequality for balanced incomplete block designs, *Ann. Math. Statist.* **20** (1949), 619–620.
51. On a problem of two-dimensional probability, *Sankhya* **10** (1950), 13–23.
52. Mathematics of factorial designs, *Proc. Intern. Congr. Math., Cambridge, Massachusetts* (1950), 543–548.
53. Review of 'some theory of sampling' by William Edwards Deming, *Intern. J. Opinion and Attitude Res.* **5** (1951), 424–425.
54. Partially balanced incomplete block designs with two associate classes involving only two replications, *Calcutta Statist. Ass. Bull.* **3** (1951), 120–125.
55. (With T. Shimamoto) Classification and analysis of partially balanced incomplete block designs with two associate classes, *J. Am. Statist. Ass.* **47** (1952), 151–184. (Inst. of Statist. Reprint Series No. 20.)
56. (With W. S. Connor) Combinatorial properties of group divisible incomplete block designs, *Ann. Math. Statist.* **23** (1952), 367–383. (Inst. of Statist. Reprint Series No. 25.)
57. (With K. A. Bush) Orthogonal arrays of strength two and three, *Ann. Math. Statist.* **23** (1952), 508–524. (Inst. of Statist. Reprint Series No. 27).
58. A note of Nair's condition for partially incomplete block designs with $k > r$, *Calcutta Statist. Ass. Bull.* **4** (1952), 123–316. (Inst. of Statist. Reprint Series No. 29.)
59. (With S. S. Shrikhande and K. N. Bhattacharya) On the construction of group divisible incomplete block designs, *Ann. Math. Statist.* **24** (1953), 167–195. (Inst. of Statist. Reprint Series No. 37).
60. (With W. H. Clatworthy and S. S. Shrikhande) Tables of partially balanced designs with two associate classes, *N. Carolina Agr. Exp. Sta. Tech. Bull.* **107** (1954). (Inst. of Statist. Reprint Series No. 50.)
61. (With S. N. Roy) Simultaneous confidence interval estimation, *Ann. Math. Statist.* **24** (1953), 513–536. (Inst. of Statist. Reprint Series No. 43.)
62. (With W. H. Clatworthy) Some classes of partially balanced designs, *Ann. Math. Statist.* **26** (1955), 212–232.
63. Paired comparison designs for testing concordance between edges, *Biometrika* **43** (1956), 113–121.
64. (With Dale M. Mesner) On linear associative algebras corresponding to association schemes of partially balanced designs, *Ann. Math. Statist.* **30** (1959), 21–38.
65. (With S. S. Shrikhande) On the falsity of Euler's conjecture about the non-existence of two orthogonal Latin squares of order $4t+2$, *Proc. Natl. Acad. Sci. U.S.A.* **46** (1959), 734–737.
66. (With S. S. Shrikhande) A note on a result in the theory of code construction, *Inform. Control* **2** (1959), 183–194.

67. (With R. L. Carter) Complex representation in the construction of rotable designs, *Ann. Math. Statist.* **30** (1959), 771–780.

68. (With Shanti S. Gupta) Moments of order statistics from a normal population, *Biometrika* **46** (1959), 433–440.

69. (With Norman R. Draper) Second order rotatable designs in three dimensions, *Ann. Math. Statist.* **30** (1959), 1097–1112.

70. On the application of finite projective geometry for deriving a certain series of balanced Kirkman arrangements, *Calcutta Math. Soc. Golden Jubilee Commemoration Volume*, Part II (1958–59), 341–356.

71. (With Roy R. Kuebler) A geometry of binary sequences associated with group alphabets in information theory, *Ann. Math. Statist.* **31** (1960), 113–139.

72. (With D. K. Ray-Chaudhuri) On a class of error correcting binary group codes, *Inform. Control* **3** (1960), 68–79.

73. (With S. S. Shrikhande) On the construction of sets of mutually orthogonal Latin squares and the falsity of a conjecture of Euler, *Trans. Am. Math. Soc.* **95** (1960), 191–209.

74. (With S. S. Shrikhande and E. T. Parker) Further results on orthogonal Latin squares and the falsity of Euler's conjecture, *Canad. J. Math.* **12** (1960), 189–203.

75. (With S. S. Shrikhande) On the composition of balanced incomplete block designs, *Canad. J. Math.* **12** (1960), 177–188.

76. On a method of constructing Steiner triple systems, *Contributions to Probability and Statistics—Essays in Honor of Harold Hotelling* (Stanford Univ. Press, Stanford, Calif., 1960), 133–141.

77. (With W. S. Connor) Analysis of fractionally replicated $2^m 3^n$ designs, *Bull. Inst. Intern. Statist.* **37** (1960), 142–160.

78. (With D. K. Ray-Chaudhuri) Further results on error correcting group codes, *Inform. Control* **3** (1960), 279–290.

79. (With I. M. Chakravarti and D. E. Knuth) On methods of constructing sets of mutually orthogonal Latin squares using a computer, I, *Technometrics* **2** (1960), 507–510.

80. (With I. M. Chakravarti and D. E. Knuth) On methods of constructing sets of mutually orthogonal Latin squares using a computer, II, *Technometrics* **3** (1961), 111–117.

81. On some connections between the design of experiments and information theory, *Bull. Inst. Intern. Statist.* **38** (1961), 257–271.

82. (With R. J. Nelson) A sorting problem, *J. Ass. Comput. Mach.* **9** (1962), 282–296.

83. (With I. M. Chakravarti) A coding problem arising in the transmission of numerical data, *Bull. Inst. Intern. Statist.* **39** (1962), 345–355.

84. (With S. Chowla) Theorems in the additive theory of numbers, *Comment. Math. Helv.* **37** (1962), 141–147.

85. Some ternary error correcting codes and the corresponding fractionally replicated designs, *Colloq. Intern. Centre Natl. Rech. Sci.* (Paris) **110** (1963), 21–32.

86. (With S. S. Shrikhande and E. T. Parker) Orthogonal Latin squares and Euler's conjecture, *Colloq. Intern. Centre Natl. Rech. Sci.* (Paris) **110** (1963).

87. (With K. R. Nair) Resolvable incomplete block designs with two replications, *Sankhya* Ser. A **24** (1962), 9–24.

88. (With R. L. Carter) Response model coefficients and the individual degrees of freedom of a factorial design, *Biometrics* **18** (1962), 160–171.
89. Strongly regular graphs, partial geometries and partially balanced designs, *Pacific J. Math.* **13** (1963), 389–419.
90. A problem of additive number theory arising in the construction of error correcting codes, *Proceedings of the Number Theory Conference*, Univ. of Colorado, Boulder, Colo. (1963), 4–11.
91. Combinatorial properties of partially balanced designs and association schemes, *Contributions to Statistics*, Presented to Professor Mahalanobis on his seventieth Birthday, Vol. 2 (1964), 21–43; *Sankhya* Ser. A **25** (1963), 109–136.
92. (With J. N. Srivastava) On a bound useful in the theory of factorial designs and error correcting codes, *Ann. Math. Statist.* **35** (1964), 408–414.
93. (With J. N. Srivastava) Mathematical theory of factorial designs; I: Analysis, II: Construction, *Bull. Inst. Intern. Statist.* **40** (1964), 786–794.
94. (With R. H. Bruck) The construction of translation planes from projective spaces, *J. Algebra* **1** (1964) 1–18.
95. (With J. N. Srivastava) Analysis of irregular factorial fractions, *Sankhya* Ser. A **26** (1964), 117–144.
96. (With J. N. Srivastava) Multidimensional partially balanced designs, *Sankhya* Ser. A **26** (1964), 145–168.
97. Samarendra Nath Roy, *Am. Statist.* **18** (1964), 26–27.
98. Magic squares, *Encyclopaedia Britannica* **14** (1965), 573–575.
99. (With J. N. Cameron) The bridge tournament problem and calibration designs for comparing pairs of objects, *J. Res. Natl. Bur. Std.* (U.S.) **69B** (1965), 323–332.
100. (With J. N. Srivastava) Economic 2^n partially balanced factorial fractions, *Ann. Inst. Statist. Math.* **18** (1966), 57–73.
101. (With R. H. Bruck) Linear representations of projective planes in projective spaces, *J. Algebra* **4** (1966), 117–172.
102. (With R. C. Burton) A characterization of flat spaces in a finite projective space, and the uniqueness of Hamming and MacDonald codes, *J. Combin. Theory*, **1** (1966), 96–104.
103. Error detecting and error correcting indexing systems for large serial numbers, *Rev. Intern. Statist. Inst.* **34** (1966), 334–340.
104. (With I. M. Chakravarti) Hermitian varieties in a finite projective space $PG(N, q^2)$, *Canad. J. Math.* **18** (1966), 1161–1182.
105. (With J. G. Caldwell) Synchronizable error corrected codes, *Inform. Control* **10** (1967), 616–630.
106. (With J. M. Cameron) Calibration designs based on solutions to the tournament problem, *J. Res. Natl. Bur. Std.* **71B** (1967), 149–160.
107. (With Renu Laskar) A characterization of tetrahedral graphs, *J. Combin. Natl. Bur. Std.* (U.S.) *Theory* **3** (1967), 366–385.
108. (With S. P. Ghosh and C. T. Abraham). File organization of records with multiple valued attributed for multi attribute queries, *Proc. Conf. on Combinatorial Mathematics and its Applications*, Chapel Hill, April 1967 (Univ. of North Carolina Press, Chapel Hill, N. Car., 1969), 277–297.
109. (With Gary G. Koch) The design of combinatorial information retrieval systems for files with multiple-valued attributes, *Siam J. Appl. Math.* **17** (1969), 1203–1214.

110. Error correcting, error detecting and error locating codes, *Essays in Probability and Statistics*, A volume dedicated to the memory of Professor S. N. Roy (Univ. of North Carolina Press, Chapel Hill, N. Car., 1970), 147–178.
111. (With Renue Laskar) Eigenvalues of the adjacency matrix of tetrahedral graphs, *Aequationes Math.* **4** (1970), 37–43.
112. (With S. S. Shrikhande) Graphs in which each pair of vertices is adjacent to the same number of other vertices, *Stud. Sci. Math. Hung.* **5** (1970), 181–196.
113. Coordinatization of a class of projective planes which can be linearly represented in a projective space of four dimensions, *Proc. Second Chapel Hill Conf. on Combinatorial Mathematics and its Applications* (Univ. of North-Carolina Press, Chapel Hill, N. Car., 1970), 37–56.
114. (With A. Barlotti) Linear representation of a class of projective planes in four dimensional projective space, *Ann. Mat. Pura. Appl.* Ser. 4, **88** (1971), 9–32.
115. Self-conjugate tetrahedra with respect to the Hermitian variety $x_0^3 + x_1^3 + x_2^3 + x_3^3 = 0$ in PG(3, 2^2) and a representation of PG(3, 3), *Proc. Symp. on Pure Math.* **19** (Am. Math. Soc., Providence, R.I., 1971), 27–37.
116. (With K. J. C. Smith) Ternary rings of a class of linearly representable semi-translation planes, *Proc. Conf. on Combinatorial Geometry and its Applications*, Perugia (Tipografia Oderisi, Gubbio, Perugia, 1971), 69–101.
117. (With R. A. Dowling) A generalization of More graphs of diameter two, *J. Combin. Theory* **11** (1971), 213–226.
118. (With S. S. Shrikhande) Some further constructions of $G_2(d)$ graphs, *Stud. Sci. Math. Hung.* **6** (1971), 127–132.
119. (With S. S. Shrikhande) Partial geometries and pseudo-geometric graphs $(q^2+1, q+1, 1)$, *J. Geometry*, to appear.
120. Some recent advances in design of experiments, *Bull. Inst. Intern. Statist.* **44** (1971), 419–431.
121. (With S. S. Shrikhande) Embedding the complement of an oval in a projective plane of even order, submitted to *Discrete Math.*
122. Characterization problems of combinatorial graph theory, *A Survey of Combinatorial Theory* (J. N. Srivastava *et al.*, eds.; North-Holland, Amsterdam, 1972), 31–51 (this volume).
123. Combinatorics and combinatorial theory. To appear in the new edition of the *Encyclopaedia Britannica*.

J. N. Srivastava et al., eds., *A Survey of Combinatorial Theory*
© North-Holland Publishing Company, 1973

CHAPTER 1

Some Classical and Modern Topics in Finite Geometrical Structures

A Brief Survey

A. BARLOTTI

Università di Perugia, Perugia, Italy

1. Introduction

In recent decades one of the fields in which Mathematics has developed in a very remarkable way is that of finite geometrical structures. A fairly complete account of this topic is given in the recent book "Finite geometries" by P. Dembowski. A rapid glance at the bibliography in this volume will show the tumultuous way in which these theories have developed, especially in the last few years.

This article will offer only a very short survey of some classical questions and draw attention to some geometrical structures, the study of which has developed only recently and appears to be interesting not only in itself but also for its applications to other fields. Far from giving a thorough account of even the main subjects we shall merely sketch some typical situations of this branch of Mathematics, "characterized by an interplay of combinatorial, geometric, and algebraic ideas" (P. Dembowski [1968], p. v).†

We are glad to dedicate this work to Professor R. C. Bose, who cultivated this field of Geometry with an intense enthusiasm obtaining many results, some of which may already be regarded as classical. He also has the high merit of being a friendly and able teacher who has been successful in transmitting his interest in, and love of science to many of his students and associates.

2. Finite projective and affine planes

2.1. Combinatorial structures related to finite planes

In this section we recall some classical results, which lately have had some interesting developments, and which may be useful in solving some important problems.

† The bibliography will refer mainly to books and general papers.

1

The following theorem states the equivalence of finite affine planes and complete sets of mutually orthogonal Latin squares, and hence is a key connecting geometry and combinatorics:

Theorem 2.1.1 (R. C. Bose [1938]): *Let $n > 2$. From any finite affine plane of order n we may construct a complete set of $n-1$ mutually orthogonal Latin squares, and vice versa.*

The same result was obtained by N. L. Stevens [1939] for Desarguesian planes only.

Properties of the set of orthogonal Latin squares may be translated into properties of the corresponding planes (see R. C. Bose and K. R. Nair [1941], and G. Pickert [1955], p. 292). There is a possibility that this topic may be developed still further leading to other notable results.

The notions of a Latin hypercube and of a set of mutually orthogonal Latin hypercubes have also been developed, and their study in connection with finite geometric structures has been initiated (see C. R. Rao [1946] and P. Hohler [1970]).

From the fact that the Galois field $GF(p^n)$ can be regarded as a vector space over $GF(p)$, interesting results in finite geometries can be obtained in a simple and elegant way.

Among these, the following theorem, which can be used to give a compact combinatorial representation of a finite plane $PG(2, p^s)$ is of great importance.

Theorem 2.1.2 (J. Singer [1938]). *Given an integer $n \geqslant 2$ of the form p^s (p being a prime) we can find $n+1$ integers*

$$(1) \qquad\qquad d_0, d_1, \ldots, d_n,$$

such that among the $n(n+1)$ differences $d_i - d_j$ $(i, j = 0, 1, \ldots, n; \ i \neq j)$ reduced modulo $n^2 + n + 1$, each of the integers $1, 2, 3, \ldots, n^2 + n$ occurs exactly once.

The points and lines of $PG(2, p^s)$ are given in the following way:

(i) the $n^2 + n + 1$ points of the plane, by the residue classes modulo $n^2 + n + 1$;

(ii) the set of $n+1$ points of a line, by the points corresponding to the integers:

$$d_0 + t, d_1 + t, \ldots, d_n + t, \qquad 0 \leqslant t \leqslant n^2 + n + 1.$$

The affine analogue of Singer's theorem is the following one:

Theorem 2.1.3 (R. C. Bose [1942]). *Given an integer $n \geqslant 2$ of the form p^s (p being a prime) we can find n integers*

$$d_1 = 1, d_2, \ldots, d_n,$$

such that among the $n(n-1)$ differences $d_i - d_j$ $(i, j, = 1, 2, \ldots, n; \ i \neq j)$ reduced modulo $n^2 - 1$, all the positive integers less than $n^2 - 1$ and not divisible by $n+1$ occur exactly once.

The set of integers (1) is called a perfect difference set of order $n+1$. The study of sets of this kind and their generalizations give rise to one of the most interesting chapters of combinatorial mathematics (for the bibliography see, e.g., H. J. Ryser [1963]).

Related to theorems 2.1.2 and 2.1.3 are remarkable properties of the group of collineations of Desarguesian planes (see P. Dembowski [1968], Section 4.4).

2.2. Classifications of projective planes

The classification of projective planes is useful for several reasons: (i) to bring some order to the large number of planes which have already been discovered, (ii) to suggest the study of some new characteristic properties of the various types of planes, (iii) to enable the discovery of new kinds of planes which would provide examples of planes belonging to some *a priori* possible class.

2.2.1. The classification based on (C, a)-transitivities

A projective plane Π is said to be (C, a)-transitive (R. Baer [1942]) if it contains a point C and a line a such that for each pair of points P, P' with $P, P' \notin a$, $P, P' \neq C$ and $CP = CP'$ there is a (C, a)-perspectivity (i.e., a perspectivity with center C and axis a) which takes P into P'.[†]

Projective planes can be classified according to the set of (C, a)-transitivities which they have. The point-line pairs (C, a) for which a plane is (C, a)-transitive form a figure $F = \{(C, a): C \text{ a point, } a \text{ a line, } \pi \text{ is } (C, a)\text{-transitive}\}$. Lenz studied the subset $\{(C, a) \in F: C \in a\}$. Later, Barlotti gave a refinement of the classification obtained by Lenz by removing the restriction $C \in a$. Both authors obtained upper bounds for the number of various types of possible figures. Results of many other authors make it possible to rule out some of these types, and we can state the following theorem (which refers to both finite and infinite planes):

Theorem. *Let F be the figure formed by all the point-line pairs (C, a) for which a projective plane is (C, a)-transitive. Then the following exhaust the possible cases:*

Class I

(I.1) F is the empty set.

(I.2) F contains a single pair (C, a), with $C \notin a$.

(I.3) F contains two pairs (C_i, a_i), $i = 1, 2$, with $C_i \in a_j$ if and only if $i \neq j$.

(I.4) F contains three pairs (C_i, a_i), $i = 1, 2, 3$, with $C_i \in a_j$ if and only if $i \neq j$.

[†] It is well known that the existence of (C, a)-transitivities in Π is closely related to certain properties of a ternary ring which is suitably chosen in Π.

(I.5) The plane contains an incident point-line pair (R, r) and a one-to-one mapping ϕ of the points of $r - \{R\}$ onto the lines passing through R and different from r. F contains all the pairs $(C, \phi(C))$ with $C \in r - \{R\}$. (The existence of planes of this type is still undetermined.)

Class II

(II.1) F contains a single pair (C, a), with $C \in a$.

(II.2) F contains two pairs $(C_i, a_i), i = 1, 2$, with $C_i \in a_1, i = 1, 2, C_1 \in a_2$, $C_2 \notin a_2$.

Class III

(III.1) There exist in the plane a line r and a point $R \notin r$ such that the figure F contains all the pairs (C, CR) for all $C \in r$.

(III.2) F contains all the pairs of the figure of type (III.1) together with the single additional pair (R, r).

Class IV_a

(IV_a.1) There exists a line r in the plane such that F contains all the pairs (C, r) for all $C \in r$.

(IV_a.2) There exist in the plane a line r and two points A and B with $A, B \in r$, such that F contains all the following pairs:

$$(C, r) \quad \text{for all } C \in r.$$
$$(A, b) \quad \text{for all } b \ni B.$$
$$(B, a) \quad \text{for all } a \ni A.$$

(IV_a.3) There exist in the plane a line r and an involutory permutation ϕ of the points of r, which fixes no point of r, such that F contains all the pairs (C, a) with $C \in r, a \ni \phi(C)$.

Class IV_b

Types (IV_b.1–3) are the duals to (IV_a.1–3).

Class V

(V.1) There exist in the plane a line r and a point $R \in r$, such that F contains all the following pairs:

$$(C, r) \quad \text{with } C \in r.$$
$$(R, a) \quad \text{with } a \ni R.$$

Class VII

(VII.1) F contains all the pairs (C, a) with $C \in a$.

(VII.2) F contains all the point-line pairs.

The analogous study of all the possible types of F^* where, for any collineation group Γ of Π, F^* is defined to be the figure $\{(C, a): C$ a point, a a line, Γ is (C, a)-transitive} has been carried out by Dembowski [1968, pp. 123–125].

The problem of the existence of planes of different types has been studied by many authors and is not yet completely solved. (See A. Barlotti [1969], P. Dembowski [1968] and J. C. D. S. Yaqub [1967].) Recently, C. Hering and W. K. Kantor have succeeded in proving the non-existence of finite planes of type III.

It is also of interest to see how the question of existence for the different classes change when some further condition is imposed (see, e.g., T. V. S. Jagannathan [1970]).

2.2.2. The classification based on (C, a, μ)-homogeneities

A projective plane is said to be (C, a, μ)-homogeneous if:

(i) it is (C, a)-transitive;

(ii) there is a correlation τ whose square is a (C, a)-perspectivity and μ is the mapping of the lines through C onto the points of a induced by τ.

This notion was used by W. Jonsson [1963, 1965] to give a refinement of the classification described in 2.2.1.

2.2.3. Translation planes

The classification based on (C, a)-transitivities is far from providing a complete order for the family of projective planes. For instance, the translation planes (types IV to VII) are, even in the finite case, a very wide class, and there is definitely a need for a further classification of them. This can be done in different ways:

(i) Since the finite translation planes can be represented in higher-dimensional space by using the notion of spread (J. André [1954]; R. H. Bruck and R. C. Bose [1964, 1966]; B. Segre [1964]), a classification can be given by classifying the spreads (R. H. Bruck [1969]).

(ii) A detailed study of finite translation planes has been made by T. G. Ostrom, and some of the notions which have been introduced by him (e.g., the notion of Desarguesian decomposition) may lead to useful classifications (T. G. Ostrom [1970, 1971]).

(iii) Attacking the problem from an analytical point of view, by attempting to classify quasifields. There is a possibility that this may be done developing the research by G. Menichetti [1968, 1969, 1970] who studied quasifields in connection with other algebraic structures or by using geometric tools.

2.2.4. Other special classifications

A classification of semi-translation planes has been given by N. L. Johnson [1969a, b].

Also the notion of k-closure (A. Barlotti [1964]) leads to a classification for a special class of infinite planes.

We would also like to note explicitly that some relations existing between two classes of planes may be used to obtain a classification of one class from a classification of the other.

2.3. Construction of planes

2.3.1. Many techniques have been developed to construct finite projective planes. Excellent accounts can be found in P. Dembowski [1968], chapter V, and in T. G. Ostrom [1970].

Here we want to point out just a few papers which have appeared recently:

(i) A generalization of Moulton's technique has been given by C. Petit [1969]: the lines of a plane are divided in "pieces" and a procedure is defined (through the ternary field) to arrange these together in a different way in order to obtain a new plane.

(ii) In a paper by R. C. Bose and A. Barlotti [1971] (see also R. C. Bose [1970] and R. C. Bose and K. J. C. Smith [1971]), a linear representation is obtained for a class of non-Desarguesian planes which are derived from the dual translation planes. Consequently, a further step has been taken towards answering the question (raised in R. H. Bruck and R. C. Bose [1966]): "can every (non-Desarguesian) projective plane be embedded (in some natural geometric fashion) in a (Desarguesian) projective space?"

(iii) In connection with the same question, a representation of free planes in projective spaces has been given by V. Chval [1970].

2.3.2. All the finite non-Desarguesian projective planes which have been constructed so far have as order the power of a prime, p^h ($h > 1$).

Whether or not there are planes whose order is not the power of a prime is as yet an unsolved mystery. It has to be noticed, in this connection, that no further result whatsoever on non-existence for finite planes of specified order has been found subsequent to the following theorem (R. H. Bruck and H. J. Ryser [1949]):

Theorem 2.3.2. *If* $n \equiv 1$ *or* 2 (mod 4) *and if* n *is not the sum of two squares, then there exists no projective plane of order* n.

Other unsolved problems are (i) whether or not there exist non-Desarguesian planes of prime order, (ii) whether or not there exist non-Desarguesian planes which can be represented by a perfect difference set.

By means of theorem 2.1.1, statements of existence or non-existence of planes of order n can be expressed in the language of combinatorial mathematics by considering the maximum number of mutually orthogonal Latin squares of order n. Thus, if the set of five mutually orthogonal Latin squares given by R. C. Bose, I. M. Chakravarti and D. E. Knuth [1960] could be completed to a set of eleven mutually orthogonal Latin squares, the existence of a projective plane of order 12 would be proved.

2.4. A final remark

There are many other topics of fundamental interest in the field of projective and affine planes which we cannot discuss for lack of space. Some examples are the problems related to the introduction of coordinates (see M. Hall [1959] and G. Pickert [1955]), the study of configurations (see A. Heyting [1963] and G. Pickert [1955]),† and the characterizations of planes of different types.

Although our exposition has been restricted to the finite planes, we sometimes also referred to the infinite case. We want to point out here explicitly that both finite and infinite cases sometimes interact with suggestions of procedures, or results, which may appear surprising. We mentioned (2.3.1(i)) for instance, that Moulton's construction, which is suggested by intuition of the real plane, can be adapted to the finite case. As an example of the opposite interaction we want to point out an interesting result given by R. P. Burn [1967]. This author proved that the fact that involutory elations and homologies cannot coexist in finite projective planes, can be extended to most of the classes of infinite projective planes. Precisely the possibility of coexistence is excluded for all classes except classes I.1 and I.2, for which the question is still open.

3. Other geometric structures

Finite projective and affine spaces are two particular types of designs. Due to reasons of space we shall not illustrate the geometric aspect of the theory of these structures in general (see, e.g., P. Dembowski [1968], chapters 2 and 7). For the same reason we do not consider many other interesting classes of incidence structures (such as "Hjelmslev planes",‡ "Generalized polygons", "Semi-planes", etc.) for which the reader is referred to P. Dembowski [1968], chapter 7.

3.1. Inversive planes

For the study of these structures see P. Dembowski [1968], chapter 6.

Here we want to point out that a generalization of these structures was presented by R. Permutti [1967, 1968]; see also W. Heise [1970].

3.2. Partial geometries

A wide class of geometric structures which has a strong connection with graph theory and with PBIB-designs is the class of partial geometries defined by R. C. Bose [1963a].

---

† Here we want to recall, among the many open questions, that no *geometrical* proof has yet been found of the well known property that a finite Desarguesian plane is also Pappian.

‡ See F. Bachmann [1971].

A partial geometry (r, k, t) is a system consisting of two undefined classes of objects called *points* and *lines*, together with a relation of *incidence*, in which the following axioms are satisfied:

A1. Any two distinct points are incident with not more than one line.

A2. Each point is incident with exactly r lines.

A3. Each line is incident with exactly k points.

A4. Given a point P not incident with a line l, there are exactly t lines ($t \geqslant 1$) which are incident with P, and intersect l.

The study of partial geometries has been carried out by R. C. Bose [1963a; see also 1963b] and by D. G. Higman [1971]. Some generalizations of this notion have been given by D. Ghinelli [1968].

3.3. Weak affine spaces

E. Sperner [1960], looking at the possibility of constructing geometric structures which are non-Desarguesian and, in some cases, can be considered of dimension greater than two, introduced the notion of weak affine space. The class of these structures includes the affine spaces of any dimension and many new classes of structures which can be considered as non-Desarguesian spaces.

A weak affine space consists of a class P of *points*, a class L of *lines*, a relation "incidence" (given in $P \times L$) and a relation "parallelism" (given in $L \times L$), with the condition that the following axioms are satisfied:

A1. Any two distinct points are incident with exactly one line.

A2. Each line contains the same cardinal $s \geqslant 2$ of points (s is called the *order* of the space).

A3. Parallelism is an equivalence relation.

A4. To each line a and each point B there exists exactly one line b such that $B \in b$ and b is parallel to a.

Many interesting problems arise from the study of the algebraic structures which can be used to coordinatize a weak affine space, and from the study of special classes of them. For a complete bibliography on this topic see E. Sperner (1971).

3.4. Quasi-translation general spaces

"General spaces" are incidence structures which satisfy axiom A1 in 3.3. G. Zappa [1964] has established a connection between special classes of general spaces (R-transitive general spaces, quasi-translation general spaces) and regular S-partitions which leads to many interesting theorems.

3.5. Pseudo-planes

R. Sandler [1966] introduced a generalization of projective planes which can be coordinatized in a way similar to projective planes. Further results in this direction have been obtained by E. H. Davis [1970] and by P. Quattrocchi [1969].

4. Galois geometries

The purpose of Galois geometries is to study the sets of points of finite spaces; particularly those which can be defined through simple geometric properties. For more extended expositions of this theory, and detailed information on the many unsolved problems, the reader is referred to the following papers by B. Segre [1960, 1961, 1965, 1967].

4.1. The $(k; n)$-arcs in a projective plane

In what follows we shall denote by $\pi(q)$ any projective plane of order q, while the symbol $PG(2, q)$ will be used only for a Desarguesian plane of order q.

A $(k; n)$-arc of $\pi(q)$ is a set of k points of $\pi(q)$ such that n is the largest number of them which are collinear. $(k; 2)$-arcs may be more simply called k-arcs, provided that the value of n is given once for all.

In a given plane a $(k; n)$-arc is said to be *complete* if there exists no $(k'; n)$-arc, with $k' > k$, which contains it.

k-arcs have been extensively studied (see, e.g., B. Segre [1961 and 1967]). We want to recall here only a few of the most interesting theorems concerning them.

Theorem 4.1.1 (B. Segre [1955]). *In* $PG(2, q)$, *if q is odd, every $(q+1)$-arc is an irreducible conic.*

This theorem opens the way to a myriad of questions concerning how to characterize algebraic varieties in finite spaces, through graphic properties (see B. Segre [1960], n. 3).

Theorem 4.1.2 (B. Segre [1957]. *In* $PG(2, 2^h)$, *for $h = 4$, $h = 5$ and for $h \geqslant 7$, there exist $(q+2)$-arcs which do not contain $q+1$ points forming a conic.*

It is not known whether the same property holds for $h = 6$. For $h = 1, 2, 3$ every $(q+2)$-arc can be obtained by adding the nucleus to the $q+1$ points of a conic.

Theorem 4.1.3 (B. Segre [1967]). *In* $PG(2, q)$, *q even, a k-arc with*

$$q - \sqrt{q} - 1 < k \leqslant q + 1$$

is incomplete.

In $PG(2, q)$, q odd, a k-arc with

$$q - \tfrac{1}{4}\sqrt{q} + \tfrac{7}{4} < k < q + 1$$

is incomplete.

The previous theorems show how the fact of the plane's being Desarguesian allows one to state properties concerning the nature of the arc (of being a conic, of being incomplete, etc.). On the other hand, the following theorem shows how a geometrical property of an arc can determine the nature of the plane:

Theorem 4.1.5 (F. Buekenhout [1966]). *If in $\pi(q)$ there is a $(q+1)$-arc K such that every hexagon whose vertices are points of K is Pascalian, then $\pi(q)$ is Pappian and K is a conic.*

Other proofs of this theorem have been given by R. Artzy [1968], J. F. Rigby [1969], and by H. Karzel and K. Sörensen [to appear]. A first characterization of this type was given by Pickert [1959] who proved that the validity of Pappus' theorem for a pair of lines implies that the plane is Pappian (see also E. Schröder [1968]).

4.2. The packing problem

In $PG(r, q), r \geq 3$, a *k-cap* is a set of k distinct points no three of which are collinear. In a given $PG(r, q)$, a k-cap is *complete* if there is no k'-cap, $k' > k$, which contains it.

We shall denote the maximum number of points in $PG(r, q)$ that belong to a k-cap if $r \geq 3$, or to a k-arc if $r = 2$, by $m(r, q)$. The knowledge of this character is very important for some applications (e.g., to coding theory and to statistics). The known values of $m(r, q)$ are given below:

$m(2, q) = q+1$ (if q is odd and the plane is Desarguesian)
               R. C. Bose [1947]

$m(2, q) = q+2$ (if q is even and the plane is Desarguesian)
               R. C. Bose [1947]

$m(3, q) = q^2+1$ (the result was found by R. C. Bose [1947] for q odd
               and by B. Qvist [1952] for q even)

$m(r, 2) = 2^r$

$m(4, 3) = 20$ G. Pellegrino [1970].

To determine the value of $m(r, q)$ in the general case seems to be a very difficult problem. A considerable amount of work has been done in recent years providing upper bounds for $m(r, q)$. A complete list of these bounds is given in a paper by B. Segre [1967, p. 166]. We shall list here only the better values which have been published to date:

$$m(4, q) < q^3 - \tfrac{1}{2}q\sqrt{q} + \tfrac{3}{4}q + 1 \qquad (q \text{ even}) \text{ B. Segre } [1967]$$

$$m(r, q) < q^{r-1} - \tfrac{1}{2}q^{r-\frac{5}{2}} + \tfrac{3}{4}q\left(\sum_{i=0}^{r-4} q^i\right) + 2 \qquad (r > 4, q \text{ even})$$

$$\text{B. Segre } [1967]$$

$$m(r, q) < q^{r-1} - \tfrac{1}{2}q^{r-\frac{5}{2}} + (\tfrac{3}{4}q - 2)\left(\sum_{i=0}^{r-4} q^i\right) + 3 \qquad (r > 3, q = 2^h, h \text{ even})$$

$$\text{J. A. Thas } [1968].$$

For q odd, the best results are "asymptotic" with respect to q, i.e., they hold for $q > q_0$ where q_0 denotes a suitable integer:

$$m(r, q) < q^{r-1} - cq^{r-\frac{3}{4}} \qquad (0 < c < \tfrac{1}{4}, q \text{ odd}, q > q_0) \quad \text{B. Segre [1967]}.$$

The best known results for $m(r, q)$ for small values of q (q odd) are still those obtained by A. Barlotti [1965b] as a small improvement on previous results by B. Segre.

The same problem arises in connection with the *k-sets of kind s*. A k-set of kind s in $PG(r, q)$ is a set of k points such that any $s + 1$ distinct points in the set are independent (in other words, there is no $(s-1)$-flat containing $s+1$ points of the set). The problem of finding the maximum number of points which can belong to a k-set of kind s is known as the "packing problem" (R. C. Bose and J. N. Srivastava [1964]; see also R. C. Bose [1967], p. 17). The connection between the packing problem and the general combinatorial problem known as the "critical problem" (H. Crapo and G. C. Rota [1970]) has been established by T. A. Dowling [1971]. Bounds for k-sets can be found (in the terminology of coding theory) in chapter 13 of the book by E. R. Berlekamp [1968].

The packing problem for $(k; n)$-arcs (A. Basile and P. Brutti [1971]) and $(k; n)$-caps seems to be even more difficult. Only cases with particular conditions have been studied to date.

References

J. André, 1954, Über nicht-Desarguessche Ebenen mit transitiver Translationsgruppe, *Math. Z.* **60**, 156–186.

R. Artzy, 1968, Pascal's theorem on an oval, *Am. Math. Monthly* **75**, 143–146.

F. Bachmann, 1971, Hjelmslev planes, *Atti Convegno Geometria Combinatoria* (Perugia), pp. 43–56.

R. Baer, 1942, Homogeneity of projective planes, *Am. J. Math.* **64**, 137–152.

A. Barlotti, 1965a, Configurazioni k-chiuse e piani k-aperti, *Rend. Sem. Mat. Padova* **35**, 56–64.

A. Barlotti, 1965b, Some topics in finite geometrical structures (Lecture Notes, Chapel Hill, N.Car.).

A. Barlotti, 1969, Classification of finite projective planes, *Combinatorial Mathematics and its Applications* (University of North Carolina Monograph Series in Probability and Statistics), chapter 25, pp. 405–415.

A. Basile, and P. Brutti, 1971, Alcuni risultati sui $\{q(n-1)+1; n\}$—archi di un piano proiettivo finito, *Rend. Sem. Mat. Univ. Padova* **46**, 107–125.

A. Basile, and P. Brutti, (b) On the completeness of regular $\{q(n-1)+m; n\}$—arcs in a finite projective plane (to appear).

W. Benz, 1960, Uber Möbiusebenen. Ein Bericht, *Jber. Deutsch. Mat.Verein.* **63**, 1–27.

W. Benz, and H. Mäurer, 1964, Uber die Grundlagen der Laguerre-Geometrie, Ein Bericht, *Jber. Deutsch. Mat.Verein.* **67**, 14–42.

E. E. Berlekamp, 1968, *Algebraic Coding Theory* (McGraw-Hill, New York).

R. C. Bose, 1938, On the application of the properties of Galois fields to the construction of hyper Graeco-Latin squares, *Sankhya* **3**, 323–338.

R. C. Bose, 1942, An affine analogue of Singer's theorem, *J. Ind. Math. Soc.* **6**, 1–15.

R. C. Bose, 1947, Mathematical theory of the symmetrical factorial design, *Sankhya* **8**, 107–166.

R. C. Bose, 1963a, Strongly regular graphs, partial geometries and partially balanced designs, *Pacif. J. Math.* **13**, 389–419.

R. C. Bose, 1963b, Combinatorial properties of partially balanced designs and association schemes, *Sankhya* **25**, 109–136.

R. C. Bose, 1967, Lectures on Coding Theory (Lecture Notes, Tata Institute Bombay).

R. C. Bose, 1970, Coordinatization of a class of projective planes which can be linearly represented in a projective space of four dimensions, *Proc. Second Chapel Hill Conference on Combinatorial Mathematics and its Applications*, pp. 37–56.

R. C. Bose, and A. Barlotti, 1971, Linear representation of a class of projective planes in a four dimensional projective space, *Ann. Mat. Pura e Appl.* **88**, 9–31.

R. C. Bose, and R. C. Burton, 1966, A characterization of flat spaces in a finite geometry and the uniqueness of the Hamming and the MacDonald codes, *J. Combin. Theory* **1**, 96–104.

R. C. Bose, I. M. Chakravarti, and D. E. Knuth, 1960, On methods of constructing sets of mutually orthogonal Latin squares using a computer, *Technometrics* **2**, 507–516.

R. C. Bose, and K. R. Nair, 1941, On complete sets of Latin squares, *Sankhya* **5**, 361–382.

R. C. Bose, and K. J. C. Smith, 1971, Ternary rings of a class of linearly representable semi-translation planes, *Atti Convegno Geometria Combinatoria* (Perugia), pp. 69–101.

R. C. Bose, and J. N. Srivastava, 1964, On a bound useful in the theory of factorial designs and error correcting codes, *Ann. Math. Statist.* **35**, 408–414.

R. H. Bruck, 1963, Existence problems for classes of finite projective planes (Lecture Notes, Canad. Math. Congress, Saskatoon).

R. H. Bruck, 1969, Construction problems of finite projective planes, *Combinatorial Mathematics and its Applications, Proc. Chapel Hill Conference* (R. C. Bose and T. A. Dowling, eds.; Univ. of North Carolina Press, Chapel Hill, N.Car.), pp. 426–514.

R. H. Bruck, and R. C. Bose, 1964, The construction of translation planes from projective spaces, *J. Algebra* **1**, 85–102.

R. H. Bruck, and R. C. Bose, 1966, Linear representation of projective planes in projective spaces, *J. Algebra* **4**, 117–172.

R. H. Bruck, and H. J. Ryser, 1949, The nonexistence of certain finite projective planes, *Canad. J. Math.* **1**, 88–93.

A. Bruen, and J. C. Fisher, 1969, Spreads which are not dual spreads, *Canad. Math. Bull.* **12**, 801–803.

F. Buekenhout, 1966, Plans projectifs à ovoides pascaliens, *Arch. Math.* **17**, 89–93.

R. J. Bumcrot, 1969, *Modern Projective Geometry* (Holt, Rinehart and Winston, New York).

R. P. Burn, 1968a, The coexistence of involutory elations and homologies, *Math. Z.* **103**, 195–200.

R. P. Burn, 1968b, Bol Quasi-fields and Pappus' Theorem, *Math. Z.* **105**, 351–364.

V. Chval, 1970, Rappresentazione dei piani liberi nello spazio proiettivo, *Rend. Trieste* **2**, 139–145.

H. Crapo, and G.-C. Rota, 1970, *Combinatorial Geometries* (M.I.T. Press, Cambridge, Mass.).

E. H. Davis, 1970, Incidence systems associated with non-planar nearfields, *Canad. J. Math.* **22**, 939–952.

P. Dembowski, 1966, Endliche Geometrien, *Math.-Phys. Semesterber.* **13**, 32–61.

P. Dembowski, 1968, *Finite Geometries* (Springer Verlag, Berlin-Heidelberg-New York).

T. A. Dowling, 1971, Codes, packings and the critical problem, *Atti Convegno Geometria Combinatoria* (Perugia), pp. 209–224.

B. R. Gulati, and E. G. Kounias, 1970, On bounds useful in the theory of symmetrical factorial designs, *J. Roy. Statist. Soc.* B **32**, 123–133.

M. Hall, Jr., 1943, Projective planes, *Trans. Am. Math. Soc.* **54**, 229–277.

M. Hall, Jr., 1959, *The Theory of Groups* (Macmillan, New York).

W. Heise, 1970, Eine neue Klasse von Möbius m-Strukturen, *Rend. Trieste*, **2**, 125–128.

C. Hering, and W. K. Kantor, 1971, On the Lenz-Barlotti classification of projective planes, *Arch. der Math.* **22**, 221–224.

A. Heyting, 1963, *Axiomatic Projective Geometry* (Noordhoff, Groningen).

D. G. Higman, 1971, Partial geometries, generalized quadrangles and strongly regular graphs, *Atti Convegno Geometria Combinatoria* (Perugia), pp. 263–293.

P. Hohler, 1970, Eine Verallgemeinerung von ortogonalen lateinischen Quadraten au, höhere Dimensionen, Dissertation Nr. 4522, Eidgenössische Tech. Hochschule, Zürich.

T. V. S. Jagannathan, 1970, The Lenz-Barlotti class I-6, *Math. Z.* **115**, 350–357.

N. L. Johnson, 1969a, A classification of semi-translation planes, *Canad. J. Math.* **21** 1372–1387.

N. L. Johnson, 1969b, Nonstrict semi-translation planes, *Arch. Math.* **20**, 301–310.

N. L. Johnson, 1970a, Derivable chains of planes, *Boll. Un. Mat. Ital.* **3**, 167–184.

N. L. Johnson, 1970b, Derivable semi-translation planes, *Pacif. J. Math.* **34**, 687–707.

W. Jónsson, 1963, Transitivität and Homogenität projektiver Ebenen, *Math. Z.* **80**, 269–292.

W. Jónsson, 1965, (C, γ, μ)-homogeneity of projective planes, *Canad. J. Math.* **17**, 331–334.

M. J. Kallaher, and T. G. Ostrom, 1971, Fixed point free linear groups, rank three planes, and Bol quasifields, *J. Algebra* **18**, 159–178.

H. Karzel, and K. Sörensen, Projektive Ebenen mit einem pascalschen Oval (to appear).

R. Lingenberg, 1969, *Grundlagen der Geometrie* (Bibliographisches Institut, Mannheim-Wien-Zürich).

H. Lüneburg, 1965, Die Suzukigruppen und ihre Geometrien Lecture Notes in Mathematics; Springer Verlag, Berlin-Heidelberg-New York).

H. Lüneburg, 1967, Gruppentheoretische Methoden in der Geometrie. Ein Bericht, *Jber. Deutsch. Mat.Verein.* **70**, 16–51.

C. Maneri, and R. Silverman, 1966, A vector-space packing problem, *J. Algebra* **4**, 321–330.

C. Maneri, and R. Silverman, 1971, A combinatorial problem with applications to geometry, *J. Combin. Theory* (to appear).

G. Menichetti, 1968, Sui quasicorpi destri di ordine finito $n = p^k$, *Le Matematiche* **23**, 262–272.

G. Menichetti, 1969, Un procedimento geometrico per determinare quasicorpi di ordine q^2, *Le Matematiche* **24**, 342–354.

G. Menichetti, 1970, Quasicorpi, di dimensione 2 sopra un campo K, associati a transformazioni quadratiche nel piano affine $A(K)$, *Le Matematiche* **25**, 117–148.

T. G. Ostrom, 1970, *Finite Translation Planes* (Springer Verlag, Berlin-Heidelberg-New York).

T. G. Ostrom, 1971, Collineation groups generated by homologies in translation planes, *Atti Convegno Geometria Combinatoria* (Perugia), pp. 351–366.

G. Pellegrino, 1970, Sul massimo ordine delle calotte in $S_{4,q}$, *Le Matematiche* **25**, 149–157.

R. Permutti, 1967, Una generalizzazione dei piani di Möbius, *Le Matematiche* **22**, 360–374.

R. Permutti, 1968, Sulle *m*-strutture ovoidali di Möbius, *Le Matematiche* **23**, 50–59.

J. C. Petit, 1969, Construction de corps ternaires par tronconnage, *Math. Z.* **10**, 127–152.

G. Pickert, 1955, *Projektive Ebenen* (Springer Verlag, Berlin-Göttingen-Heidelberg).

G. Pickert, 1959, Der Satz von Pappos mit Festelementen, *Arch. Math.* **10**, 56–61.

P. Quattrocchi, 1969, Pseudo piani finiti e piani proiettivi, *Atti Sem. Mat. Fis. Univ. Modena*, **18**, 317–331.

B. Qvist, 1952, Some remarks concerning curves of the second degree in a finite plane, *Ann. Acad. Sci.* **134**, pp. 1–27.

C. R. Rao, 1946, Hypercubes of strength *d* leading to confounded designs in factorial experiments, *Bull. Calcutta Math. Soc.* **38**, 67–78.

J. F. Rigby, 1969, Pascal ovals in projective planes, *Canad. J. Math.* **21**, 1462–1476.

H. J. Ryser, 1963, *Combinatorial Mathematics* (Wiley, New York).

R. Sandler, 1966, Pseudo planes and pseudo ternaries, *J. Algebra*, **4**, 300–316.

E. Schröder, 1968, Projektive Ebenen mit pappusschen Geradenpaaren, *Arch. Math.* **19**, 325–329.

B. Segre, 1955, Ovals in a finite projective plane, *Canad. J. Math.* **7**, 414–416.

B. Segre, 1957, Sui *k*-archi nei piani finiti di caratteristica due, *Rev. Math. Pures Appl.* **2**, 289–300.

B. Segre, 1960, On Galois geometries, *Proc. Intern. Congr. Mathematicians, 1958* (Cambridge), pp. 488–499.

B. Segre, 1961, *Lectures on Modern Geometry* (Cremonese, Roma).

B. Segre, 1964, Teoria di Galois, fibrazioni proiettive e geometrie non desarguesiane, *Ann. Mat. Pura Appl.* **64**, 1–76.

B. Segre, 1965, *Istituzioni di geometria superiore*, Vols. I, II, III (Lecture Notes, Istituto Matematico, Università di Roma).

B. Segre, 1967, Introduction to Galois geometries, *Memorie Lincei* **7** (8), 133–236.

J. Singer, 1938, A theorem in finite projective geometry and some applications to number theory, *Trans. Am. Math. Soc.* **43**, 377–385.

E. Sperner, 1960, Affine Räume mit schwacher Inzidenz und zugehörige algebraische Strukturen, *J. Reine Angew. Math.* **204**, 205–215.

E. Sperner, 1971, Zur Geometrie der Quasimoduln, *Ist. Nazl. di Alta Mat., Symp. Math.* Vol. V.

J. A. Thas, 1968, Bijdrage tot de theorie der *k*-kappen in projectieve ruimten van Galois, *Mededel. Koninkl, Vlaamse Acad. Wetenschap.* **30** (13), 15 pp.

J. A. Thas, 1968, Normal rational curves and *k*-arcs in Galois spaces, *Rend. Mat.* **1**, 331–334.

J. A. Thas, 1969, Connection between the Grassmannian $G_{k-1; n}$ and the set of the *k*-arcs of the Galois space $S_{n,q}$, *Rend. Mat.* **2**, 121–134.

R. Wille, 1970, *Kongruenzklassengeometrien* (Springer Verlag, Berlin-Heidelberg-New York).

J. C. D. S. Yaqub, 1967, The Lenz-Barlotti classification, *Proc. Projective Geometry Conf.*, Univ. of Illinois, Chicago, pp. 129–160.

G. Zappa, 1964, Sugli spazi generali quasi di translazione, *Le Mathematiche* **19**, 127–143.

J. N. Srivastava et al., eds., *A Survey of Combinatorial Theory*
© North-Holland Publishing Company, 1973

CHAPTER 2

Balanced Hypergraphs and Some Applications to Graph Theory

CLAUDE BERGE

University of Paris, Paris, France

1. Introduction

The purpose of this paper is to survey the combinatorial properties of a special kind of hypergraphs, called the balanced hypergraphs, which yield easily some new theorems of graph theory. This class of hypergraph was introduced (Berge [1969]) to generalize the class of all $(0, 1)$-matrices with the unimodular property.

2. General definitions

A *hypergraph* $H = (X, \xi)$ consists of a finite set X of n *vertices* together with a family $\xi = (E_i / i \in I)$ of m non-empty subsets of X, called the *edges*; in addition, we assume $\bigcup E_i = X$, which permits us to define also the hypergraph by $H = (E_i / i \in I)$. If $|E_i| \leq 2$ for all i, H is a *multigraph* (undirected graph). $|X|$ is the *order* of hypergraph H.

The *sub-hypergraph* of H *induced by a subset* $A \subset X$, is a hypergraph $H_A = (A, \xi_A)$, where

$$\xi_A = (E_i \cap A / i \in I, E_i \cap A \neq \emptyset).$$

The *partial hypergraph of H generated by a subset* $J \subset I$, is a hypergraph (X', ξ'), where

$$X' = \bigcup_{i \in J} E_i,$$

$$\xi' = (E_i / i \in J).$$

A *chain* is a sequence $(x_1, E_1, x_2, E_2, \ldots, E_p, x_{p+1})$ such that

   (1) the x_j's are all different vertices of H (for $j \leq p$),

   (2) the E_i's are all different edges of H (for $i \leq p$),

   (3) $x_j, x_{j+1} \in E_j$ $(j \leq p)$.

If $p > 1$, and if $x_{p+1} = x_1$, this chain is called a *cycle*.

   If H is defined by vertices $x_1, x_2, \ldots, x_n$ and by edges $E_1, E_2, \ldots, E_m$, its dual H^* is a hypergraph defined by vertices $e_1, e_2, \ldots, e_m$ (representing

15

respectively the E_i's) and by edges $X_1, X_2, \ldots, X_n$ (representing respectively the X_j's), where

$$X_j = \{e_i / i \leqslant m, E_i \ni x_j\};$$

the dual $(H^*)^*$ of H is identical to H.

The main coefficients for graphs can be extended to hypergraphs.

The *covering number* $\rho(H)$ is the smallest number of edges of H whose union is X.

The *strong stability number* $\alpha(H)$ is the largest number of vertices of a set S with $|S \cap E_i| \leqslant 1$ for all $i \leqslant p$.

The *rank* $r(A)$ of a set A is $r(A) = \max |E_i \cap A|$; the number $r(X)$ is also called the *rank of H*. If $|E_i| = r(X)$ for every i, H is a *uniform* hypergraph.

The *chromatic number* $\chi(H)$ is the smallest number of colors required to color the vertices of H such that no edge E_i with $|E_i| > 1$ is of one color (cf. Erdös and Hajnal [1966]). A q-coloration of H is a partition $(S_1, S_2, \ldots, S_p)$ such that no edge with more than one element is contained in only one class.

The *strong chromatic number* $\gamma(H)$ is the smallest number of colors required to color the vertices of H such that no color appears twice in a same edge.

A *matching* is a family of pairwise disjoint edges.

A *transversal set* is a set $T \subset X$ such that $|T \cap E_i| \geqslant 1$ for all i.

The maximum cardinality of a matching is denoted by $v(H)$, and the minimum cardinality of a transversal set is denoted by $\tau(H)$.

3. Balanced hypergraphs

A hypergraph $H = (E_i / i \in I)$ is said to be *balanced* if every odd cycle $(a_1, E_1, a_2, E_2, \ldots, E_{2p+1}, a_1)$ has an edge E_i which contains at least three of its vertices a_j. For instance, if $H = (E_i / i \in I)$ is a family of intervals of points on a straight line, it is a balanced hypergraph.

If H is a multigraph, it is a balanced hypergraph if (and only if) the multigraph is bipartite. An example of a balanced hypergraph (which is not unimodular) is given by fig. 1.

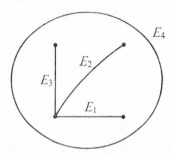

Fig. 1.

Proposition 1. *If H is a balanced hypergraph, every partial hypergraph $H' = (E_j/j \in J)$ is balanced.*

This is obvious, since if H' had an odd cycle without any edge containing three of its vertices, then this sequence would be also an odd cycle of H without any edge containing three of its vertices, and this contradicts that H is balanced.

Proposition 2. *If H is a balanced hypergraph, every sub-hypergraph H_A is balanced.*

This is obvious, since if H_A had an odd cycle without any edge containing three of its vertices, then this sequence would define in H an odd cycle without any edge containing three of its vertices.

Proposition 3. *If $H = (E_i/ i \in I)$ is a balanced hypergraph, and if A is any set disjoint of $\bigcup E_i$, the hypergraph $H' = (E_i \cup A/ i \in I)$ is also balanced.* (Same proof.)

Proposition 4. *If $H = (x_1, \ldots, x_n; E_1, \ldots, E_m)$ is a balanced hypergraph, its dual $H^* = (e_1, \ldots, e_m; X_1, \ldots, X_n)$ is a balanced hypergraph.*

By definition, we have

$$X_j = \{e_i/ i \leqslant m, E_i \ni x_j\}.$$

H^* is a hypergraph, since $X_j \neq \emptyset$ and

$$\bigcup_{j=1}^{m} X_j = \{e_1, e_2, \ldots, e_m\}.$$

Consider in H^* an odd cycle $\mu = (e_1, X_1, e_2, \ldots, e_{2p+1}, X_{2p+1}, e_1)$; it corresponds in H to an odd cycle $(x_1, E_2, x_3, \ldots, E_{2p+1}, x_{2p+1}, E_1, x_1)$.

As H is balanced, an E_i contains three of the x_j's, and therefore in H^*, an e_i belongs to three of the X_j's; or, equivalently, an X_k contains three of the e_i's.

Proposition 5. *If $H = (E_i/ i \in I)$ is a balanced hypergraph, and if $H' = (E_i/ i \in J)$ is a partial hypergraph such that $i, j \in J$ implies $E_i \cap E_j \neq \emptyset$, then*

$$\bigcap_{i \in J} E_i \neq \emptyset.$$

In other words, the edges of a balanced hypergraph have the Helly property.

Proof. (By induction on the number of edges of H'.) The statement is true for any subfamily with two edges; let us assume that it is true for every subfamily with p edges, and let us prove it for a subfamily $(E_i/ i \in J) = (E'_1, E'_2, \ldots, E'_p)$ with $p \geqslant 3$.

By the induction hypothesis, there exists for every $k \leqslant p$ a vertex a_k with

$$a_k \in \bigcap_{i \neq k} E'_i.$$

We can assume that the a_k's are all different (if not, we have proved that $\bigcap E'_i \neq \emptyset$).

2

Consider the sequence

$$\mu = (a_1, E_2', a_3, E_1', a_2, E_3', a_1)$$

If two of E_1', E_2', E_3' are equal, for instance if $E_1' = E_2'$, then a_1 belongs to $\bigcap_{i \in J} E_i$, and the proof is achieved. If no two of E_1', E_2', E_3' are equal, then μ is an odd cycle, and one of these edges, for instance E_1', contains $\{a_1, a_2, a_3\}$; hence we have

$$a_1 \in \bigcap_{j \in J} E_j,$$

which proves the proposition.

If H is a balanced hypergraph, propositions 4 and 5 show that H is a *conformal hypergraph* ("family of sets with a faithful graph representation" Gilmore [1962]).

Theorem 1. *A hypergraph $H = (X, \xi)$ is balanced if and only if for every $A \subset X$, the sub-hypergraph H_A satisfies $\chi(H_A) \leq 2$.*

The condition is necessary: If every sub-hypergraph of H admits a bicoloration, then H is balanced, because if not, there exists an odd cycle

$$(a_1, E_1, a_2, \ldots, E_p, a_1)$$

without any E_i containing three of the a_j's, and $A = \{a_1, a_2, \ldots, a_p\}$ induces a sub-hypergraph H_A with $\chi(H_A) > 2$.

The condition is sufficient: Assume that there exists a balanced hypergraph without a bicoloration, and consider a balanced hypergraph H without a bicoloration and of minimum order $n = |X|$. We shall show that this yields a contradiction.

(1) We shall show first that any vertex x_0 belongs to at least two different edges of H having exactly two elements.

The sub-hypergraph H_0 induced by $X - \{x_0\}$ is balanced (proposition 2), of order $n - 1$, and therefore it admits a bicoloration (S_1^0, S_2^0). If x_0 did not belong to an edge having two elements, $(S_1^0 \cup \{x_0\}, S_2^0)$ would be a bicoloration of H, which contradicts our assumptions; if x_0 belongs to some edges having two elements, then their traces on $X - \{x_0\}$ are not all contained in S_1^0 because $(S_1^0, S_2^0 \cup \{x_0\})$ would be a bicoloration of H. Hence, x_0 belongs to at least two edges having two elements, say $[x_0, y]$ and $[x_0, z]$, with $z \neq y$.

(2) Denote by F the partial family of edges having two elements; and consider $G = (X, F)$. As G is a balanced hypergraph (proposition 1), it is balanced, and therefore it is a bipartite graph.

Consider a connected component C of G; it is of order ≥ 3 (due to 1), and there exists at least one vertex x_1 which is not an articulation point.

(3) Consider the sub-hypergraph H_1 induced by $X - \{x_1\}$; it is balanced of order $n - 1$, and therefore admits a bicoloration (S_1, S_2). In graph G, all the vertices adjacent to x_1 have the same color; let S_1 be the class of the vertices having this color.

$(S_1, S_2 \cup \{x_1\})$ is a bicoloration of H, because every edge of H with two elements is bicolored (because it is an edge of G), and every edge of H with more than two elements is also bicolored (because its trace on $X - \{x_1\}$ is bicolored).

This contradicts the assumption that H does not admit a bicoloration.

Theorem 2. *For a balanced hypergraph $H = (E_i/i \in I)$, let $k = \min_{i \in I} |E_i|$; there exist k transversal sets of H which partition the set of all the vertices.*

Proof. Let $(S_1, S_2, \ldots, S_k)$ be a partition of X into k classes, and let $k(i)$ be the number of classes which meet edge E_i. If $k(i) = k$ for every i, then we have a partition of X into k transversal sets.

If $k(i) < k$ for an index $i = i_0$, we have $k(i_0) < |E_{i_0}|$; hence there exist two indices p and q with

$$|S_p \cap E_{i_0}| \geq 2,$$
$$|S_q \cap E_{i_0}| = 0.$$

The sub-hypergraph H' induced by $S_p \cup S_q$ is balanced, hence, by theorem 1, it admits a bicoloration (S_p', S_q').

Let $S_j' = S_j$ for $j \neq p, q$; the partition $(S_1', S_2', \ldots, S_k')$ determines new coefficients $k'(i)$, with

$$k'(i_0) = k(i_0) + 1,$$
$$k'(i) \geq k(i) \qquad (i \neq i_0).$$

Repeating with E_{i_0} this reconstruction of the partition if necessary, we can obtain a new partition $(S_1', S_2', \ldots, S_k')$ with $k'(i_0) = k$. Repeating this transformation with any edge E_i with $i \neq i_0$, if necessary, we finally obtain a partition $(T_1, T_2, \ldots, T_k)$ with $k(i) = k$ for every i, which is a partition of X into k transversal sets.

Theorem 3. *The strong chromatic number $\gamma(H)$ of a balanced hypergraph H is equal to its rank.*

Proof. Let $h = \max_{i \in I} |E_i|$ be the rank of H. Consider a hypergraph $H' = (X', \xi')$, uniform of rank h, obtained from H by adding for every i a set A_i with $|A_i| = h - |E_i|$, and by defining

$$X' = X \cup \bigcup_{i \in I} A_i,$$
$$\xi' = (E_i' = E_i \cup A_i / i \in I).$$

H' is a uniform hypergraph of rank h; let us show that it is also a balanced hypergraph.

Consider an odd cycle μ' of H'; if no vertex of μ' lies in $\bigcup_{i \in I} A_i$, then μ' induces in H an odd cycle μ with an edge containing three of its vertices, and μ' admits an edge containing three of its vertices. No vertex of μ' lies in $\bigcup_{i \in I} A_i$, this is also true. Therefore, H' is balanced.

By theorem 2 applied to H', there exists a partition $(T'_1, T'_2, \ldots, T'_h)$ of X' into h transversal sets. As $|T'_j \cap E_i| \leq 1$ for every i and every j, the sets $S_j = T'_j \cap X$ are the classes of a strong h-coloration of H, and therefore we have $\gamma(H) \leq h$. As it is obvious that $\gamma(H) \geq h$, we finally obtain $\gamma(H) = h$.

Applications. Let $G = (X, \xi)$ be a bipartite multigraph, and consider its dual hypergraph G^*.

As G is balanced, hypergraph G^* is also balanced (proposition 4). If we apply theorem 3 to G^*, we obtain a well known result: *The chromatic index of a bipartite multigraph is equal to the maximum of its degrees.*

If we apply Theorem 2 to G^*, we obtain Gupta's theorem (Gupta [1967]): *In a bipartite multigraph of minimum degree k, there exists a partition of its edges into k classes, each of them covering all the vertices.*

4. The main result

For a hypergraph $H = (X, \xi)$, a family $\xi' = (E_i / i \in J)$ is a *matching* if no two edges of ξ' intersect.

Denote by $v(H)$ the maximum number of edges in a matching, and by $\tau(H)$ the minimum cardinality of a transversal set. It is obvious that $\tau(H) \geq v(H)$, and we shall see that if H is a balanced hypergraph, then $\tau(H) = v(H)$. This result was first obtained (Berge and Las Vergnas [1970]) by using a simplified form of Ray-Chaudhuri's theorem [1963]; it can also be obtained from theorem 3, by the powerful theory of antiblocking polyhedra of Fulkerson [1970]. The combinatorial proof given here is essentially due to Lovasz [1970], and makes use only of our theorem 1.

Consider a hypergraph $H = (X, \xi)$ with m edges $E_1, E_2, \ldots, E_m$ and a mapping ϕ of X into the set of all non-negative integers. If $S \subset X$, define

$$\phi(S) \begin{cases} = \sum_{s \in S} \phi(s) & \text{if } S \neq \emptyset, \\ = 0 & \text{if } S = \emptyset. \end{cases}$$

If $t = (t_1, t_2, \ldots, t_m)$, we shall say that ϕ is *t-covering* when

$$\phi(E_i) \geq t_i \qquad (i = 1, 2, \ldots, m).$$

If $t_i = \lambda$ for all i, we shall say for short that ϕ is a *λ-covering function of H*. If we denote by ϕ_1 an arbitrary 1-covering function, we have

$$\min_{\phi_1} \phi_1(X) = \tau(H).$$

A 1-covering function ϕ_1 with $\phi_1(X) = \tau(H)$ is said to be *minimal*, and it is the characteristic function of a minimum transversal set.

Lemma 1. *If we denote by $\phi_1(x)$ a 1-covering function of a balanced hypergraph H, and by $\phi_2(x)$ a 2-covering function of H, we have*

$$\min_{\phi_2} \phi_2(X) = 2 \min_{\phi_1} \phi_1(X).$$

Proof. Let $\phi_1(x)$ be a minimal 1-covering function; as $2\phi_1(x)$ is a 2-covering function, we have

$$\min_{\phi_2} \phi_2(X) \leqslant \min_{\phi_1} [2\phi_1(X)].$$

Let us prove the inverse inequality for a balanced hypergraph H. Let $\phi_2(x)$ be a 2-covering minimal function of H. The values of $\phi(x)$ can only be 0, 1 or 2; for $j = 0, 1, 2$, let

$$S_j = \{x/\ x \in X,\ \phi_2(x) = j\};$$

(S_0, S_1, S_2) is a partition of X. The sub-hypergraph of H induced by S_1, being balanced, admits a bicoloration (S_1', S_1''). Assume for instance $|S_1'| \leqslant |S_1''|$. $S_2 \cup S_1'$ is a transversal set of H, because if an edge E_i does not meet S_2, it contains at least two vertices of S_1 (since $\phi_2(E_i) \geqslant 2$), and hence it meets S_1'.

The characteristic function ϕ of $S_2 \cup S_1'$ is then 1-covering, and we can write:

$$2 \min_{\phi_1} \phi_1(X) \leqslant 2\phi(X) = \phi_2(S_2) + 2\phi_2(S_1')$$

$$\leqslant \phi_2(S_2) + \phi_2(S_1') + \phi_2(S_1'') = \min_{\phi_2} \phi_2(X)$$

The equality follows.

Lemma 2. *For a hypergraph H, the two following statements are equivalent:*

(1) $v(H') = \tau(H')$ for every partial hypergraph H' of H.

(2) $\tau_2(H') = 2\tau(H')$ for every partial hypergraph H' of H, where $\tau_2(H)$ is the minimum of $\phi_2(X)$ for all 2-covering functions ϕ_2 of H.

Proof. (1) $\Rightarrow$ (2), because, for any hypergraph H, we have

$$2v(H) \leqslant \tau_2(H) \leqslant 2\tau(H).$$

Let us show that (2) $\Rightarrow$ (1). Let us assume that a hypergraph H satisfies (2) and $v(H) < \tau(H)$, and let us show that this leads to a contradiction. Let H' be a partial hypergraph of H with $v(H') < \tau(H')$ and with a minimal number of edges.

For every edge E_i' of H', we have

$$v(H' - E_i') = \tau(H' - E_i'),$$
$$v(H' - E_i') = v(H') \text{ or } v(H') - 1,$$
$$\tau(H' - E_i') = \tau(H') \text{ or } \tau(H') - 1.$$

Hence

$$v(H' - E_i') = v(H') \qquad (i),$$
$$\tau(H' - E_i') = \tau(H') - 1 \qquad (i).$$

Let E_1 and E_2 be two intersecting edges of H' (they exist by the above equalities). Let T_1 be a minimum transversal set of $H' - E_1$, and let T_2 be a minimum transversal set of $H' - E_2$.

If $x_0 \in E_1 \cap E_2$, denote by $\phi_0(x)$ the characteristic function of $\{x_0\}$, and by $\phi_1(x)$ and $\phi_2(x)$ the characteristic functions of T_1 and T_2 respectively. The function $\phi(x) = \phi_0(x) + \phi_1(x) + \phi_2(x)$ is a 2-covering for H'; we have

$$\tau_2(H') \leqslant \phi(X') = \phi_1(X') + \phi_2(X') + 1$$
$$= \tau(H' - E_1) + \tau(H' - E_2) + 1$$
$$= 2\tau(H') - 1.$$

This contradicts (2).

Theorem 4. *A hypergraph* $H = (X, \xi)$ *satisfies* $v(H') = \tau(H')$ *for every partial sub-hypergraph* H' *if and only if* H *is a balanced hypergraph.*

Proof. (1) If H is a balanced hypergraph, a partial hypergraph H' is also balanced, and by Lemma 1, we have

$$\tau_2(H') = 2\tau(H').$$

By Lemma 2, we obtain

$$v(H) = \tau(H).$$

This proves the sufficiency of the above condition.

(2) Let H be a hypergraph with $v(H') = \tau(H')$ for every partial sub-hypergraph. Assume that H is not balanced; there exists an odd cycle $(a_1, E_1, a_2, \ldots, a_{2p+1}, E_{2p+1}, a_1)$ without any edge containing three of the a_j' s. Let

$$A = \{a_1, a_2, \ldots, a_{2p+1}\}, \qquad \xi' = (E_1, E_2, \ldots, E_{2p+1}).$$

$H' = \xi'_A$ being a graph consisting of an odd cycle, we have $v(H') = \tau(H') - 1$, which contradicts the definition of H and proves the necessity of the condition.

Application. Let $G = (X, E)$ be a graph; let C be the family of all its maximal cliques. If $A \subset X$ and $D \subset C$, denote by $G_{A,D}$ the graph obtained from G by deleting the vertices of $X - A$ and the edges which do not belong to a clique in D. Denote by $\alpha(G)$ the stability number of G (maximum cardinality of a stable set), by $\theta(G)$ the partition number of G (smallest number of cliques which cover X), by $\gamma(G)$ the chromatic number of G, and by $\omega(G)$ the density number of G (maximum number of elements in a clique). The three following statements are equivalent:

(1) $\alpha(G_{A,D}) = \theta(G_{A,D})$ for every A and every D.

(2) $\gamma(G_{A,D}) = \omega(G_{A,D})$ for every A and every D.

(3) every odd cycle in G contains at least one edge with the property that every maximal clique containing this edge contains a third vertex of the cycle.

If we denote by $H = (X, C)$ the hypergraph of the maximal cliques of G, (3) is equivalent to

(3') H is balanced,

or to

(3″) the dual H^* of H is balanced.

(1) is equivalent to

(1′) $v(H') = \tau(H')$ for every partial sub-hypergraph H' of H^*.

(2) is equivalent to

(2′) $\gamma(H')$ is equal to the rank of H' for every partial sub-hypergraph H' of H.

By theorem 3, (2′) is equivalent to (3′).

By theorem 4, (1′) is equivalent to (3″).

This proves the equivalence between (1), (2), (3); and (3) defines a new class of perfect graphs.

References

C. Berge, 1969, The rank of a family of sets and some applications to graph theory, *Recent Progress in Combinatorics* (W. T. Tutte, ed.; Academic Press, New York), pp. 49–57.

C. Berge, 1970, *Graphes et Hypergraphes* (Dunod, Paris), ch. 20.

C. Berge and M. Las Vergnas, 1970, Sur un théorème du type König pour hypergraphes, *Intern. Conf. on Combinatorial Mathematics* (A. Gerwitz, L. Qyintas, eds.), *Ann. New York Acad. Sci.* **175**, 32–40.

P. Erdös and A. Hajnal, 1966, On chromatic number of graphs and set-systems, *Acta Math. Acad. Sci. Hungarica* **17**, 61–99.

D. R. Fulkerson, 1970, Anti-blocking polyhedra, Rand Rept. No. RM–6201.

P. Gilmore, 1962, Families of sets with faithful graph representation, I.B.M. Res. Note N.C. 184.

R. P. Gupta, 1967, A decomposition theorem for bipartite graphs, *Theorie des Graphes*, Rome I.C.C. (P. Rosenstiehl, ed.; Dunod, Paris), pp. 135–138.

L. Lovasz, October 1970, Private communication.

D. K. Ray-Chaudhuri, 1963, An algorithm for a maximum cover of an abstract complex, *Canad. J. Math.* **15**, 11–24.

J. N. Srivastava et al., eds., *A Survey of Combinatorial Theory*
© North-Holland Publishing Company, 1973

CHAPTER 3

A Strongly Regular Graph Derived from the Perfect Ternary Golay Code

E. R. BERLEKAMP†

Bell Telephone Laboratories, Inc., Murray Hill, N.J. 07974, U.S.A.

and

J. H. VAN LINT and J. J. SEIDEL

Technological University of Eindhoven, Eindhoven, The Netherlands

By use of the perfect ternary Golay code, a strongly regular graph on 243 vertices is constructed, having the property that any adjacent pair of vertices is in one triangle and that any nonadjacent pair of vertices is in one quadrangle. The graph realizes one of the five possibilities for graphs with this property. It provides a (243, 22, 2)-system on 22 and 23 in the sense of Bridges and Ryser [1969].

1. Introduction

Strongly regular graphs have been introduced by Bose [1963], as an abstraction from 2-association schemes for partially balanced incomplete block designs. They satisfy the following, essentially characteristic, properties. For any pair of adjacent vertices x and y, the number p_{11}^1 of vertices adjacent to x and to y is independent of the choice of x and y. For any pair of nonadjacent vertices u and v, the number p_{11}^2 of vertices adjacent to u and to v is independent of the choice of u and v. We shall be interested in strongly regular graphs with the special property

$$p_{11}^2 - p_{11}^1 = 1.$$

For $p_{11}^1 = 0$, this amounts to the Moore graphs of diameter 2 and girth 5, which have been discussed by Hoffman and Singleton [1960]. For $p_{11}^1 = 1$, these graphs have the property that any adjacent pair of vertices is in one triangle, and that any nonadjacent pair is in one quadrangle. In Section 2 it is shown that such a graph can exist for at most 5 values of n. An example is provided by the lattice graph on 3 symbols. In Section 4 a further such graph, on 243 vertices, is constructed in three different ways.

As a starting point for the constructions of this graph, in Section 3 the perfect ternary 2-error-correcting code is explained. This code has been

† Current address: University of California, Berkeley, Calif. 94720, U.S.A.

discovered by Golay [1949]. It was discussed by Coxeter [1958] in a geometric context, and by Bose [1961], who indicated its connection to the theory of confounding and fractional replication.

Bridges and Ryser [1969] considered yet another generalization of block designs. Their (n, k, λ)-systems on r and s are defined in terms of binary square matrices X and Y of order n satisfying the real matrix equation†

$$XY = YX = (k-\lambda)I + \lambda J, \qquad k \neq \lambda,$$

which, for $\lambda \neq 0$, implies

$$JX = XJ = rJ, \qquad JY = YJ = sJ$$

for integer r and s. In particular, (n, k, λ)-systems on k and $k+1$ are characterized by symmetric matrices C with zero diagonal and elements $+1$ and -1 elsewhere, such that

$$C^2 = (1+4(k-\lambda))I + (n-2-4(k-\lambda))J, \qquad CJ = (2k-n+1)J.$$

For special choices of the parameters, such as $n-1 = 2k$, and $n-2 = 4(k-\lambda)$, this leads to orthogonal matrices with zero diagonal, which have been discussed by Goethals and Seidel [1967]. However, there are other values of the parameters for which these systems exist. In fact, (n, k, λ)-systems on k and $k+1$ are the same objects as strongly regular graphs with $p_{11}^2 - p_{11}^1 = 1$.

2. Strongly regular graphs with $p_{11}^2 - p_{11}^1 = 1$

Strongly regular graphs on n vertices may be defined in terms of their $(1, 0)$ adjacency matrix A, and its eigenvalues k, r, s as follows (cf. Hoffman [1963], Seidel [1968, 1969]):

$$(A - rI)(A - sI) = \frac{(k-r)(k-s)}{n} J, \qquad AJ = kJ.$$

Excluding complete bipartite graphs and their complements, we take $r \geqslant 0$ and $s < 0$. For the multiplicities $1, \alpha, \beta$ of the eigenvalues k, r, s we have

$$n = 1 + \alpha + \beta, \qquad \text{tr} A = 0 = k + \alpha r + \beta s, \qquad \text{tr} A^2 = nk = k^2 + \alpha r^2 + \beta s^2,$$

whence, by elimination of α and β,

$$(k-r)(k-s) = n(k+rs).$$

By multiplying out the defining equation, we obtain

$$-r - s + p_{11}^1 = k + rs, \qquad p_{11}^2 = k + rs.$$

From now on, we restrict ourselves to strongly regular graphs with the special property

$$p_{11}^2 - p_{11}^1 = 1,$$

† Here I denotes the $n \times n$ identity matrix and J denotes the $n \times n$ matrix all of whose entries are 1.

that is, $r+s = -1$. Putting $p_{11}^2 = \lambda$, we have

$$A^2+A = (k-\lambda)I+\lambda J, \qquad AJ = kJ,$$
$$2k-(n-1)+(\alpha-\beta)(2r+1) = 0,$$
$$k^2 = (n-1)\lambda, \qquad k = \lambda+r(r+1).$$

The case $\alpha = \beta$ reduces to

$$(J-I-2A)^2 = nI-J, \qquad n = 2k+1 = 4\lambda+1 = (2r+1)^2,$$

$$C^2 = \begin{bmatrix} 0 & j^{\mathrm{T}} \\ j & J-I-2A \end{bmatrix}^2 = nI,$$

where j is $(n \times 1)$. So C, of order $n+1$, is an orthogonal matrix with zero diagonal. Such matrices, for which n must be a sum of two squares of integers, have been constructed for

$$n = p^\alpha \equiv 1 \pmod 4, \qquad p \text{ prime},$$

and for some additional orders, e.g., for $n = 225$. An easy example is C_6, with $A =$ circulant $(0, 1, 0, 0, 1)$. The smallest order for which the existence is unknown is $n = 45$. For details the reader is referred to Van Lint and Seidel [1966], and Goethals and Seidel [1967].

In the case $\alpha \neq \beta$, the relations between the parameters imply that r is an integer. Elimination of n and k yields

$$(2r+1)^4 - 2(2r+1)^2 - 16(\alpha-\beta)\lambda(2r+1) - 16\lambda^2 + 1 = 0.$$

Therefore, $2r+1$ must divide $16\lambda^2 - 1$. Thus, given λ, there are only finitely many possibilities for the parameters r, k, n.

Hoffman and Singleton [1960] considered $\lambda = 1$, which admits $r = 1, 2, 7$, with $(n, k, \lambda) = (10, 3, 1), (50, 7, 1), (3250, 57, 1)$, respectively.† The first case is realized by the Petersen graph. The second graph was constructed by Hoffman and Singleton [1960], drawn by N. Robertson [private comm.], and has an adjacency matrix which may be arranged as follows by use of the cyclic permutation matrix P of order 5:

$$\begin{bmatrix}
P+P^{-1} & 0 & 0 & 0 & 0 & I & I & I & I & I \\
0 & P+P^{-1} & 0 & 0 & 0 & I & P & P^2 & P^3 & P^4 \\
0 & 0 & P+P^{-1} & 0 & 0 & I & P^2 & P^4 & P^6 & P^8 \\
0 & 0 & 0 & P+P^{-1} & 0 & I & P^3 & P^6 & P^9 & P^{12} \\
0 & 0 & 0 & 0 & P+P^{-1} & I & P^4 & P^8 & P^{12} & P^{16} \\
I & I & I & I & I & P^2+P^{-2} & 0 & 0 & 0 & 0 \\
I & P & P^2 & P^3 & P^4 & 0 & P^2+P^{-2} & 0 & 0 & 0 \\
I & P^2 & P^4 & P^6 & P^8 & 0 & 0 & P^2+P^{-2} & 0 & 0 \\
I & P^3 & P^6 & P^9 & P^{12} & 0 & 0 & 0 & P^2+P^{-2} & 0 \\
I & P^4 & P^8 & P^{12} & P^{16} & 0 & 0 & 0 & 0 & P^2+P^{-2}
\end{bmatrix}$$

The existence of the last graph is still undecided.

We now turn to $\lambda = 2$, which admits $r = 1, 3, 4, 10, 31$, with $(n, k, \lambda) = (9, 4, 2), (99, 14, 2), (243, 22, 2), (6273, 112, 2), (494019, 994, 2)$, respectively.

† The case $r = 0$ is excluded because it leads to $n = 2$.

The first case is realized by the lattice graph of order 3, that is, the graph whose 9 vertices are the ordered pairs out of 3 symbols, any two vertices being adjacent if the corresponding ordered pairs have one coordinate equal. In Section 4 the graph with parameters (243, 22, 2) will be constructed. The existence of the other graphs remains undecided.

3. The perfect ternary Golay code

Let C_6 be the orthogonal matrix with zero diagonal of order 6 which was mentioned above. The rows of the 6×12 matrices

$$G = [I_6 \quad C_6], \qquad H = [-C_6 \quad I_6]$$

each generate a subspace of dimension 6 of the vector space $V(12, 3)$ of dimension 12 over GF(3). These subspaces are orthogonal, since $GH^T = 0$. By inspection it is observed that in H no 5 columns are linearly dependent. This implies that the subspace generated by the rows of G has the property that its vectors, apart from the zero vector, have at least 6 nonzero coordinates. By deleting any one column of G the 6×11 matrix G^* is obtained. The rows of G^* generate a subspace of dimension 6 of $V(11, 3)$, whose nonzero vectors have at least 5 nonzero coordinates. By the count

$$3^6(1 + 2 \times 11 + 2^2(\tbinom{11}{2})) = 3^{11}$$

it follows that every 11-dimensional vector over GF(3) differs from a unique vector in the row space of G^* in at most two coordinates.

In terms of coding theory (Berlekamp [1968]), we call the vectors of $V(11, 3)$ the *words*, the number of nonzero coordinates of a word its *weight*, the subspace generated by G^* a *code*, the vectors of a code its *codewords*, G^* a *generator matrix*, and a matrix H^* generating the orthogonal complement of the code a *parity check matrix* of the code. The fact that every 11-dimensional vector over GF(3) differs from a unique codeword in at most two coordinates makes the code a *perfect* code. The perfect ternary code introduced above is called the (11, 6) *ternary Golay code*, after its discoverer Golay [1949].

We remark that an alternative description for the (11, 6) ternary Golay code is given by

$$G^* = \begin{bmatrix} 1 & 1 & 1 & 1 & 1 & 1 & 1 & 1 & 1 & 1 & 1 \\ 1 & 0 & 0 & 1 & 1 & 1 & 0 & 0 & 1 & 0 & 1 \\ 1 & 1 & 0 & 0 & 1 & 1 & 1 & 0 & 0 & 1 & 0 \\ 1 & 1 & 1 & 0 & 0 & 1 & 0 & 1 & 0 & 0 & 1 \\ 1 & 1 & 1 & 1 & 0 & 0 & 1 & 0 & 1 & 0 & 0 \\ 1 & 0 & 1 & 1 & 1 & 0 & 0 & 1 & 0 & 1 & 0 \end{bmatrix},$$

and H^* the matrix consisting of the last 5 rows of G^* (cf. Van Lint [1969]).

The subspace of $V(12, 3)$ generated by the rows of the 6×12 matrix G

mentioned above is called the (12, 6) ternary Golay code. It is a special case of an extended quadratic residue code. Several general properties of such codes are summarized in Section 15.2 of Berlekamp [1968].

4. Constitution of the 243-graph

The perfect ternary Golay code partitions the vector space $V(11, 3)$ into 3^5 cosets obtained by adding a fixed word to all codewords. Each word of weight $\leqslant 2$ must be in a coset containing no other word of weight $\leqslant 2$, since the minimum nonzero weight of the code equals 5. Therefore, the 243 cosets of the code are uniquely represented by the 220 words of weight 2, the 22 words of weight 1, and the word of weight 0. Furthermore, the cosets form a linear space of dimension 5 over GF(3).

Now consider the 243 cosets as the vertices of a graph. Any two vertices are called adjacent iff the difference of the corresponding cosets is a coset of minimum weight 1. This graph possesses the triangle and the quadrangle property. Indeed, by linearity this only needs to be verified if one of the vertices is the code. Let $a, b \in \mathrm{GF}(3) - \{0\}$. The vertices represented by $(0, 0, 0, \ldots, 0)$ and $(a, 0, 0, \ldots, 0)$ are both adjacent to $(-a, 0, 0, \ldots, 0)$ only. The nonadjacent vertices $(0, 0, 0, \ldots, 0)$ and $(a, b, 0, \ldots, 0)$ are adjacent to $(a, 0, 0, \ldots, 0)$ and $(0, b, 0, \ldots, 0)$ only.

Secondly, we give the following alternative construction for the 243-graph. Let H^* be the 5×11 parity check matrix of the perfect ternary Golay code. The columns of H^* are denoted by $x_1, x_2, \ldots, x_{11}$, which are vectors of the vector space $V(5, 3)$ of dimension 5 over GF(3). There are 22 vectors of type $\pm x_i$, and 220 vectors of type $\pm x_i \pm x_j$; $i \neq j$; $i, j = 1, 2, \ldots, 11$. These vectors are pairwise distinct, since by the minimum weight 5 of the code no 4 of the vectors $x_1, x_2, \ldots, x_{11}$ are dependent. Therefore, these vectors and the zero vector represent all vectors of $V(5, 3)$.

Now consider the 3^5 vectors of $V(5, 3)$ as the vertices of a graph. Any two vertices are called adjacent if the difference of the corresponding vectors is one of $\pm x_1, \pm x_2, \ldots, \pm x_{11}$. Again, the triangle and the quadrangle property are easily verified.

Finally, we indicate a third construction of the 243-graph, similar to the construction of the 2048-graph obtained from the (24, 12) binary Golay code in Goethals and Seidel [1970]. This construction depends on the compositions of the codewords of the (12, 6) ternary Golay code. The composition of a word is the unordered set of values of its coordinates (e.g., the composition of $(0, 0, 1, 2, 1, 0, 2, 2, 1, 0, 1, 1)$ is $0^4 1^5 2^3$).

Lemma. *Every word of composition* $0^9 1^3$ *lies in some coset of the* (12, 6) *ternary Golay code which contains two words of composition* $0^9 1^3$, *two words of composition* $0^9 2^3$, *and no other words of weight* $\leqslant 3$.

This lemma may be proved by investigating the distribution of words of weight 3 in the cosets of the (12, 6) code; we shall not present the details here.

Now the construction runs as follows: It is wellknown that the (12, 6) ternary Golay code contains as a subcode the (12, 1) repetition code whose three vectors have compositions 0^{12}, 1^{12}, and 2^{12}. The 243 vertices of the graph are associated with the quotient of the (12, 6) ternary Golay code modulo its (12, 1) repetition subcode. Two vertices are adjacent iff the difference of the corresponding words contains only two elements of GF(3) in its composition. It is trivially verified that this definition of adjacency is reflective, and independent of the representative word of the vertex. The triangle and the quadrangle property are verified by use of the lemma.

References

E. R. Berlekamp, 1968, *Algebraic Coding Theory* (McGraw-Hill, New York).

R. C. Bose, 1961, On some connections between the design of experiments and information theory, *Bull. Intern. Statist. Inst.* **38** (4), 257–271.

R. C. Bose, 1963, Strongly regular graphs, partial geometries and partially balanced designs, *Pacif. J. Math.* **13**, 389–419.

W. C. Bridges, and H. J. Ryser, 1969, Combinatorial designs and related systems, *J. Algebra* **13**, 432–446.

H. S. M. Coxeter, 1958, Twelve points in *PG*(5, 3) with 95040 self-transformations, *Proc. Roy. Soc. London* A **247**, 151–165.

J. M. Goethals, and J. J. Seidel, 1967, Orthogonal matrices with zero diagonal, *Canad. J. Math.* **19**, 1001–1010.

J. M. Goethals, and J. J. Seidel, 1970, Strongly regular graphs derived from combinatorial designs, *Canad. J. Math.* **22**, 597–614.

M. Golay, 1949, Notes on digital coding, *Proc. I.R.E.* **37**, 637.

A. J. Hoffman, and R. R. Singleton, 1960, On Moore graphs with diameters 2 and 3, *I.B.M. J. Res. Develop.* **4**, 497–504.

A. J. Hoffman, 1963, On the polynomial of a graph, *Am. Math. Monthly* **70**, 30–36.

N. Robertson, (private communication).

J. J. Seidel, 1968, Strongly regular graphs with (−1, 1, 0) adjacency matrix having eigenvalue 3, *Linear Algebra Appl.* **1**, 281–298.

J. J. Seidel, 1969, Strongly regular graphs, *Recent Progress in Combinatorics* (W. T. Tutte, ed.; Academic Press, New York), pp. 185–198.

J. H. van Lint, 1969, 1967–1969 Rept. Discrete Mathematics Group, Technol. Univ. Eindhoven Rept. 69–WSK–04.

J. H. van Lint, and J. J. Seidel, 1966, Equilateral point sets in elliptic geometry, *Koninkl. Ned. Akad. Wetensch. Proc.* A, **69** (= *Indag. Math. 28*), 335–348.

J. N. Srivastava et al., eds., *A Survey of Combinatorial Theory*
© North-Holland Publishing Company, 1973

CHAPTER 4

Characterization Problems of Combinatorial Graph Theory†

R. C. BOSE

Colorado State University, Fort Collins, Colo., U.S.A.

1. Introduction

Combinatorial mathematics or combinatorial theory concerns itself with the problems of operations on or arrangements and selections from a finite or discrete set. Though mathematicians have concerned themselves with specific combinatorial problems since ancient times, the emergence of combinatorial mathematics as a distinct discipline is of comparatively recent date. Many classical results of combinatorial mathematics were obtained in connection with the development of probability theory since the times of Pascal and Fermat. Another fertile source of combinatorial problems was provided by the work of R. A. Fisher and his associates (Fisher [1925, 1936], Fisher and Yates [1938]) when they showed that the design and analysis of experiments, in which the response of an experimental unit to a treatment or treatments applied to it is subject to statistical variation, depends on the solution of certain combinatorial problems. The theory of statistically controlled experiments has had a considerable development since the pioneering work of Fisher and has required at every stage the solution of new combinatorial problems. The methods first developed in this connection have found applications in other situations, for example in the construction of error correcting codes (Bose and Ray-Chaudhuri [1960], Peterson [1961], Berlekamp [1968]). Another by-product of this work has been the study of certain types of graphs intimately connected with designs which were originally developed for the purpose of statistical experiments. It is the object of this paper to review some of the problems in this area and to indicate where further research is desirable.

2. Association schemes and partially balanced designs

Given v treatments, 1, 2, . . ., v, a relation satisfying the following conditions is said to be an m-class association scheme:

† The research in this report was supported by the National Science Foundation under Grant No. GP 30958X.

(a) Any two treatments are either 1st, 2nd, ... or mth associates, the relation of association being symmetrical, i.e., if the treatment α is the ith associate of the treatment β, then β is the ith associate of the treatment α.

(b) Each treatment α has n_i ith associates, the number n_i being independent of α.

(c) If any two treatments are ith associates then the number of treatments which are jth associates of α and kth associates of β is p^i_{jk} and is independent of the pair of ith associates α and β.

The numbers

$$v, n_i, p^i_{jk}(i, j, k = 1, 2, \ldots, m), \tag{2.1}$$

are the parameters of the association scheme.

If we have an m-class association scheme then we get a partially balanced incomplete block (PBIB) design with r replications and b blocks based on the association scheme if we can arrange the v treatments into b sets called "blocks" satisfying the following conditions:

(i) Each block contains k treatments (all different).

(ii) Each treatment is contained in r blocks.

(iii) If two treatments α and β are ith associates then they occur together in λ_i blocks, the number λ_i being independent of the pair of ith associates α and β ($i = 1, 2, \ldots, m$).

For a PBIB design based on any association scheme, the parameters of the association scheme may be called the parameters of the first kind, and the additional parameters

$$b, r, k, \lambda_i \, (i = 1, 2, \ldots, m), \tag{2.2}$$

may be called parameters of the second kind. Clearly

$$vr = bk, n_1\lambda_1 + n_2\lambda_2 + \cdots + n_m\lambda_m = r(k-1). \tag{2.3}$$

By definition, the number p^i_{jk} is independent of which pair α, β of ith associates we start with. Considering the pair β, α, we see at once that

$$p^i_{jk} = p^i_{kj}. \tag{2.4}$$

The following relations are easy to prove:

$$\sum_{i=1}^{m} n_i = v-1, \tag{2.5}$$

$$\sum_{k=1}^{m} p^i_{jk} = n_j \qquad \text{if } i \neq j \tag{2.6}$$

$$= n_j - 1 \quad \text{if } i = j,$$

$$n_i p^i_{jk} = n_j p^j_{ik} = n_k p^k_{ij}. \tag{2.7}$$

These relations were proved by Bose and Nair [1939] in their paper introducing the PBIB designs. The concept of an association scheme was first introduced by Bose and Shimamoto [1952] to clarify the analysis of the results of experiments using PBIB designs.

For the construction and combinatorial properties of PBIB designs, reference may be made to Bose and Connor [1952], Shrikhande [1952], Bose, Shrikhande and Bhattacharya [1953], Bose, Clatworthy and Shrikhande [1954], Connor and Clatworthy [1954], Bose and Clatworthy [1955], Bose [1964] and Mesner [1967].

3. Two class association schemes and strongly regular graphs

Two class association schemes are especially important. Bose and Clatworthy [1955] showed that the constancy of all the parameters p^i_{jk} is assured by assuming the constancy of the parameters p^1_{11}, p^2_{11}, n_1 and n_2 (or v). The relations connecting the parameters now become

$$p^1_{12} = n_1 - p^1_{11} - 1 = p^1_{21}, \qquad p^1_{22} = n_2 - n_1 + p^1_{11} + 1, \qquad (3.1)$$

$$p^2_{12} = n_1 - p^2_{11} = p^2_{21}, \qquad p^2_{22} = n_2 - n_1 + p^2_{11} - 1, \qquad (3.2)$$

$$v - 1 = n_1 + n_2, \qquad n_1 p^1_{12} = n_2 p^2_{11}, \qquad n_1 p^1_{22} = n_2 p^2_{12}. \qquad (3.3)$$

In this paper we shall consider only finite graphs without loops or multiple edges. A strongly regular graph G with parameters $(v, n_1, p^1_{11}, p^2_{11})$ is defined to be a graph with v vertices which is regular of degree (valence) n_1 and for which any two adjacent vertices are simultaneously adjacent to p^1_{11} other vertices, and any two non-adjacent vertices are simultaneously adjacent to p^2_{11} other vertices.

The concept of a strongly regular graph is isomorphic to the concept of a two class association scheme. Given a two class association scheme with parameters $v, n_1, p^1_{11}, p^2_{11}$, we can obtain from it a strongly regular graph with parameters $(v, n_1, p^1_{11}, p^2_{11})$ by taking the treatments of the association scheme for the vertices of a graph in which two vertices are adjacent or non-adjacent according as the corresponding treatments of the association scheme are first associates or second associates.

It is therefore convenient to call two vertices of a strongly regular graph G, first associates or second associates according as they are adjacent or non-adjacent. The parameters p^i_{jk} then have obvious meanings, and the relations (3.1), (3.2) and (3.3) hold. The concept of a strongly regular graph was first introduced by Bose [1963].

In what follows we shall assume that $p^2_{11} > 0$. Then a strongly regular graph is a connected graph. If A is the adjacency matrix of a strongly regular graph, then

$$A^2 = (p^1_{11} - p^2_{11})A + p^2_{11}J + (n_1 - p^2_{11})I, \qquad (3.5)$$

where I is the unit matrix of order v, and J is a $v \times v$ matrix with each element unity.

n_1 is a simple eigenvalue of A. There are only two other distinct eigenvalues θ_1 and θ_2 with multiplicities α_1 and α_2, given by

$$\alpha_1, \alpha_2 = \tfrac{1}{2}(v-1)\pm\frac{(2n_1-v+1)+\gamma(v-1)}{2\sqrt{\Delta}} \tag{3.6}$$

where

$$\gamma = p_{11}^1 - p_{11}^2 + 1, \qquad \beta = 2n_1 - 1 - (p_{11}^1 + p_{11}^2), \tag{3.7}$$

$$\Delta = \gamma^2 + 2\beta + 1. \tag{3.8}$$

Since the multiplicities α_1 and α_2 are necessarily integral, a necessary condition for the existence of a strongly regular graph or the corresponding two class association scheme with parameters $v, n_1, p_{11}^1, p_{11}^2$ is that α_1 and α_2 given by (3.6) are integral. For proofs of these results see Connor and Clatworthy [1954], Bose and Mesner [1959].

4. The triangular association scheme and the line graph of a complete graph

We take an $m \times m$ square $m \geqslant 3$ and fill in the $\tfrac{1}{2}m(m-1)$ positions above the leading diagonal by different treatments, taken in any order. The positions in the leading diagonal are left blank, while positions below this diagonal are filled so that the scheme is symmetrical with respect to the leading diagonal. Two treatments in the same row (or same column) are first associates. Two treatments which do not occur in the same row or same column are second associates. The treatment in the ith row and jth column may be denoted by (ij), $i \neq j$. The treatments are therefore the $\tfrac{1}{2}m(m-1)$ unordered pairs (ij), $i \neq j$ formed from the m symbols $1, 2, \ldots, m$. Two treatments are first associates or second associates according as the corresponding pairs have or do not have a common symbol.

The association scheme described above may be denoted by $T_2(m)$. It is readily seen that

$$v = \tfrac{1}{2}m(m-1), \qquad n_1 = 2(m-2), \qquad p_{11}^1 = m-2, \qquad p_{11}^2 = 4. \tag{4.1}$$

Two m-class association schemes can be defined to be isomorphic if there is a $(1, 1)$ correspondence between the treatments such that if two treatments are ith associates in the first scheme, then the corresponding treatments are ith associates in the second scheme. Thus isomorphic association schemes are essentially the same. The parameters of an association scheme are said to characterize it if there is essentially only one association scheme with these parameters. For the case $m = 2$, the parameters of a strongly regular graph may be said to characterize the graph if there is essentially one strongly regular graph with the given parameters.

Connor [1958] considered the question whether the triangular association scheme $T_2(m)$ is characterized by the parameters (4.1). He showed that the answer is in the affirmative if $m > 8$. This result can be translated into graph theoretic language.

The line graph H of a graph G is defined to be a graph whose vertices correspond to the edges of G, and in which two vertices are adjacent or non-adjacent according as the corresponding edges of G have or do not have a common vertex. Let K_m be the complete graph with m vertices which may be denoted by the symbols $1, 2, \ldots, m$. The edge joining the vertices i and j can then be denoted by the unordered pair (ij). Two edges have a common vertex if and only if the corresponding pairs have a common symbol. This shows that the line graph of K_m is isomorphic with the strongly regular graph of the triangular association scheme $T_2(m)$ and has the parameters (4.1). Hence Connor's result can be stated as follows: If $m > 8$ then a strongly regular graph with parameters (4.1) is the line graph of the complete graph K_m.

The case $m \leqslant 6$ was investigated by Shrikhande [1959a], and the case $m = 7$ by Hoffman [1960a]. They showed that in these cases too the parameters (4.1) characterize the triangular association scheme $T_2(m)$ or the line graph of the complete graph K_m.

However, it is surprising that for $m = 8$ the parameters (4.1) do not characterize the line graph of K_8. It turns out that there are three other non-isomorphic graphs. This was demonstrated by Hoffman [1960b] and Chang [1959, 1960].

Hoffman's proofs use the fact that if G is a strongly regular graph with parameters (4.1), then the eigenvalues of its adjacency matrix A are

(a) $2m-4$ with multiplicity 1 and eigenvector $(1, 1, \ldots, 1)$,

(b) $m-4$ with multiplicity $n-1$,

(c) -2 with multiplicity $v-n$.

Shrikhande's and Chang's proofs use purely combinatorial arguments.

It is convenient to denote a two class association scheme and the strongly regular graph corresponding to it by the same symbol. Thus we shall denote by $T_2(m)$ the line graph of the complete graph K_m with m vertices. A strongly regular graph with parameters (4.1) which is not isomorphic to $T_2(m)$ may be called a non-$T_2(m)$ graph.

It is of great interest to see how the three non-$T_2(8)$ graphs are related to the graph $T_2(8)$. This question has been studied by Seidel [1967]. Let G be a strongly regular graph with parameters

$$v, n_1, p_{11}^1, p_{11}^2.$$

We can obtain another graph G^* from it by the following process: Let the set of vertices V of G be divided into disjoint subsets V_1 and V_2, $V = V_1 \cup V_2$. G^* has the same set of vertices as G. Two vertices of G^* both of which belong to V_1 or to V_2 are adjacent or non-adjacent in G^* according as they are adjacent or non-adjacent in G. Two vertices of G^* one of which belongs to V_1 and the other to V_2 are adjacent in G^* if they are non-adjacent in G, and non-adjacent in G^* if they are adjacent in G. Then G^* may be said to be derived from G by complementation with respect to V_1 and V_2. If G^* is strongly regular, it is defined to be S-equivalent to G. Seidel showed that the three distinct non-$T_2(8)$ graphs are all S-equivalent to $T_2(8)$. It can be shown (Bose and Shrikhande [1970]), that the necessary and sufficient conditions for G^* to have the same parameters as G are:

(a) In G_1 each vertex in V_1 is adjacent to exactly half of the vertices in V_2, and each vertex in V_2 is adjacent to exactly half of the vertices in V_1.

(b) $p_{11}^1 + p_{11}^2 = 2n_1 - \frac{1}{2}v$.

The condition (b) is clearly satisfied if we take $m = 8$ in (4.1). The vertices of $T_2(8)$ are the 28 unordered pairs of the symbols $1, 2, \ldots, 8$. The three non-$T_2(8)$ graphs are obtained from $T_2(8)$ by complementation with respect to V_1 and $V_2 = V - V_1$, where

(i) $V_1 = \{(12), (34), (56), (78)\}$,

(ii) $V_1 = \{(12), (34), (56), (78), (13), (24), (57), (68)\}$,

(iii) $V_1 = \{(12), (34), (56), (78), (13), (24), (57), (68), (14), (23), (58), (67)\}$.

5. The $L_r(k)$ association scheme, and net graphs

Consider $v = k^2$ treatments which may be set forth in a $k \times k$ square scheme. For the case $r = 2$, we define two treatments as first associates if they occur together in the same row or same column of the square scheme and second associates otherwise. The scheme so defined may be called the $L_2(k)$ association scheme, with the same notation for the corresponding strongly regular graph. The parameters are given by

$$v = k^2, \qquad n_1 = 2(k-1), \qquad p_{11}^1 = k-2, \qquad p_{11}^2 = 2. \qquad (5.1)$$

If the treatment in the ith row and jth column is denoted by (i, j), then the k^2 treatments are denoted by the k^2 ordered pairs formed from the k symbols $1, 2, \ldots, k$. Two treatments are first associates if and only if the ordered pairs denoting them have a common coordinate.

Shrikhande [1959b] considered the question whether the parameters (5.1) characterize the $L_2(k)$ scheme and showed that the answer is in the affirmative except when $k = 4$. For the case $k = 4$ there is exactly one non-$L_2(4)$ graph which is S-equivalent to the $L_2(4)$ graph and is obtained from it by complementation with respect to V_1 and $V_2 = V - V_1$, where

$$V_1 = \{(1, 1), (2, 2), (3, 3), (4, 4)\}. \qquad (5.2)$$

In the general case $2 \leqslant r \leqslant k+1$, we take a set of $r-2$ mutually orthogonal Latin squares of order k (if such a set exists). For an $L_r(k)$ association scheme we then define two treatments to be first associates if they occur together in the same row or same column of the square scheme, or if they correspond to the same symbol of one of the Latin squares.

The parameters of the $L_r(k)$ association scheme are given by

$$v = k^2, \quad n_1 = r(k-1), \quad p_{11}^1 = (k-2)+(r-1)(r-2), \quad p_{11}^2 = r(r-1). \quad (5.3)$$

A set of $r-2$ mutually orthogonal Latin squares is isomorphic to a net (r, k) of degree r and order k. A net (r, k) is a set of k^2 points and rk lines, such that a point may or may not be incident with a line. The incidence obeys the following properties:

(1) Each point is incident with exactly r lines.

(2) Each line is incident with exactly k points.

(3) The lines fall into r parallel classes of k lines each. Distinct lines of the same parallel class have no common points and two lines of different classes have one common point.

To obtain a set of $(r-2)$ mutually orthogonal Latin squares from a net (r, k), we may proceed as follows: Let the parallel classes be denoted by $(X), (Y), (V_1), (V_2), \ldots, (V_{r-2})$. The lines of a parallel class can be assigned the numbers $1, 2, \ldots, k$ in some order or other. We now take a $k \times k$ square scheme. Then the treatment (i, j) in the ith row and jth column is made to correspond to the point of the net which is the intersection of the line i of (X) and line j of (Y). To obtain the Latin square $[L_t]$, $t = 1, 2, \ldots, r-2$, we put in the cell (i, j) of L_t the number of the line of (V_t) which passes through the point corresponding to (i, j). Hence, alternatively we may take the points of the net (r, k) as the treatments of a two class association scheme $L_r(k)$, where two treatments are first associates if and only if the corresponding points are incident with the same line of the net. The corresponding strongly regular graph may be called a net graph $L_r(k)$.

When $r > 2$, there will exist in general non-isomorphic net graphs $L_r(k)$ for given r and k.

One may ask the question whether a strongly regular graph with the parameters (5.3) is indeed a net graph $L_r(k)$. Bruck [1963] showed that the answer is in the affirmative if

$$k > \tfrac{1}{2}(r-1)(r^3-r^2+r+2). \quad (5.4)$$

We therefore have to generalize our concept of characterization. Given a certain class C of graphs, we may say that the class C is characterized by certain parameters or certain properties if any graph with those parameters or properties must belong to the class C. For example we can consider the class of $L_r(k)$ graphs with given r and k. Then if the condition (5.4) is satisfied, the parameters (5.3) characterize the class of net graphs $L_r(k)$.

The same result can be expressed in another manner. A strongly regular graph with parameters (5.3) may be called a pseudo-net graph $L_r(k)$. We can then say that a pseudo-net graph $L_r(k)$ is a net graph $L_r(k)$ provided that Bruck's condition (5.4) holds.

No general results are known for the case when (5.4) is not satisfied, but some progress has been made for small values of r and k. Thus Shrikhande and Bhat [1971] have studied the case $r = 3$, $k = 5$.

6. Embedding theorems

(A) A BIB design with parameters $(v^*, b^*, r^*, k^*, \lambda^*)$ is an arrangement of v^* treatments into b^* sets (called blocks) satisfying the following conditions:
 (i) Each block contains k^* objects (all distinct).
 (ii) Each object occurs in r^* blocks.
 (iii) Each pair of objects occurs together in exactly λ^* blocks.

A BIB design is called symmetric if $v^* = b^*$, $r^* = k^*$. We can then say that it has parameters v^*, k^*, λ^*. If D^* is a symmetric BIB design with parameters v^*, k^*, λ^*, then any two blocks of D^* intersect in exactly λ^* treatments (Bose [1939]) and by deleting one block of D^* and all the treatments contained in it we obtain another BIB design D_0, the residual of D^* which has the parameters

$$v_0 = v^* - k^*, \quad b_0 = v^* - 1, \quad r_0 = k^*, \quad k_0 = k^* - \lambda^*, \quad \lambda_0 = \lambda^* \qquad (6.1)$$

where $\lambda^*(v^* - 1) = k^*(k^* - 1)$.

It is interesting to ask under what conditions this process is reversible, i.e., given a BIB design with parameters (6.1), where $\lambda_0(v_0 - 1) = r_0(k_0 - 1)$, i.e. $\lambda^*(v^* - 1) = k^*(k^* - 1)$, can we embed it in a symmetric BIB design with parameters v^*, k^*, λ^*. The answer is in the affirmative when $\lambda^* = 1$, for in this case D_0 is isomorphic to an affine plane and D^* is isomorphic to a projective plane in which D_0 can be embedded. A counter example was given by Bhattacharya [1944] for the case $\lambda^* = 3$. He gave a design D_0 with parameters

$$v_0 = 16, \qquad b_0 = 24, \qquad r_0 = 9, \qquad k_0 = 6, \qquad \lambda_0 = 3 \qquad (6.2)$$

for which two blocks intersect in 4 treatments and which therefore cannot be embedded in a symmetric BIB design with parameters $v^* = 25$, $k^* = 9$, $\lambda^* = 3$.

The question was settled for $\lambda^* = 2$ by Hall and Connor [1953], who showed that D_0 can be embedded in a symmetric design D^* such that D_0 is the residual of D^*. Shrikhande [1960] gave an alternative proof of the Hall-Connor theorem depending on the fact that the parameters (4.1) characterize the triangular association scheme. Let D_0 be a BIB design with parameters

$$v_0 = \frac{(k^* - 1)(k^* - 2)}{2}, \qquad b_0 = \frac{k^*(k^* - 1)}{2},$$

$$r_0 = k^*, \qquad k_0 = k^* - 2, \qquad \lambda_0 = 2. \qquad (6.3)$$

It is readily proved that two blocks of D_0 intersect in either one or two treatments. If we take the blocks of D_0 as the treatments of an association scheme and call two blocks first associates if they intersect in one treatment and second associates if they intersect in two treatments then it can be shown (Shrikhande [1952], Hall and Connor [1953]) that we get a two class association scheme with parameters

$$v = v_0 = \frac{k^*(k^*-1)}{2}, \qquad n_1 = 2(k^*-2), \qquad p_{11}^1 = k^*-2, \qquad p_{11}^2 = 4 \quad (6.4)$$

Hence if $k^* \neq 8$, the association scheme must be the triangular association scheme $T_2(k^*)$. If we take k^* symbols, $1, 2, \ldots, k^*$, then to each block we can assign an unordered pair ij of symbols such that two blocks are first associates, i.e., intersect in one treatment if and only if the corresponding pairs have one symbol in common. We now take k^* new treatments. If (ij) is the pair assigned to a block B_{ij}, then we extend the block by adding two new treatments i and j. Finally we add a new block consisting of all the new treatments. Then we have $\frac{1}{2}(k^{*2}-k^*+2)$ blocks of size k^*, such that any two blocks intersect in exactly two treatments. Thus by extension we get a symmetric BIB design with parameters v^*, k^*, λ^*.

It is interesting to note that Hall and Connor's original proof does not cover the case $k^* = 8$, which is the case when the parameters (6.4) do not characterize the corresponding association scheme. Connor [1952] gave a separate proof for the non-existence of (6.3) in this case. An alternative treatment is due to Shrikhande [1967].

(B) Given a finite projective plane Π of even order q, it is well known that we can find a set of $q+2$ points in Π such that no three are collinear. They form an oval in Π. If we delete the points of the oval from Π, the incidence structure of the remaining points and lines has the following properties:

(i) There are q^2-1 points.

(ii) There are two types of lines. Lines of type I are each incident with $q+1$ points. Lines of type II are each incident with $q-1$ points.

(iii) Each point is incident $\frac{1}{2}q$ lines of type I and $\frac{1}{2}(q+2)$ lines of type II.

(iv) Any two distinct points are both incident with exactly one line which may be of type I or type II.

Let D be the incidence structure with the above properties. We may ask the question whether it is possible to embed D in a projective plane of order q by suitably extending the lines of type II.

Bose and Shrikhande [1972a] have proved that the answer is in the affirmative except possibly for the case $q = 6$. Now a projective plane of order q is non-existent by the Bruck-Ryser theorem [1949] if $q \equiv 2 \pmod 4$ and the square free part of q contains a prime factor $p \equiv 3 \pmod 4$. It follows that for these values of q (except for $q = 6$), D cannot exist. The question whether D can exist when $q = 6$ is still open.

Bose's and Shrikhande's proof essentially depends on the characterization theorem for the triangular association scheme. They prove that if an incidence structure D with the properties (i), (ii), (iii), (iv) exists, then there are $v = \frac{1}{2}(q+2)(q+1)$ lines of type II and $\frac{1}{2}q(q-1)$ lines of type I satisfying the following conditions:

(a) Any two lines of type I intersect in exactly one point.

(b) A line of type I intersects a line of type II in exactly one point.

(c_1) Given any line l of type II, there are $n_1 = 2q$ lines of type II not intersecting l. The remaining lines of type II intersect l in a single point.

(c_2) Given any two non-intersecting lines l_1 and l_2 of type II, there are exactly q lines of type II which intersect neither l_1 nor l_2.

(c_3) Given any two intersecting lines l_1' and l_2' of type II, there are exactly four lines of type II which intersect neither l_1' nor l_2'.

If we now take the lines of type II as the treatments of an association scheme and call two lines first associates if they do not intersect and second associates if they intersect in a point, then the association scheme has the parameters (4.1) with $m = q+2$. Hence it is the triangular association scheme except possibly for the case $q = 6$. Hence if $q \neq 6$ we can associate with each line of type II an unordered pair (ij) whose elements have been chosen from a set of $q+2$ symbols, $1, 2, \ldots, q+2$, such that the pairs associated to two lines of type II have one symbol or no symbol in common according as the lines are first associates or second associates. If we now choose $q+2$ new points, one corresponding to each symbol, and to the line of type II which is associated to the pair (ij) we adjoin the new points i and j, then after extension any two lines of type II intersect in one point. This taken together with (a), (b) and the properties (i), (ii), (iii), (iv) shows that the extended incidence structure is a projective plane of order q.

(C) There cannot exist more than $k-1$ mutually orthogonal Latin squares of order k. If a set of $k-1$ mutually orthogonal Latin squares of order k exists, it is called a complete set. If there exists a set of $r-2$ mutually orthogonal Latin squares of order k, then $d = k-r+1$ can be called the deficiency of the set.

It is a folk theorem that a set of mutually orthogonal Latin squares of deficiency one can always be completed. Shrikhande [1961] showed that the same holds for a set of deficiency 2 except possibly for $k = 4$. As a matter of fact, it is well known that there does not exist an orthogonal mate for a cyclic Latin square of order 4. Shrikhande's result shows that this is the only counter example for sets of deficiency 2.

Shrikhande's result was generalized by Bruck (1963), who showed that if there exists a set of mutually orthogonal Latin squares of order k and deficiency d then we can always extend it to a complete set by adjoining d new squares provided that

$$k > \tfrac{1}{2}(d-1)(d^3-d^2+d+2). \tag{6.5}$$

This result can also be stated in geometrical language. Given a net (r, k) of degree r and order k, we can define its deficiency d by setting $d = k-r+1$. If it is possible to add d new parallel classes to the existing r parallel classes, the extended net becomes an affine plane. Hence the Shrikhande-Bruck Theorem can be stated as follows: Given a net (r, k) of order k and degree r, if the deficiency $d = k-r+1$ satisfies (6.5), then the net can be embedded in an affine plane of order k.

The proofs of Shrikhande and Bruck depend essentially on the characterization theorem for net graphs. Suppose a net (r, k) is given, then the corresponding strongly regular graph G has the parameters (5.3). If $\bar{G}$ is the complementary graph, i.e. a graph with the same vertices as G such that two vertices of $\bar{G}$ are adjacent if and only if they are non-adjacent in G, then it can be proved that $\bar{G}$ is a strongly regular with parameters

$$v = k^2, \quad n_1 = d(k-1), \quad p_{11}^1 = (k-2)+(d-1)(d-2), \quad p_{11}^2 = d(d-1). \quad (6.6)$$

This shows that $\bar{G}$ is a pseudo-net graph $G_d(k)$. Hence from the characterization theorem for net graphs, $G_d(k)$ is a net graph provided (6.5) is satisfied. Let G_N and $\bar{G}_N$ be the nets corresponding to the graphs G and $\bar{G}$. They have the same k^2 points. Through any point there pass r lines of G_N (one belonging to each parallel class of G_N) and $k-r+1$ lines of $\bar{G}_N$ (one belonging to each parallel class of $\bar{G}_N$). If we now extend G_N by adjoining the lines of $\bar{G}_N$ then for the extended incidence structure Π there pass $r+1$ lines through every point. A line l of G_N and a line $\bar{l}$ of $\bar{G}_N$ cannot intersect in more than one point. In fact if l and $\bar{l}$ intersect in P_1 and P_2, then the vertices corresponding to P_1 and P_2 would be adjacent in both G and $\bar{G}$ which is a contradiction. Consider the parallel class of G_N containing l. Let its lines be $l = l_1, l_2, \ldots, l_k$. They contain between them all the points of Π. Since l has k points it must intersect each of $l_1, l_2, \ldots, l_k$ in exactly one point. We have thus shown that l and $\bar{l}$ intersect in exactly one point. Hence the lines of Π are divisible into $r+1$ parallel classes such that any two lines intersect in a point or do not intersect according as they do not belong or belong to the same parallel class. Again two points are incident with exactly one line of Π, as the corresponding vertices are adjacent in exactly one of G and $\bar{G}$. Hence Π is an affine plane.

7. Partial geometries

The results of Connor and Bruck were generalized by Bose [1963] who introduced the concept of partial geometries. A partial geometry (r, k, t) is an incidence structure with two kinds of elements, "points" and "lines", satisfying the following axioms:

   A1. A pair of distinct points is not incident with more than one line.
   A2. Each point is incident with exactly r lines.
   A3. Each line is incident with exactly k points.

A4. Given a point P not incident with a line l, there exist exactly t lines ($t \geqslant 1$) which are incident with P, and also incident with some point incident with l.

The axiom A1 clearly implies its dual:

A'1. A pair of distinct lines is not incident with more than one point.

The axiom A4 is clearly self dual. Hence the existence of the partial geometry (r, k, t) implies the existence of the dual partial geometry (k, r, t).

The number of points v, and the number of lines b in a partial geometry is determined by the parameters (r, k, t). It can be shown (Bose [1963]) that

$$v = k[(r-1)(k-1)+t]/t, \tag{7.1}$$
$$b = r[(r-1)(k-1)+t]/t. \tag{7.2}$$

The graph G of a partial geometry (r, k, t) is a graph whose vertices are the points of the geometry and for which two vertices are adjacent if and only if the corresponding points are incident with the same line. Then G is a strongly regular graph with parameters.

$$v = k[(r-1)(k-1)+t]/t, \qquad n_1 = r(k-1),$$
$$p_{11}^1 = (r-1)(t-1)+(k-2), \qquad p_{11}^2 = rt, \tag{7.3}$$

where

$$1 \leqslant t \leqslant r, \qquad 1 \leqslant t \leqslant k. \tag{7.4}$$

The multiplicities of the eigenvalues α_1 and α_2 of the adjacency matrix of G are given by (3.6). Since these are integral, it follows that a necessary condition for the existence of a partial geometry (r, k, t) is that

$$\alpha_1 = \frac{rk(r-1)(k-1)}{t(k+r-t-1)} \tag{7.5}$$

is integral.

A strongly regular graph with parameters (7.3) for which r, k, t satisfy (7.4) is called a pseudo-geometric graph (r, k, t). A pseudo-geometric graph (r, k, t) is called a geometric graph if there exists a partial geometry (r, k, t) of which it is a graph.

When G is geometric (r, k, t) there will exist in G a set Σ of distinct cliques $K_1, K_2, \ldots, K_b$ corresponding to the lines of the geometry satisfying the following axioms:

A*1. Any two adjacent vertices of G are contained in one and only one clique of Σ.

A*2. Each vertex of G is contained in r cliques of Σ.

A*4. If P is a vertex of G not contained in a clique K_i of Σ, there are exactly t vertices of K_i which are adjacent to G.

Conversely, if there exists a graph with a set of cliques Σ satisfying A*1–A*4, then it is isomorphic to a partial geometry (r, k, t), the vertices of the graph being the points of the geometry and the cliques being the lines of the geometry. The containing contained relation in G is the incidence relation in the geometry.

Bose [1963] proved the following theorem:

A pseudo-geometric graph (r, k, t) is geometric if

$$k > \tfrac{1}{2}[r(r-1)+t(r+1)(r^2-2r+2)]. \tag{7.6}$$

It is readily seen that the existence of a partial geometry $(2, k, 2)$ is equivalent to the existence of a triangular association scheme $T_2(k+1)$, and the existence of a partial geometry $(r, k, r-1)$ is equivalent to the existence of a net (r, k) of degree r and order k.

Putting $r = t = 2$, Bose's theorem reduces to Connor's [1958] result that for $m > 8$, parameters (4.1) characterize the triangular association scheme. Similarly putting $t = r-1$, we obtain Bruck's [1963] result that a pseudo-net graph (r, k) is a net graph if (5.4) is satisfied.

Given a BIB design D^* with parameters $(v^*, b^*, r^*, k^*, \lambda^* = 1)$, we can obtain the dual design D by taking for the treatments of D the blocks of D^* and for the blocks of D^* the treatments of D. Then any two blocks of D intersect in exactly one treatment. The treatments and blocks of D can then be taken as the points and lines of a partial geometry (r, k, t), where $r = k^*$ and $k = r^*$. If we call two treatments of D first associates or second associates according as they do or not occur together in the same block, we have a two class association scheme, and a corresponding strongly regular graph G with parameters

$$v = k[r(k-1)+1]/r, \qquad n_1 = r(k-1),$$
$$p_{11}^1 = (r-1)^2+(k-2), \qquad p_{11}^2 = r^2. \tag{7.7}$$

This association scheme may be called the linked block association scheme $LB_r(k)$. A strongly regular graph with parameters (7.7) is defined to be a pseudo $LB_r(k)$ graph. If

$$k > \tfrac{1}{2}r(r^3-r^2+r+1), \tag{7.8}$$

then a pseudo-$LB_r(k)$ graph must be the graph of an $LB_r(k)$ association scheme. Thus the parameters (7.7) characterize the class of linked block association schemes $LB_r(k)$. In general, there exist non-isomorphic linked block association schemes with the same parameters.

Non-isomorphic solutions of pseudo $LB_3(6)$ graphs have been studied by Shrikhande and Bhat [1970].

It is an important combinatorial problem to determine all possible non-isomorphic geometries with given parameters (r, k, t). As we have seen, the geometry (r, k, r) is isomorphic to a linked block design and its dual (k, r, r) is isomorphic to a BIB design with parameters $v^* = k(r-1)+1$, $b^* = k[r(k-1)+1]/r$, $r^* = k$, $k^* = r$, $\lambda^* = 1$. Again the partial geometry $(r, k,$

$r-1$) is isomorphic to a net (r, k), and consequently its dual $(k, r, r-1)$ is isomorphic to the dual of a net. Other known partial geometries have parameters $(q+1, q^n+1, 1)$ or $(q^n+1, q+1, 1)$, $n = 1, 2, 3$. They have been called generalized quadrangles. The cases $n = 1$ and 2, where q is a prime power, were mentioned by Bose [1963] in his paper introducing the concept of a partial geometry. They have been studied by Benson [1966], Ahrens and Szekeres [1969], Feit and Higman [1964], Hall [1972] and Payne [1970a], [1970b], [1971], [1972]. A number of non-isomorphic solutions is now known.

Recently, Thas [1972] has obtained the partial geometries $[\frac{1}{2}(q+2, q-1, \frac{1}{2}(q-2)]$ and $[\frac{1}{2}q(q-1), q, \frac{1}{2}(q-2)]$.

We may denote by $p(r, t)$ the function of r and t appearing on the right hand side in (7.6). The properties of pseudo-geometric graphs (r, k, t) for which $k < p(r, t)$ deserve study. Bose and Shrikhande [1972b] have studied pseudo-geometric graphs $(q^2+1, q+1, 1)$ and established sufficient conditions for such a graph to be geometric.

8. Characterization of some classes of edge regular, and other related graphs

(A) We will now consider characterization problems connected with a class of graphs more general than strongly regular graphs.

Consider a finite graph G without loops or multiple edges. A chain $C = (x_0, x_1, \ldots, x_t)$ of length t is a sequence of $t+1 \geq 2$ vertices x_i such that any two consecutive vertices in the chain are adjacent. C is said to join x_0 and x_t. The graph G is connected if and only if there exists a chain joining any two distinct vertices x and y. Every graph G can be uniquely partitioned into $s \geq 1$ subgraphs $G_1, G_2, \ldots, G_s$ such that each G_i is connected and no vertex in G_i is adjacent to a vertex in G_j, $i \neq j$. The subgraphs $G_1, G_2, \ldots, G_s$ are the connected components of G. For any two vertices x and y in the same connected component of G the distance $d(x, y)$ between x and y is the length of the shortest chain joining x and y. Clearly $d(x, y) = 1$ if and only if x and y are adjacent. We say that G, is regular of degree (valence) n_1 if each vertex is adjacent to exactly n_1 other vertices. We may denote by $A(x)$ the set of vertices adjacent to any given vertex, and by $|S|$ the cardinality of any set S. Then for a regular graph G of degree n_1, $|A(x)| = n_1$ for each vertex of G. We shall denote by $\Delta(x, y) = |A(x) \cap A(y)|$ the number of vertices which are simultaneously adjacent to two distinct vertices x and y of G. We say that a graph is edge regular of edge degree p_{11}^1 if for any two distinct vertices x and y which are adjacent $\Delta(x, y) = p_{11}^1$. Hence a strongly regular graph is edge regular but the converse is not necessarily true. For a strongly regular graph, $\Delta(x, y) = p_{11}^2$, if x and y are any two non-adjacent vertices.

A set K of mutually adjacent vertices of a graph G is called a clique of order $|K|$. K is said to be complete if we cannot adjoin a vertex x of G so that $x \cup K$ is a clique. Thus if K is complete then any vertex x of G not belonging to K is non-adjacent to at least one vertex of K.

Bose and Laskar [1967] proved the following theorem:

Let G be a graph satisfying the following conditions:

(a_1) G is regular with degree $n_1 = r(k-1)$.

(a_2) G is edge regular with edge degree $p_{11}^1 = k-2+\alpha$.

(a_3) If x and y are non-adjacent vertices of G, then $\Delta(x, y) \leqslant 1+\beta$, where $r \geqslant 1, k \geqslant 2, \beta \geqslant 0$ are fixed integers.

Define a grand clique K as a complete clique whose order $|K| \geqslant k-(r-1)\alpha$. If

$$k > \max[p(r, \alpha, \beta), \rho(r, \alpha, \beta)], \tag{8.1}$$

where

$$p(r, \alpha, \beta) = 1 + \tfrac{1}{2}(r+1)(r\beta - 2\alpha), \tag{8.2}$$

$$\rho(r, \alpha, \beta) = 1 + \beta + (2r-1)\alpha, \tag{8.3}$$

then

(i) each vertex of G is contained in exactly r grand cliques;

(ii) each pair of adjacent vertices is contained in exactly one clique.

(B) Consider a set S of m symbols $1, 2, \ldots, m$ where $m \geqslant 2$. We can form $\binom{m}{q}$ unordered q-plets of q distinct elements of m. Let G be a graph whose vertices are the unordered q-plets of S. Let two vertices of G be adjacent if the corresponding q-plets have exactly $q-1$ symbols of S in common. We shall call G a $T_q(m)$ graph. Clearly G has the following properties:

(b_0) The number of vertices in G is $\binom{m}{q}$.

(b_1) G is regular with degree $n_1 = q(m-q)$.

(b_2) G is edge regular with edge degree $p_{11}^1 = m-2$.

(b_3) If x and y are non-adjacent vertices of G, then $\Delta(x, y) \leqslant 4$.

We may ask conversely under what conditions a graph G possessing the properties (b_0)–(b_3) is a $T_q(m)$ graph. If we set

$$r = q, \qquad k = m-q+1, \qquad \alpha = q-1, \qquad \beta = 3,$$

then the conditions (a_1), (a_2), (a_3) of the Bose-Laskar theorem are satisfied:

$$p(r, \alpha, \beta) = 1 + \tfrac{1}{2}(q+1)(q+2),$$
$$\rho(r, \alpha, \beta) = 4 + (2q-1)(q-1),$$
$$\rho(r, \alpha, \beta) - p(r, \alpha, \beta) = \tfrac{3}{2}(q-1)(q-2) \geqslant 0.$$

The condition (8.1) now becomes

$$m > 2q(q-1)+4. \tag{8.4}$$

Also, a complete clique with

$$|K| > m-q(q-1)$$

is a grand clique. Hence if (8.4) is satisfied then each vertex of G is contained in exactly m grand cliques and each pair of adjacent vertices is contained in exactly one grand clique. Bose and Laskar [1967] showed for the case $m = 3$

and Dowling [1969] for the general case that if we take the grand cliques of G as the vertices of a new graph G^* and consider two vertices of G^* adjacent if the corresponding grand cliques have a vertex of G in common, then G^* satisfies the conditions $(b_0)-(b_3)$ if we replace m by $m^* = m-1$. Using induction, we have the following theorem:

If for a graph G the conditions $(b_0)-(b_4)$ are satisfied then G is the graph $T_q(m)$ if $m > 2q(q-1)+4$.

Notice that the case $m = 2$ gives Connor's result that if $m > 8$, then a strongly regular graph with parameters (4.1) is the $T_2(m)$ graph.

In particular, if a graph G satisfies the conditions $(b_1)-(b_4)$ with $q = 3$, then G is a $T_3(m)$ graph if $m > 16$. Aigner [1969] has shown that the same holds if $m < 9$. The question is open for $9 \leqslant m \leqslant 16$ though no exceptional cases are known. For $q > 3$, nothing is known about the case when $m \leqslant 2q(q-1)+4$.

(C) A cubic lattice graph of order m is a graph G whose vertices can be identified with the ordered triplets on m symbols so that two vertices are adjacent if the corresponding triplets have common symbols in exactly two positions. If G is a cubic lattice graph of order m then G possesses the following properties:

(c_0) G has $v = m^3$ vertices.

(c_1) G is regular of degree $3(m-1)$.

(c_2) G is edge regular with edge degree $p_{11}^1 = m-2$.

(c_3) If x and y are non-adjacent vertices of G, then $\Delta(x, y) = 2$, if $d(x, y)=2$.

Laskar [1967] and Dowling [1968] show conversely that if $m > 7$, then a graph G possessing the properties $(c_0)-(c_3)$ must be the cubic lattice graph of order m. Laskar's original characterization had an additional assumption which was eliminated by Dowling.

(D) The graph G of a **BIB** design D with parameters $(v^*, b^*, r^*, k^*, \lambda^*)$ is a graph with $v = v^*+b^*$ vertices, where each vertex corresponds to a treatment or block of D. Two vertices are adjacent if and only if one vertex corresponds to a treatment and the other to a block, and the treatment is contained in the block. Thus G is a bipartite graph in which the vertices are divisible into two subsets, one corresponding to treatments and the other to blocks. No two vertices of the same subset are adjacent.

The design D is a projective plane of order q when

$$v^* = b^* = q^2+q+1, \qquad r^* = k^* = q, \qquad \lambda^* = 1.$$

Let H be the line graph of the graph G of a projective plane of order q. Then the following properties of H are easy to check.

(d_0) H has $v = (q+1)(q^2+q+1)$ vertices.

(d_1) H is connected and regular of degree $n_1 = 2q$.

(d_3) H is edge regular with edge degree $p^1_{11} = q-1$, i.e. $\Delta(x, y) = q-1$ if $d(x, y) = 1$.

(d_4) $\Delta(x, y) = 1$ if $d(x, y) = 2$.

(d_5) If $d(x, y) = 2$, the number of vertices z such that $d(x, z) = 1$ and $d(y, z) = 2$ is at most $q-1$.

Dowling and Laskar [1967] proved that properties (d_1)–(d_5) characterize the line graph of a projective plane of order q, i.e. any graph H with the above properties must be the line graph of the graph of a projective plane. They show that the condition (d_5) is not redundant by exhibiting a graph which has the properties (d_1)–(d_4) but is not the line graph of a projective plane.

(E) Consider a symmetric BIB design D with parameters v^*, k^*, λ^*. Let G be the graph of D and let H be the line graph of G. Then H has the following properties:

(e_0) H has $v = v^*k^*$ vertices.

(e_1) H is regular and connected of degree $2k^*-2$.

(e_2) H is edge regular with edge degree $p^1_{11} = k^*-2$, i.e. $\Delta(x, y) = k^*-2$ if $d(x, y) = 1$.

(e_3) $\Delta(x, y) \leqslant 2$, if $d(x, y) > 1$.

(e_4) $|A_3(x) \cap A(y)| = k^*-\lambda$ if $d(x, y) = 2$, $\Delta(x, y) = 1$, where $A_3(x)$ denotes the set of vertices at a distance 3 from the vertex x.

(e_5) $d(x, y) \leqslant 3$ for any pair of distinct vertices.

Aigner and Dowling [1969] conversely prove that if the graph H satisfies (e_0)–(e_5) then H is the line graph of the graph of a symmetric BIB design, except in the case $v^* = 7$, $k^* = 4$, $\lambda^* = 2$ in which case there exists a unique exceptional graph which is not the line graph of the graph of a symmetric BIB design. The properties (e_0)–(e_4) are counter parts of the characterizing properties (d_0)–(d_4) of the line graph of a finite projective plane ($\lambda = 1$), except that if $\lambda > 1$ we have to assume $\Delta(x, y) < 2$ instead of $\Delta(x, y) = 1$ when $d(x, y) \geqslant 2$. The increased upper bound required for $\Delta(x, y)$ when $\lambda > 1$ necessitates the addition of (e_5).

The properties (e_4) and (e_5) are not redundant as there exist graphs possessing the properties (e_0)–(e_4) which are not the line graphs of the graph of a projective plane. The determination of all graphs of this type is an interesting problem.

(F) Consider the BIB design D with parameters
$$v^* = q^2, \quad b^* = q^2+q, \quad r^* = q+1, \quad k^* = q, \quad \lambda^* = 1.$$
Then the BIB design D is a finite affine plane of order q. Let G be the graph of D and H the line graph of G. Then H has the following properties:

(f_0) H has $q^2(q+1)$ vertices.

(f_1) H is connected and is of degree $2q-1$.

(f_2) $\Delta(x, y) \geqslant q-2$ if $d(x, y) = 1$.

(f_3) $\Delta(x, y) = 1$ if $d(x, y) = 2$.

Laskar [1970] has conversely shown that if $q \neq 2$, a graph H with the properties (f_0)–(f_3) must be the line graph of the graph of an affine plane of order q. When $q = 2$, there exist at least three exceptional graphs which are not the line graphs of the graph of an affine plane of order 2. It would be of interest to find out all exceptional cases.

9. Characterization of classes of graphs by the eigenvalues of their adjacency matrices

We have here considered the characterization of graphs by means of their geometrical properties. There is a large body of literature where classes of graphs are characterized by the eigenvalues of their matrices. A good review of this area will be found in a paper by Hoffman [1969]. We shall here give only a few illustrative examples to show the kind of results that can be obtained. For shortness, the eigenvalues of the adjacency matrix of a graph may be called the eigenvalues of the graph.

Hoffman and Ray-Chaudhuri [1965a] have proved that the line graph H of a symmetric BIB design with parameters v^*, k^*, λ^* is characterized by the eigenvalues -2, $-2k^*-2$, and $k^*-2\pm(k-\lambda)^{\frac{1}{2}}$ unless $v = 4$, $k = 3$, $\lambda = 2$, when the sufficiency of these conditions fails due to the existence of a single exceptional graph. The case $\lambda = 1$ gives the characterization of the line graph of a projective plane. This was given in an earlier paper by Hoffman [1965]. On comparing the geometrical characterization of Aigner and Dowling with the characterization of Hoffman and Ray-Chaudhuri by eigenvalues, one comes across the intriguing problem of the relationship of the eigenvalues of a graph to its geometrical properties. Apart from the fact that in the case of regular graphs the dominant eigenvalue is the degree of regularity, very little is known. An important tool is the polynomial of a graph introduced by Hoffman [1963] which for regular graphs gives an upper bound to the diameter. For example, the fact that the eigenvalues of H, other than the degree, are -2, $k^*-2\pm(k^*-\lambda)^{\frac{1}{2}}$ and the fact that there are just three immediately yield the property (e_5). Using some of the impossible subgraphs of Hoffman and Ray Chaudhuri [1965a], the fact that -2 is the minimal eigenvalue is easily shown to imply (e_3) except for one possible configuration. Hoffman and Ray-Chaudhuri directly prove edge-regularity (e_2) for $k^* > 4$. The exact nature of the relationship between the eigenvalues $k-2\pm(k-\lambda)^{\frac{1}{2}}$ and the condition (e_4) is unknown.

Again Hoffman and Ray-Chaudhuri [1965b] have shown that the line graph of an affine plane of order q is characterized by its eigenvalues

$$2q-1, \ -2, \ \tfrac{1}{2}[2q-3\pm(4q+1)^{\frac{1}{2}}], \ q-2.$$

There is no exceptional case. However, in Laskar's geometrical characterization, $q = 2$ is an exceptional case.

Ray-Chaudhuri [1967] has shown that if H is a graph such that (i) the degree of each vertex exceeds 46, (ii) the minimum eigenvalue of H is -2, (iii) if x_1 and x_2 are adjacent vertices of H then $\Delta(x_1, x_2) > \Delta(x_i)-2$, $i = 1, 2$, where $\Delta(x_i)$ denotes the degree of x_i, then H is the line graph of some graph G.

Conversely, if H is the line graph of some graph for which the degree of each vertex exceeds 3, then H satisfies the properties (i), (ii) and (iii).

Other papers which may be referred to in connection with the characterization of graphs by their eigenvalues are Benson [1966], Bose and Laskar [1970], Laskar [1969], Hoffman and Singleton [1960] and Singleton [1966].

References

M. Aigner, 1969, A characterization problem in graph theory, *J. Combin. Theory* **6**, 45–55.

M. Aigner and T. A. Dowling, 1969, A geometric characterization of the line graph of a symmetric balanced incomplete block design, Univ. of North Carolina Inst. of Statist., Mimeo Series No. 600, 5.

R. W. Ahrens and G. Szekeres, 1969, On a combinatorial generalization of 27 lines associated with a cubic surface, *J. Austral. Math. Soc.* **10**, 485–492.

C. T. Benson, 1966a, A partial geometry $(q^3+1, q^2+1, 1)$ and corresponding BIB design, *Proc. Am. Math. Soc.* **17**, 747–749.

C. T. Benson, 1966b, Minimal regular graphs of girth eight and twelve, *Canad. J. Math.* **18**, 1091–1094.

E. R. Berlekamp, 1968, *Algebraic Coding Theory* (McGraw-Hill Book Company, New York).

K. N. Bhattacharya, 1944, A new balanced incomplete block design, *Sci. Cult.* **9**, 508.

R. C. Bose, 1939, On the construction of balanced incomplete block designs, *Ann. Eugen. London* **9**, 358–399.

R. C. Bose, 1963, Strongly regular graphs, partial geometries and partially balanced designs, *Pacif. J. Math.* **13**, 389–419.

R. C. Bose, 1964, Combinatorial properties of partially balanced designs and association schemes, *Sankhya*, Ser. A, **25**, 109–136.

R. C. Bose and W. H. Clatworthy, 1955, Some classes of partially balanced designs, *Ann. Math. Statist.* **26**, 212–232.

R. C. Bose, W. H. Clatworthy and S. S. Shrikhande, 1954, Tables of partially balanced designs with two associate classes, *N. Carolina State College Exp. Stat. Tech. Bull.* **107**.

R. C. Bose and W. S. Connor, 1952, Combinatorial properties of group divisible incomplete block designs, *Ann. Math. Statist.* **30**, 367–383.

R. C. Bose and Renu Laskar, 1967, A characterization of tetrahedral graphs, *J. Combin. Theory* **3**, 366–385.

R. C. Bose and Renu Laskar, 1970, Eigenvalues of the adjacency matrix of tetrahedral graphs, *Aeq. Math.* **4**, 37–43.

R. C. Bose and D. M. Mesner, 1959, On linear associative algebras corresponding to the association schemes of partially balanced designs, *Ann. Math. Statist.* **30**, 21–38.

R. C. Bose and K. R. Nair, 1939, Partially balanced incomplete block designs, *Sankhya* **4**, 337–372.

R. C. Bose and D. K. Ray-Chaudhuri, 1960, On a class of error correcting binary group codes, *Inform. Control* **3**, 68–79.

R. C. Bose and T. Shimamoto, 1952, Classification and analysis of partially balanced incomplete block designs with two associate classes, *Ann. Statist. Assoc.* **47**, 151–184.

R. C. Bose and S. S. Shrikhande, 1970, Graphs in which each pair of vertices is adjacent to the same number of other vertices, *Studia Sci. Math. Hungarica* **5**, 181–196.

R. C. Bose and S. S. Shrikhande, 1972a, Embedding the complement of an oval in a projective plane of even order, *J. Discrete Math.*

R. C. Bose and S. S. Shrikhande, 1972b, Geometric and pseudogeometric graphs (q^2+1, $q+1$, 1), *J. Geometry*.

R. C. Bose, S. S. Shrikhande and K. N. Bhattacharya, 1953, On the construction of group divisible incomplete block designs, *Ann. Math. Statist.* **24**, 167–195.

R. H. Bruck, 1963, Finite nets II: Uniqueness and imbedding, *Pacif. J. Math.* **13**, 421–457.

R. H. Bruck and H. J. Ryser, 1949, The non-existence of certain finite projective planes, *Canad. J. Math.* **1**, 88–93.

Chang Li-chien, 1959, The uniqueness and non-uniqueness of the triangular association schemes, *Sci. Record Peking Math. New Ser.* **3**, 604–613.

Chang Li-chien, 1960, Association schemes of partially balanced designs with parameters $v = 28$, $n_1 = 12$, $n_2 = 15$ and $p_{11}^2 = 4$, *Sci. Record Peking Math. New Ser.* **4**, 12–18.

W. S. Connor, 1952, On the structure of balanced incomplete block designs, *Ann. Math. Statist.* **23**, 57–71.

W. S. Connor, 1958, The uniqueness of the triangular association scheme, *Ann. Math. Statist.* **29**, 262–266.

W. S. Connor and W. H. Clatworthy, 1954, Some theorems for partially balanced designs, *Ann. Math. Statist.* **55**, 100–112.

T. A. Dowling, 1968, Note on "A characterization of cubic lattice graphs", *J. Combin. Theory* **5**, 425–426.

T. A. Dowling, 1969, A characterization of the T_m graphs, *J. Combin. Theory* **6**, 251–263.

T. A. Dowling and Renu Laskar, 1967, A geometric characterization of the line graph of a projective plane, *J. Combin. Theory* **3**, 402–410.

W. Feit and G. Higman, 1964, The non-existence of certain generalized polygons, *J. Algebra* **1**, 114–131.

R. A. Fisher, 1925, *Statistical Methods for Research Workers* (Oliver and Boyd, Edinburgh).

R. A. Fisher, 1936, *The Design of Experiments* (Oliver and Boyd, Edinburgh).

R. A. Fisher and F. Yates, 1938, *Statistical Tables* (Oliver and Boyd, Edinburgh).

M. Hall, Jr., 1972, Affine generalized quadrangles (unpublished).

M. Hall, Jr. and W. S. Connor, 1953, An embedding theorem for balanced incomplete block designs, *Canad. J. Math.* **6**, 35–41.

A. J. Hoffman, 1960a, On the uniqueness of the triangular association scheme, *Ann. Math. Statist.* **31**, 492–497.

A. J. Hoffman, 1960b, On the exceptional case in a characterization of the arcs of a complete graph, *IBM J. Res. Develop.* **4**, 497–502.

A. J. Hoffman, 1963, On the polynomial of a graph, *Am. Math. Monthly* **70**, 30–36.

A. J. Hoffman, 1965, On the line graph of a projective plane, *Proc. Am. Math. Soc.* **16**, 297–302.

A. J. Hoffman, 1969, The eigenvalues of the adjacency matrix of a graph, *Combinatorial Mathematics and its Applications* (R. C. Bose and T. A. Dowling, eds.; Univ. of North-Carolina Press, Chapel Hill, N. Car.), ch. 32, pp. 578–584.

A. J. Hoffman and D. K. Ray-Chaudhuri, 1965a, On the line graph of a symmetric balanced incomplete block design, *Trans. Am. Math. Soc.* **116,** 238–252.

A. J. Hoffman and D. K. Ray-Chaudhuri, 1965b, On the line graph of a finite affine plane, *Canad. J. Math.* **17,** 687–694.

A. J. Hoffman and R. R. Singleton, 1960, On Moore graphs with diameters 2 and 3, *IBM J. Res. Develop.* **4,** 497–504.

Renu Laskar, 1967, A characterization of cubic lattice graphs, *J. Combin. Theory*, **3,** 386–401.

Renu Laskar, 1969, Eigenvalues of cubic lattice graphs, *Pacif. J. Math.* **29,** 623–630.

Renu Laskar, 1970, A geometric characterization of the line graph of a finite affine plane, Clemson Univ. Tech. Rept. No. 23.

D. M. Mesner, 1967, A new family of partially balanced incomplete block designs with some Latin square design properties, *Ann. Math. Statist.* **38,** 571–581.

S. E. Payne, 1970a, Affine representations of generalized quadrangles, *J. Algebra* **16,** 473–485.

S. E. Payne, 1970b, Collineations of affinely represented generalized quadrangles, *J. Algebra* **16,** 496–508.

S. E. Payne, 1971, The equivalence of certain generalized quadrangles, *J. Combin. Theory* **10,** 284–289.

S. E. Payne, 1972, Non-isormorphic generalized quadrangles, *J. Algebra.*

W. W. Peterson, 1961, *Error Correcting Codes* (M.I.T. Press, Wiley Cambridge, Mass.; New York).

D. K. Ray-Chaudhuri, 1967, Characterization of line graphs, *J. Combin. Theory* **3,** 201–214.

J. J. Seidel, 1967, Strongly regular graphs of L_2 type and triangular type, *Koninkl. Ned. Akad. Wetensch. Proc. Ser.* A **70** (=*Indag Math.* **29**), 188–196.

S. S. Shrikhande, 1952, On the dual of some balanced incomplete block designs, *Biometrics* **8,** 66–72.

S. S. Shrikhande, 1959a, On a characterization of the triangular association scheme, *Ann. Math. Statist.* **30,** 39–47.

S. S. Shrikhande, 1959b, The uniqueness of the L_2 association scheme, *Ann. Math. Statist.* **30,** 781–798.

S. S. Shrikhande, 1960, Relations between incomplete block designs, *Contributions to Probability and Statistics*, Essays in honor of Harold Hotelling (Stanford Univ. Press, Stanford Calif.), pp. 388–395.

S. S. Shrikhande, 1961, A note on mutually orthogonal Latin squares, *Sankhya* **23,** 115–116.

S. S. Shrikhande, 1967, Seidel equivalence and embedding of strongly regular graphs, MRC Tech. Sum. Rept. No. 779.

S. S. Shrikhande and V. N. Bhat, 1970, Non-isomorphic solutions of pseudo-(3, 5, 2) and pseudo-(3, 6, 3) graphs *Ann. New York Acad. Sci.* **175,** 331–350.

R. Singleton, 1966, On minimal graphs of maximum even girth, *J. Combin. Theory* **1,** 306–332.

J. A. Thas, 1972, Private Communication.

J. N. Srivastava et al., eds., *A Survey of Combinatorial Theory*
© North-Holland Publishing Company, 1973

CHAPTER 5

Line-minimal Graphs with Cyclic Group †

IZAK Z. BOUWER and ROBERTO FRUCHT

University of New Brunswick, Fredericton, New Brunswick, Canada

Universidad Técnica Santa María, Valparaiso, Chile

Abstract

Upper bounds (and in some cases the exact values) are given for the least number of lines a graph must have if its group of automorphisms is to be isomorphic to the cyclic group of order n.

1. Introduction

Only finite undirected graphs without loops or multiple lines are considered in this paper. (For other terms not defined here see Harary [1969], chapter 14.)

Let us call a graph *n-cyclic* if its group of automorphisms is isomorphic to the cyclic group of order n. Since both the complete graph K_2 and its complement $\overline{K_2}$ are 2-cyclic, interesting problems only arise for $n \geqslant 3$. One such problem was suggested by Harary and Palmer [1966]: to find for each $n \geqslant 3$ the smallest graph (with fewest points or lines) that is n-cyclic. It turns out, however, that these are really two different problems, since an n-cyclic graph with fewest points is not always minimal also with respect to the number of lines, and vice versa. (We conjecture that n-cyclic graphs minimal both as to points and lines exist only if n is a prime or $n = 4$; see Section 5.) Thus we shall have to distinguish between point-minimal and line-minimal n-cyclic graphs. For example, for $n = 2$ both K_2 and $\overline{K_2}$ are point-minimal, but only $\overline{K_2}$ is also line-minimal.

Let $\alpha(n)$ denote the number of points of a point-minimal n-cyclic graph, that is, $\alpha(n)$ is the least number of points for which an n-cyclic graph exists. In an analogous fashion define $\beta(n)$ as the number of lines of a line-minimal n-cyclic graph. Since all the values of $\alpha(n)$ have been determined by R. L. Meriwether [1963], we shall be mainly concerned with the analogous problem of finding $\beta(n)$.

† Work supported in part by National Research Council of Canada grant A7332, and in part by the Chilean government (law 11575).

One might also ask how these numbers are affected if only connected graphs are considered. It is rather obvious that the requirement of connectedness does not affect the values of $\alpha(n)$. Indeed, if a point-minimal n-cyclic graph is not connected, its complement will be connected without losing the properties of being n-cyclic and point-minimal.

The situation can be quite different for line-minimal n-cyclic graphs. For example, for $n = 2$ we already mentioned the fact that $\overline{K_2}$ is line-minimal; hence $\beta(2) = 0$. But $\overline{K_2}$ is not a connected graph (and there is of course no other 2-cyclic graph without lines). If we want a connected 2-cyclic graph, we need at least one line (graph K_2).

Accordingly, it seems natural to introduce a number $\beta_c(n)$ as the least number of lines for which a connected n-cyclic graph exists. Such a graph will of course be called a line-minimal connected n-cyclic graph.

For example, for $n = 2$ we had $\beta(2) = 0$, while $\beta_c(2) = 1$. However, it will turn out (see Section 3) that $\beta(n) = \beta_c(n)$ when n is a prime $\neq 2$ or a prime power. The problem of finding the exact value, or at least an upper bound, for $\beta_c(n)$ in the general case has led us to define still another number, called $\beta_f(n)$, as the least number of lines for which a connected n-cyclic graph with exactly one fixed point exists. For example, for $n = 2$ the path P_3 is such a line-minimal connected 2-cyclic graph with one fixed point; hence $\beta_f(2) = 2$.

Of course, we always have:

$$\beta(n) \leqslant \beta_c(n) \leqslant \beta_f(n).$$

The main aim of this paper is then to try to find the exact values of the numbers $\beta(n)$, $\beta_c(n)$, and $\beta_f(n)$, or at least upper bounds which we conjecture represent the true values, although complete proofs are not available. We have summarized those upper bounds in two tables at the end of the paper.

Finally, it should be remarked that the problem suggested by Harary and Palmer should be interpreted as asking not only for the *numbers* $\alpha(n)$, $\beta(n)$, etc., but also for the *construction* of all the existing point-minimal and line-minimal n-cyclic graphs for each $n \geqslant 3$. It seems, however, that this goal might be too ambitious because there are too many such graphs. For instance, already for $n = 4$, in which case $\alpha(4) = 10$, we found at least 12 non-isomorphic point-minimal 4-cyclic graphs with from 18 to 27 lines. One of them will be described in Section 2. It will also be shown there that already for $n = 5$ there are four non-isomorphic 5-cyclic graphs that are both point-minimal and line-minimal.

2. Line-minimal n-cyclic graphs for $n = 3, 4, 5$

Using (only partially correct, see Meriwether [1963]) values of $\alpha(n)$ found by Sabidussi [1959], Harary and Palmer [1966] exhibited three graphs M_n, $n = 3, 4, 5$ (see Fig. 2.1 for $n = 5$), which are n-cyclic and which they claimed

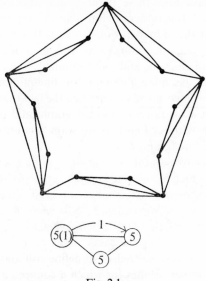

Fig. 2.1.

to be (i) minimal with respect to their number $3n$ of points, and (ii) minimal with respect to their number $5n$ of lines.

However, $\alpha(4) = 10$ so that M_4, having 12 points, cannot satisfy (i). Since we are mainly interested in line-minimal graphs, we describe only that 4-cyclic graph with 10 points having the smallest number of lines known to us:

The points of the graph are the poles N and S of a sphere and the eight vertices of a regular octagon $A_1A_2 \ldots A_8$ inscribed in the equator of the sphere; and the lines of the graph are the sides of the octagon together with the lines A_1A_5, A_3A_7, A_1S, A_2S, A_3N, A_4N, A_5S, A_6S, A_7N, A_8N; see Fig. 2.5. Since this graph has only 18 lines, it is seen that the graph M_4 also does not satisfy (ii). This erratum has already been mentioned (Gewirtz *et al.* [1969] and Bouwer [1969], p. 61).

When Harary was shown our graph (Fig. 2.5), he asked: "Can it be *proved* that it is line-minimal?" Not yet, as we shall soon see.

The claims for the graphs M_3 and M_5 (Harary and Palmer [1966]) remain unaffected. However, it should be pointed out that M_3 is the *only* 3-cyclic graph that is both point-minimal ($\alpha(3) = 9$) and line-minimal ($\beta(3) = \beta_c(3) = 15$), contrary to a statement appearing in Harary and Palmer [1966], and in a weakened form in Harary [1969], p. 170 and p. 176, Ex. 14.4.

A proof of the uniqueness of M_3 can be deduced from Frucht *et al.* [to appear], where the following more general problem is solved: for each $\alpha_0 \geqslant \alpha(3) = 9$ find a connected 3-cyclic graph with α_0 points and the smallest

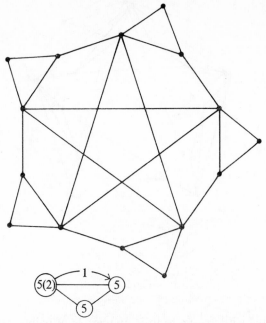

Fig. 2.2.

number of lines.† If this number turns out to be equal to $\beta_c(3) = 15$ we have a line-minimal connected 3-cyclic graph.

Thus it can be seen from Frucht *et al.* [to appear] not only that M_3 is unique but that there also exist five line-minimal connected 3-cyclic graphs that are not point-minimal, namely, two with 10 points, one of which has been known since 1949 Frucht [1949], Fig. 4, two with 12 points and one with 13 points.

Since a proof of the analogous fact that M_5 is not only point-minimal, but also line-minimal (and hence $\beta(5) = \beta_c(5) = 25$) has not been published, we sketch such a proof at the end of Section 4 in a Remark.

It should, however, be pointed out (and has already been mentioned in the Introduction) that M_5 is not the only 5-cyclic graph which is point-minimal and line-minimal at the same time. Indeed we have found three other 5-cyclic graphs with the same number of points and lines as M_5; see Figs. 2.2, 2.3, and 2.4.

That the four graphs thus obtained are non-isomorphic can readily be seen from their orbit representation‡ that is given in the same figures, except

---

† The same problem for not necessarily connected 3-cyclic graphs was solved (Frucht *et al.* [1971]).

‡ The reader not familiar with that notation should consult Frucht [1970].

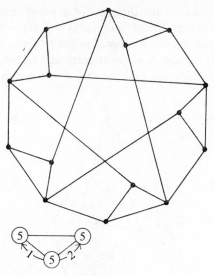

Fig. 2.3.

perhaps for the first two, whose being non-isomorphic follows more readily from the fact that M_5 (Fig. 2.1) contains 10 triangles, while the graph of Fig. 2.2 contains only 5 triangles. The proof that all four graphs are really 5-cyclic is left as an exercise to the reader; the detailed proofs given in Frucht [1949], pp. 370–372, for two similar cases might serve as a hint.

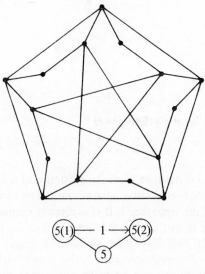

Fig. 2.4.

Finally coming back to the question of a proof that the 4-cyclic graph of Fig. 2.5 is line-minimal (and hence $\beta(4) = \beta_c(4) = 18$), it must be said that it is to be hoped that such a proof might be given along the same lines as the proof for M_3 (Frucht *et al.* [to appear]) and for M_5 below in Section 4, although it is to be expected that the case $n = 4$ is more tedious, since 4 is not a prime.

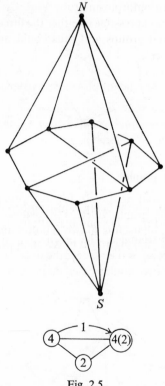

Fig. 2.5.

3. A general lemma for n a prime power

Lemma 3.1. *If* $n = 2^t$ ($t \geqslant 2$) *or* $n = p^t$ (*p an odd prime, $t \geqslant 1$), then* $\beta_c(n) = \beta(n)$.

Proof. We have to show that from the existence of a line-minimal n-cyclic graph G it follows that there exists also a connected one with the same number of lines if n satisfies the hypothesis. If G is already connected, nothing needs to be proved. If G is not connected, let $G_1, G_2, \ldots, G_k$ be its connected components ($k \geqslant 2$). No G_i can be an identity graph with lines, because such a G_i might be dropped and an n-cyclic graph with fewer lines would result. Also, at most one G_i can be an identity graph without lines (that is, an isolated point K_1). Indeed if there were m such isolated points ($m \geqslant 2$), the automorphism group of G would be a direct product of the form $H \times S_m$, and it is clear that such a direct product cannot be a cyclic group of prime power order n (the case $n = 2$ having been excluded).

Hence our line-minimal graph is either of the form $G_1 \cup G_2 \cup \ldots \cup G_s$ or of the form $G_1 \cup G_2 \cup \ldots \cup G_s \cup K_1$, where G_i are now connected graphs with nontrivial automorphism group. But if $s \geqslant 2$, the automorphism group of G would then be expressible as either the direct product or the wreath product of two nontrivial groups and thus could not be a cyclic group of prime order $n \neq 2$. Hence $s = 1$.

That means that either $G = G_1$ or $G = G_1 \cup K_1$; in both cases, G_1 is the line-minimal connected n-cyclic graph whose existence was to be shown.

Corollary 3.1. *If n satisfies the hypothesis of Lemma 3.1, then a line-minimal n-cyclic graph is either connected or is obtained from such a connected graph by adjoining an isolated point.*

4. Line-minimal p-cyclic graphs

Sabidussi [1959] found that $\alpha(p) = 2p$ for any prime $p \geqslant 7$, by exhibiting a point-minimal p-cyclic graph called $X(p)$ whose orbit representation is given in Fig. 4.1; the graph itself is shown (Harary and Palmer [1966], Fig. 1) for $p = 7$ and reproduced here in Fig. 4.2. We shall prove that the graph $X(p)$ is also line-minimal.

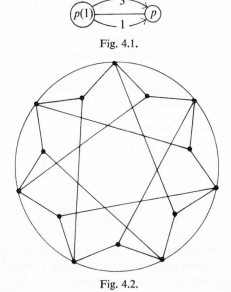

Fig. 4.1.

Fig. 4.2.

Theorem 4.1. $\beta(p) = \beta_c(p) = 4p$ *for* $p \geqslant 7$.

Proof. It follows from Lemma 3.1 that $\beta(p) = \beta_c(p)$; hence we have to show only that $\beta_c(p) = 4p$. From the fact that the point-minimal p-cyclic graph $X(p)$ has $4p$ lines it follows that $\beta_c(p) \leqslant 4p$.

On the other hand, in order to show that $4p \leqslant \beta_c(p)$, let us forget for a while the existence of Sabidussi's graph $X(p)$ and let us try to construct a line-minimal connected p-cyclic graph M beginning "from scratch".

We shall use the term "orbit" ("line orbit") of M to denote an orbit on the points (lines) of M. It is evident that the length of an orbit or line orbit of M is 1 or p, and that any line of M incident with a non-fixed point of M belongs to a line orbit of M of length p. Clearly, M must have at least two orbits of length p. On the other hand, M cannot have four orbits of length p, for the connectedness of M implies that then there would be at least $3p$ lines interconnecting the orbits, and the presence of p more lines, forming a line orbit of M, cannot eliminate all the automorphisms of order 2. Thus, the number of orbits of M of length p is either 2 or 3.

Let us first consider the case when M has exactly two orbits of order p. To avoid the existence of an automorphism of even order due to the possibility of interchanging the two orbits, there must be some distinguishing feature between the two orbits. The requirement of line-minimality leaves us with only two possibilities:

(i) one of the two orbits is formed by a p-gon, while the points of the other orbit are not joined by lines;

(ii) none of the orbits is a p-gon, but the points of one of the orbits are joined by lines to a fixed point.

Fig. 4.3 shows orbit representations of both possibilities. The corresponding graphs have $2p$ lines, but of course they are not p-cyclic, since they obviously admit automorphisms of order 2. It can now easily be checked, by exhaustion, that to destroy all those automorphisms of order 2 it is not sufficient to adjoin just one other line-orbit of length p. Therefore, our graph M must have at least $4p$ lines.

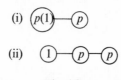

Fig. 4.3.

The reasoning in the case of three orbits of length p is quite similar; again it turns out that it is impossible to construct a graph M with less than $4p$ lines. Hence $4p \leqslant \beta_c(p)$, and the theorem is proved.

It follows from the foregoing discussion that with $4p$ lines we can obtain as line-minimal connected p-cyclic graphs either the graph $X(p)$ of Fig. 4.1 or, for $p \geqslant 11$, also those of Figs. 4.4 and 4.5. Of course, in the three figures the number 3 in the orbit representation might be replaced by some other integer between 4 and $p-2$ (not arbitrarily, since for instance $\frac{1}{2}(p+1)$ is excluded).

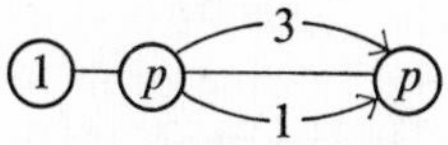

Fig. 4.4.

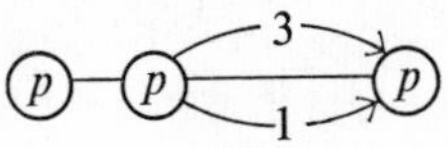

Fig. 4.5.

The case $p = 7$ is rather exceptional, since it turns out that in this case the automorphism groups of the graphs represented by the two Figs. 4.4 and 4.5 are isomorphic to the well known simple group of order 168.

Remark. When $p = 5$, neither the graph $X(p)$ nor those of Figs. 4.4 and 4.5 are 5-cyclic. By similar reasoning one obtains in this case the strict inequality $20 < \beta_c(5)$. On the other hand, the existence of the graphs of Figs. 2.1–2.4 shows us that $\beta_c(5) \leqslant 25$.

This proves that $\beta_c(5) = 25$, since the possibility that $\beta_c(5)$ is equal to 21, 22, 23 or 24 is easily ruled out by the remark that the necessary introduction of fixed points would not reduce the order of the automorphism group to 5.

5. Line-minimal n-cyclic graphs when n is a prime power

Let $n = p^t$, p a prime, $t \geqslant 2$.

Because of Lemma 3.1, we can confine ourselves to connected line-minimal graphs.

Theorem 5.1. *If p is an odd prime and $t \geqslant 2$, then*

$$\beta_c(p^t) \leqslant 2p^t + kp,$$

where

$$k = \begin{cases} 4 & \text{for } p = 3 \text{ or } 5, \\ 3 & \text{for } p \geqslant 7. \end{cases}$$

When $p = 2$, we have

$$\beta_c(2^t) \leqslant \begin{cases} 18 & \text{if } t = 2, \\ 2.2^t + 14 & \text{if } t \geqslant 3. \end{cases}$$

Proof. The first part is proved by the graphs whose orbit representation is given in Fig. 5.1. In an analogous fashion the validity of the second part is proved by Fig. 5.2 for $t \geqslant 3$; the case $t = 2$ has already been discussed in Section 2.

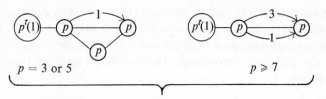

Fig. 5.1.

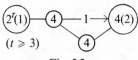

Fig. 5.2.

With the exception of this case, the graphs used in the proof are not point-minimal. Indeed our graphs have

$$\begin{cases} 2^t + 12 \text{ points} & \text{if } p = 2, t \geqslant 3, \\ p^t + 3p \text{ points} & \text{if } p = 3 \text{ or } 5, \\ p^t + 2p \text{ points} & \text{if } p \geqslant 7, \end{cases}$$

while Meriwether found the following values for $\alpha(p^t)$:

$$\begin{cases} 2^t + 6 & \text{if } p = 2, \\ p^t + 2p & \text{if } p = 3 \text{ or } 5, \\ p^t + p & \text{if } p \geqslant 7. \end{cases}$$

Nevertheless, we conjecture that the graphs of Figs. 5.1 and 5.2 are line-minimal n-cyclic graphs for $n = p^t$, or equivalently, that in Theorem 5.1 the equality sign always holds. This conjecture is, of course, based on our unsuccessful efforts to construct n-cyclic graphs with a smaller number of lines, when n is a prime power. It would follow that connected line-minimal n-cyclic graphs which are also point-minimal exist only if n either is a prime or $=4$.

6. Line-minimal n-cyclic graphs for n not a prime power

Let us now consider the case that n is not a prime power, say

$$n = p_1^{t_1} p_2^{t_2} \cdots p_r^{t_r}, \qquad r \geqslant 2,$$

the p_i being distinct primes.

Theorem 6.1. $\beta(n) \leqslant \beta(p_1^{t_1}) + \beta(p_2^{t_2}) + \cdots + \beta(p_r^{t_r})$.

Proof. If $m \geqslant 3$ let $M(m)$ be some connected line-minimal m-cyclic graph; let $M(2) = \overline{K_2}$. Then the graph

$$M(p_1^{t_1}) \cup M(p_2^{t_2}) \cup \cdots \cup M(p_r^{t_r}),$$

that is, the disjoint union of the r graphs $M(p_i^{t_i})$, has by Lemma 3.1

$$\beta(p_1^{t_1}) + \beta(p_2^{t_2}) + \cdots + \beta(p_r^{t_r})$$

lines. By virtue of a well known theorem (Harary [1969], p. 166, Corollary 14.6(a)) the group of automorphisms of that graph is the direct product ("sum" in the terminology used in Harary [1969], p. 163), of cyclic groups of order $p_1^{t_1} p_2^{t_2}, \ldots, p_r^{t_r}$, respectively. Hence that graph is n-cyclic, and the theorem follows.

Remark. A similar reasoning had been used by Meriwether to obtain an analogous formula for the least number of points of n-cyclic graphs, namely:

$$\alpha(n) \leqslant \alpha(p_1^{t_1}) + \alpha(p_2^{t_2}) + \cdots + \alpha(p_r^{t_r});$$

but Meriwether already realized that here the equality sign does not always hold, and he gave a complete list of the exceptional cases.

For example, $\alpha(3) = 9$ and $\alpha(5) = 15$, but, according to Meriwether, $\alpha(15) = 21$; indeed there is the 15-cyclic graph of Fig. 6.1.

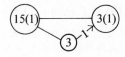

Fig. 6.1.

Since we could not find similar exceptions in the case of line-minimal n-cyclic graphs, we conjecture that the equality sign always holds in Theorem 6.1.

It should be pointed out that the graph constructed in the proof of that theorem is not connected, and therefore, for $\beta_c(n)$ we do not obtain a similar inequality. Indeed there are cases where

$$\beta_c(n) > \beta_c(p_1^{t_1}) + \beta_c(p_2^{t_2}) + \ldots + \beta_c(p_r^{t_r});$$

for example, we saw above that $\beta_c(2) = 1$ and $\beta_c(3) = 15$, but $\beta_c(6) = 17$, since it can be verified, by exhaustion, that the connected 6-cyclic graph of Fig. 6.2 is line-minimal.

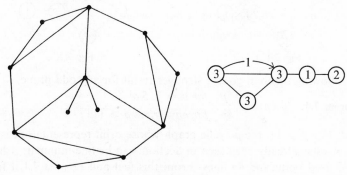

Fig. 6.2.

The method used in the construction of this example can be generalized as follows.

It is easy to see that for each $n \geqslant 2$ there exist connected n-cyclic graphs with exactly one fixed point. Select for each n one such graph with the least number of lines, call it $M_f(n)$, and let $\beta_f(n)$ be its number of lines. For example, $M_f(2)$ is the path P_3, showing us that $\beta_f(2) = 2$. More will be said about the values of $\beta_f(n)$ in the following section.

Since, by the identification of the single fixed points of the graphs $M_f(p_1^{t_1})$, $M_f(p_2^{t_2})$, ..., $M_f(p_r^{t_r})$, there results a connected n-cyclic graph with again one fixed point, it follows that

$$\beta_f(n) \leqslant \beta_f(p_1^{t_1}) + \beta_f(p_2^{t_2}) + \ldots + \beta_f(p_r^{t_r}).$$

But, of course, it is always true that $\beta(n) \leqslant \beta_c(n) \leqslant \beta_f(n)$. So we have the next result.

Theorem 6.2. $\beta_c(n) \leqslant \beta_f(n) \leqslant \beta_f(p_1^{t_1}) + \beta_f(p_2^{t_2}) + \ldots + \beta_f(p_r^{t_r})$.

Conjecture. The equality signs hold here whenever $r \geqslant 2$.

7. The numbers $\beta_f(p^t)$.

In order to make practical use of Theorem 6.2, we need more information about the numbers $\beta_f(p^t)$ defined in the foregoing section, where we already saw that $\beta_f(2) = 2$.

Lemma 7.1. $4p \leqslant \beta_f(p) \leqslant 5p$ *for any odd prime p.*

Proof. Generalizing (Frucht [1949], Fig. 4), we can construct the connected p-cyclic graph with exactly one fixed point whose orbit representation is given in Fig. 7.1. That graph has $5p$ lines; hence $\beta_f(p) \leqslant 5p$. Fig. 7.2 shows the orbit representation of a second graph with the same properties. The inequality $4p \leqslant \beta_f(p)$ follows from $\beta_c(p) \leqslant \beta_f(p)$ and the values obtained for $\beta_c(p)$ in Section 4.

Fig. 7.1.   Fig. 7.2.

The following theorem gives a stronger result for any odd prime.

Theorem 7.1. $\beta_f(p) = \begin{cases} 5p & \text{for } p = 3, 5 \text{ or } 7, \\ 4p & \text{for any prime } p \geqslant 11. \end{cases}$

Proof. For $p \geqslant 11$, the p-cyclic graph whose orbit representation is given by Fig. 4.4 has already been seen in Section 4 to be line-minimal; it has the required fixed point and $4p$ lines. From this fact and Lemma 7.1 it follows that $\beta_f(p) = 4p$ for $p \geqslant 11$.

The assertion for $p = 3$ and 5 follows from the same Lemma and the fact that in these two cases $\beta_c(p) = 5p$ (see Section 2).

It has already been pointed out in Section 4 that the graphs corresponding to Figs. 4.4 and 4.5 for $p = 7$ are not 7-cyclic, and by reasoning similar to that used there, we can rule out the possibility of the existence of a 7-cyclic graph with one fixed point and less than 35 lines. With 35 lines we have of course the graphs corresponding to Figs. 7.1 and 7.2, for $p = 7$; hence $\beta_f(7) = 35$.

Theorem 7.2. *If $t \geqslant 2$, then $\beta_f(p^t) \leqslant 2p^t + 4p$ for any odd prime p.*

Proof. If $n = p^t$ ($t \geqslant 2$, $p \geqslant 3$), the graph with the orbit representation given in Fig. 7.3 is n-cyclic, has one fixed point and $2p^t + 4p$ lines; this proves the theorem. If $p \geqslant 11$, the graph with orbit representation given in Fig. 7.4 might also be used; it has a smaller number of points than the foregoing one.

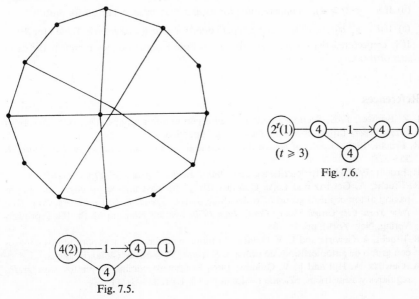

Fig. 7.3.   Fig. 7.4.

Theorem 7.3. $\beta_f(2^t) \leqslant \begin{cases} 18 & \text{for } t = 2, \\ 2(2^t + 8) & \text{for } t \geqslant 3. \end{cases}$

Proof. See Fig. 7.5 for $t = 2$ and Fig. 7.6 for $t \geqslant 3$.

Finally we conjecture that the equality sign holds also in Theorems 7.2 and 7.3.

Fig. 7.6.

Fig. 7.5.

TABLE 1

Upper bounds for:

$\beta(n)$ = least number of lines for n-cyclic graphs, and

$\beta_f(n)$ = least number of lines for connected n-cyclic graphs with exactly one fixed point

n	$\beta(n)$	$\beta_f(n)$
2	0	2
3	15	15
4	18	18
5	25	25
7	28	35
$p(\geqslant 11)$	$4p$	$4p$
$2^t \ (t \geqslant 3)$	$2(2^t+7)$	$2(2^t+8)$
$p^t(t \geqslant 2)\begin{cases} p = 3 \text{ or } 5 \\ p \geqslant 7 \end{cases}$	$\left.\begin{array}{l} 2(p^t+2p) \\ 2p^t+3p \end{array}\right\}$	$2(p^t+2p)$
$p_1^{t_1} p_2^{t_2} \ldots p_r^{t_r}, \quad (r \geqslant 2)$	$\sum_{\rho=1}^{r} \beta(p_\rho^{t_\rho})$	$\sum_{\rho=1}^{r} \beta_f(p_\rho^{t_\rho})$

It is proved in the paper that these upper bounds represent the exact values of $\beta(n)$ and $\beta_f(n)$ if n is a prime; it is conjectured that the same is true if n is not a prime.

TABLE 2

Upper bounds for:

$\beta_c(n)$ = least number of lines for a connected n-cyclic graph:

(a) If n is a prime:

$$\beta_c(p) = \begin{cases} 1 & \text{if } p = 2, \\ \beta(p) & \text{if } p \geqslant 3. \end{cases}$$

(b) If $n = p^t \ (t \geqslant 2)$, an upper bound for $\beta_c(n)$ is that given in Table 1 for $\beta(n)$.

(c) If $n = p_1^{t_1} p_2^{t_2} \ldots p_r^{t_r} (r \geqslant 2)$, an upper bound for $\beta_c(n)$ is that given in Table 1 for $\beta_f(n)$.

It is conjectured that in the cases (b) and (c) these upper bounds represent the exact values of $\beta_c(n)$.

References

I. Z. Bouwer, 1969, Section graphs for finite permutation groups, *The Many Facets of Graph Theory* (Springer-Verlag, New York), pp. 55–61.

R. Frucht, 1949, Graphs of degree three with a given abstract group, *Canad. J. Math.* **1**, 365–378.

R. Frucht, 1970, How to describe a graph, *Ann. New York Acad. Sci.* **175**, 159–167.

R. Frucht, A. Gewirtz and L. V. Quintas, 1971, The least number of edges for graphs having automorphism group of order three, *Recent Trends in Graph Theory, Proc. First New York City Graph Theory Conf., June 1970*, Lecture Notes in Math. 186 (Springer-Verlag, New York), pp. 95–104.

R. Frucht, A. Gewirtz and L. V. Quintas, El número mínimo de líneas para grafos conexos con grupo de automorfismos de orden 3, *Scientia* (Valparaiso, Chile) (to appear).

A. Gewirtz, A. Hill and L. V. Quintas, 1969, El número mínimo de puntos para grafos regulares y asimétricos, *Scientia* (Valparaiso, Chile) **138**, 103–111.

F. Harary, 1969, *Graph Theory* (Addison-Wesley, Reading, Mass.).

F. Harary and E. M. Palmer, 1966, The smallest graph whose group is cyclic, *Czech. Math. J.* **16**, 70–71.

R. L. Meriwether, 1963, unpublished; see G. Sabidussi, 1967, *Math. Rev.* **33** (2563).

G. Sabidussi, 1959, On the minimum order of graphs with given automorphism group, *Monatsh. Math.* **63**, 124–127.

J. N. Srivastava et al., eds., *A Survey of Combinatorial Theory*
© North-Holland Publishing Company, 1973

CHAPTER 6

Circle Geometry in Higher Dimensions

R. H. BRUCK

University of Wisconsin, Madison, Wisc. 53706, U.S.A.

1. Introduction

In this note I propose a definition of a d-dimensional circle-geometry for each integer $d \geqslant 2$. The definition is framed in such a manner that the circle-geometries of dimension 2 are precisely the inversive planes and that the circle-geometries of dimension d embrace (in the sense of isomorphism) all regular spreads (see Bruck and Bose [1966]) of $(d-1)$-dimensional projective subspaces of a $(2d-1)$-dimensional projective space. (The members and reguli of such a spread constitute the points and circles, respectively, of a d-dimensional circle-geometry.) My concern for the theory of circle geometries might be sufficiently indicated by Bruck [1969].

In the classical affine model of a (Miquelian) inversive plane we adjoin a single point, ∞, to the points of a (Desarguesian) affine plane A. For circles we use the lines of A (considered as circles through ∞) together with a suitable collection of conics in A. (See Dembowski [1968] for other representations.) By contrast, for $d > 2$, the affine plane A is merely replaced by a d-dimensional affine space (necessarily Desarguesian) but the collection of conics is replaced by a suitable collection of "twisted, unicursal" curves of degree d. (The latter form interesting patterns on certain algebraic surfaces of degree d, but we shall omit this topic here. There is a connection with Bruck [1970].)

It seems to me that higher-dimensional circle-geometries deserve a careful study, beginning with those of dimension 3. May I point out why they are not as familiar in classical geometry as inversive planes; the reason is simple: In an affine 3-space over the field of real or complex numbers, every twisted cubic curve must meet the plane at infinity. And the analogous difficulty occurs for twisted unicursal curves of odd degree d in real or complex affine d-space. By contrast, there is no such difficulty either in the finite case or when the underlying field is the field of rationals.

In Section 2 we present our axioms. In Section 3 we show that they are satisfied by a wide variety of circle-geometries. Although we do not examine the point in the paper, these examples are all isomorphic to circle-geometries of regular spreads.

Nothing is said in the axioms about the higher-dimensional analogs of inversions. It may be deduced from Section 3 that such analogs can exist (corresponding to field automorphisms); but they appear to be less interesting than inversions, since they do not have order two.

2. The axioms

Consider a system $(\mathscr{P}, \mathscr{C})$ consisting of a set $\mathscr{P}$, the elements of which we shall call *points*, and of a set $\mathscr{C}$ of subsets, called *circles*, of $\mathscr{P}$. For any point P of $\mathscr{P}$ we define an *induced system*, $(\mathscr{P}(P), \mathscr{C}(P))$, as follows: The elements of $\mathscr{P}(P)$ are called *points*, the elements of $\mathscr{C}(P)$ are called *lines*. $\mathscr{P}(P)$ is the set of all points of $\mathscr{P}$ distinct from P. For each circle C in $\mathscr{C}$ which contains P, $C(P)$ is the set obtained from C by deleting P; and $\mathscr{C}(P)$ is the set of all such lines $C(P)$.

A wide variety of systems $(\mathscr{P}, \mathscr{C})$ satisfy the following set (A) of axioms:

(A) **Axioms of existence, uniqueness, equi-cardinality.**

(A.1) *$\mathscr{P}$ and $\mathscr{C}$ are non-empty. In addition:*

(A.2) *There exists a (finite or infinite) cardinal number n, called the order of* $(\mathscr{P}, \mathscr{C})$, *such that* (a) $n \geqslant 2$ *and* (b) *each circle in $\mathscr{C}$ contains precisely $n+1$ distinct points of $\mathscr{P}$.*

(A.3) *Each (unordered) set of three distinct points of $\mathscr{P}$ is contained in exactly one circle of $\mathscr{C}$.*

The variety of systems is sharply reduced when we add:

(B) **Axiom of dimension.** *There exists a positive integer d such that* (a) $d \geqslant 2$ *and* (b) *for each point P in $\mathscr{P}$, the induced system $A(P) = (\mathscr{P}(P), \mathscr{C}(P))$ is a d-dimensional affine geometry.*

Note that, for $(\mathscr{P}, \mathscr{C})$ finite and subject to (A) and (B), $(\mathscr{P}, \mathscr{C})$ has prime-power order q and each $A(P)$ is the affine space over $\mathrm{GF}(q)$, except perhaps when $d = 2$.

In the presence of (A), (B), consider a point P of $\mathscr{P}$ and a circle D of $\mathscr{P}$ which does not contain P. It is natural to inquire into the nature of D as a "curve" in the affine geometry $A(P)$. I offer the following set of axioms for the case that d is an odd prime, and leave the question of redundancy (and, indeed, of technical adequacy) to the reader. (There is a similar but more complex axiom-set for the case that d is not a prime.)

(C) **Axioms for odd prime dimension.** *Let $(\mathscr{P}, \mathscr{C})$ satisfy (A), (B) and have dimension d, where d is an odd prime. Then, for each point P in $\mathscr{P}$ and each circle D in $\mathscr{C}$ which does not contain P, the following holds in the affine geometry $A(P) = (\mathscr{P}(P), \mathscr{C}(P))$:*

(C.1) *For $1 \leqslant i \leqslant d-1$, no i-dimensional affine subspace of $A(P)$ contains more than $i+1$ distinct points of D.*

(C.2) *To each point Q of D there corresponds at least one collection $\{T_i | 1 \leqslant i \leqslant d-1\}$, called a nest of tangent spaces to D at Q, with the following properties:*

(C.2a) *For $1 \leqslant i \leqslant d-1$, T_i is an i-dimensional affine subspace of $A(P)$.*

(C.2b) *For $1 \leqslant i \leqslant d-2$, $T_i \subset T_{i+1}$. Also T_1 contains Q; and T_{d-1} contains no point of D other than Q.*

(C.2c) *For $1 \leqslant i < j \leqslant d-1$, each j-dimensional affine subspace of $A(P)$ which contains T_i contains at most $j-i+1$ distinct points of D (including Q).*

Because (C) is somewhat complicated, it seems worthwhile to restate it for the case that $d = 3$. Then D is a "curve" in affine 3-space with the following properties: (C.1) Each line has at most two distinct points of D, and each plane has at most three distinct points of D. (C.2) If Q is a point of D then Q is contained in at least one "tangent line" T_1, and T_1 is contained in at least one "osculating plane" T_2 such that T_2 (and hence T_1) contains no point of D besides Q and such that no plane through T_1 contains more than two points of D (including Q).

In other words, if we construct the (quotient) projective plane π over Q, with the lines and planes through Q as its points and lines, respectively, then T_1 and the lines joining Q to the other points of D become the points of an oval $\mathcal{O}$ in π, and T_2 becomes a tangent line to $\mathcal{O}$ at T_1. This remark should explain why we do not claim uniqueness for T_1 and T_2. Note that, in the case of affine 3-space over $GF(q)$ for q odd, the Theorem of Segre [1955a] makes $\mathcal{O}$ a conic and (for $q \geqslant 5$) ensures the uniqueness of T_1 and T_2. Segre [1955b] is also pertinent here.

3. A class of circle-geometries

We begin with a pair of fields F, K, with F a subfield of K. At some points we shall want to assume all of the following:

(I) K has finite dimension d over F, where $d \geqslant 2$.

(II) K is Galois (that is, separable and normal) over F.

(III) d is an odd prime.

(IV) F has at least d distinct elements.

Note that if
$$F = GF(q), \qquad K = GF(q^d),$$
where q is any prime-power and $d \geqslant 2$ any positive integer, then (I), (II) hold. If, in addition, d is an odd prime and $q \geqslant d$, then (I)–(IV) hold. We can also take F to be the field of rationals, d to be any odd prime, and find K in many ways so that (I)–(IV) hold. But, for example, if (I)–(IV) hold, F cannot be the real or complex field.

To begin with, let F, K satisfy (I). We represent the projective line $L = PG(1, K)$ over K as a two-dimensional vector space V over K. A point of L is a one-dimensional space $\langle v \rangle$, $v \in V$, $v \neq 0$, over K.

The following notation will be convenient. Call an ordered triple

$$\{e_1, e_2, e_3\} \tag{3.1}$$

of three (distinct) vectors e_1, e_2, e_3 of V a canonical triple provided (a) e_1, e_2 are linearly independent over K and (b) $e_1 + e_2 + e_3 = 0$. Note that each of the six ordered sets obtained from a canonical triple (3.1) by permutation of the vectors is also canonical.

Lemma 3.1. *If P_1, P_2, P_3 are three distinct points of $PG(1, K)$, there exists a canonical triple (3.1), unique to within multiplication by a nonzero element of K, such that*

$$P_i = \langle e_i \rangle, \qquad i = 1, 2, 3. \tag{3.2}$$

Proof. For $i = 1, 2, 3$, let u_i be any basis element of P_i. If $i \neq j$, then $P_i \neq P_j$ and hence u_i, u_j are linearly independent over K and form a basis for V over K. Since u_1, u_2 form such a basis,

$$au_1 + bu_2 + u_3 = 0$$

for unique elements a, b of K. Clearly a, b are nonzero. Thus (3.2) holds for a canonical triple (3.1) if and only if

$$e_1 = kau_1, \qquad e_2 = kbu_2, \qquad e_3 = ku_3$$

for a nonzero element k of K. This proves Lemma 3.1.

Given three distinct points P_1, P_2, P_3 of $PG(1, K)$, we define the *circle* $C = (P_1, P_2, P_3)$ to be the set of all points $\langle v \rangle$ of $PG(1, K)$ such that v is expressible in the form

$$v = xe_1 + ye_2 + ze_3 \neq 0, \qquad x, y, z \in F, \tag{3.3}$$

with respect to a canonical triple (3.1) for which (3.2) holds. Since (3.3) is equivalent to

$$kv = x(ke_1) + y(ke_2) + z(ke_3) \neq 0$$

for each nonzero k in K, the point-set C depends only on P_1, P_2, P_3. Note that C contains the P_i. Note also that (3.3) holds if and only if it holds with, say, $z = 0$.

Lemma 3.2. *If P_1, P_2, P_3 are three distinct points of $PG(1, K)$ and if Q_1, Q_2, Q_3 are three distinct points of (P_1, P_2, P_3), then*

$$(P_1, P_2, P_3) = (Q_1, Q_2, Q_3). \tag{3.4}$$

Proof. We may suppose that (3.2) holds for a canonical triple (3.1) and that

$$Q_i = \langle v_i \rangle, \qquad i = 1, 2, 3, \tag{3.5}$$

where each v_i is expressible in the form (3.3) and therefore in the form (3.3) with $z = 0$. Thus

$$v_i = x_{i1}e_1 + x_{i2}e_2, \qquad i = 1, 2, 3, \tag{3.6}$$

where the x_{ij} are in F. Since the Q_i are distinct, each two of v_1, v_2, v_3 are linearly independent over K. In particular, the square matrix

$$M = \begin{pmatrix} x_{11} & x_{12} \\ x_{21} & x_{22} \end{pmatrix}$$

is nonsingular and has an inverse with elements in F. The equation

$$av_1 + bv_2 + v_3 = 0$$

holds for a, b in K if and only if

$$(a, b)M + (x_{31}, x_{32}) = 0.$$

This shows that a, b are in F. Hence, after replacing v_1, v_2 by av_1, bv_2, respectively, we may assume (3.5), (3.6), where $\{v_1, v_2, v_3\}$ is a canonical triple. Then it is clear that (Q_1, Q_2, Q_3) is contained in (P_1, P_2, P_3). Since any two of equations (3.6) allows us to solve for e's in terms of v's with coefficients in F, it is equally clear that (P_1, P_2, P_3) is contained in (Q_1, Q_2, Q_3). This proves Lemma 3.2.

Now we are ready to construct a circle-geometry $(\mathscr{P}, \mathscr{C})$. For $\mathscr{P}$ we take the set of all points of $\mathrm{PG}(1, K)$. For $\mathscr{C}$ we take the set of all circles (P_1, P_2, P_3). From the definitions and from Lemma 3.2, $(\mathscr{P}, \mathscr{C})$ satisfies the axioms (A) of Section 2.

To consider axiom (B), we take any point P of $\mathscr{P}$ and choose three distinct points P_1, P_2, P_3 of $\mathscr{P}$ such that $P_1 = P$. We may assume (3.2), where (3.1) is a canonical triple. We define

$$P(\infty) = P = P_1 = \langle e_1 \rangle, \qquad P(k) = \langle ke_1 + e_2 \rangle, \qquad k \in K, \qquad (3.7)$$

for every k in K. Thus the points of $\mathscr{P}(P)$ are the points $P(k)$, $k \in K$. The circles containing $P = P(\infty)$ are those of form

$$C = (P(\infty), P(a), P(b)), \qquad a \neq b, \qquad (3.8)$$

where a, b are distinct elements of K. If

$$v_1 = ae_1 + e_2, \qquad v_2 = -be_1 - e_2, \qquad v_3 = (b-a)e_1,$$

then $\{v_1, v_2, v_3\}$ is a canonical triple with $P(a) = \langle v_1 \rangle, P(b) = \langle v_2 \rangle, P(\infty) = \langle v_3 \rangle$. The points of C are those of form $\langle w \rangle$, where

$$w = xv_1 + yv_2 + zv_3 \neq 0, \qquad x, y, z \in F.$$

We may assume $z = 0$. Then $\langle w \rangle = P(\infty)$ if $x - y = 0$. If $x - y \neq 0$, we may as well take $x - y = 1$, $y = x - 1$. Thus, for C given by (3.8),

$$C(P) = \{P(xa + (1-x)b) \mid x \in F\}. \qquad (3.9)$$

If we represent the point $P(k)$ of $\mathscr{P}(P)$ by the element k of K, then the line of $\mathscr{C}(P)$ joining two distinct points a, b is the set of all points of the form

$$xa + (1-x)b = b + x(a-b), \qquad x \in F.$$

Since K is a d-dimensional vector space over F, clearly $(\mathscr{P}(P), \mathscr{C}(P))$ is the affine geometry $\mathrm{AG}(d, F)$. This proves (B).

Now consider a circle
$$D = (Q_1, Q_2, Q_3) \tag{3.10}$$
which does not contain $P = P(\infty)$. We may assume that
$$Q_i = \langle w_i \rangle, \qquad i = 1, 2, 3, \tag{3.11}$$
for a canonical triple $\{w_1, w_2, w_3\}$ of the form
$$w_i = a_i e_1 + b_i e_2, \qquad i = 1, 2, 3. \tag{3.12}$$
Here the a_i, b_i are in K and, in particular, the b_i are all nonzero. Indeed,
$$Q_i = P(a_i b_i^{-1}), \qquad i = 1, 2, 3. \tag{3.13}$$
Every point of D has the form $\langle w \rangle$, where
$$w = x w_1 + y w_2 \neq 0, \qquad x, y \in F.$$
In detail,
$$w = (x a_1 + y a_2) e_1 + (x b_1 + y b_2) e_2.$$
Since D does not contain $P(\infty)$, b_1 and b_2 must be linearly independent over F. Thus

$$D \text{ consists of the points } P(k), \text{ where } k \text{ has the form} \tag{3.14a}$$

$$k = \frac{x a_1 + y a_2}{x b_1 + y b_2}, \qquad x, y \in F, (x, y) \neq (0, 0). \tag{3.14b}$$

Before we consider the axioms (C), let us take a moment to see how nicely the family of circles D not containing $P = P_1 = P(\infty)$ may be represented in the *projective* space $\mathrm{PG}(d, F)$. First we pick a basis $E_1, E_2, \ldots, E_d$ of K over F. Then we adjoin $2d$ commutative and associative indeterminates
$$X_1, \ldots, X_d, \qquad Y_1, \ldots, Y_d$$
to F and observe that if
$$X = \sum_{i=1}^{d} X_i E_i, \qquad Y = \sum_{i=1}^{d} Y_i E_i, \tag{3.15}$$
then
$$N(X) Y = X \sum_{i=1}^{d} P_i(X; Y) E_i \tag{3.16}$$
where $N(X)$ and the $P_i(X; Y)$ are in the polynomial ring
$$P[X_1, \ldots, X_d, Y_1, \ldots, Y_d].$$
Here $N(X)$ is the *norm polynomial*, homogeneous of degree d in $X_1, \ldots, X_d$. Also, for each $i = 1, 2, \ldots, d$, $P_i(X; Y)$ is linear in $Y_1, \ldots, Y_d$ and homogeneous of degree $d-1$ in $X_1, \ldots, X_d$.

If $u = \Sigma u_i E_i$, $v = \Sigma v_i E_i$ are elements of K with components u_i, v_i in F, the meaning of $N(u)$, $P_j(u; v)$ should be clear. With this understood, we see that D, given by (3.14), consists of all elements k of the form

$$k = \frac{1}{N(x b_1 + y b_2)} \sum_{i=1}^{d} P_i(x b_1 + y b_2; x a_1 + y a_2) E_i, \qquad x, y \in F, (x, y) \neq 0. \tag{3.17}$$

Here the denominator $N(xb_1 + yb_2)$ is nonzero unless $x = y = 0$. If we embed $AG(d, F)$ in $PG(d, F)$ in the usual way by introducing $d+1$ homogeneous coordinates, D can be considered to consist of all points of form

$$(N(xb_1 + yb_2), \ P_1(xb_1 + yb_2; xa_1 + ya_2), \ldots, P_d(xb_1 + yb_2; xa_1 + ya_2)),$$
$$x, y \in F, \qquad (x, y) \neq (0, 0). \tag{3.18}$$

Clearly D is a unicursal algebraic curve of degree d in $PG(d, F)$ which has no point in common with the hyperplane at infinity for $AG(d, F)$.

Note that we have given a uniform representation of the family of all circles in D. (There are, in fact, more elegant representations in $AG(d, F)$ which show the interrelations of the curves more plainly, but we shall not discuss these here.)

In proving axioms (C), we can specialize a little. We choose the distinct points P_1, P_2, P_3 so that $P_1 = P = P(\infty)$. In dealing with the circle D, given by (3.10), (3.14), which does not contain P_1, there is no loss of generality in choosing P_2, P_3 so that $P_2 = Q_1, P_3 = Q_2$. This means that in (3.14b) we have $a_1 = 0$, $a_2 = b_2$, where b_1, b_2 are linearly independent over K. If we define $c = b_1 b_2^{-1}$, then D consists of 0 and the points $k(f), f \in F$, where

$$k(f) = \frac{1}{c+f}, \qquad f \in F. \tag{3.19}$$

Here the elements $1, c$ of K are linearly independent over F. At the same time, for the purposes of axiom (C.2) we can assume that $Q = Q_1 = P_2$. Thus we need only consider affine subspaces of $AG(d, F)$ $F)$ which contain 0. These are vector subspaces of K over F.

Here, for the first time, we use the assumption (III) that d is a prime (an odd prime if we want anything new.) Thus the element c of K, since it is not in F, must have degree d over F. For each integer n in the range $0 \leqslant n \leqslant d$, we define a vector subspace $T(n)$ of K over F in terms of a basis as follows:

$$T(0) = \{0\}; \qquad T(n) = \langle 1, c, \ldots, c^{n-1} \rangle, \qquad 1 \leqslant n \leqslant d. \tag{3.20}$$

Here, since c has degree d over K, $T(n)$ is n-dimensional for each n, and $T(d) = K$. Clearly

$$0 = T(0) \subset T(n) \subset T(n+1), \qquad 0 \leqslant n \leqslant d-1. \tag{3.21}$$

The set $\{T(n)| \ 1 \leqslant n \leqslant d-1\}$ is to be the nest of tangent spaces (to D at $Q = 0$) mentioned in (C.2). We prove both parts of (C) by showing that if n, m are integers with $n \geqslant 0$, $m > 0$ and $n+m \leqslant d$, a vector space spanned by $T(n)$ and m distinct vectors $k(f)$ has dimension $n+m$ (and hence contains precisely $m+1$ points of D, including 0).

If the last statement as to dimension is false, then there exist distinct elements $f_1, f_2, \ldots, f_m$ of F and elements $x_1, x_2, \ldots, x_m$ of F, not all zero, such that the element

$$x_1 k(f_1) + x_2 k(f_2) + \cdots + x_m k(f_m) \tag{3.22}$$

is in $T(n)$. To see that this is false, let X be an indeterminate over F and, for $i = 1, 2, \ldots, m$, define the polynomial $Q_i(X)$ to be the product of the $m-1$ factors $X+f_j, j \neq i$ (or to be 1 in case $m = 1$.) Consider the polynomial

$$P(X) = x_1 Q_1(X) + x_2 Q_2(X) + \cdots + x_m Q_m(X) \qquad (3.23)$$

(which is in $F[X]$) and note that, for $i = 1, 2, \ldots, m$,

$$P(-f_i) = x_i Q_i(-f_i), \qquad Q_i(-f_i) \neq 0. \qquad (3.24)$$

Since, by assumption, the x_i are not all zero, $P[X]$ is nonzero. Since each $Q_i(X)$ has degree $m-1$, $P(X)$ has degree at most $m-1$. Next note that if the element (3.22) is in $T(n)$ then there exists a unique polynomial $R(X)$ in $F[X]$ such that the element (3.22) is equal to $R(c)$ and such that either (a) $R(X)$ is the zero polynomial or (b) $R(X) \neq 0$, $R(X)$ has degree at most $n-1$, and (of course) $n \geqslant 1$. After multiplying (3.22) by

$$(c+f_1)(c+f_2) \cdots (c+f_m)$$

and using the definitions of $P(X)$, $R(X)$, we see that the polynomial

$$S(X) = P(X) - (X+f_1)(X+f_2) \cdots (X+f_m)R(X)$$

is in $F[X]$ and has c as a zero. If $R(X) = 0$, then $S(X) = P(X)$ is nonzero and $S(X)$ has degree at most $m-1 \leqslant m+n-1 \leqslant d-1$. If $R(X) \neq 0$, then $S(X)$ is nonzero (since the degree of P is less than that of the second term in S) and $S(X)$ has degree $\leqslant m+(n-1) \leqslant d-1$. In either case we have a contradiction to the fact that c has degree d over F. This proves axioms (C).

The reader may perhaps be mystified by the unexplained choice of the spaces $T(n)$. At the risk of mystifying him further I will parrot the theory of complex variables as follows: If a point $k(f)$ is to be "near" 0, f must be "large". Hence we expand $k(f)$ in powers of f^{-1}:

$$k(f) = f^{-1} - f^{-2}c + f^{-3}c^2 - \ldots,$$

and "deduce" that the spaces $T(1) = \langle 1 \rangle$, $T(2) = \langle 1, c \rangle, \ldots$ fit the curve closely "near" 0. Obviously, we could introduce some curves of degrees $2, 3, \ldots, d-1$ which "osculate" D at 0.

We have not used assumptions (II), (IV) explicitly. Note that, if (IV) is false, the axiom-set (C) loses some of its force. As for (II), it is of importance for further study; in connection, for example, with the collineation group of the circle-geometry, or with the projective representation (3.18) of the family of circles D.

References

R. H. Bruck, 1969, Construction problems of finite projective planes, *Combinatorial Mathematics and its Applications*, (R. C. Bose and T. A. Dowling, eds.; Univ. of North Carolina Press, Chapel Hill, N.Car.), ch. 27, pp. 426–514.

R. H. Bruck, 1970, Some relatively unknown ruled surfaces in projective space, *Arch. Inst. Grand-Ducal Luxembourg, Sect. Sci. Nat. Phys. Math., N.S.*

R. H. Bruck and R. C. Bose, 1966, Linear representations of projective planes in projective spaces, *J. Algebra* **4**, 117–172.

Peter Dembowski, 1968, *Finite Geometries* (Springer Verlag, New York).

B. Segre, 1955a, Ovals in a finite projective plane, *Canad. J. Math.* **7**, 414–416.

B. Segre, 1955b, Curvi razionale normali e *k*-archi negli spazi finiti, *Ann. Mat. Pura Appl.* **39**, 357–379.

J. N. Srivastava et al., eds., *A Survey of Combinatorial Theory*
© North-Holland Publishing Company, 1973

CHAPTER 7

Bose as Teacher—The Early Years

K. A. BUSH

Washington State University, Pullman Wash., U.S.A.

Professor R. C. Bose arrived at Chapel Hill early in 1949, and his first students were S. S. Shrikhande, William S. Connor, Willard Clatworthy, and myself. Connor and I had arrived in the fall of 1948, Clatworthy came the following year, but Shrikhande had begun his work somewhat earlier and was already demonstrating remarkable insight and substantial knowledge. The rest of us were not so well endowed.

Both Connor and I had been trained as economists and were not proficient mathematically, and I was well past thirty when I arrived at North Carolina. Since our time at the University overlapped by only one year, I had fewer contacts with Clatworthy, but I had the general impression that he was almost in the same straits as Connor and myself.

How, then, does one explain the paradox of a group of Ph.D. candidates ostensibly below average later developing into mathematicians very much above average? I claim, if I may coin a phrase for the educationalists, that he made us over-achievers. His methods were essentially based on two factors.

In the first place, he set highly imaginative but quite difficult topics for our investigation. In the second place, the student had to supply substantial originality. Bose seemed totally unaware of the modern trend where the thesis advisor shows the student how to solve part of the problem and then suggests that he carry it further with some inconsequential generalization. Instead he expected students to discover appropriate theorems on their own and develop sound techniques for their proof. At this stage, he would read every word, frequently rewrite entire sections to improve the exposition, and sometimes find important refinements or extensions of the results. While this was helpful, the really important element is that he forced us to do genuine research from the outset.

In retrospect, the way he achieved this objective is essentially uncomplicated. In some strange way he made us feel that his overwhelming desire was to see each of us succeed. Nothing else seemed to matter to him. His kindliness, his evident deep concern for our welfare, and his unfailing interest in our progress drove us to an almost Bacchanalian frenzy of activity. It was no longer a question of getting the degree that was paramount, it was the question of trying to meet the expectations of a devoted friend.

And the method worked with surprising celerity. By the late spring of 1950, Shrikhande had discovered a number of exciting theorems, and his thesis was nearly complete. Perhaps spurred on by Shrik's (as we called him) accomplishments, I began working night and day on my own topic completing the research in less than eight weeks. Just as this work was drawing to a close, Connor made an important discovery that formed the basis of his distinguished thesis. It was indeed an electrifying summer for us, an exhausting one for Bose.

At the conclusion of the summer, he and I drove from Chapel Hill to Cambridge where he was to present an invited address before the International Congress, a somewhat belated major recognition. With characteristic generosity, he referred to some of my new results in his talk. I heard him again at the International Congress in 1970 speaking with the same clarity and enthusiasm that I so vividly remembered. Somehow it did not seem possible that two decades had so swiftly slipped by.

To this brief account of the early days, I append a personal homage to my teacher. Even my papers in classical analysis, which I regard as my best accomplishment, are dependent upon him, for the techniques I used in them are really only readaptations of things he taught me. Thus my association with Bose lasting less than twenty months has influenced me for more than twenty years.

J. N. Srivastava et al., eds., *A Survey of Combinatorial Theory*
© North-Holland Publishing Company, 1973

CHAPTER 8

Construction of Symmetric Hadamard Matrices

K. A. BUSH

Washington State University, Pullman, Wash., U.S.A.

1. Introduction

In preparing an essay for this volume, the thought occurred to me that it would be particularly appropriate if I could present material with some connection with the investigations I undertook years ago when I wrote under Professor Bose. As it turns out, this became a relatively easy task, for in recent work (Bush [1971a,b,c]) I had used more complex methods than were necessary and thus failed to achieve the best theorems. This note will fill that gap. It also complements the interesting discoveries of Seidel and Goethals [1970]. Among other things, they succeeded in constructing symmetric Hadamard matrices with constant diagonal and fixed row sum. We shall do the same, but we will be able to specify the form with additional restrictions. They obtained the cases 36 and 100. We are unable to obtain the latter case, but we can assert a general theorem.

Theorem. *If* $t-2$ *mutually orthogonal Latin squares of order* $2t$ *exist, then a symmetric Hadamard matrix of order* $4t^2$ *exists with the property that it can be partitioned into* $2t \times 2t$ *submatrices such that* (i) *the submatrices on the main diagonal have* $+$ *on the diagonal and* $-$ *elsewhere, and* (ii) *the off-diagonal submatrices have* $-$ *on the principal diagonal, and each row and column contains* $t+1$ *elements that are* $+$.

Proof. We display the $t-2$ mutually orthogonal Latin squares in an orthogonal array (OA). The first two rows of the OA will be in the standard form indicated below:

$$
\begin{array}{ccccccc}
0\,0\cdots\;0 & 1\,1\cdots\;1 & \cdots & 2t-1 & 2t-1 & \cdots & 2t-1 \\
0\,1\cdots 2t-1 & 0\,1\cdots 2t-1 & \cdots & 0 & 1 & \cdots & 2t-1.
\end{array}
$$

The remaining rows are completed by adding the row consisting of the first row followed by the second row ⋯ followed by the last row of the first Latin square for the first adjoined row. We do the same to adjoin the next row associated with the second orthogonal Latin square, and so on until we have adjoined $t-2$ rows in this way. This array will then have the property

4   81

that between any two rows each of the possible ordered pairs will occur exactly once.

Denoting the Hadamard matrix by $H = (h_{ij})$, we set $h_{ii} = 1$, and we set $h_{ij} = -1$ in case the ith and jth columns of the OA have an element in common. Otherwise, we set $h_{ij} = +1$. By a common element we mean that some row of the OA has the same element in the ith and jth columns. It is clear that two columns cannot have two elements in common, for then this pair would be repeated twice. Considering the ith column, we note that associated with each row there are $2t-1$ columns with the same element so that each row of the matrix contains $2t^2 - t$ negative entries, and H is surely symmetric. Furthermore, from the first row of the OA displayed above, we secure the main diagonal submatrices described in the statement of the theorem. From the second row of the OA, we secure the result that the off-diagonal submatrices each have -1 on the main diagonal. It is also clear that an off-diagonal submatrix is generated by the columns in two separate divisions of the OA indicated in our display by separating the columns into groups of columns, each group consisting of $2t$ columns. Except for the first row of the OA, in each such group we have a permutation of the symbols. Thus any column will have precisely $t-1$ columns in another group of coincidences of common elements. Therefore each off-diagonal submatrix will have $t-1$ negative entries in each row and column.

It remains to establish orthogonality. Consider any two columns with respective entries $(i_1, i_2, \ldots, i_t)$ and $(j_1, j_2, \ldots, j_t)$ in the OA. These columns will generate (say) the ith and jth rows of H. These two rows will have $h_{ik} = h_{jk} = -1$ if and only if the kth column of the OA has a common element with each of these two columns. We seek the frequency of this combination of $-$ versus $-$. There will be such a combination in case the uth row of the column contains i_u and the vth row contains the element j_v. There is exactly one such column. If the ith and jth column have no common element, then for each i_u there are precisely $t-1$ columns that contain i_u in the uth row and j_v in the vth row, $v \neq u$. Thus there are $t^2 - t$ such pairs. If the two columns contain a common element, then there are $2t-2$ columns remaining with this common element. There are $(t-1)(t-2)$ remaining columns that generate the pairing minus versus minus.

Having located the position of the negative entries, we thus have the following frequencies of sign combinations between two rows of H:

$$
\begin{array}{cccc}
+ & + & - & - \\
+ & - & + & - \\
t^2+t & t^2 & t^2 & t^2-t
\end{array}
$$

establishing orthogonality.

By using the type of argument given above, we have the further result:

Theorem. *If* $t-1$ *mutually orthogonal Latin squares of order* $2t$ *exist, then a symmetric Hadamard matrix exists of order* $4t^2$ *such that it can be partitioned into* $2t \times 2t$ *submatrices such that* (i) *the submatrices on the main diagonal consist exclusively of* $+$ *elements and* (ii) *the off-diagonal submatrices have* -1 *on the main diagonal, and each row and column contains* t $+$ *elements.*

Proof. We proceed exactly as in the proof of the preceding theorem with one exception. We disregard the first row of the OA and locate the $-$ elements using the remaining t rows of the OA exactly as we did before. Then the negative elements all occur in the off-diagonal matrices, but the number in each row of H remains the same as above with $2t^2 - t$ negative entries in each row and column of H. By the same argument as given above, this leads to t negative entries in each row of each off-diagonal submatrix.

References

K. A. Bush, 1971a, An inner orthogonality of Hadamard matrices, *Austral. J. Math.*

K. A. Bush, 1971b, Hadamard matrices and finite projective planes of even order, *J. Comb. Theory.*

K. A. Bush, 1971c, Forms of Hadamard matrices induced by finite projective planes, *Atti Convegno Geomtria Combinatoria* (Perugia).

J. M. Goethals and J. J. Seidel, 1970, Strongly regular graphs derived from combinatorial designs, *Canad. J. Math.* **22,** 597–614.

J. N. Srivastava et al., eds., *A Survey of Combinatorial Theory*
© North-Holland Publishing Company, 1973

CHAPTER 9

Cayley Diagrams and Regular Complex Polygons

H. S. M. COXETER

University of Toronto, Toronto, Ontario, Canada

Abstract

Section 1 generalizes a technique of R. Frucht for constructing symmetrical graphs as Cayley diagrams for some interesting finite groups, such as the linear groups SL(2, 3) of order 24, and GL(2, 3) of order 48.

Section 2 relates these graphs to certain configurations in the unitary plane. Such configurations are "regular polygons" $p\{q\}r$ according to a natural generalization of the ordinary regular q-gon $2\{q\}2$, and their symmetry groups $p[q]r$ generalize the dihedral group $D_q \cong 2[q]2$, which is the symmetry group of the q-gon.

Section 3 indicates how the regular complex polygons can be represented in a real 4-space and orthogonally projected onto a real plane. The real figures so obtained provide an agreeably symmetrical way to draw the Cayley diagrams.

I am grateful to Dr. B. B. Phadke for making these figures.

1. Cayley diagrams

R. Frucht [1955], p. 11, observed that, in many cases, a symmetrical graph of valency r can be obtained as a Cayley diagram (Burnside [1911], pp. 423–427) for a group whose presentation

$$T_\mu^2 = 1, \qquad f(T_1, T_2, \ldots) = f(T_2, T_3, \ldots) = \ldots = f(T_r, T_1, \ldots) = 1$$
$$(\mu = 1, 2, \ldots, r)$$

reveals an automorphism cyclically permuting the r generators. (Actually he considered only $r = 3$.) For instance, the strongly regular Thomsen graph (Fig. 1), with each triad of "parallel" edges colored alike, provides a Cayley diagram for the dihedral group D_3 of order 6:

$$T_1^2 = T_2^2 = T_3^2 = T_1T_2T_1T_3 = T_2T_3T_2T_1 = T_3T_1T_3T_2 = 1.$$

These six relations can easily be contracted to three, namely

$$T_1^2 = 1, \qquad T_1T_2 = T_2T_3 = T_3T_1.$$

The three generators T_ν are represented by the three "colors". Starting from any vertex, we can take a path representing the word T_1T_2, and see that the destination is the same as if we had taken the path T_2T_3 or T_3T_1. Other true relations, such as $(T_1T_2)^3 = 1$, can be deduced algebraically or read off from the diagram.

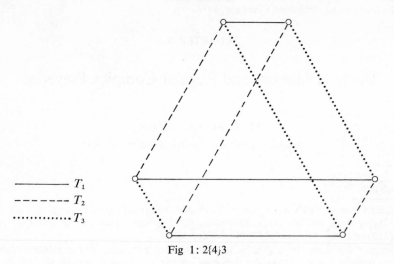

Fig 1: 2{4}3

When $r = 3$ (as in this example), the graph is *2-regular* in the sense of Tutte [1947], p. 459: its group of automorphisms is transitive on the pairs of adjacent edges. For greater values of r, the graph is only 1-regular (meaning that its group is transitive on the vertices and on the edges).

Other 1-regular graphs are obtained by writing $T_\mu^p = 1$ instead of $T_\mu^2 = 1$. Then the Cayley diagram is a directed graph of girth p. For instance, Fig. 2 shows another strongly regular graph. If we assume the edges to be directed so as to provide a positive sense round each (equilateral) triangle, this graph (with its two colors) serves as a Cayley diagram for the non-cyclic group of order 9:

$$T_1^3 = T_2^3 = 1, \qquad T_1T_2 = T_2T_1.$$

The last relation is represented by the rhombi.

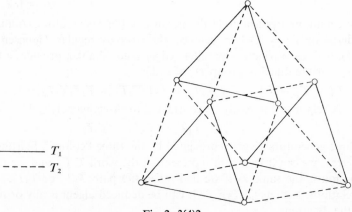

Fig. 2: 3{4}2

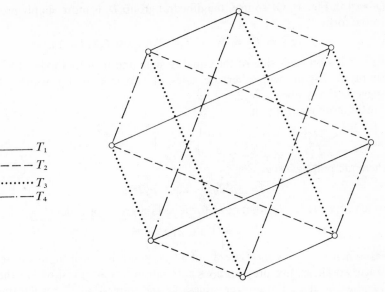

$$\begin{aligned}&\underline{\hspace{3cm}} \; T_1\\ &------ \; T_2\\ &\cdots\cdots\cdots \; T_3\\ &-\cdot-\cdot- \; T_4\end{aligned}$$

Fig. 3: 2{4}4

More generally, there is a graph $2\{4\}r$, of valency r and girth 4, representing the dihedral group D_r:

$$T_\mu^2 = 1, \qquad T_1T_2 = T_2T_3 = \ldots = T_{r-1}T_r = T_rT_1$$

($r = 4$ in Fig. 3). And there is a graph $p\{4\}2$, of valency 4 and girth p, representing $C_p \times C_p$:

$$T_1^p = T_2^p = 1, \qquad T_1T_2 = T_2T_1$$

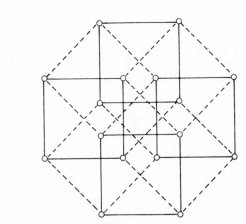

$$\begin{aligned}&\underline{\hspace{3cm}} \; T_1\\ &------ \; T_2\end{aligned}$$

Fig. 4: 4{4}2

($p = 4$ in Fig. 4). Of course, the dihedral group D_r is more simply presented in the form

$$T_1^2 = T_2^2 = 1, \qquad T_1T_2T_1 \ldots = T_2T_1T_2 \ldots,$$

with r T's on each side of the equality sign, and then its Cayley diagram is simply a $2r$-gon (with alternate sides of two colors), for which analogy suggests the symbol $2\{2r\}2$.

Still more generally, if

$$\frac{1}{p} + \frac{1}{q} + \frac{1}{r} > 1,$$

there are graphs

$$p\{2q\}r \quad \text{and} \quad p\{2r\}q$$

which serve as Cayley diagrams for a certain group of order

$$\frac{4}{qr}\left(\frac{1}{p} + \frac{1}{q} + \frac{1}{r} - 1\right)^{-2}$$

in its two alternative presentations

$$T_\mu^p = 1, \qquad T_1T_2 \ldots T_q = T_2 \ldots T_qT_{q+1} = \ldots = T_rT_1 \ldots T_{q-1}$$
$$(T_{\mu+r} = T_\mu; \mu = 1, 2, \ldots, r) \qquad (1.1)$$

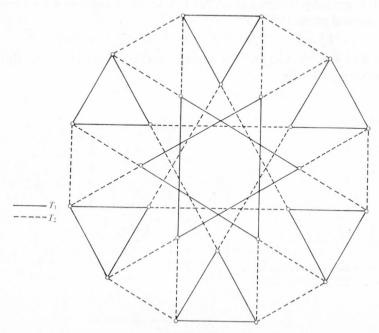

Fig. 5: $3\{6\}2$

and

$$U_v^p = 1, \qquad U_1 U_2 \ldots U_r = U_2 \ldots U_r U_{r+1} = \ldots = U_q U_1 \ldots U_{r-1}$$
$$(U_{v+q} = U_v; v = 1, 2, \ldots, q).$$

In particular, Fig. 5 and Fig. 6 are Cayley diagrams for the binary tetrahedral group 3[3]3 $\cong$ $SL(2, 3)$ of order 24 (Coxeter [1959], pp. 97, 99) in its two presentations

$$T_1^3 = 1, \qquad T_1 T_2 T_1 = T_2 T_1 T_2$$

and

$$T_1^3 = 1, \qquad T_1 T_2 = T_2 T_3 = T_3 T_1.$$

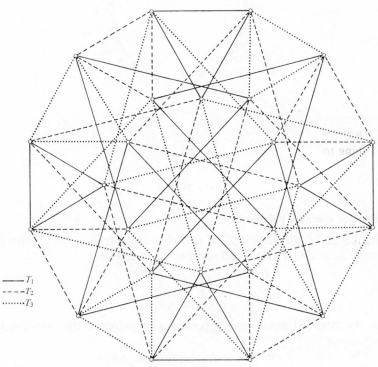

Fig. 6: 3{4}3

Fig. 7 is a Cayley diagram for $GL(2, 3)$, of order 48:

$$U_1^2 = 1, \qquad U_1 U_2 U_3 U_1 = U_2 U_3 U_1 U_2 = U_3 U_1 U_2 U_3.$$

As matrices modulo 3, we may write

$$T_1 = \begin{bmatrix} 1 & 1 \\ 0 & 1 \end{bmatrix}, \qquad T_2 = \begin{bmatrix} 0 & 1 \\ -1 & -1 \end{bmatrix}, \qquad T_3 = \begin{bmatrix} -1 & 1 \\ -1 & 0 \end{bmatrix},$$
$$U_1 = \begin{bmatrix} 0 & 1 \\ 1 & 0 \end{bmatrix}, \qquad U_2 = \begin{bmatrix} 1 & 1 \\ 0 & -1 \end{bmatrix}, \qquad U_3 = \begin{bmatrix} -1 & 1 \\ 0 & 1 \end{bmatrix}.$$

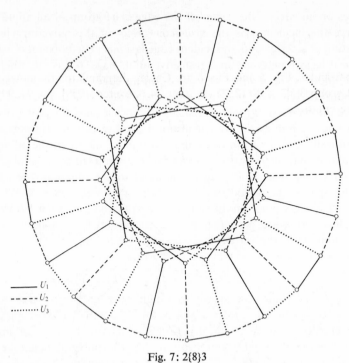

Fig. 7: 2{8}3

2. Regular complex polygons

Adjoining to 1.1 an element S, of period r, which transforms T_μ into $T_{\mu-1}$, we write $T_\mu = S^{\mu-1}T_1 S^{1-\mu}$, and observe that the relation

$$T_1 T_2 \ldots T_q = T_2 \ldots T_q T_{q+1}$$

yields

$$(T_1 S)^q = (ST_1)^q.$$

Thus the extended group, which permutes the edges of the graph, has the presentation

$$T_1^p = S^r = 1, \qquad (T_1 S)^q = (ST_1)^q$$

(Coxeter [1965], p. 76). This serves to identify it with the symmetry group $p[2q]r$ of the regular complex polygon $p\{2q\}r$, which can be described as follows.

In the complex affine plane, we define a *polygon* to be a finite configuration consisting of points (not all on one line) called *vertices*, and lines (not all through one point) called *edges*, such that every edge is incident with at least two vertices, every vertex with at least two edges, and any two vertices belong to a chain of successively incident vertices and edges. The centroid O of all the vertices is the *center* of the polygon; similarly, the centroid of the vertices on an edge is the *center* of the edge. The group of all affinities that preserve the incidences is the *symmetry group* of the polygon. The figure consisting of a vertex and an incident edge is a *flag*. The polygon is said to be *regular* if its symmetry group is transitive on its flags. If g is the order of this group, and p is the number of vertices on an edge, and r is the number of edges through a vertex, there are g flags, g/p edges, and g/r vertices.

Since the symmetry group consists of affinities leaving the center invariant, it is a linear group and leaves invariant a positive definite Hermitian form. We naturally use this form to establish a unitary metric in the affine space.

It follows that the p vertices on an edge are permuted by a cyclic group C_p of unitary *reflections* whose mirror joins O to the center of the edge. (In complex spaces, the period of a "reflection" may well exceed 2. We merely require that every point on the mirror is left invariant.) Similarly, the r edges through a vertex are permuted by a cyclic group C_r of reflections whose mirror joins O to the vertex. If T generates the C_p and S generates the C_r, the whole symmetry group is generated by T and S. The polygon is denoted by

$$p\{q\}r,$$

where q is a certain function of the angle between the two mirrors. For instance, the ordinary regular q-gon is $2\{q\}2$, because its symmetry group D_q is generated by reflections, of period 2, in mirrors forming an angle π/q.

If we ignore the slight complication that the polygon might be a *star* polygon (Shephard [1952], p. 92) such as the pentagram $2\{\frac{5}{2}\}2$, we find that the symmetry group of the polygon $p\{q\}r$ (and of its reciprocal $r\{q\}p$) is the group $p[q]r$:

$$T^p = S^r = 1, \qquad TST\ldots = STS\ldots,$$

where the last relation has q factors on each side of the equals sign (Coxeter [1962], pp. 90, 96). The order of this group is

$$g = \frac{8}{q}\left(\frac{1}{p} + \frac{2}{q} + \frac{1}{r} - 1\right)^{-2}.$$

3. Real representations

When the points $(u, v) = (x+yi, z+ti)$ of the unitary plane are represented by the points (x, y, z, t) of a real 4-space, the g/p edges of $p\{q\}r$ appear as real regular p-gons $2\{p\}2$. For, the coordinate axes can be chosen so that the p vertices on an edge are

$$(l\,e^{2v\pi i/p}, v) \qquad (v = 0, 1, \ldots, p-1),$$

yielding the real p-gon

$$\left(l\cos\frac{2v\pi}{p}, l\sin\frac{2v\pi}{p}, z, t\right).$$

Thus the real representation of $p\{q\}r$ is, in general, a set of g/r points and g/p p-gons, so arranged that each of the points is a common vertex of r p-gons. However, if $p = 2$, the p-gons reduce to edges. In other words, the figure in real 4-space is a graph having g/r vertices. If $p = 2$, the graph has $g/2$ edges, valency r, and girth $2q$. But if $p > 2$, it has g edges, valency $2r$, and girth p.

This description remains valid when the 4-dimensional figure is ortho-gonally projected onto a suitable plane. In saying "suitable", we mean chiefly that we wish to avoid having any of the p-gons or edges foreshortened into a mere point. It is, perhaps, a pleasant surprise to find that all the p-gons remain regular. This happens because, instead of going into real 4-space and then projecting onto a plane, we can just as well work first in the original unitary plane (with its two complex coordinates): project the complex polygon orthogonally onto a line, and then represent the resulting set of points on the complex line by an Argand diagram. Each complex edge yields a set of p points whose abscissae are

$$l'\, e^{2v\pi i/p} \qquad (v = 0, 1, \ldots, p-1)$$

for some $l' \leqslant l$.

We thus see how it happens that the triangles in Figs. 2, 5 and 6 are all equilateral, and that the significant quadrangles in Fig. 4 are squares, all positively oriented. (In these cases, where the generators are not involutory, the Cayley diagrams are *directed* graphs. The positive orientation of the regular p-gons indicates how the edges are to be directed.) For the same reason, the midpoints of the three edges at one vertex in Figs. 1 or 7 form an equilateral triangle, and the midpoints of the four edges at one vertex in Fig. 3 form a square.

The results of Sections 1 and 2 may now be summarized as follows:

$p\{2q\}2$ and $p\{4\}q$ are Cayley diagrams for $p[q]p$.

$\qquad$ $2\{6\}3$ is a Cayley diagram for $\langle 2, 2, 2\rangle_2$:

$$T_\mu^2 = 1, \qquad T_1 T_2 T_3 = T_2 T_3 T_1 = T_3 T_1 T_2.$$

$2\{8\}3$ and $2\{6\}4$ are Cayley diagrams for GL(2, 3).

$2\{10\}3$ and $2\{6\}5$ are Cayley diagrams for $\langle 5, 3, 2\rangle_2$

(Coxeter and Moser [1972], pp. 71, 134).

References

W. Burnside, 1911, *Theory of Groups of Finite Order* (Univ. Press, Cambridge).

H. S. M. Coxeter, 1959, Factor groups of the braid group, *Proc. Fourth Canad. Math. Congress, Toronto.*

H. S. M. Coxeter, 1962, The symmetry groups of the regular complex polygons, *Arch. Math.* **13,** 86–97.

H. S. M. Coxeter and W. O. J. Moser, 1972, *Generators and Relations for Discrete Groups* (Springer Verlag, Berlin).

R. Frucht, 1955, Remarks on finite groups defined by generating relations, *Canad. J. Math.* **7**, 8–17, 413.

G. C. Shephard, 1952, Regular complex polytopes, *Proc. London Math. Soc.* (3), **2**, 82–97.

G. C. Shephard and J. A. Todd, 1954, Finite unitary reflection groups, *Canad. J. Math.* **6**, 274–304.

W. T. Tutte, 1947, A family of cubical graphs, *Proc. Cambridge Philos. Soc.* **43**, 459–474.

J. N. Srivastava et al., eds., *A Survey of Combinatorial Theory*
© North-Holland Publishing Company, 1973

CHAPTER 10

Combinatorial Problems in Finite Abelian Groups

GEORGE T. DIDERRICH and HENRY B. MANN

University of Arizona, Tucson, Ariz. B5721, U.S.A.

Let $S = \{a_1, a_2, \ldots, a_m\}$ be a sequence of elements from a group G written additively. We consider the set $\Sigma(S)$ consisting of all elements of G which can be expressed as the sum of a subsequence of S:

$$\Sigma(S) = \{\varepsilon_1 a_1 + \cdots + \varepsilon_m a_m\}, \tag{1}$$

where ε_i takes the value 0 or 1 for $i = 1, \ldots, m$ but not all ε_i are 0.

Knowing m and perhaps some properties of the sequence S, we wish to infer properties of $\Sigma(S)$.

Two problems have received particular attention.

Problem 1. *What is the smallest integer $s(G)$ such that*

$$m = |S| = s(G) \to \Sigma(S) \ni 0.$$

Problem 2. *If the a_i are distinct and not 0 what is the smallest integer $c(G)$* *such that*

$$|S| \geqslant c(G) \to \Sigma(S) = G.$$

Considerable progress has been made in recent years as regards Problem 1, notably due to the efforts of J. E. Olson [1969a, b], and to a group of mathematicians at the Mathematisch Centrum in Amsterdam (Van Emde Boas [1969], Van Emde Boas and Kruyswijk [1969]).

Let G be an Abelian group with invariants $n_1 | n_2 | \cdots | n_t$ and set

$$s^* = s^*(G) = \sum n_i - t + 1. \tag{2}$$

Olson [1969a, b] and Kruyswijk proved independently that $s^*(G) = s(G)$ if G is a p-group or if $t = 2$. An impressive number of other groups for which $s^*(G) = s(G)$ was found by P. C. Baayen, J. H. van Lint and P. van Emde Boas. An extensive list of references will be found in (Van Emde Boas [1969], Van Emde Boas and Kruyswijk [1969]. These results made it seem reasonable to conjecture that $s^*(G) = s(G)$ for all Abelian groups. However, in 1969 P. C. Baayen found a counterexample in the group G_n with $4n$ invariants equal to 2 and one equal to $4n+2$ (Van Emde Boas [1969]). The number $s^*(G)$ computed from (2) is $8n+2$. Represent the elements of G_n by points $(x_1, x_2, \ldots,$

x_{4n+1}, x_{4n+2}), where $x_1, \ldots, x_{4n+1}$ are residues mod 2 and x_{4n+2} is a residue mod $2n+1$. Consider the array

$$
\begin{array}{ccccccccc}
1 & 0 & \cdots & 0 & \quad & 0 & 1 & \cdots & 1 \\
0 & 1 & \cdots & 0 & \quad & 1 & 0 & \cdots & 1 \\
\cdot & & \cdot & & & \cdot & & & \cdot \\
\cdot & & & \cdot & & & \cdot & & \cdot \\
\cdot & & & \cdot & & & \cdot & & \cdot \\
0 & 0 & \cdots & 1 & \quad & 1 & 1 & \cdots & 0 \\
1 & 1 & \cdots & 1 & \quad & 1 & 1 & \cdots & 1
\end{array}
\tag{3}
$$

where the columns represent elements of G. If we take c_1 columns from the first $4n+1$ columns and c_2 columns from the last $4n+1$ columns and if the sum of these columns is the 0 column, then $c_1 + c_2$ can only take the values $2n+1$, $4n+2$ or $6n+3$. However, a moment's reflection will show that $c_1 = c_2$ if c_2 is even and $c_1 = 4n+1-c_2$ if c_2 is odd and therefore either $c_1 + c_2 = 4n+1$ or $c_1 + c_2 \equiv 0 \pmod 4$. Hence no subset of the columns of (3) can add up to 0.

The smallest number of generators obtainable in this way is 5. Later, Van Emde Boas and Kruyswijk showed that $s(G) > s^*(G)$ for $G = (3, 3, 3, 6)$. It is not known if $s(G) = s^*(G)$ for all groups with 3 generators.

As regards Problem 2, an important result was obtained by Erdös and Heilbronn [1964], who proved

$$
\sqrt{4p+5}-2 < c(G) \leqslant 2\sqrt{6p}+1
$$

for $G = (p)$, p a prime, and conjectured $c(G) \leqslant 2\sqrt{p}+1$. This conjecture was proved by Olson [1968], who even improved it to $c(G) \leqslant \sqrt{4p-3}+1$ which is within at most 2 of the best possible value.

For groups of type (p, p), Mann and Olson [1967] showed

$$
2p-3 < c(G) \leqslant 2p-1,
$$

with the upper bound reached for $p = 3$. This result was generalized by George T. Diderrich (forthcoming) to $G = (p, q)$, where p and q are primes and $p+q-3 < c(G) \leqslant p+q-1$. Mann and Olson [1967] also showed $\Sigma(S) \ni 0$ if $|S| = 2p-2$, $G = (p, p)$ and the elements of S are distinct and non-zero. It is not known for what values of p, if any, we have $c(G) = 2p-2$ for $G = (p, p)$.

In this paper we shall prove:

Theorem. *Let G be a group of order $2l$ ($l > 1$). If G possesses a maximal subgroup of order l, then $l \leqslant c(G) \leqslant l+1$. Moreover, if G is Abelian, then*
   (i) $c(G) = l$ *if* $l \geqslant 5$ *or* $G = (2, 2, 2)$.
   (ii) $c(G) = l+1$ *otherwise*.

Let $S = \{a_1, \ldots, a_m\}$ be a set of m elements in a group G. Let S_t denote the set of elements of G which can be written as a sum of t distinct elements $a_{i_1}, \ldots, a_{i_t}$, $0 < t \leqslant m$. If $t > m$, let $S_t = \emptyset$.

We first prove two lemmas.

Lemma 1. *Let G be an Abelian group and let $S = \{a_1, \ldots, a_m\}$ be a set of m distinct elements of G.*

(i) *If m is odd and $m \geqslant 3$ then $|S_2| \geqslant m$.*

(ii) *If m is even $m \geqslant 6$ then $|S_2 \cup S_4| \geqslant m$.*

(iii) *If $m = 4$ and $|S_2 \cup S_4| = 3$ then G contains a subgroup H of type $(2, 4)$ and*

$$S = \{a, b+a, 3a, b+3a\},$$

where $H = (b, a)$ and b is of order 2, a of order 4.

Proof. (i). If $a_{i_1} + a_{i_2} = a_{j_1} + a_{j_2}$ then the pairs (i_1, i_2), (j_1, j_2) are either disjoint or identical. Hence any element in S_2 can have at most $\frac{1}{2}(m-1)$ distinct representations of the form $a_{i_1} + a_{i_2}$. Hence

$$|S_2| \frac{m-1}{2} \geqslant m \frac{m-1}{2},$$

$$|S_2| \geqslant m.$$

(ii). Now let m be even, $m \geqslant 6$. By the counting argument used in the proof of (i) we see that either $|S_2| \geqslant m$ or every element of S_2 has $\frac{1}{2}m$ representations as a sum of two elements of S. In the latter case, if v and w are in S_2 ($v = w$ not excluded) then

$$v = x_1 + x_2, \qquad w = x_3 + x_4, \tag{4}$$

where x_1, x_2, x_3, x_4 are four distinct elements of S. Therefore either $S_4 \not\subseteq S_2$ and $|S_2 \cup S_4| \geqslant m$ (since $|S_2| \geqslant m-1$ in any case) or S_2 is a subgroup H of G. In the latter case, if x_1, x_2, x_3 are any distinct elements of S, we have $x_1 + x_2 \in H$, $x_1 + x_3 \in H$, hence

$$x_2 \in H + x_3.$$

This means that S is contained in a coset mod H and therefore

$$|H| = |S_2| \geqslant S = m.$$

This proves (ii).

(iii). Now let $m = 4$ and assume $|S_2 \cup S_4| = 3$. Then we must have

$$\begin{aligned}
a_1 + a_2 &= a_3 + a_4, \\
a_1 + a_3 &= a_2 + a_4, \\
a_1 + a_4 &= a_2 + a_3.
\end{aligned} \tag{5}$$

Setting $a_i - a_1 = b_i$ for $i = 2, 3, 4$, we get

$$\begin{aligned}
b_2 &= b_3 + b_4, \\
b_3 &= b_2 + b_4, \\
b_4 &= b_2 + b_3.
\end{aligned} \tag{6}$$

By summing pairs of the equations (6), we get

$$2b_2 = 2b_3 = 2b_4 = 0.$$

Hence $(0, b_2, b_3, b_4)$ is a subgroup of G of type $(2, 2)$. Now $a_1 + a_2 + a_3 + a_4 \in S_2$ (since $|S_2 \cup S_4| = |S_2| = 3$). Hence we may arrange the notation so that

$$a_1 + a_2 + a_3 + a_4 = a_1 + a_2 = a_3 + a_4 = 0. \tag{7}$$

Summing over (5) and adding a_1 to both sides gives

$$4a_1 = a_1 + a_2 + a_3 + a_4 = 0. \tag{8}$$

From (7) and (8) we get

$$a_2 = 3a_1, \qquad b_2 = 2a_1 \neq 0, \qquad a_3 = b_3 + a_1, \qquad a_4 = -a_3 = b_3 + 3a_1$$

This completes the proof of (iii) and of Lemma 1.

Lemma 2. *Let G be an Abelian group and let $S = (a_1, \ldots, a_m)$ be a set of $m \geqslant 3$ distinct non 0 elements of G. Then either $S \cup 0$ is a subgroup of G or $S_1 \cup S_2 \cup S_3$ contains at least $m+1$ non 0 elements.*

Proof. If $S_2 \cup S_3 \nsubseteq S_1 \cup 0$ then the lemma is true. Hence we may assume

$$S_2 \cup S_3 \subset S_1 \cup 0. \tag{9}$$

We now consider first the case that m is even. The element 0 has at most $\frac{1}{2}m$ representations as a sum of two elements of S and elements of S can have at most $\frac{1}{2}(m-2)$ such representations. If S_2^* is the set of non 0 elements of S_2 we must have

$$|S_2^*| \frac{m-2}{2} + \frac{m}{2} \geqslant m \frac{m-1}{2} \tag{10}$$

and this yields $|S_2^*| = |S_1| = m$. We therefore have for each i

$$a_i = a_j + a_k$$

for some j, k. Hence

$$2a_i = a_i + a_j + a_k \in S_3$$

and on account of (9), $2a_i \in S_1 \cup 0$; since also $S_2 \subset S_1 \cup 0$ we see that $S_1 \cup 0$ is a group.

Now let m be odd, $m \geqslant 5$. If $a_i \in S_2$ for all i, we show as before that $2a_i \in S_1 \cup 0$ and that $S_1 \cup 0$ is a group. Otherwise let $a_1 \notin S_2$. Set $S' = (a_2, \ldots, a_m)$. If $S' \cup 0$ is a group then $S' + a_1 \subset S_2$, $(S' + a_1) \cap S' = \emptyset$ and $S_2 \nsubseteq S_1 \cup 0$ contradicting (9). If $S' \cup 0$ is not a group then $S_1' \cup S_2' \cup S_3'$ contains at least m non-zero elements by the preceding result. If $a_1 \notin S_1' \cup S_2' \cup S_3'$ then $S_1 \cup S_2' \cup S_3'$ contains $m+1$ non 0 elements. Hence we may assume $a_1 \in S_1' \cup S_2' \cup S_3'$, and since $a_1 \notin S_1' \cup S_2'$, we must have

$$a_1 = a_i + a_j + a_k,$$

where i, j, k are distinct from 1. Since $a_1 \notin S_2$ but $S_2 \in S_1 \cup 0$, we must have

$$a_i + a_j = a_k,$$
$$a_i + a_k = a_j, \tag{11}$$
$$a_j + a_k = a_i.$$

Summing the equations (11) gives

$$a_1 = a_i + a_j + a_k = 0,$$

a contradiction. This proves the theorem for $m \geqslant 4$. It may be left to the reader to verify Lemma 2 for $m = 3$.

We are now prepared to prove the theorem.

Let G be any group with a subgroup H of index 2; $G = \{H, H+g\}$. Let $S = \{a_1, \ldots, a_{l+1}\}$. Suppose

$$\{a_1, a_2, \ldots, a_u\} \subset H, \qquad \{a_{u+1}, \ldots, a_{u+v}\} \subset H+g.$$

$u+v = l+1$. Then $a_i + a_{u+j} \in H+g$, $1 \leqslant i \leqslant u$, $1 \leqslant j \leqslant v$. By theorem 1.1 of Mann [1967] we have $H+g \subset S_2$. Now $a_{u+1} + a_{u+j} \in H$ for $2 \leqslant j \leqslant v$. Set $A = \{0, a_1, \ldots, a_u\}$, $B = \{a_{u+1} + a_{u+j}, j = 2, \ldots, v\}$. Then

$$|A| = u+1, \qquad |B| = v-1,$$

and $|A| + |B| = u+v = l+1 > |H|$; hence $A+B = H$. Combining these results we see that $S_2 \cup S_3 = G$.

We proceed to prove statement (i) of our theorem.

Let G be Abelian and either $l \geqslant 5$ or $G = (2, 2, 2)$. Let $S = \{a_1, \ldots, a_l\}$, and let u and v be defined as before except that $u+v = l$. Let $A = \{a_1, \ldots, a_u\}$, $B = \{a_{u+1}, \ldots, a_{u+v}\}$. Then $a_{u+j} = h_j + g$ with $h_j \in H$, $|0, A| + |B| > |H|$ and

$$(0, A) + B \subset H+g.$$

Hence by theorem 1.1 of Mann [1967] we have $H+g \subset S_1 \cup S_2$.

It remains to show that $H \subset \Sigma(S)$.

If $|B_2 \cup B_4| \geqslant v$ or if $A_1 \cup A_2 \cup A_3$ contains $u+1$ non 0 elements then statement (i) follows from theorem 1.1 of (Mann [1967]). By lemma 1 we have to consider only the cases $v = 1$, $v = 2$ and the case that $v = 4$ and G contains a subgroup of type $(2, 4)$. By lemma 2 we have to consider only the cases $v = 1$, $v = 2$ and the case that $u+1|l$. Also $u+v = l \geqslant 4$. This leaves only the following cases:

Case 1: $v = 1$, $u = l-1$.

Case 2: $v = 2$, $u = 2$, $l = 4$.

Case 3: $v = 4$, G contains a subgroup of type $(2, 4)$ and $u+1 = l-3|l$.

Case 1. In this case all non 0 elements of H are in S. If one element of H is distinct from its inverse then $0 \in S_2$. If all elements of H have order 2 then if a, b are distinct and not 0, we have $a+b = c \neq 0$ and $a+b+c = 0 \in S_3$. Hence $G \subset S_1 \cup S_2 \cup S_3$.

Case 2. For $l = 4$ we have to consider only the case $G = (2, 2, 2)$. Hence $H = (2, 2)$, and in this case $A_1 \cup A_2$ has 3 non 0 elements of H.

Case 3. In this case we have $G \neq (2, 2, 2)$. Hence $l \geqslant 5$ and $l \equiv 0 \pmod 4$ and so $l \geqslant 8$. But $u + 1 = l - 3$ and $l - 3 | l$ only for $l = 4$ and $l = 6$. This completes the proof of statement (i) of the theorem.

It is easy to verify statement (ii) of the theorem and this verification may be left to the reader.

References

P. Erdös and H. Heilbronn, 1964, On the addition of residue classes mod p, *Acta Arithmetica*, 149–159.

P. Erdös, A. Ginsburg and A. Ziv, 1961, A Theorem in the additive number theory, *Bull. Res. Council Israel* **10F**, 41–43.

H. B. Mann, 1965, *Addition Theorems* (Wiley, New York).

H. B. Mann, 1967, Two addition theorems, *J. Combin. Theory* **3**, 233–235.

H. B. Mann and J. E. Olson, 1967, Sums of sets in the elementary Abelian group of type (p, p), *J. Combin. Theory* **2**, 275–284.

J. E. Olson, 1968a, An addition theorem for the elementary Abelian group, *J. Combin. Theory* **5**, 53–58.

J. E. Olson, 1968b, An addition theorem modulo p, *J. Combin. Theory* **5**, 45–52.

J. E. Olson, 1969a, A combinatorial problem on finite Abelian groups I, *J. Number Theory* **1**, 8–11.

J. E. Olson, 1969b, A combinatorial problem on finite Abelian groups II, *J. Number Theory* **1**, 195–199.

P. van Emde Boas, 1969, A combinatorial problem on finite Abelian groups II, Z.W. 1969–007 (Math. Centrum, Amsterdam).

P. van Emde Boas and D. Kruyswijk, 1969, A combinatorial problem on finite Abelian groups III, WN31 1969–008 (Math. Centrum, Amsterdam).

J. N. Srivastava et al., eds., *A Survey of Combinatorial Theory*
© North-Holland Publishing Company, 1973

CHAPTER 11

A q-Analog of the Partition Lattice †

T. A. DOWLING

University of North Carolina, Chapel Hill, N. Car., U.S.A.

1. Introduction

The set of all partitions of a finite set, when ordered by refinement, is a well-known geometric lattice enjoying a number of structural properties. Every upper interval of a partition lattice is a partition lattice, and every interval is a direct product of partition lattices. The partitions with a single non-trivial block form a Boolean sublattice of modular elements, and the Whitney numbers are the familiar Stirling numbers. Because of these and other structural properties, the partition lattices occupy a middle ground between the highly structured modular geometric lattices (projective geometries), and arbitrary geometric lattices (combinatorial geometries), exhibiting some of the consequences of the departure from modularity while still retaining enough of the structure to facilitate their study.

We describe in this article for any prime power q a class of geometric lattices, here called q-partition lattices, which share a number of the properties of partition lattices. There is a natural order- and rank-preserving map from the q-partition lattice to the partition lattice of the same rank, which reduces to an isomorphism when $q = 2$. We examine the interval structure of the q-partition lattice and obtain a representation of it as the lattice of a subgeometry of a projective geometry over the q-element field. The characteristic polynomial and Möbius function are obtained, and a Stirling-like identity and recursion derived for the Whitney numbers of the q-partition lattice. We conclude with an application to the enumeration of factorial designs using the geometrical formulation of the design problem developed by Professor Bose in his classical paper [1947].

2. Preliminaries

We summarize in this section a number of results and definitions needed later (Crapo and Rota [1970], Rota [1964]).

A (*partially*) *ordered set* $(P, \leqslant)$ is a set P together with a reflexive,

† The research reported here was partially supported by the Air Force Office of Scientific Research under Contract AFOSR-68-1415.

anti-symmetric, transitive relation on P, written $x \leqslant y$. When the order relation is implicit, we write simply P for $(P, \leqslant)$. An ordered set $(P, \leqslant)$ is *finite* if P is a finite set. All ordered sets considered here are finite. The *direct product* of two ordered sets P, Q is the set $P \times Q$ with order $(u, v) \leqslant (x, y)$ iff $u \leqslant x$ in P and $v \leqslant y$ in Q. An *interval* $[x, y]$ of an ordered set P is the ordered subset $[x, y] = \{z \in P \mid x \leqslant z \leqslant y\}$ (with the order relation of P), defined for all $x, y \in P$ such that $x \leqslant y$. An element y *covers* x in P if $y > x$ and $[x, y]$ consists only of the two elements x, y. A subset $C = \{x_i \mid i \in [0, n]\}$ of P is a *chain* if it is *totally ordered*: $x_0 < x_1 < \cdots < x_n$. The *length* of C is n, one less than its cardinality. C is a *maximal chain* in $[x, y]$ iff $x_0 = x$, $x_n = y$, and x_i covers x_{i-1}, $i \in [1, n]$. P satisfies the *chain condition* if all maximal chains in any interval $[x, y]$ are of the same length. If P has a *zero element* 0 $(0 \leqslant x$ for all $x \in P)$, and satisfies the chain condition, the *rank* of an element $x \in P$, denoted $r(x)$, is the length of a maximal chain in $[0, x]$. If P has a unit element 1 $(1 \geqslant x$ for all $x \in X)$, and satisfies the chain condition, the *corank* of $x \in P$ is the length of a maximal chain in $[x, 1]$.

Let P have a 0 and 1. An *atom* (*coatom*) is an element covering 0 (covered by 1). P is a *lattice* iff any two elements x, y have a unique minimal upper bound $x \vee y$, called their *join*, and a unique maximal lower bound $x \wedge y$, called their *meet*.

If P, Q are ordered sets, a function $\phi: P \to Q$ is *order-preserving* iff $x \leqslant y$ implies $\phi(x) \leqslant \phi(y)$. If both P, Q satisfy the chain condition, ϕ is *rank-preserving* if $r(\phi(x)) = r(x)$ for all $x \in P$. P and Q are *isomorphic*, written $P \cong Q$, iff there is a bijection $\phi: P \to Q$ such that ϕ, ϕ^{-1} are both order-preserving. If Q is a lattice and $\phi: P \to Q$ is an isomorphism of ordered sets, then P is a lattice and $\phi(x \vee y) = \phi(x) \vee \phi(y)$, $\phi(x \wedge y) = \phi(x) \wedge \phi(y)$, i.e. ϕ is a *lattice isomorphism*. An *anti-isomorphism* of two ordered sets P, Q is a bijection $\phi: P \to Q$ such that both ϕ, ϕ^{-1} are *order-inverting*.

A (finite) lattice L is *geometric* when y covers x iff $y = x \vee p$ for some atom p of L. A geometric lattice satisfies the chain condition, and its rank function satisfies the *semimodular inequality*: $r(x \vee y) + r(x \wedge y) = r(x) + r(y)$. Elements of rank $1, 2, r(1) - 2, r(1) - 1$, are called *points*, *lines*, *colines*, *copoints*, respectively.

A (finite) *combinatorial geometry* $G = G(S)$ is a finite set S together with a closure operator $A \to \bar{A}$ on S satisfying the *exchange property*: $a \in \overline{A \cup b}$, $a \notin \bar{A}$ implies $b \in \overline{A \cup a}$, and such that the empty set and all elements of S, called *points*, are closed. An *independent set* is a set $B \subseteq S$ such that $\overline{B - b} \neq \bar{B}$ for all $b \in B$. All maximal independent subsets of any set $A \subseteq S$, called *bases* of A, have the same cardinality, called the *rank* of A. A basis of G is a basis of S, and the rank of G is the rank of S. A *subgeometry* H of G is a subset T of S with closure operator $A \mapsto \bar{A} \cap T$. A set $A \subseteq T$ is independent in H iff it is independent in G.

If $G(S)$ is a combinatorial geometry, the set of closed sets of G, ordered by inclusion, is a geometric lattice $L(G)$ whose points are the elements of S.

Conversely, every geometric lattice L defines a combinatorial geometry $G(S)$ on its set S of points by $A \rightarrow \bar{A} = \{p \in S \mid p \leqslant \vee_{a_i \in A} a_i\}$. If $H = H(T)$ is a subgeometry of $G = G(S)$, the lattice $L(H)$ consists of all elements $x \in L(G)$ such that $x = \vee_{a_i \in A} a_i$ for some subset $A \subseteq T$.

3. The lattice of partitions

A *partition* of a finite set X with n elements is a set $\pi = \{X_i \mid i \in [1, k]\}$ of disjoint, nonempty subsets of X, such that $X = \bigcup_{i=1}^k X_i$. The subsets X_i are the *blocks* of π. There is an obvious correspondence between partitions of X and equivalence relations defined on X, wherein the blocks of the partition are the equivalence classes.

The set Π_n of all partitions of X may be ordered by *refinement*: $\pi \leqslant \sigma$ iff each σ-block is a union of π-blocks. With this order, Π_n is a lattice, with zero element the partition consisting of n singleton blocks, corresponding to the identity relation on X, and with unit element the partition 1 consisting of the single block X, corresponding to the universal relation on X.

The join and meet in Π_n of two partitions $\pi = \{X_i \mid i \in [1, k]\}$, $\tau = \{Y_j \mid j \in [1, l]\}$ may be easily found from the *intersection graph* $I(\pi, \sigma)$, a bipartite graph with vertices X_i, Y_j ($i \in [1, k]$, $j \in [1, l]$) and edges $X_i Y_j$ iff $X_i \cap Y_j$ is nonempty. A block of $\pi \wedge \sigma$ is an intersection $X_i \cap Y_j$, for each edge $X_i Y_j$, and a block of $\pi \vee \sigma$ is a union $\bigcup X_i$ of all X_i in a connected component of $I(\pi, \sigma)$.

The partition lattice is a geometric lattice, with rank function $r(\pi) = n - k(\pi)$, where $k(\pi)$ is the number of blocks of π.

4. The q-partition lattice Q_n

Let $\mathbb{P}_{n-1}(F)$ be a projective geometry of dimension $n-1$ over a finite field $F = GF(q)$, and let S be the set of points of $\mathbb{P}_{n-1}$. Given any independent set B in S, we may associate with each point $a \in \bar{B}$, the subspace spanned by B, a unique subset $B(a)$ of B, defined by

$$B(a) = \{b \in B \mid a \notin \overline{B-b}\}.$$

$B(a)$ is the minimal subset of B whose closure contains a. We call $B(a)$ the *B-support* of a, $a \in \bar{B}$.

Let $\mathbb{P}_{n-1}$ be coordinatized over F, i.e., each point $a \in S$ is represented by a non-zero list $\alpha = (\alpha_1, \alpha_2, \ldots, \alpha_n) \in F^n$ so that a subset $B = \{b_i \mid i \in [1, k]\}$ of S is independent iff the corresponding set $\{\beta_i \mid i \in [1, k]\}$ of coordinate lists is linearly independent in F^n. We follow the usual convention of associating with every point $a \in S$ the set of all non-zero scalar multiples of its coordinate list. Thus every list $\alpha \in F^n - \{0\}$ represents a point of S, with two lists representing the same point iff they are scalar multiplier of each other.

Then if $B = \{b_i | i \in [1, k]\}$ is an independent set, with $\{\beta_i | i \in [1, k]\}$ a set of coordinate lists of the b_i, $\bar{B}$ is the set of all points $a \in S$ such that for some $\lambda_i \in S$, $i \in [1, k]$, $\alpha = \sum_{i=1}^k \lambda_i \beta_i$, where α is a coordinate list of a. We shall find it convenient to denote this relationship by writing $a = \sum_{i=1}^k \lambda_i b_i$, with the understanding implicit in all such expressions that a set of coordinate lists of the b_i is fixed, so that the λ_i are uniquely determined up to a constant scalar multiple. Thus $a = \sum_{i=1}^k \lambda_i b_i$, $a' = \sum_{i=1}^k \lambda_i' b_i$ implies $a = a'$ iff there is a $\kappa \in F^* = F - \{0\}$ such that $\lambda_i' = \kappa \lambda_i$ for all $i \in [1, k]$. In particular, $a = \kappa a$, $\kappa \in F^*$. Then $a = \sum_{i=1}^k \lambda_i b_i$ is a point of $\bar{B}$, and $B(a) = \{b_i \in B | \lambda_i \neq 0\}$. We shall frequently find it convenient to write the expression $\sum_{i=1}^k \lambda_i b_i$ as $\sum \lambda_i b_i$ $(b_i \in B(a))$ so that all the coefficients λ_i are non-zero. More generally, if B_1 is a nonempty subset of $B(a)$, then $\sum \lambda_i b_i$ $(b_i \in B_1)$ is the point $\sum \kappa_i b_i$, where $\kappa_i = \lambda_i$, $b_i \in B_1$, and $\kappa_i = 0$, $b_i \notin B_1$.

Let $X = \{x_i | i \in [1, n]\}$ be a basis of $\mathbb{P}_{n-1}$. We define a q-partition of X (briefly, a q-partition) as a set A of points for which the X-supports $X(a)$, $a \in A$, are disjoint sets. Thus if $A = \{a_i | i \in [1, k]\}$ is a q-partition of X, and α_i is a coordinate list of a_i, then no two α_i have non zero elements in the same position. Equivalently, the matrix with rows α_i is a $k \times n$ column-monomial matrix over F, with no zero rows. Two $k \times n$ column-monomial matrices represent the same q-partition A iff one is obtainable from the other by a permutation of rows and multiplication of rows by non-zero scalars, i.e. by premultiplication by a $k \times k$ monomial matrix.

It is clear that every q-partition A of X is an independent set, and that any subset of a q-partition is one also. In particular, X is a q-partition, as is the empty set. We denote by Q_n the set of all q-partitions of X. For $A \in Q_n$, let $k(A)$ be the cardinality of A, and let

$$X(A) = \bigcup_{a \in A} X(a), \qquad X_0(A) = X - X(A). \tag{1}$$

Then $\{X(a) | a \in A\}$ is a partition of $X(A)$ into $k(A)$ blocks. However, $X_0(A)$ may be empty. We may remedy this by adjoining to $X_0(A)$ the empty subspace z (or any element not in x). Then $A \in Q_n$ yields a partition

$$\pi(A) = \{X(a) | a \in A\} \cup \{X_0(A) \cup z\} \tag{2}$$

of $X \cup z$ into $k(A) + 1$ blocks.

Proposition 1. *Let* $\sigma: Q_n \to L_n$ *be the map* $A \mapsto \bigvee_{a_i \in A} a_i$ *from the set of q-partitions of X to the lattice of subspaces of* $\mathbb{P}_{n-1}$. *Then the relation*

$$A \geqslant B \quad \text{iff} \quad \sigma(A) \leqslant \sigma(B) \tag{3}$$

defines an order on Q_n. *So ordered,* Q_n *satisfies the chain condition with rank function* $r(A) = n - k(A)$. *The map* $\pi: Q_n \to \Pi_{n+1}$, *defined by* (2), *is surjective and preserves order and rank. If $q = 2$, π is an isomorphism,*

Proof. Suppose $\sigma(A) \leqslant \sigma(B)$. Then $A \subseteq \bar{B}$, so every $a \in A$ can be written $a = \sum \lambda_i b_i$ $(b_i \in B(a))$, hence $X(a) = \bigcup X(b_i)$ $(b_i \in B(a))$. The remaining block $X_0(A) \cup z$ of $\pi(A)$ must then be the union of $X_0(B) \cup z$ and the blocks $X(b_i)$ contained in no $X(a)$. Thus $\pi(B) \leqslant \pi(A)$, so $\sigma(A) = \sigma(B)$ implies $\pi(A) = \pi(B)$, and each $B(a)$ is a singleton, i.e. $A = B$. σ is therefore injective, so the relation defined by (3) is anti-symmetric, hence an order on Q_n. It is evident from the above that $A > B$ in Q_n iff A can be obtained from Q_n by a sequence of single point deletions and/or replacement of two points b_i', b_j' of B' $(B \leqslant B' < A)$ by a third point $b_i' + \lambda b_j'$ on the line joining b_i' and b_j'. Each such operation decreases cardinality by one, so the length of all maximal chains in $[B, A]$ is $k(B) - k(A)$. Thus Q_n satisfies the chain condition and has rank function $r(A) = k(X) - k(A) = n - k(A)$. Since $\pi(A)$ has $k(A) + 1$ blocks, the rank of $\pi(A)$ is $(n+1) - (k(A)+1) = n - k(A)$, so π preserves rank. Given any partition $\tau = \{X_0 \cup z, X_1, \ldots, X_k\}$ of $X \cup z$, let $a_i = \sum x_j$ $(x_j \in X_i)$, for each $i \in [1, k]$. Then $\pi(A) = \tau$, where $A = \{a_i | i \in [1, k]\}$, so π is surjective. If $q = 2$, A is the only preimage of τ, hence π is a bijection. Clearly π^{-1} is order-preserving.

Remark. One could define a G-partition of a basis X of a combinatorial geometry G in the analogous way. It can be shown that σ is injective, so (3) defines an order, while the chain condition holds iff G has no trivial lines.

Corollary 1. *A covers B in Q_n iff*
$$A = B - b_i$$
or
$$A = B - \{b_i, b_j\} \cup \{b_i + \lambda b_j\}$$
for some $b_i, b_j \in B$, $\lambda \in F^*$.

Corollary 2. *Each element of rank $n-k$ in Q_n is covered by*
$$\binom{k}{1} + (q-1)\binom{k}{2}$$
elements of rank $n-k+1$.

Remark. The partial order defined by (3) can easily be described in terms of representative matrices of q-partitions. If M_A, M_B are representative matrices of A, B, respectively, then $A \geqslant B$ iff there is a $k(B) \times k(A)$ column-monomial matrix P such that $M_A = P M_B$. An equivalent definition, independent of dimension, can be obtained by adding $n - k(A)$ zero rows in arbitrary positions to any representative matrix of each element $A \in Q_n$. The $n \times n$ column-monomial matrices over F form a semigroup S_n under multiplication, containing as a maximal subgroup the group M_n of monomial matrices. Then $A \geqslant B$ iff there is a $P \in S_n$ such that $N_A = P N_B$, i.e. iff $N_A \in S_n N_B$, so $A \geqslant B$ if $S_n N_A \subseteq S_n N_B$. Two matrices $N_1, N_2 \in S_n$ generate the same principal left ideal of S_n iff there is an $M \in M_n$ such that $N_1 = M N_2$. But this is precisely the requirement that N_1, N_2 represent the same q-partition, so Q_n is *anti-isomorphic* to the set of principal left ideals of S_n, ordered by inclusion.

Our next two propositions concern the interval structure of Q_n.

Proposition 2. *If $B \in Q_n$ is of cardinality k, then $[B, 1] \cong Q_k$.*

Proof. In Q_n, $A \geqslant B$ iff $A \subseteq \bar{B}$. But $A \in Q_n$, $A \subseteq \bar{B}$ imply that the subsets $B(a)$, $a \in A$, are disjoint; so A is a q-partition of B. Conversely, every q-partition of B is a q-partition of X in $[B, 1]$. The closure in $\bar{B}$ is that of $\mathbb{P}_{n-1}$, so this correspondence preserves order, and the subgeometry on $\bar{B}$ is isomorphic to $\mathbb{P}_{k-1}$.

Proposition 3. *Let $A = \{a_l| \ l \in [1, k]\} \in Q_n$, and let $|X_0(A)| = n_0$, $|X(a_l)| = n_l$, $l \in [1, k]$. Then*
$$[0, A] \cong Q_{n_0} \times \Pi_{n_1} \times \Pi_{n_2} \times \cdots \times \Pi_{n_k}.$$

Proof. Let $X_0 = X_0(A)$, $X_l = X(a_l)$, $l \in [1, k]$. Suppose $a_l = \sum \kappa_i x_i$ $(x_i \in X_l)$. Then $B \leqslant A$ iff $\pi(B) \leqslant \pi(A)$ and $a_l \in \bar{B}_l$ for all $l \in [1, k]$, where $B_l = \{b \in B| \ X(b) \subseteq X_l\}$. But given only the $\pi(B)$-blocks not in $X_0 \cup z$, the subsets B_l, $l \in [1, k]$, are uniquely determined by the a_l. Namely, if X_{lm} are the $\pi(B)$-blocks contained in X_l, then $b_{lm} = \sum \kappa_i x_i$ $(x_i \in X_{lm})$. To the set $\bigcup B_l$ $(l \in [1, k])$, as determined by the blocks of $\pi(B)$ not in $X_0 \cup z$, an arbitrary q-partition B_0 of X_0 can be added to obtain a $B \in [0, A]$. The order in $[0, A]$ is clearly the product of the orders in the blocks X_l, $l \in [1, k]$, and the set X_0. It is evident from the above that the order in X_l is that of Π_{n_l}, while that of X_0 is Q_{n_0}.

Corollary 1. *Let $B \leqslant A$ in Q_n, where $A = \{a_l| \ l \in [1, k]\}$. Let $|B_0(A)| = m_0$, $|B(a_l)| = m_l$, $l \in [1, k]$. Then*
$$[B, A] \cong Q_{m_0} \times \Pi_{m_1} \times \Pi_{m_2} \times \cdots \times \Pi_{m_k}.$$

Corollary 2. *If $a_l = \sum \kappa_i x_i$ $(x_i \in X_l)$, $A = \{a_l| \ l \in [1, k]\}$, the atoms of $[0, A]$ are*
$$X - \{x_i, x_j\} \cup \{\kappa_i x_i + \kappa_j x_j\}$$
for all $x_i, x_j \in X_l$, $l \in [1, k]$, and
$$X - x_i, \ X - \{x_i, x_j\} \cup \{x_i + \lambda x_j\}$$
for all $x_i, x_j \in X_0$, $\lambda \in F^*$.

Corollary 3. *Let $A = \{a\}$ be a copoint of Q_n. If $X(a) = X$, then $[0, A] \cong \Pi_n$, while if $X(a) = \{x_i\}$, then $[0, A] \cong Q_{n-1}$.*

We prove in Section 5 that Q_n is isomorphic to the lattice of a subgeometry of $\mathbb{P}_{n-1}$. The next proposition (and its corollary) is not required for that proof, but we include it to describe the nature of joins and meets in Q_n.

Proposition 4. *Q_n is a lattice.*

Proof. Let A, $B \in Q_n$. Since π preserves order, $\pi(C) \geqslant \pi(A) \vee \pi(B)$ and $\pi(D) \leqslant \pi(A) \wedge \pi(B)$ for any upper bound C and lower bound D of A, $B \in Q_n$.

Let $\pi(A) \vee \pi(B) = \{X_l | l \in [1, m]\} \cup \{X_0 \cup z\}$. Suppose $\{X_l | l \in [1, p]\}$, where $p \leq m$, are the blocks of $\pi(A) \vee \pi(B)$ for which there exists a point $c_l \in \bar{A}_l \cap \bar{B}_l$ with X-support X_l, where A_l, B_l are the subsets of A, B, respectively, with $X(A_l) = X_l$, $X(B_l) = X_l$. Then clearly $C = \{c_l | l \in [1, p]\}$ is a minimal upper bound of A, B in Q_n. To show that $A \vee B$ exists, we must prove that the c_l are uniquely defined. For a fixed $l \in [1, p]$, let $A_l = \{a_i | i \in [1, r]\}$, $B_l = \{b_j | j \in [1, s]\}$. Suppose there are two points c_l, c_l' in $\bar{A}_l \cap \bar{B}_l$ with X-support X_l. Then

$$c_l = \sum_{i=1}^{r} \kappa_i a_i = \sum_{j=1}^{s} \lambda_j b_j, \qquad c_l' = \sum_{i=1}^{r} \kappa_i' a_i = \sum_{j=1}^{s} \lambda_j' b_j,$$

say, where all coefficients are non-zero. Let $a_i = \sum \alpha_{iu} x_u$, $b_j = \sum \beta_{jv} x_v$. If $x \in X_l$, say $x = x_t$, and x_t is contained in the X-supports of $a_i \in A_l$, $b_j \in B_l$, then the coefficient of x_t in c_l, c_l' is

$$\kappa_i \alpha_{it} = \lambda_j \beta_{jt}, \qquad \kappa_i' \alpha_{it} = \lambda_j' \beta_{jt},$$

respectively. Thus $\kappa_i / \lambda_j = \kappa_i' / \lambda_j'$, so $\kappa_i' / \kappa_i = \lambda_j' / \lambda_j$ whenever $X(a_i) \cap X(b_j)$ is nonempty. Since X_l is a block of $\pi(A) \vee \pi(B)$, the intersection graph of the $X(a_i)$, $i \in [1, r]$, and the $X(b_j)$, $j \in [1, s]$ is connected. Thus all the ratios κ_i' / κ_i, λ_j' / λ_j are equal, so $c_l = c_l'$ and $A \vee B = C$.

The blocks of $\pi(A) \wedge \pi(B)$ are the nonempty intersections of the blocks of $\pi(A)$ with the blocks of $\pi(B)$. Let $a = \sum \kappa_i x_i$ $(x_i \in X(a))$, $b = \sum \lambda_i x_i$ $(x_i \in X(b))$ be arbitrary points of A, B, respectively, and let Y_l be a block of $\pi(A) \wedge \pi(B)$ such that $z \notin Y_l$. We define a q-partition D of X as follows. If $Y_l = X_0(A) \cap X(b)$, let D contain the point $\sum \lambda_i x_i$ $(x_i \in Y_l)$, while if $Y_l = X(a) \cap X_0(B)$, let D contain the point $\sum \kappa_i x_i$ $(x_i \in Y_l)$. Finally, if $Y_l = X(a) \cap X(b)$, then partition Y_l into the blocks Y_{lm} defined by the equivalence relation E: $x_i E x_j$ iff $\lambda_i / \kappa_i = \lambda_j / \kappa_j$. For each Y_{lm} let D contain the point $\sum \lambda_i x_i$ $(x_i \in Y_{lm})$. It is clear (see the proof of Proposition 3) that D is the unique maximal lower bound of A, B in Q_n, i.e. $D = A \wedge B$.

Corollary. Q_n is a geometric lattice.

Proof. By Corollary 1 of Proposition 1, A covers B iff $A = B - b_r$, or $A = B - \{b_r, b_s\} \cup \{b_r + \lambda b_s\}$. In the first case, $A = B \vee (X - x_i)$ for any $x_i \in X(b_r)$, while in the second case $A = B \vee (X - \{x_i, x_j\} \cup \{x_i + \lambda_{ij} x_j\})$ for any $x_i \in X(b_r)$, $x_j \in X(b_s)$, where $\lambda_{ij} = \lambda \kappa_j / \kappa_i$, $b_r = \sum \kappa_i x_i (x_i \in X(b_r))$, $b_s = \sum \kappa_j x_j (x_j \in X(b_s))$.

5. Representation of Q_n

In the lattice L_n of subspaces of $\mathbb{P}_{n-1}$, let $X^* = \{x_i^* | i \in [1, n]\}$ be the set of copoints defined by $x_i^* = \vee x_j$ $(x_j \in X - x_i)$. X^* is the dual basis of X; every copoint $c \in L$ may be written as $c = \sum \kappa_i x_i^*$ such that if $a = \sum \lambda_i x_i$ is a point, then $a \leq c$ iff $\sum \kappa_i \lambda_i = 0$. The dual of a point $a = \sum \kappa_i x_i$ is the copoint

$a^* = \sum \kappa_i x_i^*$, and the dual of a copoint $c = \sum \lambda_i x_i^*$ is the point $c^* = \sum \lambda_i x_i$. The pair of bijections $a \mapsto a^*$, $c \mapsto c^*$ extend to an anti-isomorphism *: $L_n \to L_n$;

$$(\vee a_i)^* = \wedge a_i^*, \qquad (\wedge c_i)^* = \vee c_i^*.$$

Recall the injection $\sigma: Q_n \to L_n$ defined by $\sigma(A) = \vee a_i \, (a_i \in A)$. The elements of Q_n are independent sets, so

$$r_L(\sigma(A)) = k(A) = n - r_Q(A).$$

Since $A \geqslant B$ in Q_n iff $\sigma(A) \leqslant \sigma(B)$, the image of Q_n under σ is anti-isomorphic to Q_n. Thus the image R_n of the composite map $\sigma^* = {}^* \circ \sigma: Q_n \to L_n$ is isomorphic to Q_n. Further,

$$r_L(\sigma^*(A)) = n - r_L(\sigma(A)) = r_Q(A),$$

so the rank of an element in R_n is its rank in L_n. We now prove that R_n is the lattice of a subgeometry of $\mathbb{P}_{n-1}$.

Proposition 5. Q_n *is a geometric lattice, isomorphic to its image* R_n *under the map* $\sigma^*: Q_n \to L_n$. R_n *is the lattice of the subgeometry of* $\mathbb{P}_{n-1}$ *on the point set*

$$S_2 = \{a \in S | \, |X(a)| \leqslant 2\},$$

consisting of X and all points $x_i + \lambda x_j$ on the lines joining two points of X.

Proof. We have only to show that S_2 is the σ^*-image of the atoms of Q_n, and that a subspace U is in R_n iff it contains a basis in S_2.

The σ^*-image of $A = \{a_l | \, l \in [1, k]\} \in Q_n$ is $\vee b_r \, (r \in [1, n-k])$ where $\{b_r | \, r \in [1, n-k]\}$ is any independent set such that $b_r \leqslant a_l^*$ for all $r \in [1, n-k]$, $l \in [1, k]$. Consider now the image of the atoms of Q_n. The dual of x_i is x_i^* and that of $x_i + \lambda x_j$ is $x_i^* + \lambda x_j^*$. A point $a = \sum \kappa_l x_l$ is the σ^*-image of $X - x_i$ iff $\kappa_l = 0$ for all $l \neq i$. Thus

$$\sigma^*(X - x_i) = x_i. \tag{4}$$

The point $a = \sum \kappa_l x_l$ is the σ^*-image of $X - \{x_i, x_j\} \cup \{x_i + \lambda x_j\}$ iff $\kappa_l = 0$ for all $l \neq i, j$ and $\kappa_i + \lambda \kappa_j = 0$. Hence

$$\sigma^*(X - \{x_i, x_j\} \cup \{x_i + \lambda x_j\}) = x_i - \lambda^{-1} x_j, \tag{5}$$

so S_2 is the set of points of R_n.

Let $A = \{a_l | \, l \in [1, k]\} \in Q_n$, where $a_l = \sum \kappa_i x_i \, (x_i \in X_l = X(a_l))$. By Corollary 1 of Proposition 3, (4) and (5), the points of $[0, \sigma^*(A)]$ in R_n are

$$\kappa_i^{-1} x_i - \kappa_j^{-1} x_j \qquad (x_i, x_j \in X_l), \tag{6}$$

$$x_i, \qquad x_i + \lambda x_j \qquad (x_i, x_j \in X_0 = X_0(A)). \tag{7}$$

Fix a point, say x_l, in X_l, for each $l \in [1, k]$, and let

$$T_l = \{x_l - \kappa_l \kappa_j^{-1} x_j | \, x_j \in X_l, x_j \neq x_l\}.$$

Then $T = T_1 \cup T_2 \cup \cdots \cup T_k \cup X_0$ is independent and of cardinality $n - k = r_L(\sigma^*(A))$, so T is a basis of $\sigma^*(A)$ in S_2.

It remains to show that for every $U \in L_n$, $U \cap S_2$ is of the form given by (6) and (7). Let U be a subspace, and let $X_0 = \{x_i | x_i \in U\}$. Then $x_i + \lambda x_j \in U$ for all $\lambda \in F^*$, $x_i, x_j \in X_0$. If $X_0 = X$, $U = 1 = \sigma^*(1)$, so assume $X - X_0$ is nonempty. Let $x_i \in X - X_0$. If $x_i + \lambda x_j \in U$, and $x_j \in X_0$, then $x_i \in U$, so $x_i \in X_0$, a contradiction. Hence every point $x_i + \lambda x_j$ in U has either $x_i, x_j \in X_0$ or $x_i, x_j \in X - X_0$. If $x_i + \lambda x_j$, $x_i + \kappa x_j \in U$, $\lambda \neq \kappa$, then $x_i, x_j \in U$, so $x_i, x_j \in X_0$. We conclude that for every pair $x_i, x_j \in X - X_0$, there is at most one point $x_i + \lambda x_j$ in U on the line joining x_i and x_j.

Define now a directed graph† D with vertex set $X - X_0$ such that for each point $x_i + \lambda x_j$ in U, D has one edge $x_i \xrightarrow{\lambda} x_j$ labelled λ and one edge $x_j \xrightarrow{\lambda^{-1}} x_i$ labelled λ^{-1}. Then any two vertices in D are joined by no edges or exactly two, with opposite orientations, and representing the same point. Suppose that

$$x_l \xrightarrow{\lambda} x_j \xrightarrow{\kappa} x_i$$

is a path of length two in D. Then $(x_l + \lambda x_j) - \lambda(x_j + \kappa x_i) = x_l - \lambda \kappa x_i \in U$, so there is an edge $x_l \xrightarrow{-\kappa\lambda} x_i$ in D. We conclude that every connected component D_l, $l \in [1, k]$, say, is a complete graph. Let X_l be the vertex set of D_l. To complete the proof, we need only verify that the points $x_i \in X_l$ can be assigned labels $\kappa_i^{-1} \in F^*$ so that $x_i \xrightarrow{\lambda} x_j$ in D_l implies $\lambda = -\kappa_i \kappa_j^{-1}$ (see (6)). Fix a point, say x_l, in X_l, and assign the label 1 to x_l. To each $x_i \in X_l - x_l$ assign the label, say $-\kappa_i^{-1}$, of the edge $x_l \to x_i$. Then

$$x_l \xrightarrow{-\kappa_i^{-1}} x_i \xrightarrow{\lambda} x_j \qquad \text{and} \qquad x_l \xrightarrow{-\kappa_j^{-1}} x_j$$

imply $\kappa_j^{-1} = -\kappa_i^{-1}\lambda$, so $\lambda = -\kappa_i \kappa_j^{-1}$, as required.

Remark. With the exception of Corollary 2 of Proposition 1, the assumed finiteness of F has not been used. Hence all our results to this point hold for an arbitrary field.

6. The characteristic polynomial and Whitney numbers of Q_n

A *modular* element of a geometric lattice L with rank function r is an element $x \in L$ such that $r(x \vee y) + r(x \wedge y) = r(x) + r(y)$ for all $y \in L$. If x is a modular element, the map $z \mapsto x \vee z$ is an isomorphism $[x \wedge y, y] \cong [x, x \vee y]$ with inverse $w \mapsto w \wedge y$, for any $y \in L$. Every point of a geometric lattice is a modular element.

Proposition 6. *In the geometric lattice Q_n of q-partitions of X, the subset $M = \{A | A \subseteq X\}$ is a sublattice anti-isomorphic to the Boolean algebra $B(X)$. Every element of M is modular in Q_n.*

† This idea, although similar to our original proof, is suggested in Rota, Doubilet and Stanley [1971].

Proof. If $A \in M$, $B \in Q_n$, the blocks of $\pi(A) \wedge \pi(B)$ not containing z are

$$\{a_i\}, a_i \in A,$$
$$X(b) \cap (X - A), b \in B, X(b) \nsubseteq A, \qquad (8)$$

and the blocks of $\pi(A) \vee \pi(B)$ not containing z are

$$X(b), \qquad b \in B, X(b) \subseteq A. \qquad (9)$$

It is clear (see the proof of Proposition 4) that $\pi(A \wedge B) = \pi(A) \wedge \pi(B)$ and $\pi(A) \vee \pi(B) = \pi(A) \vee \pi(B)$. The total number of blocks in (8) and (9) is $k(A) + k(B)$, so $k(A) + k(B) = k(A \vee B) + k(A \wedge B)$. Thus since $r(C) = n - k(C)$, A is modular. If also $B \in M$, then $X(b) = \{b\}$, so $A \wedge B = A \cup B \in M$, $A \vee B = A \cap B \in M$. Thus M is a sublattice, and $A \mapsto X - A$ is an anti-isomorphism of M to $B(X)$.

The *Möbius function* $\mu: L \times L \to Z$ of a finite partially ordered set L is defined recursively by $\mu(x, x) = 1$, $\mu(x, y) = 0$ if $x \nleqslant y$, and $\mu(x, y) = -\sum_{z:x \leqslant z < y} \mu(x, z)$ if $x \leqslant y$ (Rota [1964]). If L is a geometric lattice of rank n with rank function r, the *characteristic polynomial* of L is

$$p_L(v) = \sum_{x \in L} \mu(0, x) v^{n - r(x)}.$$

The characteristic polynomial extends to geometric lattices the notion of the chromatic polynomial of a graph. In particular, if L is the lattice of contractions (Rota [1964]) of a linear graph G with k components, then the chromatic polynomial of G is $\chi(v) = v^k p(v)$. The partition lattice Π_{n+1} is the lattice of contractions of the complete graph K_{n+1} with chromatic polynomial $v(v-1) \cdots (v-n)$, so the characteristic polynomial of Π_{n+1} is $(v-1)(v-2) \cdots (v-n) = (v-1)_{(n)}$. We may obtain the characteristic polynomial $p_n(v)$ of Q_n with the aid of the following special case of a theorem of Crapo [1968], Th. 6, Cor. 5: If L is a finite geometric lattice of rank n and c is a copoint of L, then

$$v p_{[0,c]}(v) = \sum_{x: x \wedge c = 0} p_{[x,1]}(v), \qquad (10)$$

where $p_{[a,b]}(v)$ is the characteristic polynomial of the interval $[a, b]$ of L.

Proposition 7. *The characteristic polynomial of Q_n is*

$$p_n(v) = \prod_{i=0}^{n-1} (v - 1 - (q-1)i) = (q-1)^n \left(\frac{v-1}{q-1}\right)_{(n)}. \qquad (11)$$

Proof. We take as our c in (10) the copoint $C = \{x_1\}$ of Q_n. By Corollary 2 of Proposition 3, $[0, C] \cong Q_{n-1}$. Since C is modular, $B \wedge C = 0$ iff $B = 0$ (i.e. $B = X$) or B is an atom of Q_n not in $[0, C]$. The number of such atoms is $1 + (q-1)(n-1)$. By Proposition 2, $[B, 1] \cong Q_{n-1}$, for every atom $B \in Q_n$. Thus

$$p_n(v) = (v - 1 - (q-1)(n-1)) p_{n-1}(v). \qquad (12)$$

Since $p_1(v) = v - 1$, we obtain (11).

Remark. Stanley [to appear] has recently investigated the class of geometric lattices containing a maximal chain $0 = x_0 < x_1 < \cdots < x_n = 1$ of modular elements. Such lattices, called *supersolvable* lattices, have the property that all zeros of the characteristic polynomial are positive integers, namely,

$$p(v) = (v - \alpha_1)(v - \alpha_2) \cdots (v - \alpha_n),$$

where α_i is the number of atoms of $[0, x_i]$ not in $[0, x_{i-1}]$. By Proposition 6, Q_n is supersolvable, with $\alpha_i = 1 + (q-1)(i-1)$.

Corollary 1. Let μ be the Möbius function of Q_n, and let $\mu_n = \mu(0, 1)$. Then

$$\mu_n = (-1)^n \prod_{i=0}^{n-1} (1 + (q-1)i)$$

$$= (-(q-1))^n (1/(q-1))^{(n)},$$

where $x^{(k)} = x(x+1) \cdots (x+k-1)$.

Proof. Set $v = 0$ in (11).

When $q = 2$, $\mu_n = (-1)^n n! = \mu'_{n+1}$ (say), where $\mu'_{n+1} = \mu(0, 1)$ for the partition lattice Π_{n+1}. Since the Möbius function is multiplicative on direct products, we obtain from Corollary 1 of Proposition 3:

Corollary 2. Let $B \leqslant A$ in Q_n, where $A = \{a_i | i \in [1, k]\}$, $B = \{b_i | j \in [1, m]\}$. Let $m_0 = |B_0(A)|$, $m_i = |B_0(a_i)|$, $i \in [1, k]$. Then

$$\mu(B, A) = \mu_{m_0} \mu'_{m_1} \mu'_{m_2} \cdots \mu'_{m_k}$$

$$= (-1)^{m-k} (q-1)^{m_0} (1/(q-1))^{(m_0)} \prod_{i=1}^{k} (m_i - 1)!.$$

The *Whitney numbers* of a finite geometric lattice L of rank n are defined by

$$w(n, k) = \sum_{x:r(x)=n-k} \mu(0, x) \quad \text{(first kind)}, \tag{13}$$

the coefficient of v^k in the characteristic polynomial, and

$$W(n, k) = \sum_{x:r(x)=n-k} 1 \quad \text{(second kind)}, \tag{14}$$

the number of elements of corank k. Some classical examples are the following: If $L = B_n$, the lattice of subsets of an n-set,

$$w(n, k) = (-1)^{n-k} \binom{n}{k}, \qquad W(n, k) = \binom{n}{k}.$$

If $L = L(V_n) \cong L(\mathbb{P}_{n-1})$, the lattice of subspaces of a vector (projective) space over $GF(q)$, then

$$w(n, k) = (-1)^{n-k} q^{\binom{n-k}{2}} \binom{n}{k}_q, \qquad W(n, k) = \binom{n}{k}_q,$$

where $\binom{n}{k}_q$ are the Gaussian coefficients (Goldman and Rota [1970]),

$$\binom{n}{k}_q = \frac{(q^n - 1)(q^{n-1} - 1) \cdots (q^{n-k+1} - 1)}{(q^k - 1)(q^{k-1} - 1) \cdots (q - 1)}. \tag{15}$$

Finally, if $L = \Pi_{n+1}$, the lattice of partitions of an n-set,

$$w(n, k) = s(n+1, k+1), \qquad W(n, k) = S(n+1, k+1),$$

the Stirling numbers of the first and second kind, respectively. All of these, as well as the q-partition lattices Q_n, are classes of lattices which satisfy the hypotheses of the following proposition. Here $\delta(a, b) = 1$ or 0 according as $a = b$ or $a \neq b$.

Proposition 8. *Let* $\{P_n| n = 0, 1, \ldots\}$ *be a class of finite geometric lattices with the property that* P_n *is of rank n and* $[x, 1] \cong P_k$ *for all* $x \in P_n$ *of corank k, $k \in [0, n]$, $n \in [0, \infty)$, Let $w(n, k)$, $W(n, k)$ be the Whitney numbers of P_n, $k \in [0, n]$. Then*

$$\sum_k W(n, k)w(k, m) = \delta(n, m), \tag{16}$$

and

$$\sum_k w(n, k)W(k, m) = \delta(n, m). \tag{17}$$

The numbers $W(n, k)$, $w(n, k)$ *then satisfy the inverse relations*

$$a_n = \sum_k W(n, k)b_k, \qquad b_n = \sum_k w(n, k)a_k. \tag{18}$$

Proof. We use the identities $\delta(0, y) = \sum_{x \leqslant y} \mu(x, y) = \sum_{x \leqslant y} \mu(0, x)$. Then

$$\sum_k W(n, k)w(k, m) = \sum_{x \in P_n} \sum_{y:y \geqslant x} \mu(x, y)\delta(m, n-r(y))$$

$$= \sum_{y \in P_n} \delta(m, n-r(y)) \sum_{x:x \leqslant y} \mu(x, y)$$

$$= \sum_{y \in P_n} \delta(m, n-r(y))\delta(0, y)$$

$$= \delta(n, m).$$

Similarly,

$$\sum_k w(n, k)W(k, m) = \sum_{x \in P_n} \mu(0, x) \sum_{y:y \geqslant x} \delta(m, n-r(y))$$

$$= \sum_{y \in P_n} \delta(m, n-r(y)) \sum_{x:x \leqslant y} \mu(0, x)$$

$$= \sum_{y \in P_n} \delta(m, n-r(y))\delta(0, y)$$

$$= \delta(n, m).$$

Then if $a_n = \sum_k W(n, k)b_k$,

$$\sum_k w(n, k)a_k = \sum_k w(n, k) \sum_m W(k, m)b_m$$

$$= \sum_m b_m \sum_k w(n, k)W(k, m)$$

$$= \sum_m b_m \delta_{nm}$$

$$= b_n.$$

The converse is proved analogously.

Corollary. *The Whitney numbers* $T(n, k)$, $t(n, k)$ *of the q-partition lattice* Q_n *satisfy the inverse relations*

$$(q-1)^n \binom{v-1}{q-1}_{(n)} = \sum_k t(n, k)v^k,$$

$$v^n = \sum_k T(n, k)(q-1)^k \binom{v-1}{q-1}_{(k)}.$$

Proof. Set $a_n = v^n$ in (18). Then b_n is the characteristic polynomial (11).

Note that on setting $q = 2$ and multiplying both equations above by v, we obtain the defining relations of the Stirling numbers.

Proposition 9. *The numbers* $T(n, k)$, $t(n, k)$ *satisfy the recursions*

$$T(n, k) = T(n-1, k-1) + (1 + (q-1)(k-1))T(n-1, k), \tag{19}$$

$$t(n, k) = t(n-1, k-1) - (1 + (q-1)(n-1))t(n-1, k). \tag{20}$$

Proof. Every q-partition of X of size k is obtainable either from a unique q-partition of $X - x_n$ of size $k-1$ by adding the point x_n, or from a unique q-partition B of $X - x_n$ of size k by replacing some $b_i \in B$ by a point $b_i + \lambda x_n$ ($\lambda \in F^*$), or else is equal to a q-partition of $X - x_n$ of size k. This proves (19), while (20) follows from a comparison of the coefficients of v^k in (12).

7. An application to design

We conclude with an application to a problem of design in statistics. Consider an experiment in which n factors are to be observed, each factor at q levels, where q is a prime power. The q^n different combinations of levels of the factors are to be partitioned into q^{n-r} blocks of size q^r, in such a way that no t-factor or lower interaction is confounded with blocks. As Bose [1947, 1961] has shown, the problem may be represented geometrically by representing the main effects of the n factors by the points of a basis X of $\mathbb{P}_{n-1}$ over $GF(q)$. Each t-factor interaction is then represented by a point of X-weight (cardinality of its X-support) t. The design is specified by the choice of a subspace U of $\mathbb{P}_{n-1}$ of dimension $n-r-1$, in which the points of U represent the interactions confounded with blocks in the design. Thus the design will confound no interactions of t or fewer factors iff U contains no points of

$$S_t = \{a \in S \mid |X(a)| \leqslant t\}.$$

For given n, t, it is desirable to maximize the dimension $n-r-1$ of U so as to minimize the block size q^r. Bose [1961] observed that the problem is equivalent to that of finding an $(n, n-r)$ linear code over $GF(q)$ with minimum distance at least $t+1$.

The problem of determining the maximum k may be considered as a special case of the *critical problem of combinatorial geometry* (Crapo and Rota

5

[1970]). It is shown by Crapo and Rota [1970] that the number of lists $(C_1, C_2, \ldots, C_r)$ of projective hyperplanes such that no point of a given spanning set T of $\mathbb{P}_{n-1}$ is in every C_i is $(q-1)^{-r}p(q^r)$, where $p(v)$ is the characteristic polynomial of the lattice $L(T)$ of closed sets in the subgeometry of $\mathbb{P}_{n-1}$ on the point set T. This implies in particular that $p(q^r) = 0$ iff $r < c$, where $n - c - 1$ is the maximum dimension of a subspace of $\mathbb{P}_{n-1}$ containing no points of T. The integer c is called the critical exponent of T. We show in Dowling [1971] by an application of Möbius inversion that the number N_{n-r} of $(n - r - 1)$-dimensional projective subspaces not meeting T is then given by

$$\left(\prod_{i=0}^{r-1}(q^r - q^i)\right)N_{n-r} = \sum_{i=0}^{r}(-1)^i q^{\binom{i}{2}}\binom{r}{i}_q p(q^{r-i}), \tag{21}$$

where $\binom{r}{i}_q$ is the Gaussian coefficient (15). The critical exponent of the set S_2 is well known to be the smallest r such that $n \leqslant \binom{r}{1}_q = (q^r - 1)/(q - 1)$. We may, in addition, obtain the numbers N_{n-r} in this case, since $Q_n \cong L(S_2)$. We state this result as

Proposition 10. *The number N_{n-r} of (q^n, q^r) factorial designs (i.e., n factors at q levels each, in q^{n-r} blocks of size q^r) such that no main effects or two-factor interactions are confounded is*

$$N_{n-r} = \frac{\displaystyle\sum_{i=0}^{r-1}(-1)^i q^{\binom{i}{2}}\binom{r}{i}_q (q-1)^n \left(\frac{q^r-1}{q-1}\right)_{(n)}}{\displaystyle\prod_{i=0}^{r-1}(q^r - q^i)}.$$

Note that $N_{n-r} = 0$ iff $n > (q^r - 1)/(q - 1)$. For the critical exponent c, we have the

Corollary. *The number of (q^n, q^c) designs with minimum block size q^c, such that no main effects or two factor interactions are confounded, is*

$$N_{n-c} = \frac{(q-1)^n \left(\dfrac{q^c-1}{q-1}\right)_{(n)}}{\displaystyle\prod_{i=0}^{c-1}(q^c - q^i)}.$$

References

R. C. Bose, 1947, Mathematical theory of the symmetrical factorial design, *Sankhya* **8**, 107–166.

R. C. Bose, 1961, On some connections between the design of experiments and information theory, *Bull. Intern. Statist. Inst.* **38**, 257–271.

H. Crapo, 1968, Möbius inversion in lattices, *Arch. Math.* **19**, 595–607.

H. Crapo and G.-C. Rota, 1970, *Combinatorial Geometries* (MIT Press, Cambridge, Mass.).

T. A. Dowling, 1971, Codes, packings and the critical problem, *Proc. Conf. on Combinatorial Geometry and its Applications* (A. Barlotti, ed.; Perugia, Italy).

J. Goldman and G.-C. Rota, 1970, On the foundations of combinatorial theory. IV: Finite vector spaces and eulerian generating functions, *Stud. Appl. Math.* **49**, 239–258.

G.-C. Rota, 1964, On the foundations of combinatorial theory. I: Theory of Möbius functions, *Z. Wahrscheinlichkeitstheorie und verw. Gebiete* **2**, 340–368.

G.-C. Rota, 1964, The number of partitions of a set, *Am. Math. Monthly* **71**, 498–504.

G.-C. Rota, P. Doubilet and R. Stanley, 1971, The idea of generating function, *Proc. Sixth Berkeley Symp. on Mathematical Statistics and Probability*.

R. Stanley, 1971, Modular elements of geometric lattices, *Algebra Universalis* **1/2**, 214–217.

R. Stanley, (to appear), Supersolvable semimodular lattices.

J. N. Srivastava et al., eds., *A Survey of Combinatorial Theory*
© North-Holland Publishing Company, 1973

CHAPTER 12

Problems and Results on Combinatorial Number Theory

P. ERDÖS

Hungarian Academy of Sciences, Budapest, Hungary

I will discuss in this paper number theoretic problems which are of combinatorial nature. I certainly do not claim to cover the field completely and the paper will be biased heavily towards problems considered by me and my collaborators. Combinatorial methods have often been used successfully in number theory (e.g. sieve methods), but here we will try to restrict ourselves to problems which themselves have a combinatorial flavor. I have written several papers in recent years on such problems and in order to avoid making this paper too long, wherever possible, will discuss either problems not mentioned in the earlier papers or problems where some progress has been made since these papers were written.

Before starting the discussion of our problems I give a few of the principal papers where similar problems were discussed and where further literature can be found.

I. P. Erdös, On unsolved problems, *Publ. Math. Inst. Hung. Acad.* **6** (1961) 229–254; Some unsolved problems, *Michigan Math. J.* **4** (1957) 291–300.

II. P. Erdös, Remarks on number theory IV and V. Extremal problems in number theory I and II *Mat. Lapok* **13** (1962) 28–38; **17** (1966) 135–166. See also: P. Erdös, *Proc. Symp. Pure Math.*, vol. 8 (Am. Math. Soc., Providence, R.I.), pp. 181–189.

III. P. Erdös, Some recent advances and current problems in number theory, *Lectures on modern mathematics*, vol. III (L. Saaty, ed.), pp. 196–244.

IV. P. Erdös, Some extremal problems in combinatorial number theory, *Math. Essays dedicated to A. J. Macintyre* (H. Shankar, ed.; Ohio Univ. Press, Athens, Ohio, 1971), pp. 123–133.

V. P. Erdös, Some problems in number theory, *Computers in Number Theory* (Academic Press, London, 1971), pp. 406–414.

VI. H. Halberstam and K. F. Roth, *Sequences* (Oxford Univ. Press, London, 1966).

VII. A. Stöhr, Gelöste und ungelöste Fragen über Basen der natürlichen Zahlenreihe I and II, *J. Reine Angew. Math.* **194** (1955) 40–65, 111–140.

Many interesting unsolved problems of a combinatorial and number theoretic nature are mentioned in the proceedings of the meetings on number theory held in Boulder, Colorado, in 1959 and 1963. See also a forthcoming book of Croft and Guy.

1. Denote by $r_k(n)$ the maximum number of integers not exceeding n which do not contain an arithmetic progression of k terms. The first publication on $r_k(n)$ was due to Turán and myself (Erdös and Turán [1936]). We were lead to this problem by the following two facts: $r_k(n) < \frac{1}{2}n$ for $n > n_0(k)$ would immediately imply the well known theorem of Van der Waerden that if we split the integers into two classes, then at least one class contains arbitrarily long arithmetic progressions. If we could prove $r_k(n) < \pi(n)$ ($\pi(n)$ is the number of primes not exceeding n) for every k if $n > n_0(k)$, we would obtain that there are arbitrarily long arithmetic progressions among the primes. Unfortunately, none of these results have been proved so far. The best inequalities for $r_3(n)$ are due to Behrend [1946] and Roth [1953]:

$$n^{1-c_1/(\log n)^{\frac{1}{2}}} < r_3(n) < c_2 n/\log\log n.$$

A weaker lower bound has been proved earlier by Salem and Spencer.

L. Moser constructed an infinite $a_1 < \cdots$ of integers not containing an arithmetic progression of three terms so that for every n

$$\sum_{a_i \leqslant n} 1 > n^{1-c/(\log n)^{\frac{1}{2}}}.$$

Moser also raised the following interesting problem: Denote by $f_3(n)$ the largest integer so that one can find $f(n)$ lattice points in the n-dimensional cube

$$\{x_1^{(r)}, \ldots, x_n^{(r)}\}, x_i^{(r)} \text{ is } 0, 1 \text{ or } 2, 1 \leqslant i \leqslant n; 1 \leqslant r \leqslant f_3(n)$$

no three of which are on a straight line. Clearly $f_3(n) \geqslant r_3(3^n)$. Moser showed that

$$f_3(n) > c3^n/\sqrt{n} \tag{1.1}$$

and asks: is it true that $f_3(n) = o(3^n)$? (1.1) has never been improved. It is easy to see that $\lim f_3(n)/3^n$ exists. One can easily generalize this question when the x_i can take k integral values, but nothing seems to be known.

The following problem is due to Straus: Let $a_1 < \cdots < a_k \leqslant x$ be such that no a_i is the arithmetic mean of any subset of the a's consisting of two or more elements. Put max $k = F(x)$. We have

$$\exp(2 \log x)^{\frac{1}{2}} < F(x) < cx^{\frac{2}{3}}. \tag{1.2}$$

The lower bound in (1.2) is due to Straus [1967] and the upper bound to Erdös and Straus [1970] (these papers contain other related problems and results on combinatorial number theory). Straus conjectures that in (1.2) the lower bound is the correct one, but even $F(x) = o(x^\varepsilon)$ seems very difficult to obtain.

Recently, Szemerédi [1969] (see also Behrend [1946]) proved $r_4(n) = o(n)$; his very complicated proof is a masterpiece of combinatorial reasoning.

Recently, Roth [1970] obtained a more analytical proof of $r_4(n) = o(n)$. $r_5(n) = o(n)$ remains undecided. Very recently, Szemerédi proved $r_5(n) = o(n)$.

Rankin [1962] proved that for every $k \geqslant 3$ ($\exp z = e^z$)

$$r_k(n) > n \exp \left(-c(\log n)^{1/(k-1)} \right)$$

and also investigated the question of the densest sequence of integers which do not contain a geometric progression of k terms. Riddell [1969] obtains sharper results, but many interesting questions remain unsettled.

Riddell [1969] defines $g_k(n)$ as the largest integer so that among any n real numbers one can always find $g_k(n)$ of them which do not contain an arithmetic progression of k terms. It is not hard to see that without loss of generality we can always assume that the real numbers are in fact positive integers. $r_3(5) = 4$, but Riddell showed $g_3(5) = 3$ (from the set 1, 3, 4, 5, 7 one can select only 3 integers not containing an arithmetic progression of three terms). $r_3(14) = 8$, but Riddell recently showed $g_3(14) \leqslant 7$ (he feels that $g_3(14) = 7$). Riddell, with the help of a computer, showed that any subset of 8 elements of

$$0, 2, 3, 4, 6, 7, 8, 10, 11, 12, 14, 15, 16, 18$$

contains an arithmetic progression of three terms, thus $g_3(14) \leqslant 7$.

Trivially, $g_k(n) \leqslant r_k(n)$, and we cannot disprove that for large n, $g_3(n) = r_3(n)$. Riddell proved very simply that $g_3(n) > c\sqrt{n}$ and for $k > 3$, $g_k(n) > cn^{1-(2/k)}$. Riddell and I recently slightly improved this, e.g., we showed $g_4(n) > cn^{\frac{2}{3}}$. It seems certain that $g_3(n) > n^{1-\varepsilon}$ for every ε if $n > n_0(\varepsilon)$, but even the proof of $\lim g_3(n)/n^{\frac{1}{2}} = \infty$ seems to present difficulties. (Szemerédi just writes that he proved $g_3(n) > n^{1-\varepsilon}$.)

The following problem might be of interest in this connection: Let $f_k(n)$, $k > 3$, be the largest integer so that there is a sequence of integers $a_1 < \cdots < a_n$ which contains $f_k(n)$ arithmetic progressions of three terms but no progression of k terms. It is easy to see that $f_4(3^n) \geqslant 5^{n-1}$ and G. Simmons considerably improved this estimate. We proved that for every $k > 3$

$$\lim_{n \to \infty} \log f_k(n)/\log n = c_k$$

exists, but we do not know if $c_k < 2$. In fact, we do not know if $f_4(n) = o(n^2)$.

Kleitman and I observed that 1, 2, 3 and 1, 3, 4, 5, 7 are essentially the only sets of integers where every pair is contained in a three term arithmetic progression. It is not clear to us at this moment if there are sequences $a_1, \ldots, a_n$ which do not form an arithmetic progression but where every pair is contained in some arithmetic progression other than arithmetic progressions of even length and even difference with the middle integer (this example is due to Jeffrey Lagaris).

Denote by $P(n, k)$ (Riddell [1969]) the largest integer so that amongst n points in k-dimensional space one can always find $P(n, k)$ which do not contain an isosceles triangle. Clearly $P(n, 1) = g_3(n)$. It is not hard to prove by induction with respect to k that

$$P(n, k) > n^{\varepsilon_k},$$

but it is not easy to determine (or estimate) the best value of ε_k. I expect $P(n, 2) < n^{1-c}$. In fact, it seems probable that amongst the lattice points (x, y), $0 \leqslant x, y \leqslant n$, x, y integer, one cannot select $n^{2-\varepsilon}$ of them which do not determine an isosceles triangle. A technique used by Guy and myself (Erdös and Guy [1959]) seems to give that one can give cn such points — I would guess that one can give more than n^{1+c} for some $c > 0$.

As far as I know, questions of the following type have not yet been investigated: Let there be given cn^2 lattice points (x_i, y_i) $(0 \leqslant x_i, y_i \leqslant n)$. Is it true that they determine four vertices of a square whose sides are parallel to the axes? Clearly many generalizations are possible which we leave to the reader. Improving a previous result of Rennie Abbott and Hanson recently proved that one can give $n^{\log 5/\log 3}$ lattice points (x_i, y_i), $0 \leqslant x_i, y_i \leqslant n$ which do not contain four vertices of a square whose sides are parallel to the axes. Very recently Aytai found $n^{2-\varepsilon}$ such lattice points.

Before ending this chapter, I mention a problem considered in Riddell [1969]. Let $a_1 < \cdots < a_n$ be any set of real numbers. Denote by $l(n)$ the largest integer so that one can always find $l(n)$ of them, $a_{i_1}, \ldots, a_{i_r}$, $r \geqslant l(n)$ so that all the sums $a_{i_{j_1}} + a_{i_{j_2}}$, $1 \leqslant j_1 \leqslant j_2 \leqslant r$ are distinct. It is known that

$$cn^{\frac{1}{3}} < l(n) \leqslant (1 + o(1))n^{\frac{1}{2}}. \tag{1.3}$$

In (1.3), probably the upper bound is the right one. Szemerédi and Komlós proved a slightly weaker upper bound $cn^{\frac{1}{2}}$. Many generalizations are possible for more than two summands or vectors in higher dimensions.

2. The theorem of Van der Waerden states that there is a smallest integer $f(n)$ so that if we split the integers from 1 to $f(n)$ into two classes, at least one contains an arithmetic progression of n terms. Van der Waerden obtains a very poor upper bound for $f(n)$ and it would be very desirable to obtain a more reasonable upper bound for it. The best lower bound for $f(n)$ is due to Berlekamp [1968] who proved $f(p) > p2^p$, sharpening previous results of Rado, Schmidt and myself. It would be very interesting to decide whether $f(n) < c^n$ holds for a certain constant c.

Perhaps the following modification of the problem is more amenable to attack: Denote by $f(c, n)$, $\frac{1}{2} < c \leqslant 1$, the smallest integer so that if we split the integers from 1 to $f(c, n)$ into two classes, there is an arithmetic progression of n terms so that at least cn of its terms belong to the same class. $f(1, n)$ clearly equals $f(n)$. By probability methods it is not hard to show that for every $c > \frac{1}{2}$, $f(c, n) > (1+\varepsilon_c)^n$. I never could get a good upper bound for $f(c, n)$. Perhaps $f(c, n) < c^n$, at least if c is sufficiently close to $\frac{1}{2}$ (Erdös [1963]).

A related problem was considered by Roth [1964, 1967]. Let $g(n) = \pm 1$ be an arbitrary number theoretic function. Put

$$F(x) = \min_{g(n)} \max |\sum g(a+kd)|,$$

where the maximum is to be taken over all arithmetic progressions whose terms are positive integers not exceeding x and the minimum is to be taken over all the functions $g(n) = \pm 1$. Roth proved that

$$F(x) > cx^{\frac{1}{4}}$$

and conjectured $F(x) > x^{\frac{1}{2}-\varepsilon}$ for every $\varepsilon > 0$ if $x > x_0(\varepsilon)$.

Y. Spencer recently proved

$$F(x) < cx^{\frac{1}{2}} \frac{\log \log x}{\log x},$$

he uses probabilistic methods, his proof will be published soon. These results will also be treated in the forthcoming booklet of Y. Spencer and myself on probabilistic methods in combinatorial analysis.

Many years ago, Cohen asked the following question. Determine or estimate a function $f(d)$ so that if we split the integers into two classes, at least one class contains for infinitely many values of d an arithmetic progression of length $f(d)$. I showed $f(d) < cd$. To see this, let α be a quadratic irrationality, say $\sqrt{5}$. n belongs to the first class if the fractional part of $n\alpha$ is less than $\frac{1}{2}$ and in the second class otherwise. From the well known fact that $|\alpha - p/q| > c_1/q^2$, it easily follows that $f(d) < cd$. I have not been able to show that $f(d) < \varepsilon d$ for sufficiently small ε and I have not succeeded in getting a lower estimation for $f(d)$. Van der Waerden's theorem certainly implies that $f(d) \to \infty$.

Let $g(n) = \pm 1$ be an arbitrary number theoretic function. Cantor, Schreiber, Straus and I [II] proved that there is such a $g(n)$ for which

$$\max_{a, m, 1 \leqslant b \leqslant d} \left| \sum_{k=1}^{m} g(a+kb) \right| < h(d)$$

for a certain function $h(d)$. Van der Waerden's theorem implies $h(d) \to \infty$ as $d \to \infty$. We showed $h(d) < (cd)!$. We have no good lower bound for $h(d)$ and are not sure how good our upper bound is. As far as I know the following related more general question is still unsolved: Let

$$A_k = \{a_1^{(k)} < \cdots\}, k = 1, 2, \ldots$$

be infinitely many infinite sequences of integers. Does there exist a function $F(d)$ (which depends on the sequences A_k) so that for a suitable $g(n) = \pm 1$

$$\max_{m, 1 \leqslant k \leqslant d} \left| \sum_{i=1}^{m} g(a_i^{(k)}) \right| < F(d)?$$

It seems certain that the answer is affirmative.

I conjectured more than thirty years ago that if $f(n) = +1$ and $f(n)$ is multiplicative then

$$\lim_{n=\infty} \frac{1}{n} \sum_{k=1}^{n} f(k)$$

exists and is 0 if and only if $\sum_{f(p)=-1} 1/p = \infty$. Wintner observed that the conjecture fails if we only assume $|f(n)| = 1$ and $f(n)$ is multiplicative. Wirsing [1967] proved (and generalized) my conjecture and Halász [1968] obtained a still more general result.

Finally, I would like to mention an old conjecture of mine: let $f(n) = \pm 1$ be an arbitrary number theoretic function. Is it true that to every c there is a d and an m so that

$$\left| \sum_{k=1}^{m} f(kd) \right| > c?$$

I have made no progress with this conjecture.

Sanders and Folkman proved the following result (which also follows from earlier results of Rado [1933]): For every n there is a $g(n)$ so that if we split the integers not exceeding $g(n)$ into two classes, there always is a sequence $a_1 < \cdots < a_n$ so that all the sums

$$\sum_{i=1}^{n} \varepsilon_i a_i, \ \varepsilon_i = 0 \text{ or } 1 \text{ (not all } \varepsilon_i = 0)$$

belong to the same class. As far as I know there are no good upper or lower bounds for $g(n)$. The result of Sanders and Folkman also follows from the general theorems of Graham and Rotschild.

Graham and Rotschild ask the following beautiful question: split the integers into two classes. Is there always an infinitive sequence so that all the finite sums

$$\sum \varepsilon_i a_i, \ \varepsilon_i = 0 \text{ or } 1 \text{ (not all } \varepsilon_i = 0) \tag{2.1}$$

all belong to the same class? It is not even known if there is an infinite sequence so that all the sums (2.1) with $\sum \varepsilon_i = k$, $k = 1, 2, \ldots$, belong to the same class where the class may depend on k.

This problem seems very difficult. As far as I know, the following simpler question is also unsolved: Does there exist an infinite sequence $a_1 < \cdots$ so that all the numbers

$$a_i, i = 1, 2, \ldots \quad \text{and} \quad a_i + a_j, 1 \leqslant i < j < \infty$$

all belong to the same class? Galvin recently asked the following question: Does there exist a sequence $a_1 \cdots a_n$, $a_1 \leqslant n$ so that all the numbers $a_1 < \cdots < a_n$ and $a_i + a_j$, $1 \leqslant i < j \leqslant n$ belong to the same class?

3. Denote by $f(k, x)$ the largest integer r so that there is a sequence $a_1 < \cdots < a_r \leqslant x$ no $k+1$ of which are pairwise relatively prime. It seems certain that we obtain $f(k, x)$ by considering the set of multiples of the first r primes. The proof seems to present unexpected difficulties — recently, Szemerédi states that he proved this conjecture for sufficiently large r.

A more general conjecture would be: Denote by $f(p_s, k, x)$ the largest integer r so that there is a sequence $a_1 < \cdots < a_r \leqslant x$ all prime factors of the a's are $\geqslant p_s$ and no $k+1$ of them are relatively prime. One would expect to obtain $f(p_s, k, x)$ by considering the set of integers not exceeding x of the form $p_{s+i}t$, $1 \leqslant i \leqslant k$, where all prime factors of t are $\geqslant p_{s+i}$. I have not even been able to show this for $k = 1$ and all s.

A problem of Graham and myself states: Let $a_1 < \cdots < a_k = n$, $(a_i, a_j) = 1$. What is the maximum of k? A reasonable guess seems to be that max k either equals n/p where p is the smallest prime factor of n or it is the number of integers of the form $2t$, $t \leqslant \frac{1}{2}n$, $(t, n) = 1$. (See a forthcoming paper of Graham and myself in *Acta Arithmetica*.)

4. Let $a_1 < \cdots < a_k \leqslant x$. Assume that no r $(r \geqslant 3)$ a's have pairwise the same greatest common divisor. Put max $k = f_r(x)$. I proved $f_r(x) < x^{\frac{3}{4}+\varepsilon}$ (Erdös [1964a]). This was improved to $x^{\frac{1}{2}+\varepsilon}$ by Abbott and Hanson [1970]. In Erdös [1964a], I showed that

$$f_3(x) > \exp\left(c_1 \log x/\log \log x\right)$$

and I stated in Erdös [1964a] that it is likely that

$$f_3(x) < \exp\left(c_2 \log x/\log \log x\right). \tag{4.1}$$

(4.1) is intimately connected with the following purely combinatorial problem of Erdös and Rado [1960]: Let $g_r(n)$ be the smallest integer with the following property: Let $|A_i| = n$, $1 \leqslant i \leqslant g_r(n)$; then there are always r A's which have pairwise the same intersection. Rado and I proved $g_r(n) < c_r^n n!$ and conjectured

$$g_r(n) < c_r^n. \tag{4.2}$$

(4.2) would have many applications in combinatorial number theory. It is easy to see that (4.2) if true is best possible apart from the value of c_r. Abbott [1966] improved our upper and lower bounds for $g_r(n)$, but no real progress has been made with the conjecture (4.2).

I stated in Erdös [1964a] that (4.2) would imply (4.1). Abbott pointed out to me that (4.2) does not seem to suffice. The following slightly stronger conjecture is easily seen to imply (4.1): Let $g'_r(n)$ be the smallest integer so that if u_i, $1 \leqslant i \leqslant g_r(n)$, are integers satisfying

$$u_i = \prod_j p_j^{\alpha_j}, \quad \sum \alpha_j = n \quad (p_j \text{ prime}, \alpha_j \geqslant 0 \text{ integer})$$

then there are always r u's, say $u_{i_1}, \ldots, u_{i_r}$, which have pairwise the same greatest common divisor d and $(u_{i_j}/d, d) = 1$, $1 \leqslant j \leqslant r$. The method of Rado and myself gives $g'_r(n) < c_r^n n!$, and it seems likely that

$$g'_r(n) < c_r^n. \tag{4.3}$$

Let $a_1 < \cdots < a_k \leqslant n, k > cn$. Is it true that for $n > n_0(c)$ there are always three a's which have pairwise the same least common multiple? I do not know the answer to this question, but showed that there do not have to be four a's which have pairwise the same least common multiple [IV].

Let $a_1 < \cdots < a_k < n, k > cn$. Is it true that there always is an m so that $pa_i = m$ (p prime) has at least three solutions? If the answer would be yes then the least common multiple of the three a's would be m (since it is easy to see that one could assume $(p, a_i) = 1$). I. Ruzsa (a 16-year-old Hungarian mathematician) found the following simple construction of a sequence $a_1 < \cdots < a_k \leqslant n, k > cn$ so that the equation $pa_i = m$ has at most two solutions. Consider the set of all squarefree numbers of the form

$$q_1 q_2 \cdots q_r, q_{i+1} > 2q_i, i = 1, \ldots, r-1; \qquad r = 1, 2, \ldots \qquad (4.4)$$

It is easy to see that the density of the integers (4.4) is positive. Therefore there are cn of them in the interval $(\tfrac{1}{2}n, n)$ and it is easy to see that for this set of integers $pa_i = m$ has at most two solutions.

Assume that $pa_i = m$ has at most r solutions. Then clearly

$$\sum_{a_i \leqslant n} \frac{1}{a_i} \sum_{p \leqslant n} \frac{1}{p} \leqslant r \sum_{m=1}^{n^2} \frac{1}{m} < cr \log n$$

or

$$\sum_{a_i \leqslant n} \frac{1}{a_i} < \frac{c_1 r \log n}{\log \log n} \qquad (4.5)$$

I do not know whether (4.5) can be improved.

Let $a_1 < \cdots < a_k \leqslant n$ be such that for every m, $pa_i = m$ has at most one solution (i.e., the numbers $\{pa_i\}$ are all distinct). It can be shown that there is a c so that

$$\max k = n \exp(-(1+o(1))c(\log n \log \log n)^{\frac{1}{2}}).$$

5. R. L. Graham posed the following interesting problem: Let $1 \leqslant a_1 < \cdots < a_n$ be n integers. Prove

$$\max_{1 \leqslant i,j \leqslant n} \frac{a_j}{(a_i, a_j)} \geqslant n. \qquad (5.1)$$

Szemerédi proved that (5.1) holds if $n = p$. It is easy to see that in this case either $a_i \equiv a_j \pmod{p}$, or $a_i \equiv 0 \pmod{p}$ and $a_j \not\equiv 0 \pmod{p}$ for two indices $i \neq j$ (we can of course assume that not all the a's are multiples of p). (5.1) now follows easily. For composite n, the proof of (5.1) seems to present difficulties.

Winterle [1970] proved (5.1) if a_1 is a prime. Marcia and Schönheim [1969] proved that if the a's are squarefree then there are at least n distinct ratios of the form $a_j/(a_i, a_j)$, thus (5.1) follows.

Denote by $h(n)$ the greatest integer so that there are at least $h(n)$ distinct ratios of the form (5.1). Szemerédi and I showed

$$n^{\frac{1}{4}} < h(n) < n^{1-c_1}. \tag{5.2}$$

It would be interesting to improve (5.2). The determination of

$$\lim_{n=\infty} \frac{\log h(n)}{\log n}$$

will perhaps not be too difficult.

6. On covering congruences. A system of congruences

$$a_i(\bmod n_i), n_1 < \cdots < n_k \tag{6.1}$$

is called covering if every integer satisfies at least one of the congruences (6.1). The simplest covering system is: 0 (mod 2), 0 (mod 3), 1 (mod 4), 5 (mod 6), 7 (mod 12). I asked if for every n_1 there is a covering system (6.1). $n_1 = 20$ is the largest number for which this is known. This is an unpublished result of Choi. An affirmative answer to my question would imply that for every k there is an arithmetic progression no term of which is of the form $2^r + u$, where u has at most k distinct prime factors.

It is easy but not quite trivial to prove that for a covering congruence $\sum_{i=1}^{k} 1/n_i > 1$ (Erdös [1950]; L. Mirsky and D. Newman). It is easy to see that this is best possible if $n_1 = 3$ or $n_1 = 4$. Selfridge and I feel that for $n_1 > 4$, $\sum_{i=1}^{k} 1/n_i > 1 + c_{n_1}$ where $c_{n_1} \to \infty$ as n_1.

It is not known if there is a covering system where all the moduli are odd. As far as I know, it has never been proved that for every c there is an m so that $o(m)/m > c$, but one can not form a covering system from the divisors of m. It would be nice to have a usable necessary and sufficient condition on the sequence $n_1 < \cdots < n_k$ which would decide whether they can be the moduli of a covering set, but perhaps this is too much to hope for. Selfridge informs me that it is easy to see that the n_k can not all be squarefree integers having at most two prime factors and very likely the same result holds for three prime factors.

Selfridge and Dewar investigated the following problems: An infinite system

$$a_i \ (\bmod n_i), n_1 < \cdots \tag{6.2}$$

is called covering if every integer satisfies at least one of the congruences (6.2) and the density of integers which do not satisfy any of the first k congruences of (6.2) goes to 0 as $k \to \infty$. This can clearly be done if $\sum 1/n_i = \infty$. A system (6.2) is called perfect if every integer satisfies exactly one of the congruences (6.2).

Recently, several interesting results were obtained on covering congruences by Burshtein and Schónheim [1970] and Znám [1969], some of which are not yet published. The recent thesis of C. E. Krukenberg (Urbana, Illinois) also contains many interesting results and also many numerical examples of covering systems and a fairly complete literature on the subject. Several problems and results on this subject are also stated in [V]. Here I just want to state one more problem which is also stated in [V] but which is still far from being completely solved; so perhaps it deserves to be restated.

A system of congruences (6.1) is called disjoint if no integer satisfies two of them. Let $n_k \leqslant x$ and put max $k = f(x)$. Stein and I conjectured that $f(x) = o(x)$; Szemerédi and I proved this. In fact we showed

$$x \exp\left(-c_1(\log x \log\log x)^{\frac{1}{2}}\right) < f(x) < \frac{x}{(\log x)^{c_2}}. \tag{6.3}$$

In the proof of the lower bound, Stein collaborated. The lower bound seems to give the true order of magnitude of $f(x)$, but by our method the upper bound can not be improved (Erdös and Szemerédi [1968]).

7. Heilbronn and I conjectured that if n is any integer and $a_1, \ldots, a_k$, $k > c\sqrt{n}$, are k distinct residues mod n then

$$\sum_{i=1}^{k} \varepsilon_i a_i \equiv 0 \,(\text{mod } n), \qquad \varepsilon_i = 0 \text{ or } 1 \text{ (not all } \varepsilon_i \text{ are 0)}$$

is always solvable. This conjecture has recently been proved by Szemerédi [1970]. The right value of c is perhaps $\sqrt{2}$. Szemerédi's proof works for Abelian groups having n elements. The result may hold for non-Abelian groups too, but this is not yet settled.

A theorem of Ginsburg, Ziv and myself states (Mann [1967]): Let G_n be an Abelian group of n elements and let $a_1, \ldots, a_{2n-1}$ be $2n-1$ elements of G_n (of course they are not all distinct). Then the 0 element of G_n can be represented in the form

$$\sum_{i=1}^{2n-1} \varepsilon_i a_i, \qquad \sum_{i=1}^{n} \varepsilon_i = n, \qquad \varepsilon_i = 0 \text{ or } 1.$$

This result holds perhaps for non-Abelian groups too, but this has not been settled.

I would like to mention two interesting problems of Graham: Let $a_1, \ldots, a_p$ be p not necessarily distinct residues mod p. Assume that if

$$\sum_{i=1}^{p} \varepsilon_i a_i \equiv 0 \,(\text{mod } p), \qquad \varepsilon_i = 0 \text{ or } 1$$

then $\sum_{i=1}^{p} \varepsilon_i = r$. Does it then follow that there are at most two distinct residues amongst the a's?

Let $a_1, \ldots, a_k$ be k distinct residues mod p, $k < p$. Is it true that there is a permutation $a_{i_1}, \ldots, a_{i_k}$ so that none of the sums $a_{i_1} + \cdots + a_{i_r}$, $1 \leqslant r \leqslant k$ are $\equiv$ (mod p)? Graham proved this if $k = p-1$, but the general case is not yet settled.

Rényi and I proved the following result (Erdös and Rényi [1965]): Let G_n be an Abelian group of n elements (n large). Let $k > \log n/\log 2 + c \log \log n$. Then for all but $o(\binom{n}{k})$ choices of k elements $a_1, \ldots, a_k$ of G_n all elements of G_n can be written in the form $\sum_{i=1}^{k} \varepsilon_i a_i$, $\varepsilon_i = 0$ or 1. It seems likely that the summand $c \log \log n$ cannot be replaced by $o(\log \log n)$, but as far as I know, nothing has been done in this direction.

We also proved that if

$$k > 2 \frac{\log n}{\log 2} + c$$

then for all but $o(\binom{n}{k})$ choices of $a_1, \ldots, a_k$ the number of representations of every element of G in the form $\sum_{i=1}^{k} \varepsilon_i a_i$, $\varepsilon_i = 0$ or 1 is $(1+o(1))2^k/n$. It is not impossible that this result remains true for $k > (1+o(1)) \log n/\log 2$.

Some progress in this direction has been made by Miech [1967] and Bognár [1972]. Rényi and I proved (unpublished) that if $k > (1+o(1)) \log n/\log 2$ then for all but $o(\binom{n}{k})$ choices of $a_1, \ldots, a_k$ every element has at least $(1+o(1))2^k/n$ representations in the form $\sum_{i=1}^{k} {}_i a_i$, but we have not succeeded in getting an upper bound.

For further problems and results of this kind see [V] and [VI]; also a forthcoming paper of Eggleton and myself (*Acta Arithmetica*) and for a comprehensive treatment Mann [1965].

8. Some problems in additive number theory. Not very much progress has been made on these problems and they have been published several times, but because of their attractiveness it is worthwhile to repeat them (see [I], [III]).

1. Let $0 < a_1 < \cdots < a_r \leqslant 2^k$ be a sequence of integers so that all the sums $\sum_{i=1}^{r} \varepsilon_i a_i$, $\varepsilon_i = 0$ or 1, are distinct. L. Moser and I both asked: Can r be greater than $k+1$. This was answered affirmatively by Conway and Guy for every $k > 21$. It is not known if $r = k+3$ is possible.

Let now $0 < a_1 < \cdots < a_r \leqslant x$, and $\sum_{i=1}^{r} \varepsilon_i a_i$, $\varepsilon_i = 0$ or 1, are all distinct. Put $f(x) = \max r$. Is it true that

$$f(x) = \frac{\log x}{\log 2} + o(1)? \tag{8.1}$$

It is trivial that

$$f(x) < \frac{\log x}{\log 2} + \frac{\log \log x}{\log 2} + o(1).$$

L. Moser and I (Erdös [1955]) proved that

$$f(x) < \frac{\log x}{\log 2} + \frac{\log \log x}{2 \log 2} + o(1).$$ (8.2)

I offered (and still offer) 300 dollars for a proof or disproof of (8.1). I would pay something for any improvement of (8.2). Graham recently asked: Does (8.1) remain true if we only require the a_i to be positive,

$$1 \leqslant a_1 < \cdots < a_r \leqslant x, \ a_{i+1} - a_i \geqslant 1,$$

and any two of the sums $\sum_{i=1}^r \varepsilon_i a_i$, $\varepsilon_i = 0$ or 1 differ by at least one? (8.2) remains true.

2. Let $a_1 < \cdots < a_k \leqslant x$ be a sequence of integers. Assume that all the sums

$$\sum_{i=1}^r \varepsilon_i a_{j_i}, \qquad \varepsilon_i = 0 \text{ or } 1, \qquad 1 \leqslant j_1 < \cdots < j_r \leqslant k,$$

are all distinct. Put max $k = f_r(x)$. Turán and I proved $f_2(x) < x^{\frac{1}{2}} + cx^{\frac{1}{4}}$, and recently Lindstrom proved (Krückeberg [1961])

$$f_2(x) \leqslant x^{\frac{1}{2}} + x^{\frac{1}{4}} + 1.$$

Recently, Szemerédi proved $f_2(x) < x^{\frac{1}{2}} + o(x^{\frac{1}{4}})$. That $f_2(x) > (1-\varepsilon)x^{\frac{1}{2}}$ for every $\varepsilon > 0$ if $x > x_0(\varepsilon)$ easily follows from the classical result of Singer on difference sets (as observed by Chowla and myself). Turán and I conjectured

$$f_2(x) = x^{\frac{1}{2}} + o(1).$$ (8.3)

I offer 250 dollars for the proof or disproof of (8.3).

Bose and Chowla proved that for every r

$$f_r(x) \geqslant (1 + o(1))x^{1/r},$$

and they conjecture that

$$\lim_{x = \infty} f_r(x)/x^{1/r} = 1$$ (8.4)

(8.4) has never been proved for $r > 2$ and is a very attractive conjecture. I offer 100 dollars for a proof or disproof.

Let now $a_1 < \cdots$ be an infinite sequence so that all the sums $a_i + a_j$ are distinct (i.e., A is a B_2 sequence of Sidon). It is easy to see that there is a B_2 sequence for which $a_k < ck^3$ for every k. It seems certain to me that there is a B_2 sequence $a_1 < \cdots$ satisfying $a_k < k^{2+\varepsilon}$ for every ε if $k > k_0$, but as far as I know, nobody constructed a B_2 sequence satisfying $a_k = o(k^3)$. (I offer 25 dollars for this and 50 for $a_k < k^{2+\varepsilon}$.) Rényi and I proved by probabilistic methods that to every ε there is a c so that there is a sequence $a_k < k^{2+\varepsilon}$ so that the number of solutions of $n = a_i + a_j$ is less than c_ε (see [VI]).

There is a B_2 sequence for which (Krückeberg [1961])

$$\liminf_{k \to \infty} \frac{a_k}{k^2} \leqslant \sqrt{2}.$$ (8.5)

It is not known if $\sqrt{2}$ can be decreased, perhaps it can be replaced by 1. On the other hand, for a B_2 sequence [VII]

$$\limsup_{h \to \infty} \frac{a_k}{k^2 \log k} > 0.$$

Let $a_1 < \cdots$ be any sequence of integers. Denote by $f(n)$ the number of solutions of $n = a_i + a_j$. Turán and I conjectured that if $f(n) > 0$ for all n then

$$\limsup_{n \to \infty} f(n) = \infty. \tag{8.6}$$

Perhaps $a_k < ck^2$ suffices to imply (8.6). I offer 250 dollars for (8.6). The multiplicative analogue of (8.6) I succeeded to prove (Erdös [1964b]).

Let $a_1 < \cdots$ be an infinite sequence of integers. Assume that no a is the distinct sum of other a's. Then the a's have density 0 and $\sum 1/a_i < \infty$ (Erdös [1962]).

Let $a_1 < \cdots < a_n$ be n distinct numbers; L. Moser and I proved that the number solutions of (see [II])

$$t = \sum_{i=1}^{n} \varepsilon_i a_i, \qquad \varepsilon_i = 0 \text{ or } 1, \tag{8.7}$$

is less than $c2^n (\log n)^{\frac{1}{2}}/n^{\frac{3}{2}}$. We conjectured that it is in fact less than $c2^n/n^{\frac{3}{2}}$ (which apart from the value of c is best possible). Sárközi and Szemerédi [1965] proved this conjecture. It seems that the number of solutions of

$$t = \sum_{i=1}^{n} \varepsilon_i a_i, \qquad \sum_{i=1}^{n} \varepsilon_i = l, \tag{8.8}$$

is less than $c2^n/n^2$ (where c is an absolute constant independent of t, l, n and our sequence). (8.8) has never been proved.

It is likely that for $n = 2m + 1$ the number of solutions of (8.7) is largest when the a's are the integers in $(-m, +m)$, but this has never been proved (Van Lint [1967]).

9. I first mention a few special problems considered in [II], especially those where some progress has been made. Let $a_1 < \cdots < a_n$ be n real numbers all different from 0. Denote by $f(n)$ the largest integer so that for every sequence $a_1, \ldots, a_n$ one can always select $k = f(n)$ of them, $a_{i_1}, \ldots, a_{i_k}$, so that

$$a_{i_{j_1}} + a_{i_{j_2}} \neq a_{i_{j_3}}, \qquad 1 \le j_1 \le j_2 < j_3 \le k. \tag{9.1}$$

It is not hard to see that $f(n) \ge \frac{1}{3}n$. This is almost certainly not best possible but Klarner and Hilton showed $f(n) < \frac{1}{2}n$ even if we exclude $j_1 = j_2$.

Independently of this, Diananda, Yap, Rhentulla and Street considered in several papers the problem of determining in an Abelian group of order n the maximal sum free set, i.e., the largest set for which (9.1) holds. The most difficult case is when all prime factors of n are $\equiv 1 \pmod 3$, and in this case there are several unsolved problems. For a complete literature on this subject, see the two forthcoming papers of H. P. Yap, 'Maximal sum free sets in finite abelian groups I and II,' *Bull. Austral. Math. Soc.*

Denote by $g(n)$ the largest number such that from every sequence of n numbers one can always select $g(n)$ of them with the property that no sum of two distinct integers of this subsequence belongs to the original sequence. It is known that

$$c \log n < g(n) < n^{2/5+\varepsilon}. \tag{9.2}$$

The lower bound is due to Klarner, the upper bound to S. L. G. Choi. (Choi's paper is not yet published but will soon appear.)

The lower bound in (9.2) can probably be improved very much.

In Choi's paper the following interesting problem is raised: A set C of natural numbers is said to be admissible relative to a set of natural numbers B if the sum of two distinct elements of C is always outside B. Let B be any set of integers in $(2n, 4n)$ and let C be a maximal admissible subset of $(n, 2n)$ relative to B. Put

$$f(n) = \min_{B} (|C| + |B|).$$

Choi conjectures $f(n) < n^{\frac{1}{2}+\varepsilon}$, but can only show $f(n) < cn^{\frac{3}{4}}$. Choi's conjecture perhaps could be proved by probabilistic arguments, but I have not succeeded in this.

Denote by $h(n)$ the largest integer so that from any set of n integers one can always find a subset of $h(n)$ integers with the property that any two sums formed from the elements of the subset are equal only if they have the same number of summands. We have

$$c_1 n^{\frac{1}{3}} < h(n) < c_2 n^{\frac{1}{2}}$$

The upper bound is due to Straus [1966]. Recently, Choi proved $h(n) > c(n \log n)^{\frac{1}{3}}$; his proof will soon appear.

Denote by $l(n)$ the largest integer so that from any set $a_1, \ldots, a_n$ of real numbers one can always select $l(n)$ of them, $a_{i_1}, \ldots, a_{i_k}, k \geqslant l(n)$, so that no a_{i_j} is the distinct sum of other a_{i_r}'s. I observed $l(n) \geqslant \sqrt{(\frac{1}{2}n)}$; this was improved by Choi to $l(n) > (1+c)\sqrt{n}$. Probably $l(n)/\sqrt{n} \to \infty$, but Choi's method does not even seem to give $l(n) > 2\sqrt{n}$. I claimed $l(n) = o(n)$, but have difficulties in reconstructing my proof. Probably $l(n) < n^{1-c}$ holds for some $c > 0$.

Several very interesting problems on additive number theory are discussed in the papers of Rohrbach and Stöhr [VII]. Here I would like to mention one problem of Rohrbach: Let $0 \leqslant a_1 < \cdots < a_k \leqslant n$ be a sequence of integers so that every integer $0 \leqslant m \leqslant n$ can be written in the form $a_i + a_j$. Put $g(n) = \min k$. Rohrbach observed:

$$\sqrt{2n} \leqslant g(n) \leqslant 2\sqrt{n}.$$

He proved $g(n) > (1+\varepsilon)\sqrt{2n}$ for some $\varepsilon > 0$; Moser improved this result but his ε is still very small. Rohrbach conjectured $g(n) = 2\sqrt{n} + o(1)$. We are very far from being able to prove this.

10.   Let $1 \leqslant a_1 < \cdots < a_k \leqslant x$ be a sequence of integers so that the product of any two integers $a_i a_j$ is distinct. Then

$$\pi(x) + \frac{c_2 x^{\frac{3}{4}}}{(\log x)^{\frac{3}{2}}} < \max k < \pi(x) + \frac{c_1 x^{\frac{3}{4}}}{(\log x)^{\frac{3}{2}}}. \tag{10.1}$$

Perhaps

$$\max k = \pi(x) + \frac{c x^{\frac{3}{4}}}{(\log x)^{\frac{3}{2}}} + o\left(\frac{x^{\frac{3}{4}}}{(\log x)^{\frac{3}{2}}}\right) \tag{10.2}$$

for a certain constant $c > 0$, but I have not been able to prove (10.2).

Assume that $1 \leqslant a_1 < \cdots < a_k \leqslant x$ is such that all products $a_{i_1} \cdots a_{i_r}$ are distinct. Perhaps in this case

$$\max k < \pi(x) + c x^{\frac{1}{2}(1 + 1/r)}, \tag{10.3}$$

but I could prove this only for $r = 2$.

Now let $1 \leqslant a_1 < \cdots < a_k \leqslant x$ so that all the products

$$\prod_{i=1}^{k} a_i^{\varepsilon_i}, \qquad \varepsilon_i = 0 \text{ or } 1,$$

are distinct. Then

$$\max k \leqslant \pi(x) + c \frac{x^{\frac{1}{2}}}{\log x};$$

perhaps

$$\max k = \pi(x) + \pi(\sqrt{x}) + o\left(\frac{x^{\frac{1}{2}}}{\log x}\right).$$

All these questions become very much more difficult if the a's do not have to be integers. Let, e.g., $1 \leqslant a_1 < \cdots < a_k \leqslant x$ be a sequence of real numbers and assume that $|a_i a_j - a_r a_s| \geqslant 1$. Does (10.1) remain true? I can not even prove $k = o(x)$, though this may be simple, and perhaps I overlook a simple idea.

An old conjecture of mine states: Let $1 \leqslant a_1 < \cdots < a_k \leqslant x$, $1 \leqslant b_1 < \cdots < b_l \leqslant y$ be two sequences of integers. Assume that the products $a_i b_j$ are all distinct. Is it true that

$$kl < \frac{c x^2}{\log x}? \tag{10.4}$$

It is easy to see that if true, (10.4) is best possible. The weaker result

$$kl < \frac{x^2}{(\log x)^\alpha} \text{ for some } \alpha > 0$$

is not very hard to prove. (10.4) was recently proved by Szemerédi.

Is it true that to every $\varepsilon > 0$ there is an infinite sequence of integers of density $> 1 - \varepsilon$ so that two products $a_{i_1} \cdots a_{i_r} = a_{j_1} \cdots a_{j_s}$ can only hold if $r = s$? Selfridge constructed such a sequence of density $1/e$. Is it true that one can give $x - o(x)$ such integers not exceeding x? By taking the integers not exceeding x having a prime factor $> x^{\frac{1}{2}}$, it is easy to see that one can give $x \log 2$ such integers not exceeding x and that the constant $\log 2$ can be slightly improved.

For the literature on these questions, see [II] and Erdös [1968, 1964].

11. A sequence of integers is called primitive if no one divides any other. Chapter 5 of [VI] is devoted to the study of primitive sequences. Sárközi, Szemerédi and I wrote about ten papers on primitive sequences and related questions (see our paper at the Debrecen meeting of the Bólyai Math. Soc. 1968). The following question which I formulated nearly forty years ago is still unsolved:

Let $1 \leqslant a_1 < \cdots$ be a sequence of positive numbers. Assume that for every integer i, j and k

$$|ka_i - a_j| \geqslant 1. \tag{11.1}$$

Is it then true that

$$\sum_i \frac{1}{a_i \log a_i} < \infty \tag{11.2}$$

and

$$\sum_{a_i < x} \frac{1}{a_i} < \frac{c \log x}{(\log \log x)^{\frac{1}{2}}}? \tag{11.3}$$

If the a's are integers then (11.1) means that no a divides any other, and in this case (11.3) is an old result of Behrend and (11.2) is an old result of mine (see [VI]). But in the general case I can not even prove that (11.1) implies $\lim \inf A(x)/x = 0$ $(A(x) = \sum_{a_i < x} 1)$. Recently, Haight [unpublished] proved that if the a's are rationally independent then (11.1) implies $\lim A(x)/x = 0$. An old result of Besicovitch states that if the a's are integers then (11.1) does not imply $\lim A(x)/x = 0$ [VI].

Let $a_1 < \cdots$ be an infinite sequence of integers where no a_i divides the sum of two greater a's. Sárközi and I proved that the a's then have density

0 and this result is best possible (Erdös and Sárközi [1970]). Probably $\sum 1/a_i$ $< \infty$ holds.

Let $a_1 < \cdots < a_k \leqslant x$ be a sequence of integers where no a divides the sum of two larger a's. Probably

$$\max k = x/3 + o(1).$$

12. Let $a_1 < a_2 < \cdots$ be an infinite sequence of integers. Straus and I conjectured that there is a sequence of density 0, $b_1 < \cdots$ so that every integer is of the form $a_i + b_j$. Lorentz [1954] proved this conjecture. In fact, he showed that $b_1 < \cdots$ can be chosen so that for every x

$$B(x) < c \sum_{k=1}^{x} \frac{\log A(k)}{A(k)}. \tag{12.1}$$

(12.1) is surprisingly close to being best possible (Erdös [1954]). We will call $b_1 < \cdots$ the complementary sequence to $a_1 < \cdots$

Lorentz observed that if the a's are the primes then (12.1) gives $B(x) < c(\log x)^3$. I proved that this can be improved to $B(x) < c(\log x)^2$ (Erdös [1954]). Clearly every complimentary sequence to the primes must satisfy

$$\liminf \frac{B(x)}{\log x} \geqslant 1. \tag{12.2}$$

I am certain that (12.2) can be improved, but I could not even show

$$\limsup \frac{B(x)}{\log x} > 1.$$

In the other direction I could not find a complementary sequence to the primes satisfying $B(x) = o((\log x)^2)$. Further I could not decide whether there is a complementary sequence to the primes $b_1 < \cdots$ for which the number of solutions of $n = p + b_i$ is bounded.

I asked: Let $a_k = 2^k$. Is there a complementary sequence for which $B(x) < cx/\log x$? The 17-year-old Ruzsa gave a very ingenious proof that the answer is affirmative and he also observed that for his sequence the number of solutions of $2^k + b_i = n$ is bounded. Clearly

$$B(x) \geqslant (1 + o(1)) \frac{x \log 2}{\log 2}.$$

It seems certain that

$$B(x) > (1 + c) \frac{x \log 2}{\log 2} \tag{12.3}$$

must hold for a complementary sequence of the powers of 2, but this has never been proved. Ruzsa's proof will appear in the *Bull. Canad. Math.*

Soc. Ruzsa also finds a sequence $a_1 < \cdots$ with $A(x) > c \log x$ so that for every complementary sequence

$$B(x) > \frac{cx \log \log x}{\log x};$$

or, (12.1) is best possible in this case. It is not clear that if $a_k = r^k$, then there is a complementary sequence satisfying $B(x) < cx/\log x$. (By the way, earlier I referred to Ruzsa as being 16 years old; this is no contradiction, since he did the other work one year earlier.)

Complementary sequences of the rth powers were studied by L. Moser [1965], but several interesting unsolved problems remain.

13. Let $a_1 < \cdots$; $b_1 < \cdots$ be two infinite sequences of integers. Assume that every sufficiently large integer is of the form $a_i + b_j$. Clearly

$$\liminf \frac{A(x)B(x)}{x} \geqslant 1$$

and Hanani conjectured that

$$\limsup \frac{A(x)B(x)}{x} > 1. \qquad (13.1)$$

Narkiewicz [1960] proved that (13.1) holds under fairly general conditions, but Danzer [1964] disproved Hanani's conjecture. Danzer and I then conjectured that if every $n > n_0$ is of the form $a_i + b_j$ and

$$\lim \frac{A(x)B(x)}{x} = 1 \qquad (13.2)$$

then

$$\lim (A(x)B(x) - x) \to \infty \qquad (13.3)$$

(It is easy to see that (13.3) does not hold in general). Sárközi and Szemerédi recently proved (13.3); their proof is not yet published. It is not clear how fast $A(x)B(x) - x$ must tend to infinity if (13.2) holds.

14. Before finishing this report, I would like to mention a few miscellaneous problems and results of a combinatorial flavor. No doubt I will omit many very interesting questions, but this is inevitable since both space and time and my memory and judgement are limited.

1. Let $a_1 < \cdots < a_k \leqslant n$ be a sequence of integers satisfying

$$[a_i, a_j] > n, \ 1 \leqslant i < j \leqslant k. \qquad (14.1)$$

In other words, no $m \leqslant n$ is divisible by two or more a's.

I conjectured that

$$\max k = (1 + o(1))\frac{3}{2\sqrt{2}} n^{\frac{1}{3}}$$

and that the extremal sequence is given by the numbers $1 \leqslant i \leqslant (\frac{1}{2}n)^{\frac{1}{2}}$, $(\frac{1}{2}n)^{\frac{1}{2}} \leqslant 2j \leqslant (2n)^{\frac{1}{2}}$. Perhaps these conjectures are trivially true or false and I overlook an obvious idea.

I further conjectured that (13.1) implies

$$\sum_{i=1}^{k} \frac{1}{a_i} \leqslant \frac{31}{30},$$ (14.2)

with equality only if $n = 5$, $a_1 = 2$, $a_2 = 3$, $a_3 = 5$. Schinzel and Szekeres proved this conjecture. I thought that (14.1) implies the existence of an absolute constant c so that there are cn integers $m \leqslant n$ which do not divide any of the a's. To my great surprise this was disproved by Schinzel and Szekeres. It is probable that (14.1) implies for $n > n_0(\varepsilon)$, $\sum_{i=1}^{k} 1/a_i < 1+\varepsilon$.

Let $a_1 < \cdots < a_k \leqslant n$ satisfying $\sum_{i=1}^{k} 1/a_i < c_1$. It is true that there is a c_2 so that there are at least $n/(\log n)^{c_2}$ integers m not exceeding n which are not divisible by any of the a's. The example of Schinzel and Szekeres [1959] shows that apart from the value of c_2 this is best possible if true.

Assume now

$$\sum_{i=1}^{k} \frac{1}{a_i} < c_1, \qquad (a_i, a_j) = 1, \qquad 1 < a_i \leqslant n.$$ (14.3)

For what choice of the a's satisfying (14.3), the number of integers $m \leqslant n$ not divisible by any a is minimal? Let q_1 be the greatest prime not exceeding n and $q_1 > q_2 > \cdots$ the consecutive primes in decreasing order. Put

$$\sum_{i=1}^{j} \frac{1}{q_i} \leqslant c_1 < \sum_{i=1}^{j+1} \frac{1}{q_i}.$$ (14.4)

The q's defined by (14.4) satisfy (14.3) and it seems to me that (14.4) either gives the extremal sequence (or at least nearly gives the minimum). I made no progress with this question.

2. A sequence $n_1 < n_2 < \cdots$ is called an essential component if for any sequence $1 = a_1 < \cdots$ of Schnirelman density α, the Schnirelman sum of the two sequences $\{a_i+n_j\}$ always has density greater than α. I conjectured that if $n_{i+1}/n_i > c > 1$ then the sequence is never an essential component.

Essential components have been investigated a great deal for the older literature; see [VI] and Edrös [1961]. Also in the *J. Reine Angew. Math.*, several recent interesting papers appeared, e.g., Tülnnecke [1960].

3. Let $1 = a_1 < \cdots < a_{\phi(n)} = n-1$ be the set of integers relatively prime to n. An old conjecture of mine states that

$$\sum_{i=1}^{\phi(n)-1} (a_{i+1}-a_i)^2 < c\frac{n^2}{\phi(n)}.$$ (14.5)

C. Hooley made some progress towards the proof of (14.5), but at the moment (14.5) is not yet settled. It seems certain that for every k

$$\sum_{i=1}^{\phi(n)-1} (a_{i+1} - a_i)^k < c_k \frac{n^k}{\phi(n^{k-1})}. \tag{14.6}$$

Hooley in fact proved (14.6) for $k < 2$. I conjectured (14.5) and (14.6) more than thirty years ago and never expected it to be so difficult.

4. Let $f(n)$ be the largest integer so that for every $1 \leq i \leq f(n)$ there is a $p_i | n + i$, $p_{i_1} \neq p_{i_2}$ for $1 \leq i_1 < i_2 \leq f(n)$.

Grimm conjectured (*Am. Math. Monthly*, Dec. 1969) that for every j, $f(p_j) > p_{j+1} - p_j$. Selfridge and I proved that for all n

$$f(n) \geq (1 + o(1)) \log n \tag{14.7}$$

and for infinitely many n

$$f(n) < \exp c \, (\log n \log \log \log n / \log \log n)$$

It would be interesting to find out more about $f(n)$. Grimm's conjecture if true will be very hard to prove. Ramachandra just informed me that he improved (14.7) to

$$f(n) > c \log n \, (\log \log n)^{\frac{1}{2}} / (\log \log \log n)^{\frac{1}{2}}.$$

5. A problem in set theory lead R. O. Davies and myself to the following question: Denote by $f(n, k)$ the largest integer so that if there are given in k dimensional space n points which do not contain the vertices of an isosceles triangle, then they determine at least $f(n, k)$ distinct distances. Determine or estimate $f(n, k)$. In particular, is it true that

$$\lim_{n = \infty} \frac{f(n, k)}{n} = \infty ? \tag{14.8}$$

(14.8) is unproved even for $k = 1$. Straus observed that if $2^k \geq n$ then $f(n, k) = n - 1$.

References

H. L. Abbott, 1966, Some remarks on a combinatorial theorem of Erdös and Rado, *Canad. Math. Bull.* **9**, 155–160.

H. L. Abbott and B. Gardiner, 1967, *Canad. Math. Bull.* **10**, 173–177.

H. L. Abbott and D. Hanson, 1970, An extremal problem on number theory, *Bull. London Math. Soc.* **2**, 324–326. *See also* Abbott and Gardiner [1967].

F. Behrend, 1946, On sets of integers which contain no three terms in an arithmetic progression, *Proc. Natl. Acad. Sci. USA* **32**, 331–332.

E. R. Berlekamp, 1968, A construction for partitions which avoid long arithmetic progressions, *Canad. Math. Bull.* **11**, 409–414.

K. Bognár, On a problem of statistical group theory, *Studia Sci. Math. Hung.*

N. Burshtein and J. Schönheim, 1970, On a conjecture concerning exactly covering systems of congruences, *Israel J. Math.* **8**, 28–29.

L. Danzer, 1964, Über eine Frage von Hanani aus der additiven Zahlentheorie, *J. Reine Angew. Math.* **214/215**, 392–394.

P. Erdös, 1950, On integers of the form $2^n + p$ and some related problems, *Summa Brasil Math.* **2**, 113–123.

P. Erdös, 1952, On a problem concerning congruence systems, *Mat. Lapok* **3**, 122–128 (in Hungarian).

P. Erdös, 1954, Some results on additive number theory, *Proc. Am. Math. Soc.* **5**, 847–853.

P. Erdös, 1955, Problems and results in additive number theory, *Coll. sur la theorie des nombres*, (Bruxelles), pp. 127–137.

P. Erdös, 1957, On some geometrical problems, *Mat. Lapok* **8**, 86–92 (MR 20 6056 [1959]) (in Hungarian).

P. Erdös, 1961, Über einige Probleme der additiven Zahlentheorie, *J. Reine Angew. Math.* **206**, 61–66.

P. Erdös, 1962, Remarks on number theory III, *Mat. Lapok* **13**, 28–38 (in Hungarian).

P. Erdös, 1963, On some combinatorial theorems connected with the theorems of Ramsey and Van der Waerden, *Mat. Lapok* **14**, 29–37 (in Hungarian).

P. Erdös, 1964a, On a problem in elementary number theory and on a combinatorial problem, *Math. Comp.* **18**, 644–646.

P. Erdös, 1964b, On the multiplicative representation of integers, *Israel J. Math.* **2**, 251–261.

P. Erdös, 1968, On some applications of graph theory to number theoretic problems, *Bull. Ramanujan Inst.* **1**.

P. Erdös and R. Guy, 1970, Distinct distances between lattice points, *Elemente Math.* **25**, 121–123.

P. Erdös and R. Rado, 1960, Intersection theorems for systems of sets, *J. London Math. Soc.* **35**, 85–90.

P. Erdös and A. Rényi, 1965, Probabilistic methods in group theory, *J. Analyse Math.* **14**, 127–138.

P. Erdös and A. Sárközi, 1970, On the divisibility properties of sequences of integers, *Proc. London Math. Soc.* **21**, 97–101.

P. Erdös and E. G. Straus, 1970, Non-averaging sets II, *Combinatorial Theory and its Applications* (P. Erdös et al., eds.; North-Holland, Amsterdam), pp. 405–411.

P. Erdös and E. Szemerédi, 1968, On a problem of P. Erdös and S. Stein, *Acta Arithmetica* **15**, 85–90.

P. Erdös and P. Turán, 1936, On some sequences of integers, *J. London Math. Soc.* **11**, 261–264.

G. Halász, 1968, Über Mittelwerte multiplikativer zahlentheoretischer Funktionen, *Acta Math. Acad. Sci. Hung.* **19**, 365–403.

F. Krückeberg, 1961, B_2 Folgen und verwandte Zahlenfolgen, *J. Reine Angew. Math.* **206**, 53–60.

G. G. Lorentz, 1954, On a problem of additive number theory, *Proc. Am. Math. Soc.* **5**, 838–841.

H. B. Mann, 1965, *Addition Theorems*, Tracts in Math. **18** (Wiley-Interscience, New York).

H. B. Mann, 1967, Two addition theorems, *J. Combin. Theory* **3**, 233–235.

J. Marcia and J. Schönheim, 1969, Difference of sets and a problem of Graham, *Canad. Math. Bull.* **12**, 635–637.

R. J. Miech, 1967, On a conjecture of Erdös and Rényi, *Illinois J. Math.* **11**, 114–127.

L. Moser, 1965, On the additive completion of sets of integers, *Proc. Symp. Pure Math.* **8** (Am. Math. Soc. Providence, R.I.), pp. 175–180.

W. Narkiewicz, 1960, Remarks on a conjecture of Hanani in additive number theory, *Colloq. Math.* **7**, 161–165.

R. Rado, 1933, Studien über Kombinatorik, *Math. Z.* **36**, 429–480.

R. A. Rankin, 1962, Sets of integers containing not more than a given number of terms in arithmetical progression, *Proc. Roy. Soc. Edinburgh* **A65**, 332–344.

J. Riddell, 1969, On sets of numbers containing no *l* terms in arithmetic progression, *Nieuw Archief voor Wiskunde*, **17**, 204–209.

K. F. Roth, 1953, On certain sets of integers, *J. London Math. Soc.* **28**, 104–109.

K. F. Roth, 1964, Remark concerning integer sequences, *Acta Arithmetica* **9**, 257–260.

K. F. Roth, 1967, Irregularities of sequences relative to arithmetic progressions, *Math. Ann.* **169**, 1–25.

K. F. Roth, 1970, Irregularities of sequences relative to arithmetic progressions III, *J. Number Theory* **2**, 125–142.

A. Sárközi and Z. Szemerédi, 1965, Über ein Problem von Erdös und Moser, *Acta Arithmetica* **11**, 205–208.

A. Schinzel and G. Szekeres, 1959, Sur un problème de M. P. Erdös, *Acta Sci. Math. Szeged* **20**, 221–229.

E. G. Straus, 1966, On a problem in combinatorial number theory, *J. Math. Sci.* **1**, 77–80.

E. G. Straus, 1967, Non averaging sets, *Proc. Symp. Pure Math.* (Am. Math. Soc., Providence, R.I.).

E. Szemerédi, 1969, On sets of integers containing no four elements in arithmetic progression, *Acta Math. Acad. Sci. Hung.* **20**, 89–104.

E. Szemerédi, 1970, On a conjecture of Erdös and Heilbronn, *Acta Arithmetica* **17**, 227–229.

J. H. van Lint, 1967, Representation of 0 as $\sum_{k=-N}^{N} \varepsilon_k k$, *Proc. Am. Math. Soc.* **18**, 182–184.

H. Tülnnecke, 1960, Über die Dichte der Summen zweier Mengen deren eine die Dichte 0 hat, *J. Reine Angew. Math.* **205**, 1–20.

B. L. van der Waerden, 1927, Beweis einer Baudetschen Vermutung, *Nieuw Archief voor Wiskunde* (2) **15**, 212–216. *See also* Rado [1933].

A. Winterle, 1970, A problem of R. L. Graham in combinatorial number theory, *Proc. Louisiana Conf. on Combinatorics, Graph Theory and Computing* (Louisiana State Univ., Baton Rouge, La.), pp. 357–361.

E. Wirsing, 1967, Das asymptotische Verhalten von Summen über multiplikative Funktionen, *Acta Math. Acad. Sci. Hung.* **18**, 411–467.

S. Znám, 1969, On exactly covering systems of arithmetic sequences, *Math. Ann.* **180**, 227–232.

J. N. Srivastava et al., eds., *A Survey of Combinatorial Theory*
© North-Holland Publishing Company, 1973

CHAPTER 13

Dissection Graphs of Planar Point Sets

P. ERDÖS, L. LOVÁSZ, A. SIMMONS and E. G. STRAUS†

University of California, Los Angeles, Calif. 90024, U.S.A.

1. Introduction

Let S be a set of n points in general position (no three collinear) in the plane. For any two points $p, q \in S$, the directed line $\overrightarrow{pq}$ has a certain number, $N(\overrightarrow{pq})$, of points of S on its positive side, that is, the open half plane to the right of $\overrightarrow{pq}$. We are interested in the directed k-*graphs*, G_k, of S whose edges are the segments $\overrightarrow{pq}$ with $N(\overrightarrow{pq}) = k$ ($k = 0, 1, \ldots, n-2$). Since clearly $G_{n-k-2} = -G_k$, that is, the k-graph with all orientations reversed, it suffices to consider the cases $k \leqslant (n-2)/2$. If n is even, then the bigraph $B = G_{(n-2)/2}$ is of special interest since each edge occurs in both orientations and it can therefore be considered as an undirected graph.

This case has been studied in several previous papers.

In Section 2, we discuss some general properties of the graphs G_k. In Section 3, we answer the relatively easy question concerning the upper and lower bounds on the number of vertices of G_k and the lower bound on the number of edges of G_k. In Section 4, we tackle the far more difficult problem of the upper bound $e_{n,k}$ on the number of edges in G_k. We obtain upper bounds of the form $cn\sqrt{k}$ and lower bounds of the form $cn \log n$ for $e_{n,rn-1}$, where r is a rational number, $0 < r < 1$, and rn is an integer. Finally in Section 5 we discuss new problems and generalizations.

2. Some structural properties of k-graphs

We can construct the graph G_k as follows. Let l be any oriented line containing no points of S and having $k+1$ points of S on its positive side. Translate l to its left until it meets a point p_1 of S. Call this line $l(0)$. Now rotate $l(0)$ counterclockwise by θ about p_1 into line $l(\theta)$ until it meets a second point p_2 of S at $l(\theta_1) = l_1$. Now rotate counterclockwise about p_2 until $l(\theta)$ meets a point p_3 of S at $l(\theta_2) = l_2$, etc. We thus get a sequence of (not necessarily

† The last author was supported in part by NSF Grant #GP-28696.

distinct) points $p_1, p_2, \ldots, p_N$ of S with $p_{N+1}=p_1$, $p_{N+2} = p_2$ and a sequence of directed lines $l_1, l_2, \ldots, l_N, l_{N+1}$ with $l_{N+2} = l_1$.

Theorem 2.1. *The graph G_k consists of those vertices p_i and those edges $\overrightarrow{p_{i+1}p_i}$ for which the orientation $\overrightarrow{p_ip_{i+1}}$ is opposite to that of the line l_i.*

Proof. Clearly the number $N(\theta)$ of points on the positive side of $l(\theta)$ remains constant in any interval which does not contain one of the angles θ_i.

If $\overrightarrow{p_ip_{i+1}}$ is in the direction of l_i then for small $\varepsilon > 0$ we have $N(\theta_i-\varepsilon) = N(\theta_i) = N(\theta_i+\varepsilon)$, since the points p_i, p_{i+1} are either on or to the left of $l(\theta)$ for $\theta_i-\varepsilon \leqslant \theta \leqslant \theta_i+\varepsilon$. If $\overrightarrow{p_ip_{i+1}}$ is in the direction opposite to l_i then $N(\theta_i) = N(\theta_i-\varepsilon)-1 = N(\theta_i+\varepsilon)-1$ for small $\varepsilon > 0$, since one of p_i, p_{i+1} is to the right of $l(\theta)$ in $\theta_i-\varepsilon \leqslant \theta \leqslant \theta_i+\varepsilon$ except for $\theta = \theta_i$ when both are on l_i.

Thus we have $N(\theta) =$ constant $= k+1$ for all $\theta \neq \theta_i$ and $N(\theta_i) = k+1$ or k according as $\overrightarrow{p_ip_{i+1}}$ is in the direction of l_i or not.

Finally we can see that all edges of G_k are included in the lines $l(\theta)$ since any line l' not included in $l(\theta)$ is like-directed to a line $l(\theta')$ so that $N(l')-N(\theta')$ $\neq 0$ since $l(\theta')$ contains a point of S. If $\theta' = \theta_i$ and $\overrightarrow{p_ip_{i+1}}$ is in the direction opposite to l_i, this proves that $N(l') \neq k$. Otherwise $N(\theta') = k+1$ and l' passes through two points on the right of $l(\theta')$ so that $N(l') \leqslant k-1$.

Theorem 2.2. *Given a line L containing no points of the set S so that L divides S into two sets S_1 and S_2 with $|S_1| = m \leqslant n-m = |S_2|$. Then L intersects $m_0 = \min\{m, k+1\}$ edges of G_k going from S_1 to S_2 and m_0 edges of G_k going from S_2 to S_1.*

Proof. Since a small perturbation of L does not affect the hypotheses we may assume that L is not parallel to any line pq joining two points $p, q \in S$. We may therefore pick the point $p_1 \in S_2$ and the directed line $l(0)$ of the family defined in Theorem 2.1 through p_1; S_1 lies on the negative side of $l(0)$. As θ increases from 0 to π, the number $N(\theta, S_1)$ of points of S_1 on the positive side of $l(\theta)$ increases from 0 to $m_0 = \min\{k+1, m\}$. This increase is monotonic if we ignore the values $\theta = \theta_i$.

Now the number $N(\theta, S_1)$ is clearly constant in any interval which does not contain a θ_i. If both points p_i, p_{i+1} of $l(\theta_i)$ are in S_2 then $N(\theta_i-\varepsilon, S_1) = N(\theta_i, S_1) = N(\theta_i+\varepsilon, S_1)$ for small $\varepsilon > 0$. Similarly, if $\overrightarrow{p_ip_{i+1}}$ is in the direction of $l(\theta_i)$ then $N(\theta_i-\varepsilon, S_1) = N(\theta_i, S_1) = N(\theta_i+\varepsilon, S_1)$ for small $\varepsilon > 0$.

If p_i, p_{i+1} are both in S_1 and $\overrightarrow{p_ip_{i+1}}$ is opposite directed to $l(\theta_i)$ then the point p_{i+1} is to the right of $l(\theta_i-\varepsilon)$ and p_i is to the left of $l(\theta_i-\varepsilon)$ for small $\varepsilon > 0$, while p_{i+1} is to the left of $l(\theta_i+\varepsilon)$ and p_i is to the right of $l(\theta_i+\varepsilon)$ for small

$\varepsilon > 0$. Thus we have $N(\theta_i - \varepsilon, S_1) = N(\theta_i + \varepsilon, S_1)$ in this case. Finally if p_i, p_{i+1} are in opposite sides of L and $\overrightarrow{p_i p_{i+1}}$ is opposite directed to $l(\theta_i)$ for $0 < \theta_i < \pi$ then $p_i \in S_1$, $p_{i+1} \in S_2$ and $N(\theta_i + \varepsilon, S_1) = N(\theta_i, S_1) + 1 = N(\theta_i - \varepsilon, S_1) + 1$ for small $\varepsilon > 0$, since the point p_i is to the right of $l(\theta_i + \varepsilon)$ but on $l(\theta_i - \varepsilon)$. We have thus shown that $N(\theta, S_1)$ increases by one in the interval $0 < \theta < \pi$ whenever L is intersected by a segment $\overrightarrow{p_{i+1} p_i}$ of G_k going from S_2 to S_1. Since $N(\theta, S_1)$ increases to m_0 there must be m_0 such segments of G_k.

As θ increases from π to 2π, the number $N(\theta, S_1)$ decreases from m_0 to 0 and in a manner entirely analogous to that used above we see that $N(\theta, S_1)$ decreases by one whenever L is intersected by a segment $\overrightarrow{p_{i+1} p_i}$ of G_k going from S_1 to S_2. Thus there must be m_0 such segments of G_k.

Theorem 2.3. *If n is odd then $G_{(n-3)/2}$ is connected.*

Proof. In this case each of the segments $\overrightarrow{p_{i+1} p_i}$ in Theorem 2.1 is part of the graph. This is certainly the case if $\overrightarrow{p_{i+1} p_i}$ has the orientation of l_i since in that case there are $(n-3)/2$ points to the right of l_i, but it is also the case if $\overrightarrow{p_{i+1} p_i}$ has the opposite orientation of l_i since in that case there are $(n-1)/2$ points to the right of l_i and hence $(n-3)/2$ points to the left of l_i. Thus the construction at the beginning of this section yields a closed oriented Euler path through the graph $G_{(n-3)/2}$.

Except for the trivial case of G_0, which is the positively oriented boundary of the convex hull of S, this is the only case in which G_k must be connected.

Theorem 2.4. *For any n and any k with $0 < k \leqslant (n-2)/2$, $k \neq (n-3)/2$ there exist sets S with n elements so that G_k is not a connected graph.*

Proof. We have either $n = 2k+2$ or $n \geqslant 2k+4$. In the first case we let S consist of the vertices of a convex n-gon and G_k consists of the diagonals joining diametrically opposite vertices.

In the second case let S consist of the vertices of a regular $(2k+4)$-gon K with the remaining $n-2k-4$ points situated closer to the center of K than any of the non-diametric diagonals of K. Then $G_k(S) = G_k(K)$ consists of two closed $(k+2)$-gons obtained by joining each vertex of K to the one following it by $k+1$ steps in the counter-clockwise direction. All lines which pass through a point of $S \backslash K$ contain at least $k+1$ points of K on each side. Thus no point of $S \backslash K$ is a vertex of $G_k(S)$.

Theorem 2.5. *If we order the oriented lines of the edges of G_k at a vertex, v, in counterclockwise order, then between any two lines containing outgoing*

*edges there is a line containing an incoming edge, and between any two lines
containing an incoming edge there is a line containing an outgoing edge.*

Proof. Let l_1 and l_2 be successive oriented lines through v containing
outgoing edges of G_k. Then as l rotates from l_1 to l_2 we have $k+1$ points of S
on the positive side of l for l near to l_1 and k points of S on the positive side
of l for l near to l_2. Since the number of points of S on the positive side of l
increases by one each time l passes through a point p of S in the oriented
angle $\not\angle$ (l_1, l_2) and decreases by one each time l passes through a point p of
S in the opposite vertical angle $\not\angle$ $(-l_2, -l_1)$, it follows that at some stage
of the rotation the number of points on the positive side of l decreases from
$k+1$ to k so that l contains an incoming edge $\overrightarrow{pv}$ of G_k. The argument for
successive lines containing incoming edges is entirely analogous.

Corollary 2.6. *At each vertex of G_k, the number of incoming edges is equal
to the number of outgoing edges. Thus each component of G_k has an oriented
Euler circuit.*

*In the (unoriented) bigraph $(k = (n-2)/2)$ the number of edges at each
vertex is odd.*

Theorem 2.7. *Let G' be a component of G_k and let S' be the set of vertices
of G'. Then there exists a k', $k' \leqslant k$, so that $G' = G_{k'}(S')$.*

Proof. The directed lines $l'(\theta)$ which contain the edges of G' form a subset
of the lines $l(\theta)$ constructed at the beginning of this section. Let $N'(\theta)$ denote
the number of points of S' on the positive side of $l(\theta)$. We first show that
$N'(\theta)$ is constant, $k'+1 \geqslant 1$, for all values of θ which do not correspond to
edges of G' where $N'(\theta) = k'$. Clearly $N'(\theta)$ can change only when $l(\theta)$
contains two points of S and if at least one of these points is in S'. Now let
$p_i, p_{i+1} \in l(\theta_i) \cap S$; if p_i, p_{i+1} are both in S' then $\overrightarrow{p_{i+1}p_i}$ is not an edge of G_k
and hence $l(\theta_i)$ has the direction $\overrightarrow{p_ip_{i+1}}$. Thus neither p_i nor p_{i+1} is on the
positive side of $l(\theta)$ for $\theta_i - \varepsilon < \theta < \theta_i + \varepsilon$ for sufficiently small $\varepsilon > 0$ and $N'(\theta)$
is constant in this range. If both p_i and p_{i+1} are in S' but $l(\theta_i)$ has direction
$\overrightarrow{p_ip_{i+1}}$ then the same argument applies to keep $N'(\theta)$ constant. Finally, if
$\overrightarrow{p_{i+1}p_i}$ is an edge of G' then p_{i+1} is on the positive side of $l(\theta_i - \varepsilon)$ and p_i is on
the positive side of $l(\theta_i + \varepsilon)$ for small $\varepsilon > 0$. Thus $N'(\theta_i - \varepsilon) = N'(\theta_i + \varepsilon) =
N'(\theta_i) + 1 = k' + 1$.

We have thus shown that $G' \subset G_{k'}(S')$. The argument that $G' \supset G_{k'}(S')$
is exactly as in the proof of Theorem 2.1.

Corollary 2.8. *Theorem 2.2 applies to each component G' of G_k. That is,
if a directed line L which contains no vertices of G' divides the vertices of G'*

into sets of m' and $n'-m'$ vertices with $m' \leqslant n'-m'$, then L intersects min $\{m', k'+1\}$ edges of G' crossing L from right to left and the same number crossing L from left to right.

Corollary 2.9. Let G', G'', ..., $G^{(r)}$ be the components of G_k and set $G^{(i)} = G_{k_i}(S^{(i)})$. Then $k = k_1 + \ldots + k_r + r - 1$ and each directed line containing an edge of $G^{(i)}$ is crossed by k_j edges of $G^{(j)}$ from left to right and k_j edges of $G^{(j)}$ from right to left for each $j \neq i$.

Each union

$$G^{(i_1)} \cup G^{(i_2)} \cup \ldots \cup G^{(i_s)} = G_t(S^{(i_1)} \cup \ldots \cup S^{(i_s)}),$$

where $t = k_{i_1} + \ldots + k_{i_s} + s - 1$.

Proof. As shown in the proof of Theorem 2.7, an edge of $G^{(i)}$ is contained in a line whose positive side contains k_i points of $S^{(i)}$ and k_j+1 points of $S^{(j)}$ for each $j \neq i$. Thus $k = k_1 + \ldots + k_r + r - 1$.

3. On the number of vertices of G_k.

Lemma 3.1. A point p of S is a vertex of $G_k(S)$, $k \leqslant (n-2)/2$, if and only if there exists a directed line through p whose positive side contains no more than k points of S.

Proof. The necessity is obvious since any line of an edge of G_k with vertex p has that property. The sufficiency follows from the fact that, as a directed line l is rotated around p, the number of points on its positive side ranges through all values from the minimum m to the maximum M. We have $m \leqslant k$ and $M = n - m - 1 \geqslant n - k - 1 \geqslant k + 1$. Thus there must be an instant at which the number of points on the positive side of l changes from k to $k+1$. This can only happen when l contains an edge $\overrightarrow{pq}$ of G_k.

Corollary 3.2. All points on the convex hull of S are vertices of every $G_k(S)$. In particular, if all points of S are on its convex hull then $G_k(S)$ has n vertices.

Theorem 3.3. Every point of S is a vertex of $G_{[(n-2)/2]}$. If $k < [(n-2)/2]$ then G_k has at least $2k+3$ vertices, and for every v with $2k+3 \leqslant v \leqslant n$ there exists an S so that $G_k(S)$ has exactly v vertices.

Proof. The first part of the theorem is an immediate consequence of Lemma 3.1. If $k < [(n-2)/2]$ then choose a point p of S which lies on its convex hull and a line l through p which contains at least $[(n-1)/2]$ points of S on each side and is not parallel to any line joining two points of S. Then there exist two oppositely directed lines l_1, l_2 on either side of l parallel to l through points of S so that their positive sides, which exclude l, contain exactly k points of S each. According to Lemma 3.1, each of the $2k+2$ points to the right of or on one l_i ($i = 1, 2$) and the point p is a vertex of G_k so that G_k has at least $2k+3$ vertices.

Finally, if $2k+3 \leqslant v \leqslant n$, let S consist of the vertices of a regular v-gon inscribed in the unit circle K and of $n-v$ points located so near to the center of K that a line through one of them cuts K in arcs exceeding $2\pi(k+1)/v$. Thus by Lemma 3.1 and Corollary 3.2, the vertices of G_k are exactly those of the regular v-gon.

4. On the number of edges of G_k

The lower bound on the number of edges is easily settled by the result of the preceding section. According to Corollary 2.6, each vertex of G_k is incident to at least two directed edges, so that the number of its edges can be no less than the number of its vertices.

Theorem 4.1. *The graph $G_{[(n-2)/2]}$ has at least n directed edges. In particular, if n is even, the bigraph $B = G_{(n-2)/2}$ has at least $n/2$ undirected edges. If $k < [(n-2)/2]$ then G_k has at least $2k+3$ edges. For every number e with $2k+3 \leqslant e \leqslant n$ there is a set S so that $G_k(S)$ has exactly e edges.*

Proof. By Theorem 2.2, every vertex of the convex hull is incident to exactly two edges of G_k. Thus the theorem follows from the constructions made in the proof of Theorem 3.3.

Using Lemma 3.1, we can get an upper bound on the valence of a vertex of G_k.

Theorem 4.2. *The valence of the vertices of G_k does not exceed $2k+2$.*

Proof. Let p be a vertex of G_k, and assume $k \leqslant (n-3)/2$. The edges of G_k through p clearly lie on the directed lines joining p to other points of S whose positive side contains no more than $(n-3)/2$ points of S. There are no more than $n-1$ such lines and if n is even there are no more than $n-2$ such lines.

By Lemma 3.1, the point p is also a vertex of $G_{k+1}, G_{k+2}, \ldots, G_{[(n-3)/2]}$ all of which are edge-disjoint from each other and from G_k. Since the valence v_p of p is not less than 2 in any of these graphs, we get for odd n

$$v_p \leqslant n-1-2\left(\frac{n-3}{2}-k\right) = 2k+2$$

and for even n

$$v_p \leqslant n-2-2\left(\frac{n-4}{2}-k\right) = 2k+2.$$

Since the vertices of G_k include those of $G_0, G_1, \ldots, G_{k-1}$, it follows that all these vertices have valences less than $2k+2$, including vertices of valence 2 which are points on the boundary of the convex hull of S. We could use this to get a poor upper bound for the number of edges of G_k. A better upper bound is obtained through the use of Theorem 2.2.

Theorem 4.3. *The number of edges of G_k, $k \neq (n-2)/2$, is less than*

$$E_{n,k} = 4\sqrt{(k+1)}\sqrt{(n-k-1)}\sqrt{n}.$$

Proof. Pick a direction which is not parallel to any line joining two points of S and draw $n-1$ parallel lines in this direction separating the points of S. According to Theorem 2.2, the total number of intersections of these lines with the edges of G_k is

$$N = 2(2+4+ \ldots +2k)+2(n-1-2k)(k+1)$$
$$= 2n(k+1)-2(k+1)^2.$$

Now we choose an integer α, to be determined later, and divide the edges of G_k into two classes according to whether they intersect at least α of the parallel lines or not. The first class clearly contains no more than N/α elements. The second class contains no more elements than there are pairs of points of S separated by fewer than α of the parallel lines, that is

$$2[(n-2)+(n-3)+ \ldots +(n-\alpha)] = 2(\alpha-1)n-\alpha(\alpha+1)+2.$$

Thus the number of edges of G_k is certainly less than $2\alpha n+N/\alpha$ which is minimal when we choose $\alpha = \sqrt{(N/2n)}$. Let α be the integer just above $\sqrt{(N/2n)}$; then the number of edges is less than $2\sqrt{(2nN)} = E_{n,k}$.

Definition 4.4. Let $e_{n,k}$ denote the maximal number of edges of a graph $G_k(S)$ where S contains n points. For (undirected) bigraphs we write $e_n = \frac{1}{2}e_{2n,n-1}$.

Theorem 4.5. *If $k \neq (n-2)/2$ then*

$$e_{2n,2k+1} \geqslant 2e_{n,k}+n.$$

Proof. We first show that a $G_k(S)$ with a maximal number of edges must have n vertices. For, assume that there is a point $p \in S$ which is not a vertex of G_k. According to Lemma 3.1, this means that p is on the negative side of all the edges of G_k. If we move p across the line of an edge of G_k then that edge is removed from the graph but there will be at least two new edges incident to p, contrary to the maximality assumption for G_k.

Now associate an outgoing edge e_p to each vertex p of G_k and construct a set S' with $2n$ points by splitting each point p into two points p' and p'' at a small distance ε from p, with $\overrightarrow{p'p}$ and $\overrightarrow{pp''}$ in the direction of e_p. Consider $G_{2k+1}(S')$. First for each $p \in S$ we get $p'p''$ as an edge of G_{2k+1} since each point on the positive side of e_p has become two points; and, if $e_p = \overrightarrow{pq}$, then exactly one of the two points q', q'' is on the positive side of e_p. In addition, both $\overrightarrow{p''q'}$ and $\overrightarrow{p''q''}$ are edges of G_{2k+1} whose positive sides contain the points arising from those on the positive side of e_p and, respectively, the point p' or that point q', q'' which lies on the positive side of e_p.

6

Finally, if $\overrightarrow{pq}$ is an edge of G_k other than e_p, then the edge which joins the point p' or p'' on the negative side of $\overrightarrow{pq}$ to the point q' or q'' on the positive side of $\overrightarrow{pq}$ as well as the edge which joins the point p' or p'' on the positive side of $\overrightarrow{pq}$ to the point q' or q'' on the negative side of $\overrightarrow{pq}$ are edges of G_{2k+1}.

Thus in this splitting process each edge of G_k yields two edges of G_{2k+1} and the n edges e_p of G_k yield an additional edge $\overrightarrow{p'p''}$. Thus

$$e_{2n,2k+1} \geqslant e(G_{2k+1}) = 2e_{n,k} + n.$$

In order to get an analog to Lemma 4.5 for bigraphs B, we must avoid the possibility of associating the same edge in its two orientations with both of its endpoints, that is, $e_p = \overrightarrow{pq}$ and $e_q = \overrightarrow{qp}$. Since $e_1 = 1$, $e_2 = 3$, such associations cannot be avoided for bigraphs on 2 or 4 vertices, but they can be avoided for $2n \geqslant 6$.

Lemma 4.6. $e_{n+1} \geqslant e_n + 2$.

Proof. Let B be a bigraph of S with $2n$ vertices and e_n unoriented edges. By an affine transformation we can assume that all points of S are close to the x-axis and that all edges of B make small angles with the x-axis. Now we add two points p, q to S, where p has sufficiently large positive y-coordinate and q has sufficiently large negative y-coordinate and both have, say, x-coordinates smaller than the x-coordinates of the points of S; then all the edges of $B(S)$ are also edges of $B(S')$, where $S' = S \cup \{p, q\}$, and since pq is not an edge of $B(S')$, there are two new edges incident to p and q, respectively. Thus $e_{n+1} \geqslant e(B(S')) = e_n + 2$.

Corollary 4.7. For $n \geqslant 3$, we have $e_n \geqslant 2n$.

Lemma 4.8. If B is a bigraph of S with $2n$ vertices and e_n unoriented edges, $n \geqslant 3$, then each component of B has at least 6 vertices.

Proof. By Corollary 2.9, each union of components of B is itself a bigraph. Thus if B has a component B' with no more than 4 vertices, we can write $B = B' \cup B''$ where B' has $2i \leqslant 4$ vertices and therefore $\leqslant 2i - 1$ edges and B'' has $2n - 2i$ vertices and therefore $\leqslant e_{n-i}$ edges. This implies $e_n \leqslant e_{n-i} + 2i - 1$, in contradiction to Lemma 4.6.

Lemma 4.9. For a bigraph B with $2n$ vertices and a maximal number e_n of unoriented edges, $n \geqslant 3$, it is possible to associate to each vertex p an edge e_p of B so that $e_p \neq e_q$ whenever $p \neq q$.

Proof. By Lemma 4.8 and Corollary 4.7, each component B' of B has at least as many edges as it has vertices. It therefore contains a circuit C. If we arrange the vertices of C in cyclic order $p_1, p_2, \ldots, p_s$ with $p_{s+1} = p_1$ then we can associate with each p_i the edge $p_i p_{i+1}$. If there are vertices B' not included in C then there exist immediate neighbors, q, of vertices p_i in C. To each such neighbor we associate the edge qp_i. If this still does not exhaust the vertices of B' we get additional vertices joined to the immediate neighbors of C etc.

Theorem 4.10. *For* $n \geqslant 2$, *we have* $e_{2n} \geqslant 2e_n + 2n$.

Proof. For $n = 2$, we do this by inspection since $e_2 = 3$ is obtained whenever the vertices form a non-convex quadruple and $e_4 = 9$ by a modified splitting procedure. For $n \geqslant 3$, we use Lemma 4.9 to associate to each vertex a different incident edge of B and then employ the splitting process used in the proof of Theorem 4.5 to complete the proof.

By iterated application of Theorems 4.5 and 4.10, we get

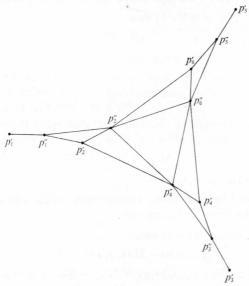

Fig. 1.

Corollary 4.11. (1) $e_{2^m n, 2^m(k+1)-1} \geqslant 2^m e_{n,k} + m2^{m-1}n$.
For $n \geqslant 2$, (2) $e_{2^m n} \geqslant 2^m e_n + m2^m n$.

Theorem 4.12. *For all* n, *we have* $e_n > \frac{1}{2}n \log_2 (2n/3)$.

Proof. Apply Corollary 4.11 to get

$$e_{3 \cdot 2^m} \geqslant 2^m e_3 + 3m \cdot 2^m = 3 \cdot 2^m(2+m). \tag{4.13}$$

Now choose m so that $3 \cdot 2^m \leqslant n < 3 \cdot 2^{m+1}$, then $m+1 > \log_2 (n/3)$ and $3 \cdot 2^m > \frac{1}{2}n$, so that (4.13) yields

$$e_n \geqslant e_{3 \cdot 2^m} > \frac{1}{2}n \log_2 (2n/3).$$

If we modify Theorem 4.3 to include the unoriented bigraph, we replace n by $2n$, k by $n-1$ and N by $\frac{1}{2}N$ so that the proof of Theorem 4.3 yields

Theorem 4.14. *For all n, $e_n < (2n)^{\frac{3}{2}}$.*

We can generalize the lower bound of Theorem 4.12 to arbitrary $e_{n,k}$. For convenience, we restrict our attention to the cases $n \geqslant 3(k+1)$.

Lemma 4.15. *For all $m \geqslant 2$, we have*

$$e_{m,0} = m;$$

and hence for all $m \geqslant 3$ and all $l \geqslant 0$,

$$e_{2^l m, 2^l - 1} \geqslant (l+2)2^{l-1}m. \tag{4.16}$$

Proof. The first part follows from the fact that $G_0(S)$ consists of the boundary of the convex hull of S traversed in the counter-clockwise direction. The second part then follows through l-fold application of Theorem 4.5 to $e_{m,0}$.

Lemma 4.17. *If $n \geqslant 3(2^l - 1)$ then*

$$e_{n,2^l - 1} \geqslant (l+2)2^{l-1}[n/2^l].$$

Proof. Choose $m \geqslant n/2^l$. Then according to Lemma 4.15 and its proof there exists a set S with $2^l m$ points whose vertices consist of clusters of 2^l points in arbitrarily small neighborhoods of vertices of a convex (say, a regular) m-gon C so that $e(G_{2^l-1}(S)) \geqslant (l+2)2^{l-1}m$. Now we can place $n - 2^l m$ points so near to the centroid of C that every line through one of these points has at least 2^l points of S on either side. Thus if the new set is called S' then $e(G_{2^l-1}(S')) = e(G_{2^l-1}(S)) \geqslant (l+2)2^{l-1}m = (l+2)2^{l-1}[n/2^l]$.

Lemma 4.18. $e_{n+2,k+1} \geqslant e_{n,k} + 4$.

Proof. The proof follows from a construction which is entirely analogous to that used in the proof of Lemma 4.6.

Theorem 4.19. *If $n \geqslant 3(k+1)$ then*

$$e_{n,k} \geqslant \tfrac{1}{2}(n - 3k) \log_2 (2k+2).$$

Proof. Write $k = 2^l - 1 + a$, where $0 \leqslant a < 2^l$, and $n = 2^l m + 2a + b$, where $0 \leqslant b < 2^l$. Then $2^{l+1} > k+1$, so that $l+2 \geqslant \log_2 (2k+2)$ and $2^l m = n - 2a - b \geqslant n - 3(2^l - 1) \geqslant n - 3k$. Now, combining Lemmas 4.17, 4.18 and 4.19, we get

$$e_{n,k} \geqslant (l+2)2^{l-1}m \geqslant \tfrac{1}{2}(n - 3k) \log_2 (2k+2).$$

5. Generalizations and problems

There are several obvious generalizations of the graphs G_k considered above. For example, if we have a set S of n points in general position in E^d (no $d+1$ on a hyperplane) then we can consider the hypergraph G_k^d consisting of oriented hyperplanes containing d points of S and having k points of S on their positive sides. By symmetry we may again assume $k \leqslant (n-d)/2$. The elements of the graph are now vertices and hyperplanes (d-tuples of vertices). However, the construction in Section 2 is no longer applicable for the construction of G_k^d since the spatial rotations do not form a one-dimensional group.

It is now clear that every $(d-1)$-tuple of points of S, through which there passes a hyperplane containing no more than k points of S on one side, is contained in at least 2 faces of G_k^d. We must get a trivial lower estimate for $e_{n,k}^d$, the maximal number of faces of G_k^d, obtained whenever the points of S are vertices of its convex hull,

$$e_{n,k}^d \geqslant 2\binom{n}{d-1}. \tag{5.1}$$

The other lower bound estimates, such as Theorems 3.3 and 4.1 can also be generalized without difficulty. However, the more difficult estimates based on Theorems 2.2 and 4.5 do not generalize as easily. In particular, if we try to extend the splitting process of Theorem 4.5, we would turn each point into d points and the distributions on opposite sides of the dividing planes would not be easy to establish.

Another generalization would be to use other classes of dividing curves and surfaces instead of straight lines and planes. For example, in the plane we could use circles through 3 points or conic sections through 5 points, etc.

It appears likely that the lower bound obtained for e_n is closer to the truth than the upper bound.

Conjecture 5.2. *The lower bound $e_n > cn \log n$ obtained in Theorem 4.12 cannot be substantially improved. In particular, we conjecture that $e_n = o(n^{1+\varepsilon})$ for all $\varepsilon > 0$.*

Finally, the upper bound construction in Theorem 4.3 leads to the following

Problem 5.3. *Given any graph with n vertices in general position in the plane (the graph need not be planar, so its edges are permitted to intersect), what is the minimal number $e = f(n, k)$ of edges that guarantees that there exists a straight line intersecting at least k of the edges?*

Obviously, $f(n, 1) = 1$, $f(n, 2) = 2$ while $f(n, 3) = n+1$ since a convex n-gon has n edges no three of which are intersected by a straight line. In the same manner we get

$$f(2ml, 2l^2+1) > ml(2l-1)$$

since we can place complete graphs on $2l$ vertices in small neighborhoods of the vertices of a regular m-gon. Then no straight line can intersect more than two of these complete graphs and therefore no straight line intersects more than $2l^2$ edges of the graph. This leads us to a final conjecture:

Conjecture 5.4. *If a graph with n vertices in general position in the plane has more than nk edges, then there exists a straight line which intersects k^2 edges.*

J. N. Srivastava et al., eds., *A Survey of Combinatorial Theory*
© North-Holland Publishing Company, 1973

CHAPTER 14

Some Combinatorial Problems and Results in Fractional Replication

W. T. FEDERER,

Department of Plant Breeding and Biometry, Cornell University, Ithaca, N.Y., U.S.A.

U. B. PAIK,

Department of Statistics, Korea University, Seoul, Korea

B. L. RAKTOE,

Department of Mathematics and Statistics,
University of Guelph, Guelph, Ontario, Canada
and
L. H. WERNER

Department of Social Sciences, University of Wisconsin,
Milwaukee, Wisc. 53201, U.S.A.

1. Introduction

In a mimeographed paper (distributed at the Spring Statistical Meetings at Tallahassee in 1965), Raktoe (see Federer and Raktoe [1965]) and, in his Ph.D. dissertation, Paik [1968] threw new light on the combinatorial structure of saturated main effect plans from an s^m factorial experiment. As a consequence, many combinatorial problems became evident. In a series of papers by Paik and Federer [1966, 1970a, b, 1972], Raktoe and Federer [1970a, b, c, 1971, 1972], and Werner [1970], several combinatorial structures and problems were investigated and formulated. Some of these are discussed below.

If we designate the s^m factorial single-degree-of-freedom effects by an $s^m \times 1$ column vector $\boldsymbol{\beta}$, the $s^m \times s^m$ columnwise-orthogonal design matrix of coefficients by X, and the $s^m \times 1$ vector of $N = s^m$ observations by $\mathbf{Y}$, then the expected value of the observation vector is $E(\mathbf{Y}) = X\boldsymbol{\beta}$. If $p < N$ observations from the complete set of N observations are selected and if p of the N parameters are to be estimated, a saturated fractional replicate of the s^m factorial results. If we let $\mathbf{Y}_p$ be the $p \times 1$ column vector of observations, $\boldsymbol{\beta}_p$ be the $p \times 1$ column vector of parameters to be estimated, and $\boldsymbol{\beta}_{N-p}$ be the $(N-p) \times 1$ column vector of parameters aliased with those in $\boldsymbol{\beta}_p$, then we may write the p observational equations for the fractional replicate, omitting the error vector, in the form $X_{11}\boldsymbol{\beta}_p + X_{12}\boldsymbol{\beta}_{N-p} = \mathbf{Y}_p$, where

$$X = (X_1 \mid X_2) = \left(\begin{array}{c|c} X_{11} & X_{12} \\ \hline X_{21} & X_{22} \end{array}\right).$$

Here X_1 is an $N \times p$ matrix of coefficients corresponding to the parameters in β_p, X_2 is an $N \times (N-p)$ matrix of coefficients corresponding to the parameters in β_{N-p}, X_{11} is a $p \times p$ matrix constructed by taking any p rows of X_1 and the corresponding observations from $\mathbf{Y}$ to form $\mathbf{Y}_p$, X_{12} is a $p \times (N-p)$ matrix determined by the selection of the rows in X_{11}, X_{21} is an $(N-p) \times p$ matrix of the remaining rows in X_1 after X_{11} has been formed, and X_{22} is an $(N-p) \times (N-p)$ matrix resulting from X_2 after the p rows in X_{12} have been removed.

The combinatorial problems considered are concerned with the character-ization of the matrix X_{11}. In particular, some of the following questions arise about the nature of X_{11} for a given β_p.

(i) How many possible plans of X_{11} exist?

(ii) Can any of these plans be generated from a given plan, i.e., is there a set of generators which can be used to generate all possible plans?

(iii) For any given s^m factorial, how many generators and how many plans per generator are there?

(iv) What values are possible for the determinant of X_{11}? More gener-ally, what are the possible spectra of $X'_{11}X_{11}$?

(v) Are there only n values of $\| X_{11} \|$, the absolute value of the deter-minant, for the 2^n factorial for $n > 2$?

(vi) How many plans are associated with each value of $\| X_{11} \|$?

(vii) What is the nature of the aliasing structure of $X_{11}^{-1}X_{12}$, given that X_{11}^{-1} is either an inverse or a generalized inverse of X_{11}, for each generator? And for each value of $\| X_{11} \|$?

(viii) What are the variances for the estimates of the parameters in β_p, given that X_{11}^{-1} exists, for all the plans derivable from a given generator or from a given value of $\| X_{11} \|$?

(ix) How many plans are associated with singular X_{11}?

(x) How many plans are associated with the maximum value of $\| X_{11} \|$?

(xi) Is there an algorithm for deriving all generators from any given generator for a specified value of $\| X_{11} \|$?

(xii) Is there an algorithm for obtaining all generators from any specified generator?

(xiii) Given that results are available for the 2^n factorial, how do they extend to the s^m and the $q \times r \times t \ldots$ factorials?

(xiv) Is there a transformation of X which will aid in characterizing X_{11}?

(xv) What is the frequency distribution of plus ones in X_{11} for fractional replicates from the 2^n factorial?

(xvi) What is the frequency distribution of plus ones in X_{11} for the generators of fractional replicates in the 2^n factorial?

(xvii) How does the frequency of plus ones in X_{11} for fractional replicates of the 2^n factorial relate to values of $\| X_{11} \|$?

(xviii) Since the columnwise-orthogonal matrix X has rank N, what are the possible ranks for the matrix X_{11} and its corresponding matrix X_{22}?

Some of the above questions have been answered or partially answered but some appear to be exceedingly difficult. In order to complete the theory for saturated fractional replicates from an s^m factorial, it is necessary to have answers to all of the above questions. Since the $(1, -1)$ coefficient matrix can be transformed to a $(0, 1)$ matrix (see, e.g., Raktoe and Federer [1970a]), answers to the above questions also extend the theory of $(0, 1)$ and $(1, -1)$ matrices. Since the rows of X, and consequently of X_1, are all distinct, there are $\binom{N}{p}$ ways of forming saturated fractional replicate plans. Paik [1968] and Paik and Federer [1970a] have shown that generators can be produced, each of which will generate s^m plans, s a power of a prime, provided that the generator does not form a group.[1] Only for the case that $p = s^k, k < m$, is it possible for a generator to form a group. Raktoe and Federer [1970c] have determined the number of generators for $p = m(s-1)+1$, s a power of a prime number, and for saturated main effect plans from an s^m factorial.[2]

Paik and Federer [1966, 1970a] and Paik [1968] have found that $\| X_{11} \|$, the absolute value of $| X_{11} |$, can take on the following values for specified factorials:

	Possible values of $\| X_{11} \|$
2^2	$2^2, 0\dagger$
2^3	$2(2^3), 2^3, 0$
2^4	$3(2^4), 2(2^4), 2^4, 0$
3^2	$2(2^2)(3^2), 2^2(3^2), 0$
3^3	$4(2^3)(3^3), 3(2^3)(3^3), 2(2^3)(3^3), 2^3(3^3), 0$

† occurs with zero frequency.

From the above, Paik and Federer [1970a] were led to the conjecture that $\| X_{11} \|$ can take the values $[s(s-1)(s-2) \ldots 1]^m[m(s-1)-i]$ for $i = s-1, s, s+1, \ldots, m(s-1)$. Since Paik has produced a counterexample to the above conjecture for $s = 5$, it may hold only for $s = 2$ and/or 3 except possibly for $m(s-1)+1 = s^k$ for $k \geqslant 3$. The fact that $\| X_{11} \|$ takes only a relatively small number of values appears to be unknown to most writers of statistical literature.

Only Webb [1965] appears to have been aware of this fact. Writers of mathematical literature also appear to be unaware of this. A proof of the number of values and the values that $\| X_{11} \|$ can take would be an important development in mathematics and in the theory of fractional replication.

[1] See note added in proof (1).   [2] See note added in proof (2).

Metropolis and Stein [1967] obtained a lower bound on the number of singular (0, 1) matrices in the class having distinct rows (i.e., those from the 2^n factorial). Raktoe and Federer [1970b] obtained a lower bound on the number of singular saturated main effect plans from an s^m factorial. The general problem is formulated precisely in a mathematical form, and these two bounds are discussed and compared in the second section of this paper. Thus, some characterizations of X_{11} have been made when X_{11} is singular. In addition, Raktoe and Federer [1970a] have given some characterizations of X_{11} when $\| X_{11} \|$ achieves its maximal value. It appears that mathematicians have been mostly concerned with studying X_{11} when $\| X_{11} \|$ achieves its maximum value. The study of $\|X_{11}\| \neq 0$ or a maximum has been neglected; knowledge of other values of $\| X_{11} \|$ is important in selecting fractional replicates with desirable aliasing properties (see Paik and Federer [1970a]). Some information on the aliasing structure of $X_{11}^{-1}X_{12}$ and on the variances of the estimated parameters in β_p has been obtained by Paik and Federer [1972].

Werner [1970] has obtained the frequency distribution of ones in (0, 1) matrices with distinct rows and of ones in the generators for saturated main effect plans from the 2^n factorial. Some of her results are presented in the third section of this paper.

2. Some combinatorial problems of singular saturated main effect plans of the 2^n factorial

In this section, some additional unsolved problems and first attempts at solving them are presented. The results are associated with the enumeration of saturated main effect plans of the 2^n factorial. Special attention is devoted to the class of singular saturated plans, and two lower bounds to its cardinality are discussed.

First, let us consider the general combinatorial problem. Let G be the set of n-tuples of the form $(x_1, x_2, \ldots, x_n)$, with $x_i \in \{0, 1\}$. Further let A_{n+1} be the set of $(n+1) \times (n+1)$ matrices of the form $[1 : D]$, where 1 is a column of $+1$'s and the rows of D are $n+1$ distinct elements of G. Note that the cardinality of A_{n+1} is $\binom{2^n}{n+1}$. If det $[1 : D]$ denotes the determinant of $[1 : D]$, then an unsolved problem of considerable complexity is the determination of the precise range of det $[1 : D]'[1 : D]$ or of $|$ det $[1 : D] |$ for arbitrary n. A second and related problem is then to determine how many matrices belong to each possible value of $|$ det $[1 : D] |$. These two problems can be considered as the general combinatorial problem of saturated main effect plans of the 2^n factorial. (See also Paik and Federer [1970a] and Raktoe and Federer [1970b].)

The above general problem was first enumerated by Paik [1968] and Paik and Federer [1970a] for $n = 2, 3$ and 4.[3] In a mimeographed note distributed

[3] Wells [1971] has given the enumeration for $n \leqslant 7$.

at the Joint Statistical Meetings of August 1969 in New York City, the general problem was pointed out to the audience by one of the authors and possible methods of attack were indicated. This could be done because the problem can be treated in terms of $(-1, 1)$-matrices, $(0, 1)$-matrices, polytopes, etc.

Possibly, the simplest problem to be solved is to find the number of matrices in A_{n+1} having determinant equal to zero, i.e., the determination of the cardinality of the singular class in A_{n+1}. Let us denote this singular class by $A_{n+1,0}$ and let $\delta_{n+1,0}$ be its cardinality. In a recent attempt to determine $\delta_{n+1,0}$ for arbitrary n, Raktoe and Federer [1970b] considered a subclass of A_{n+1}, this subclass being of the form $[1 : B]$, where the $n+1$ rows of B were considered to be $(n+1)$-subsets of the points belonging to a $(n-k)$-flat of EG$(n, 2)$, k being the largest positive integer such that $n \leqslant (2^{n-k}-1)$ for given n. (For further details concerning the more general setting of this subset, the reader is directed to Raktoe and Federer [1970b].) Denoting this subset by E_{n+1} and its cardinality by ε_{n+1}, the authors showed that

$$\varepsilon_{n+1} = a(n, k) \cdot 2^k \cdot \binom{2^{n-k}}{n+1}, \tag{2.1}$$

where

$$a(n, k) = \prod_{i=0}^{k-1} [(2^{n-i}-1)(2^{k-i}-1)^{-1}]. \tag{2.2}$$

Denoting by $E_{n+1,0}$ the singular subclass of E_{n+1} and letting $\varepsilon_{n+1,0}$ be the cardinality of $E_{n+1,0}$, the authors further showed (via a theorem first proved by Dowling [1970]) that

$$\varepsilon_{n+1,0} \geqslant [a(n, k)-b(n, k) \cdot [c(n, k)]^{-1}] \cdot 2^k \binom{2^{n-k}}{n+1} = \varepsilon^*_{n+1,0}, \tag{2.3}$$

where

$$b(n, k) = \prod_{i=0}^{n-1} [(2^{n-k}-1)-i], \tag{2.4}$$

$$c(n, k) = 2^{(n-k)(n-k-1)2^{-1}} \prod_{i=0}^{n-k-1} (2^{n-k-i}-1). \tag{2.5}$$

The number $\varepsilon^*_{n+1,0}$ now can be used as a lower bound to $\delta_{n+1,0}$.

A different and interesting subset of A_{n+1} has also been investigated by Metropolis and Stein [1967]. Putting their work in our context, their subclass, call it A_n, consisted of matrices of the form

$$\left[1 \begin{array}{c} 0' \\ \hline D^* \end{array}\right],$$

where D^* is an $n \times n$ matrix with the rows being distinct non-zero elements of G and $0'$ is the zero element of G. Note that the cardinality δ_n of A_n is equal to $\binom{2^n-1}{n}$.

As pointed out by these two authors, the determination of the cardinality $\delta_{n,0}$ of the singular subclass $A_{n,0}$ of A_n is still unresolved. They obtained a lower bound to $\delta_{n,0}$ (their derivations are complex and would require at least a couple of pages, so that the reader is advised to read the paper) and they compared it with $\delta_{n,0}$ as enumerated by Wells [1972] for $n = 2,3,4,5,6,7$. Denoting the Metropolis-Stein lower bound by $\varepsilon_{n,0}$, we have the following illustrative table:

TABLE 2.1.
Lower bounds to the number of singular plans for $n = 2, 3, 4, 5, 6$.

	Raktoe-Federer		Metropolis-Stein	
n	$\delta_{n+1,0}$	$\varepsilon_{n+1,0}$	$\delta_{n,0}$	$\varepsilon_{n,0}$
2	0	0	0	0
3	12	12	6	6
4	1,360	1,120	425	350
5	–	15,680	65,625	43,260
6	–	86,080	27,894,671	14,591,171

From Table 2.1 it can be observed that the Raktoe-Federer bound becomes crude for $n > 4$ and noting the fact that A_n is contained in A_{n+1} we see that the Metropolis-Stein bound is much sharper for these values. There is hope that further work will improve the Raktoe-Federer bound.

3. Frequency distribution of ones in (0, 1)-matrices having distinct rows

Let D be an $(n+1) \times n$ matrix composed of zeros and ones such that the rows are all distinct. One problem related to characterizing D is to find the frequency distribution of ones in all possible fractional replicates with $n+1$ observations from the 2^n factorial. This problem is resolved here. Before presenting this result, we state four lemmas without proof.

Lemma 3.1. *Given a* (0, 1)-*matrix D with T equal to the number of ones in the matrix, then the range of T is from n to n^2 for a total of $n^2 - n + 1$ values.*

Lemma 3.2. *Given the $\binom{2^n}{n+1}$ possible matrices D and letting the number of matrices having exactly T ones be denoted by* $\mathrm{df}(T)$*, then* $\mathrm{df}(n+k) = \mathrm{df}(n^2 - k)$ *for $k = 0, 1, \ldots, \frac{1}{2}(n^2 - n)$; that is, $\mathrm{df}(T)$ is symmetric around the median value of T.*

Let D^* be an $n \times n$ (0, 1)-matrix having all rows distinct and no row containing all zeros.

Lemma 3.3. *Given a* (0, 1)-*matrix D^* with T^* equal to the number of ones in the matrix, then the range of T^* is from n to $n^2 - n + 1$ for a total of $n^2 - 2n + 2$ values.*

Lemma 3.4. *Given the $\binom{2^n - 1}{n}$ possible matrices D^* and letting the number of matrices having exactly T^* ones be denoted by* $d^*(T^*)$*, then* $d^*(n+k) = d^*(n^2 - n - k)$ *for $k = 0, 1, \ldots, \frac{1}{2}(n^2 - 2n - 1)$ if n is odd and $k = 0, 1, 2, \ldots,$ $\frac{1}{2}n(n-2)$ if n is even; that is with the exclusion of $T^* = n^2 - n + 1$, $d^*(T^*)$ is symmetric around the median value of T^*.*

We are now ready to state the main result concerning the frequency distribution of ones in (0, 1)-matrices having distinct rows:

Theorem 3.1. *Given* $(0, 1)$-*matrices* D *of size* $(n+1) \times n$ *with* T *ones and letting* $\mathrm{df}(T)$ *equal the number of matrices* D *having exactly* T *ones, then*

(i) *for* $T = n$, $\mathrm{df}(n) = 1$

(ii) *for* $T = n+1$, $\mathrm{df}(n+1) = \binom{c_0}{1}\binom{c_1}{n-1}\binom{c_2}{1}$, *and*

(iii) *for* $T = n+k$, $k = 0, 1, \ldots, \frac{1}{2}(n^2-n)$,

$$\mathrm{df}(T) = \sum_{\substack{\sum_{i=0}^{n} x_i = n+1 \\ \sum_{i=0}^{n} ix_i = n+k}} \prod_{2=0}^{n} \binom{c_j}{x_j},$$

where $c_j = \binom{n}{j} \equiv$ *the number of unique* $1 \times n$ *row vectors with* n *ones and* $x_j = 0, 1, \ldots, \binom{n}{j} \equiv$ *the number of* $1 \times n$ *row vectors having* j *ones in the matrix* D.

Proof. Consider the matrix D as being composed of $n+1$ row vectors of size $1 \times n$ with each vector containing from zero to n ones with all elements not one being zero. Every row of D is unique (distinct) and the total number of ones, T, in D is equal to $n+k$, $k = 0, 1, \ldots, n^2-n$ by Lemma 3.1. Let t equal the smallest number of ones in any row vector of the matrix and let r equal the number of row vectors with t ones. When there are r row vectors with t ones, the remaining $n+1-r$ row vectors must contain $n+k-rt$ ones. If matrices are formed by starting with r row vectors with t ones in each and if

$$n+1-r > n+k-rt, \tag{3.1}$$

then no matrix exists because the number of row vectors left to choose is greater than the number of ones left to distribute. This results from the facts that

$$n+1-r = \sum_{i=t+1}^{n} x_i$$

and

$$n+k-rt = \sum_{i=t+1}^{n} ix_i,$$

when $n+1-r > n+k-rt$; this means that

$$\sum_{i=t+1}^{n} x_i > \sum_{i=t+1}^{n} ix_i,$$

i.e., that

$$0 > \sum_{i=t+1}^{n} (i-1)x_i.$$

But, by definition, $0 \leqslant x_i \leqslant \binom{n}{i}$ and $t \geqslant 0$; therefore, the strict inequality can never hold.

If there are $n+k-rt$ ones left to distribute and if each remaining row vector that can be chosen has at least $t+1$ ones (by the definition of t), then the number of ones left must be equal to or greater than the number of ones required such that every remaining row vector has $t+1$ ones, i.e.,

$$n+k-rt \geqslant (t+1)(n+1-r). \tag{3.2}$$

The inequality in (3.2) results from the fact that every row vector still to be selected has $t+1$ or more ones.

In proving the theorem, we proceed part by part. First we shall prove part (i). Since there is only $\binom{n}{0} = 1$ possible unique zero row vector and since there are $\binom{n}{1} = n$ possible unique row vectors with a single 1 each, then there is only one unique combination which will form an $(n+1) \times n$ matrix with only n ones. The formula is $\binom{c_0}{1}\binom{c_1}{n} = \binom{1}{1}\binom{n}{n} = 1$, which proves part (i).

To prove part (ii), note that if the zero row vector is selected, there are n row vectors left to choose and $n+1$ ones left to distribute. Therefore, if r row vectors with t ones each are selected from the remaining n row vectors, the following must hold (from 3.2)):

$$n+1-rt \geqslant (t+1)(n-r),$$

or

$$r \geqslant tn-1.$$

Since only n row vectors remain to be selected, $r \leqslant n$.

If $t > 1$, r becomes too large. When $t = 1$, r can either be n or $n-1$. For $r = n$, $\sum_{i=1}^{n} ix_i = 0(1)+1(n) \neq n+1$. If $r = n-1$, the following combinations are suitable:

$$\binom{c_0}{1}\binom{c_1}{n-1}\binom{c_2}{1}. \tag{3.3}$$

If the zero row vector is not selected as one of the rows of D, there are $n+1$ row vectors to be selected and $n+1$ ones to distribute. Therefore, from (3.2), the inequality

$$r \geqslant t(n+1)$$

must hold in order for any possible matrices to exist. If $t = 1$, then $r = n+1$, but there are only n unique row vectors with one 1. When $t > 1$, $r > n+1$ and therefore there are no possibilities except (3.3), which proves part (ii).

Admissible matrices D occur when the number of row vectors selected equals $n+1$ and when the number of ones totals $n+k$. If a row vector with i ones is needed, there are $\binom{n}{i} = c_i$ combinations from which to choose. If x_i row vectors with i ones are needed, there are $\binom{c_i}{x_i}$ possible unique combinations. In our notation, this is simply

$$df(T) = \sum_{\substack{\sum_{i=0}^{n} x_i = n+1 \\ \sum_{i=0}^{n} ix_i = n+k}} \prod_{j=0}^{n} \binom{c_j}{x_j}. \tag{3.4}$$

In order to perform the actual computations and in order to find all possibilities (as was done for part (ii)), the following breakdown of (3.4) is used:

$$\mathrm{df}(T) = \binom{c_0}{1} \sum_{\substack{\sum_{i=1}^{n} x_i = n \\ \sum_{i=1}^{n} i x_i = n+k}} \prod_{j=1}^{n} \binom{c_j}{x_j} + \sum_{\substack{\sum_{i=1}^{n} x_i = n+1 \\ \sum_{i=1}^{n} i x_i = n+k}} \prod_{j=1}^{n} \binom{c_j}{x_j}. \tag{3.5}$$

The above expression allows consideration of whether or not the zero row vector is selected.

Once r row vectors with t ones are selected, the first part of (3.5) may be rewritten as

$$\sum_{t=1}^{n} \sum_{r=tn+k}^{c_t \text{ or } n+1} \binom{c_t}{r} \sum_{\substack{\sum_{i=t+1}^{n} x_i = n-r \\ \sum_{i=t+1}^{n} i x_i = n+k-rt}} \prod_{j=t+1}^{n} \binom{c_j}{x_j}. \tag{3.6}$$

If $n-r > n+k-rt$ (from (3.1)), there are no possibilities, and only if $n+k-rt \geq (t+1)(n-r)$, from (3.2), are there any possibilities. In the first case, the summation is zero and the next value of r and t can be tried. For the second case, the expression.

$$\sum_{\substack{\sum_{i=t+1}^{n} x_i = n-r \\ \sum_{i=t+1}^{n} i x_i = n+k-rt}} \prod_{j=t+1}^{n} \binom{c_j}{x_j}$$

may be partitioned in the same manner as the first term of (3.5) was to obtain (3.6).

If r_1 row vectors with t_1 ones each are selected, where t_1 is the second smallest number of ones in any row vector, the number of ones left is $n+k-rt-r_1 t_1$ and the number of row vectors left to be chosen is $n+1-r-r_1$. To continue with this procedure, the following must hold:

$$n+k-rt-r_1 t_1 \geq (n+1-r-r_1)(t_1+1),$$

or

$$r_1 \geq t_1(n+1)+(1-k)+r(t-t_1);$$

if

$$n+1-r-r_1 > n+k-rt-r_1 t_1,$$

then no possibilities exist and a new path with a larger value for t must be tried. Whenever (3.1) holds, the entire path is wrong and the first choice of t must be increased if possible. When t can no longer be increased, there are no more possibilities for choosing the zero row vector. This same procedure can be used for the second part of (3.5) and hence to complete the proof of part (iii). This completes the proof of the theorem and illustrates the procedure for computing $\mathrm{df}(T)$.

The generators developed by Paik [1968] have a zero row vector in the first row. The remaining part of the matrix D is then D^*. The frequency distribution of ones in the matrix D^* is given in the following corollary.

Corollary 3.1. *Given $(0, 1)$-matrices D^* composed of n unique $(0, 1)$ row vectors with none of them being the zero row vector and letting $\mathrm{df}^*(T^*)$ be the number of matrices D^* having a total of T^* ones, then*

(i) *for $T^* = n$, $\mathrm{df}^*(n) = 1$,*

(ii) *for $T^* = n+1$, $\mathrm{df}^*(n+1) = \binom{c_1}{n-1}\binom{c_2}{1}$, and*

(iii) *for $T^* = n+k$, $k = 0, 1, \ldots, n^2 - 2n + 2$,*

$$\mathrm{df}^*(T^*) = \sum_{\substack{\sum_{i=1}^{n} x_i = n \\ \sum_{i=1}^{n} i x_i = n+k}} \prod_{j=1}^{n} \binom{c_j}{x_j}.$$

The proof of the corollary follows that given for Theorem 3.1.

Note added in proof

Pesotan, Raktoe and Federer [1972] have shown:

(1) that the generators produce less than s^m plans if and only if the generator forms a subgroup or is the union of cosets of a fixed nonzero subgroup.

(2) that the Raktoe and Federer [1971] formula holds if and only if $m(s-1)+1$ and the underlying prime of s are relatively prime, or $s^m = 2^3$ or equal to the underlying prime, and

(3) that these results may be extended for arbitrary factorials and for arbitrary p.

References

T. A. Dowling, 1970, private communication.

W. T. Federer and B. L. Raktoe, 1965, Some remarks on the generalized construction and analysis of fractional replicates, *Biometrics* **21**, 763–764.

N. Metropolis and P. R. Stein, 1967, On a class of (0, 1)-matrices with vanishing determinant, *J. Combin. Theory* **3**, 191–198.

U. B. Paik, 1968, Analysis of non-orthogonal n-way classifications and fractional replication, Ph.D. Thesis, Cornell Univ., Ithaca, N.Y.

U. B. Paik and W. T. Federer, 1966, On the construction of fractional replicates with special reference to saturated designs, Biometrics Unit Mimeograph Series No. BU-224-M, Cornell Univ., Ithaca, N.Y.

U. B. Paik and W. T. Federer, 1970a, On construction of fractional replicates and on aliasing schemes, Math. Res. Center Tech. Sum. Rept. No. 1029, U.S. Army and Univ. of Wisconsin, Madison, Wisc.

U. B. Paik and W. T. Federer, 1970b, A randomized procedure of saturated main effect fractional replicates, *Ann. Math. Statist.* **41**, 369–375.

U. B. Paik and W. T. Federer, 1972, On a randomized procedure for saturated fractional replicates of the 2^n-factorial, *Ann. Math. Statist.* **43**(4), to appear.

H. Pesotan, B. L. Raktoe and W. T. Federer, 1972, Complexes of Abelian groups with applications to fractional factorial designs, Biometrics Unit Mimeo Series BU-423-M, Cornell Univ., Ithaca, N.Y.

B. L. Raktoe and W. T. Federer, 1970a, Characterization of optimal saturated main effect plans of the 2^n factorial, *Ann. Math. Statist.* **41**, 203–206.

B. L. Raktoe and W. T. Federer, 1970b, A lower bound for the number of singular saturated main effect plans of an s^m factorial, *Ann. Inst. Statist. Math.* **22**, 519–525.

B. L. Raktoe and W. T. Federer, 1970c, A theorem on saturated plans and their complements, *Ann. Math. Statist.* **41**, 2184–2185.

B. L. Raktoe and W. T. Federer, 1971, On the number of generators of saturated main effect fractional replicates, *Ann. Math. Statist.* **42**, 1758–1760.

B. L. Raktoe and W. T. Federer, 1972, Construction of confounded mixed factorial and mixed lattice designs, *Austral. J. Statist.* **14**(1), to appear.

S. Webb, 1965, Design, testing and estimation in complex experimentation, 1: Expansible and contractible factorial designs and the application of linear programming to combinatorial problems, Rept. No. ARL 65–116, Office of Aerospace Research, U.S. Air Force.

M. B. Wells, 1971, *Elements of Combinatorial Computing* (Pergamon Press, New York).

L. H. Werner, 1970, The distribution of ones in a class of seminormalized (0, 1) matrices with applications to fractional replication in the 2^n factorial, M.S. Thesis, Cornell University, Ithaca, N.Y.

J. N. Srivastava et al., eds., *A Survey of Combinatorial Theory*
© North-Holland Publishing Company, 1973

CHAPTER 15

The Bose-Nelson Sorting Problem †

ROBERT W. FLOYD and DONALD E. KNUTH

Stanford University, Stanford, Calif. 94305, U.S.A.

A typical "sorting network" for four numbers is illustrated in Fig. 1; the network involves five "comparators", shown as directed wires connecting two lines. Four numbers are input at the left, and as they move towards the right each comparator causes an interchange of two numbers if necessary so that the larger number appears at the point of the arrow. At the right of the network the numbers have been sorted into nondecreasing order from top to bottom; it is easy to verify that this will be the case no matter what numbers are input, since the first four comparators select the smallest and the largest elements and the final comparator ranks the middle two.

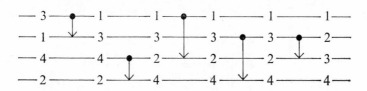

Fig. 1. A sorting network.

Sorting networks were originally constructed prior to 1957 by R. J. Nelson, who developed special networks for eight or less elements. Nelson also showed that n more comparators always suffices to go from n elements to $n+1$ (see O'Connor and Nelson [1962]).

In 1960–61, he and R. C. Bose constructed n-element sorting networks which were considerably more economical as $n \to \infty$ (see Bose and Nelson [1962]). The *Bose-Nelson sorting problem* is the problem of determining $S(n)$, the minimum number of comparators needed in an n-element sorting network.

Bose and Nelson gave an upper bound for $S(n)$, and conjectured that their method actually gave $S(n)$ exactly; but subsequent constructions have shown that their upper bound can be improved for all $n > 8$ (see Floyd and Knuth [1967], Batcher [1968]). In this paper we develop a few aspects of the theory, and prove that Bose and Nelson's conjecture was correct for $n \leqslant 8$.

† The preparation of this report has been supported in part by the National Science Foundation, and in part by the Office of Naval Research.

TABLE 1
Bounds for $S(n)$

Approx. date		$n=1$	2	3	4	5	6	7	8	9	10	11	12	13	14	15	16	Asymptotic
1954	Nelson	0	1	3	5	9	12	18	19	27	36	46	57	69	72	86	101	$\frac{1}{2}n^2$
1960	Bose and Nelson	0	1	3	5	9	12	16	19	27	32	38	42	50	55	61	65	$n^{1.6}$
1964	Floyd and Knuth	0	1	3	5	9	12	16	19	25	31	37	41	49	54	60	64	$n^{1+c/\sqrt{\log n}}$
1964	Batcher	0	1	3	5	9	12	16	19	26	31	37	41	48	53	59	63	$\frac{1}{4}n(\log_2 n)^2$
1969	Best upper bounds known	0	1	3	5	9	12	16	19	25	29	35	39	46	51	56	60	$\frac{1}{4}n(\log_2 n)^2$
1971	Best lower bounds known	0	1	3	5	9	12	16	19	23	27	31	35	39	43	47	51	$n\log_2 n$

Table 1 outlines some of the early work on the Bose-Nelson sorting problem and summarizes its current status; see Knuth [1972] for further details of recent constructions, due to M. W. Green, A. Waksman and G. Shapiro. The upper bounds listed for $n \leqslant 12$ are probably exact.

In order to study the problem in detail, it is convenient to introduce a few notational conventions. Let $x = \langle x_1, \ldots, x_n \rangle$ and $y = \langle y_1, \ldots, y_n \rangle$ be sequences of n real numbers; x is said to be *sorted* if $x_1 \leqslant x_2 \leqslant \cdots \leqslant x_n$. We define two operators on such sequences, the *exchange* operation (ij) and the *comparator* operation $[ij]$, for $1 \leqslant i, j \leqslant n, i \neq j$, as follows:

$$x(ij) = y \quad \text{iff} \quad y_i = x_j, \quad y_j = x_i, \quad y_k = x_k \quad \text{for} \quad i \neq k \neq j; \tag{1}$$

$$x[ij] = x \quad \text{if} \quad x_i \leqslant x_j, \quad x[ij] = x(ij) \quad \text{if} \quad x_i > x_j. \tag{2}$$

Thus $x[ij] = y$ iff $y_i = \min(x_i, x_j)$, $y_j = \max(x_i, x_j)$, and $y_k = x_k$ for $i \neq k \neq j$. It is clear that, when i, j, k, l are distinct, we have (see Fig. 2)

$$(ij)[ij] = [ij], \qquad [ij](ij) = [ji]; \tag{3}$$

$$(ij)[jk] = [ik](ij), \qquad (ij)[kj] = [ki](ij), \qquad (ij)[kl] = [kl](ij). \tag{4}$$

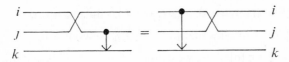

Fig. 2. A proof that $(ij)[jk] = [ik](ij)$.

A *comparator network* α is a sequence of zero or more exchange and/or comparator operations; a *sorting network* α is a comparator network such that $x\alpha$ is sorted for all x. We write $\alpha\beta$ for the network consisting of α followed by β; and we say that

$$\alpha \subseteq \beta \quad \text{iff} \quad U\alpha \subseteq U\beta, \qquad \alpha \equiv \beta \quad \text{if} \quad U\alpha = U\beta, \tag{5}$$

where U is the set of all sequences $\langle x_1, \ldots, x_n \rangle$. Figure 1 illustrates the sorting network [12][34][13][24][23]. Clearly $\alpha \subseteq \beta$ implies that $\alpha\gamma \subseteq \beta\gamma$. Furthermore, if β is a sorting network and $\alpha \subseteq \beta$ we must have $\alpha \equiv \beta$; in fact, $x\alpha = x\beta$ for all x in this case, since $x\alpha$ must be sorted.

Sorting networks can also be interpreted in a more general way, if we allow m numbers to be contained in each line for some fixed $m \geqslant 1$. If $x_1, \ldots, x_n$ are *multisets* (i.e. sets with the possibility of repeated elements), containing m elements each, we can redefine the comparator $[ij]$ to be the operation of replacing x_i and x_j by the smallest and largest m elements, respectively, of the original $2m$ elements in x_i and x_j. See Fig. 3, which illustrates the case $m = 2$. Our first result gives a basic property of this general interpretation.

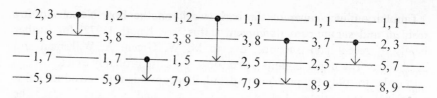

Fig. 3. Another interpretation of the network in Fig. 1.

Theorem 1. *Let* α *be a comparator network for n elements, and let i and j be indices such that* $(x\alpha)_i \not\leq (x\alpha)_j$ *for some x and for some* $m \geq 1$; *in other words, the m elements of the multiset* $(x\alpha)_i$ *are not all less than or equal to the m elements of* $(x\alpha)_j$. *Then there is a sequence* $y = \langle y_1, \ldots, y_n \rangle$ *of zeroes and ones such that* $(y\alpha)_i = 1$ *and* $(y\alpha)_j = 0$.

Proof. Let $\alpha = f_1 \cdots f_t$, where each f_s is an exchange or a comparator. Let u be the smallest element of $(x\alpha)_j$; we shall use the name A to stand for any number $\leq u$, and B for any number $> u$. By hypothesis, at least one element of $(x\alpha)_i$ is a B. We shall define a sequence $y^{(s)}$ of zeroes and ones, for $0 \leq s \leq t$, such that

$$y_p^{(s)} = 0 \text{ implies that } (xf_1 \cdots f_s)_p \text{ contains an } A, \tag{6}$$

$$y_p^{(s)} = 1 \text{ implies that } (xf_1 \cdots f_s)_p \text{ contains a } B, \tag{7}$$

for $1 \leq p \leq n$. First for $s = t$ we define $y_i^{(t)} = 1, y_j^{(t)} = 0$, and other elements $y_k^{(t)}$ are defined in any manner consistent with the above conventions (6), (7).

Assuming $y^{(s)}$ has been defined for some $s \geq 1$, we define $y^{(s-1)}$ as follows:

Case 1: $f_s = (pq)$. Then $y^{(s-1)} = y^{(s)}(pq)$.

Case 2: $f_s = [pq]$ and $y_p^{(s)} = y_q^{(s)}$. Then $y^{(s-1)} = y^{(s)}$. This fulfills the above conditions, since $y_q^{(s)} = 0$ implies that $(xf_1 \cdots f_s)_q$ contains at least one A, hence $(xf_1 \cdots f_s)_p$ contains *all* A's, hence there are more than m A's in all; some A's must be present in both $(xf_1 \cdots f_{s-1})_p$ and $(xf_1 \cdots f_{s-1})_q$. Similarly $y_p^{(s)} = 1$ implies that $(xf_1 \cdots f_{s-1})_p$ and $(xf_1 \cdots f_{s-1})_q$ both contain at least one B.

Case 3: $f_s = [pq]$ and $y_p^{(s)} \neq y_q^{(s)}$. Then $(y_p^{(s-1)}, y_q^{(s-1)})$ are defined to be either $(0, 1)$ or $(1, 0)$, in any manner consistent with the above conventions; and $y_r^{(s-1)} = y_r^{(s)}$ for $p \neq r \neq q$. This definition of $y^{(s-1)}$ is justified because $(xf_1 \cdots f_{s-1})_p$ and $(xf_1 \cdots f_{s-1})_q$ are not both all A's or both all B's. Note that $y_q^{(s)} = 0$ is impossible, since it implies as in case 2 that $(xf_1 \cdots f_s)_p$ is all A's, contradicting our convention; thus $y_p^{(s)} = 0, y_q^{(s)} = 1$.

According to this definition, $y^{(s-1)}f_s = y^{(s)}$, hence $y^{(0)}\alpha = y^{(t)}$; therefore $y = y^{(0)}$ satisfies the conditions of the theorem.

When $m = 1$, Theorem 1 implies that a network will necessarily sort all possible inputs if we can prove that it sorts the 2^n sequences of zeroes and ones;

Corollary 1. *A comparator network is a sorting network if and only if it sorts all sequences of zeroes and ones.*

Corollary 2. *Let $M(m)$ be the minimum number of comparators needed to merge two sets of m elements, i.e., to sort all sequences $\langle x_1, \ldots, x_{2m} \rangle$ such that $x_1 \leqslant \cdots \leqslant x_m$ and $x_{m+1} \leqslant \cdots \leqslant x_{2m}$. Then*

$$S(mn) \leqslant nS(m) + M(m)S(n), \tag{8}$$

$$M(mn) \leqslant M(m)M(n). \tag{9}$$

Proof. Replace each line in an n-element sorting network by m parallel lines, and replace each comparator by $M(m)$ comparators which merge the $2m$ lines corresponding to the original 2 lines. Append n m-element sorting networks at the left, in order to sort each of the groups; this yields a sorting network for mn elements having $M(m)S(n) + nS(m)$ comparators. If we start with a sorting network that was constructed in this way for $n = 2$, the right-hand part of the network has $M(m)$ comparators; expanding each line to m' lines makes the $M(m)M(m')$ comparators of the right-hand part capable of merging two ordered groups of mm' elements.

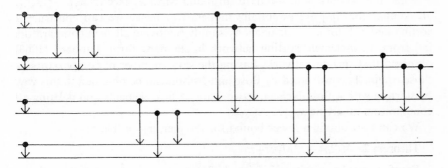

Fig. 4. An 8-element sorting network, constructed from Fig. 1 and Corollary 2 ($m = 2$).

An example of the construction in Corollary 2 appears in Fig. 4. Bose and Nelson proved Corollary 2 in the special case of binary merging, $n = 2$; this shows that $M(2^n) \leqslant 3^n$, and $S(2^n) \leqslant 3^n - 2^n$. When $S(n) \approx n^\beta - n$ and $M(n) \approx n^\beta$, the inequalities in Corollary 2 do not allow us to lower the exponent β; and in fact these inequalities do not lead to an especially efficient way to construct sorting networks, compared to other known methods. Yet the special case $m = 7$, $n = 3$ shows that 21 elements can be sorted with one less comparator than predicted by the Bose-Nelson conjecture, and this is what first showed us that the conjecture was false in general (see Floyd [1964]). We went on to find that the conjecture is false for all $n > 8$. (This is, perhaps, poetic justice, since Bose made the conjecture shortly after he had helped to disprove Euler's famous Latin-squares conjecture, for all

$n > 6$, after having first disproved it for $n = 50$! And our own sorting networks have by now been shown to be nonoptimal for all $n > 9$.)

Let us now examine the properties of comparator networks a little more closely. In the first place, we can use identities (3), (4) to transform any comparator network so that all comparators precede all exchanges, and so that all comparators $[ij]$ have $i < j$. (Working from left to right, we replace $[ji]$ by $[ij](ij)$ when $i < j$, and we permute exchanges with comparators. This process clearly converges in a finite number of steps.) In this way, a network α is transformed into $\alpha'\alpha''$, where α' has only "downward" comparators and α'' has only exchanges. If α is a sorting network, we can see by considering the effect of α on $\langle 1, 2, \ldots, n \rangle$ that α'' must be the identity transformation, so $\alpha \equiv \alpha'$.

Let us say that a sorting network is in *standard form* if it consists entirely of comparators $[ij]$ with $i < j$, and no exchanges. We have proved

Lemma 1. *Every sorting network is equivalent to a network in standard form, having the same number of comparators.*

When the network is in standard form and when x_n exceeds $x_1, \ldots, x_{n-1}$, all comparators $[in]$ are essentially inoperative; hence we can construct a sorting network for $n-1$ elements by simply removing all such comparators $[in]$ from an n-element sorting network in standard form. Hibbard [1963] observed that sorting networks having the same number of comparators as those originally constructed by Bose and Nelson can be obtained in this way by starting with a Bose-Nelson network for $2^k > n$ elements and deleting all comparators involving $x_{n+1}, \ldots, x_{2^k}$.

We can now obtain a lower bound for the merging problem:

Theorem 2. $M(2n) \geqslant 2M(n)+n$.

Proof. Consider a network with $M(2n)$ comparators in standard form, which sorts $\langle x_1, \ldots, x_{4n} \rangle$ whenever $x_1 \leqslant x_3 \leqslant \cdots \leqslant x_{4n-1}$ and $x_2 \leqslant x_4 \leqslant \cdots \leqslant x_{4n}$. We separate the comparators $[ij]$ into three types,

$$A: i \leqslant 2n, j \leqslant 2n;$$
$$B: i \leqslant 2n, j > 2n;$$
$$C: i > 2n, j > 2n.$$

Since $x_{2n+1}, \ldots, x_{4n}$ may be very large, there must be at least $M(n)$ comparators of type A; similarly there must be at least $M(n)$ of type C. And since we might have $x_i = 1$ when i is odd, 0 when i is even, there must be at least n comparators of type B in order to let n zeros rise to the top half of the diagram.

A similar proof shows that $M(2n+1) \geqslant M(n)+M(n+1)+n$; and the same relations also hold with S in place of M. It follows that $M(n) \geqslant \frac{1}{2}n \log_2 n + O(n)$ for all n; this is why sorting networks based recursively on binary merging involve the order of $n(\log n)^2$ comparators, at least.

The best sorting networks known for $n > 8$ do not use binary merging, so Theorem 2 does not give us useful information about lower bounds for $S(n)$. When n is comparatively small, exact lower bounds can be found, as we shall now see. First we shall examine a general commutativity condition:

Lemma 2. *If* $\alpha = [i_1 j_1][i_2 j_2] \cdots [i_t j_t]$, *where* $\{i_1, i_2, \ldots, i_t\} \cap \{j_1, j_2, \ldots, j_t\} = \emptyset$, *and if* β *is any rearrangement of the comparators of* α, *then* $\alpha \equiv \beta$.

Proof. We shall show that $U\alpha$ is the set $S(\alpha)$ of all vectors $\langle x_1, \ldots, x_n \rangle$ such that $x_{i_s} \leqslant x_{j_s}$ for $1 \leqslant s \leqslant t$. All $x \in S(\alpha)$ satisfy $x\alpha = x$, hence $S(\alpha) \subseteq U\alpha$.

Conversely, suppose that $x\alpha \notin S(\alpha)$, i.e., $(x\alpha)_{i_s} > (x\alpha)_{j_s}$ for some s. Clearly s must be less than t; let $\alpha' = [i_1 j_1] \cdots [i_{t-1} j_{t-1}]$. By induction on t, we have $(x\alpha')_{i_s} \leqslant (x\alpha')_{j_s}$, hence $[i_t j_t]$ has either increased the i_s component or decreased the j_s component of $x\alpha'$. This means that $i_s = j_t$ or $j_s = i_t$, contradicting the hypothesis.

Lemma 3. *Let* α *be a sorting network for* 3 *or more elements. Then there is a sorting network, with no more comparators than* α, *in which the first three operations are either*

$$\text{Case A,} \quad [12][13][23];$$
$$\text{or Case B,} \quad [12][13][45];$$
$$\text{or Case C,} \quad [12][34][13];$$
$$\text{or Case D,} \quad [12][34][14].$$

Proof. Clearly α includes at least three comparators, or it could not sort. Since $(ij)\alpha \equiv \alpha$, we may use (3) and (4) to transform α into a sorting network α' in which the first operation is [12]. For example, if $\alpha = [47]\beta$, we may take $\alpha' = (14)(27)\alpha = [12](14)(27)\beta$; and if $\alpha = [21]\beta$ we may take $\alpha' = \alpha = [12](12)\beta$. Similarly we may assume that the second operation is either [12] or [13] or [23] or [34]; and since [12][12] = [12], we may rule out the case [12][12]. If the second operation is [23], we may observe that whenever $\alpha = [i_1 j_1] \cdots [i_t j_t]$ is a sorting network, so is the "dual" network $\alpha' = [j_1 i_1] \cdots [j_t i_t]\tau$, where τ is the sequence of exchanges which transforms $\langle x_1, x_2, \ldots, x_n \rangle$ into $\langle x_n, \ldots, x_2, x_1 \rangle$. Hence, when $\alpha = [12][23]\beta$, we may consider the sorting network $\alpha' = [21][32]\beta'\tau = [12][13](13)(12)\beta'\tau$. Therefore we may assume that the first two operations are [12][13] or [12][34].

Proceeding in this way, we can analyze the possibilities for the third comparator, as follows:

[12][13][12] = [12][13].
[12][13][23] = A.
[12][13][14] ⊇ [12][34][13].
[12][13][24] → [13][12][34](23) ≡ [12][13][34](23).
[12][13][34] → [12][14][34] ≡ [12][34][14].

$[12][13][45] = B.$
$[12][34][13] = C.$
$[12][34][14] = D.$
$[12][34][23] \rightarrow [34][12][41](13)(24) = [12][34][14](14)(13)(24).$
$[12][34][24] \sim [21][43][42] = [12][34][13](13)(34)(12).$
$[12][34][15] = [12][15][34] \rightarrow [12][13][45](45).$
$[12][34][25] = [12][25][34] \sim [21][52][43] \rightarrow [21][32][45](35)$
$$= [12][13][45](13)(12)(35).$$
$[12][34][35] = [34][35][12] \rightarrow [12][13][45](45)(35)(24)(13).$
$[12][34][45] \rightarrow [34][12][25](24)(13) = [12][34][25](24)(13).$

Here "$\rightarrow$" denotes an appropriate left-multiplication by one or more exchanges, "$\equiv$" denotes an application of Lemma 2, and "$\sim$" denotes dualization as above.

Finally, if the first three operations are $[12][34][56]$, we may consider the first comparator which has an index in common with a previous one; this will reduce to a case already considered.

The exhaustive method in this proof can be extended to show that there are essentially eleven ways to choose the first four comparators, when $n \geqslant 4$, namely

A1. $[12][13][23][24].$ C1. $[12][34][13][24].$
A2. $[12][13][23][45].$ C2. $[12][34][13][35].$
                                  C3. $[12][34][13][45].$
B1. $[12][13][45][14].$ C4. $[12][34][13][56].$
B2. $[12][13][45][46].$
B3. $[12][13][45][56].$ D1. $[12][34][14][35].$
                                  D2. $[12][34][14][56].$

Details are omitted here, since we shall not need this fact.

Theorem 3. $S(n) \geqslant S(n-1)+3$, for $n \geqslant 5$.

Proof. There is a sorting network with n comparators, in standard form, having one of the four forms stated in Lemma 3. If we suppress all comparators $[ij]$ with $i = 1$ we have a sorting network for $x_2, \ldots, x_n$, so we must show that at least three comparators have $i = 1$. This is obvious, since in each case we already know two of the comparators, and at least one more is required to bring the smallest element to the required position.

Theorem 3 probably possesses the unique property that it has exactly two applications, no more and no less! Once $S(5)$ has been shown to equal 9, we can use Theorem 3 to show that $S(6) = 12$; and $S(7) = 16$ will imply that $S(8) = 19$. Besides these results, the theorem appears to be quite useless.

We always have $S(n) \geqslant \log_2 n!$ by an elementary information-theoretic argument, hence the values of $S(1)$, $S(2)$, $S(3)$, $S(4)$ are immediately established. But information theory tells us only that $S(5) \geqslant 7$, and Theorem 3 shows that $S(5) \geqslant 8$; the following theorem shows how to strengthen Theorem 3 when $n = 5$.

Theorem 4. $S(5) = 9$.

Proof. We need only show that $S(5) \geqslant 9$, in view of Bose and Nelson's construction. Proceeding as in Theorem 3, if the sorting network begins as in case D, we may permute the lines so that the first three comparators are [14][25][15]. Then we must have at least $S(3)$ more comparators $[ij]$ with $i < j \leqslant 3$, and at least $S(3)$ more with $3 \leqslant i < j$, to complete the sort. This makes 9 comparators.

For cases A, B and C we may permute the lines to obtain a sorting network in standard form in which the first three comparators are respectively

$$[12][15][25] \text{ in case A,}$$
$$[13][14][25] \text{ in case B,}$$
$$[14][25][12] \text{ in case C.}$$

Applying these to all 32 combinations $\langle x_1, x_2, x_3, x_4, x_5 \rangle$ of zeroes and ones (cf. Corollary 1), then replacing all zeroes at the left and all ones at the right by asterisks, discarding all duplicates and all sequences which are nothing but asterisks, we obtain the 5-tuples

$$
\begin{array}{ccccc}
* & * & * & 1 & 0 \\
* & * & 1 & 0 & 0 \\
* & * & 1 & 1 & 0 \\
* & * & 1 & 0 & * \\
* & 1 & 0 & 0 & * \\
* & 1 & 1 & 0 & * \\
* & 1 & 0 & * & * \\
\end{array}
\tag{10}
$$

plus the "special" 5-tuples

$$
\begin{array}{lll}
1\ 1\ 0\ 0\ *, & 1\ 1\ 0\ *\ *, & 1\ 1\ 1\ 0\ * \quad \text{in case A,} \\
1\ 0\ 1\ 1\ 0, & 1\ 0\ *\ *\ * & \qquad\qquad\quad \text{in case B,} \\
1\ 1\ 0\ *\ *, & *\ 1\ 0\ 1\ 0, & *\ 1\ 1\ 1\ 0 \quad \text{in case C.}
\end{array}
\tag{11}
$$

In order to sort (10), we need at least $S(3)$ comparators with $2 \leqslant i < j \leqslant 4$ and $S(3)$ with $3 \leqslant i < j \leqslant 5$; and there must also be another with $i = 1$. The only way to do this with five more comparators is to use the sequence [34][23][45][34] or [34][45][23][34], with an additional $[1j]$ inserted somewhere. But then it is not difficult to verify that the special 5-tuples in (11) cannot all be sorted.

Theorem 5. $S(7) = 16$.

Proof. This theorem was proved by exhaustive enumeration on a CDC G-21 computer at Carnegie Institute of Technology in 1966. The program was written by Mr. Richard Grove, and its running time was approximately

20 hours. The algorithm consisted of constructing a set S_t of sequences such that, for all α of the form $[i_1 j_1] \cdots [i_t j_t]$, there exist permutations π and ρ with $\pi \alpha \rho \supseteq \beta$ for some $\beta \in S_t$. The sets S_t were generated successively for $t = 1, 2, \ldots, 16$, taking care to keep each set rather small; for this purpose, a 128-bit vector was maintained for each element of S_t, characterizing those 7-tuples of zeroes and ones which are output by the network. Most of the computation (about 13 hours) was spent in the cases $t = 8$ and 9, since S_9 had 729 elements. None of the six elements in S_{15} was a sorting network.

The methods of proof used to establish these lower bounds on $S(n)$ are of course quite unsatisfactory for larger values of n. We have no idea how to prove that $S(n)$ grows as $cn(\log n)^2$, although the best upper bounds known to date have this asymptotic behavior. Van Voorhis has recently generalized Theorem 3, proving that $S(n) \geqslant S(n-1) + [\log_2 n]$.

References

K. E. Batcher, 1964, A new internal sorting method, Rep. No. GER-11759, Goodyear Aerospace Corporation, September 1964.

K. E. Batcher, 1968, Sorting networks and their applications, *Proc. AFIPS Spring Joint Comp. Conf.* **32,** 307–314.

R. C. Bose and R. J. Nelson, 1962, A sorting problem, *J. Assoc. Comp. Mach.* **9,** 282–296.

R. W. Floyd, 1964, A minute improvement in the Bose-Nelson sorting procedure, Memorandum, Computer Associates, Inc., Wakefield, Massachusetts, February 1964.

R. W. Floyd and D. E. Knuth, 1967, Improved constructions for the Bose-Nelson sorting problem, *Notices Am. Math. Soc.* **14,** 283.

T. N. Hibbard, 1963, A simple sorting algorithm, *J. Assoc. Comp. Mach.* **10,** 142–150.

D. E. Knuth, 1972, Sorting and searching, *The Art of Computer Programming* 3 (Addison-Wesley, Reading, Mass.).

D. G. O'Connor and R. J. Nelson, 1962, Sorting system with *n*-line sorting switch, U.S. Patent 3029413, issued April 10, 1962.

D. C. van Voorhis, 1971, An improved lower bound for the Bose-Nelson sorting problem, Technical Note No. 7, Digital Systems Laboratory, Stanford Univ., Stanford, Calif.

J. N. Srivastava et al., eds., *A Survey of Combinatorial Theory*
© North-Holland Publishing Company, 1973

CHAPTER 16

Nombres d'Euler et Permutations Alternantes

D. FOATA†
University of Florida, Gainesville, Fla., U.S.A.
et
M.-P. SCHÜTZENBERGER
Université de Paris VII et I.R.I.A., France

1. Introduction

Les entiers apparaissant dans le développement de Taylor des fonctions élémentaires de l'analyse classique sont souvent susceptibles d'interprétations combinatoires, qui donnent une signification géométrique à certaines de leurs propriétés. Réciproquement, de nombreux problèmes d'énumération d'objets rudimentaires conduisent à des fonctions génératrices remarquables, et il parait utile d'explorer systématiquement ces liaisons.

Le présent travail est la première partie d'une étude sur les nombres D_n ($n \in \mathbf{N}$) définis par le développement

$$\sum_{1 \leqslant n} D_n u^n / n! = D(u)$$

de la fonction

$$D(u) = \int_0^u (\operatorname{tg} u + (\cos u)^{-1}) \, du.$$

Pour $n = 2m$ pair, on retrouve donc les *nombres tangents* D_{2m}

$$D_{2m} = 2^{2m-1}(2^{2m}-1)m^{-1}B_m,$$

où

$$B_m = 2\zeta(2m)(2\pi)^{-2m}(2m)!$$

est le $2m$-ième *nombre de Bernoulli*. Pour $n = 2m+1$, les nombres D_{2m+1} sont les *nombres sécants*, dits aussi *nombres d'Euler*. Ces nombres ont fait l'objet de très nombreuses études arithmétiques dont un exposé systématique d'ensemble a été donné par Nielsen [1923] dans son *Traité élémentaire des nombres de Bernoulli*. D'autre part, la table la plus complète des premières valeurs de ces nombres apparaît dans Buckholtz et Knuth [1967].

Les relations

$$\exp D(u) = D''(u) = (\tfrac{1}{2})(1 + D'^2(u)), \tag{1.1}$$

$$D(0) = 0 \tag{1.2}$$

entraînent

$$D'(u) = (1 + tg \tfrac{1}{2}u)(1 - tg \tfrac{1}{2}u)^{-1} \tag{1.3}$$

† On leave from the Université de Strasbourg, 1971–72.

et les identités

$$\exp D_{(2)} = D_{(1)'}, \tag{1.4}$$

$$D_{n+3} = \sum_{0 \leqslant i \leqslant n} [^n_i] D_{i+1} D_{n+2-i}, \tag{1.5}$$

$$2D_{n+2} = \sum_{0 \leqslant i \leqslant n} [^n_i] D_{i+1} D_{n-i+1} + \delta_{n,0}, \tag{1.6}$$

$$D_{2n+1} = \sum_{0 \leqslant i \leqslant n-1} [^{2n-1}_{2i}] D_{2i+1} D_{2n-2i}, \tag{1.7}$$

où, dans la première, l'on a posé

$$D_{(2)} = \sum_{1 \leqslant n} (u^{2n}/(2n)!) D_{2n} \left(= \int_0^u \operatorname{tg} u \, du \right)$$

et

$$D_{(1)} = \sum_{0 \leqslant n} (u^{2n+1}/(2n+1)!) D_{2n+1} = D - D_{(2)}.$$

A leur tour, ces identités fournissent les congruences élémentaires suivantes valables pour tout nombre premier impair p

$$D_{p+3} \equiv D_{p+2} + D_{p+1}; \tag{1.8}$$

$$D_{p+2} \equiv D_{p+1} \equiv D_p + 1. \tag{1.9}$$

D'un point de vue combinatoire, André [1879, 1881] a montré que D_{n+1} est le nombre des *permutations alternantes* sur [n], c'est-à-dire des permutations $x_1 x_2 \ldots x_n$ des éléments de [n] $= \{1, 2, \ldots, n\}$ telles que x_{2j} soit à la fois inférieur à x_{2j-1} et à x_{2j+1} pour tout entier j tel que $0 < 2j < n$ et, en plus, si n est pair, telles que $x_n < x_{n-1}$.

De façon indépendante, Kermack et McKendrick [1938] ont étudié une distribution qui équivaut à celle des "pics" et des "creux" sur les permutations du groupe symétrique sur [n]: étant donnée une telle bijection $f : [n] \to [n]$, un pic (resp. creux) est une valeur $j \in [n]$ telle que jf soit plus *grande* (resp. *petite*) que les deux valeurs adjacentes $(j-1)f$ et $(j+1)f$. Sous cette forme, les calculs de Kermack et McKendrick ont été repris par David et Barton [1962] dans leur ouvrage *Combinatorial Chance*. Les fonctions génératrices associées sont des polynômes qui se rencontrent aussi dans la représentation des polynômes eulériens comme sommes de monômes $t^p(1+t)^m$ à coefficients entiers non négatifs (cf. Foata et Schützenberger [1970]). Une transformation simple les ramène à des polynômes à deux variables $D_n(s, t)$ prenant la valeur D_n pour $s = t = 1$. Ce sont ces derniers que nous appellerons *polynômes d'André* et dont nous nous proposons ici d'aborder l'étude.

Dans la section 2 suivante nous établissons les principales formules concernant les polynômes d'André. En particulier, une formule explicite pour leur fonction génératrice est donnée. Cette section est de nature purement analytique.

Ainsi qu'il se produit souvent dans ce domaine, une meilleure compréhension de objets est atteinte en opérant dans une algèbre non commutative.

Nous introduisons donc dans la section 3 les polynômes d'André *non commutatifs*, qui sont eux susceptibles de plusieurs interprétations. La place nous a manqué pour donner toutes celles qu'exigerait une vérification géométrique des identités données à la deuxième section. Nous nous sommes donc bornés à discuter ce que nous appelons les "permutations d'André", grâce auxquelles diverses identités binomiales au sens de Mullin et Rota [1970] s'expliquent en termes de variation. Un article ultérieur traitera des "complexes d'André" qui permettent de retrouver certaines propriétés fondamentales de symétrie.

Il est clair que la plupart de nos énoncés pourraient aussi bien être présentés dans le langage statistique qui fut celui d'une grande partie de notre carrière. Notre choix d'une formulation moins spéciale est un hommage à notre Maître Bose dont l'oeuvre a tant illustré les enrichissements mutuels de la mathématique et de ses applications.

Nous remercions notre ami John Riordan de nous avoir signalé l'article de Kermack et McKendrick [1938] et d'avoir bien voulu relire et commenter une première version de cet article.

2. Les polynômes d'André

2.1. Définition et propriétés élémentaires

Soit D une fonction réelle de la variable u, analytique à l'origine et satisfaisant l'équation différentielle

$$D'' = t \exp D, \tag{2.1}$$

avec les conditions initiales

$$0 = D(0), \qquad s = D'(0), \tag{2.2}$$

où s et t sont des constantes. En raison de $0 = D(0)$, la relation (2.1) est équivalente à

$$D''' = D'D''. \tag{2.3}$$

Nous posons

$$D = \sum_{0 \leqslant n} (u^n/n!)D_n,$$

où, d'après (2.1) et (2.2), $D_0 = 0$, $D_1 = s$, $D_2 = t$. Considérant s et t comme des paramètres, les relations (2.1) et (2.3) déterminent de façon univoque par récurrence sur n les D_n comme polynômes en s et t. Ce sont eux que nous appellerons *polynômes d'André* et que nous désignerons dans cette section par $D_n = D_n(s, t)$ ($n \geqslant 0$). La liste des premiers d'entre eux est la suivante:

$$D_0 = 0, \qquad D_1 = s, \qquad D_2 = t, \qquad D_3 = st,$$
$$D_4 = s^2t + t^2, \qquad D_5 = s^3t + 4st^2,$$
$$D_6 = s^4t + 11s^2t^2 + 4t^3, \qquad D_7 = s^5t + 26s^3t^2 + 34st^3.$$

Les valeurs $D_n(1, 1)$ sont entières et sont bien les coefficients de la fonction

$D(u)$ présentée dans l'introduction puisque celle-ci était définie par l'équation différentielle
$$D'' = \exp D$$
avec les valeurs initiales
$$D(0) = 0, \qquad D'(0) = 1(= s)$$
et que l'on avait donc
$$D''(0) = \exp 0 = 1(= t).$$

Soit maintenant l'opérateur
$$\Delta = st\frac{\partial}{\partial t} + t\frac{\partial}{\partial s}.$$
On a
$$\Delta D_1 = t = D_2 \qquad \text{et} \qquad \Delta D_2 = st = D_3.$$
Observant que (2.3) équivaut à l'identité binomiale
$$D_{n+3} = \sum_{0 \leqslant j \leqslant n} \left[{}^n_j\right] D_{j+1} D_{n+2-j} \qquad (n \geqslant 0), \tag{2.4}$$
on en conclut que
$$D_{n+1} = \Delta D_n \qquad (n \geqslant 1), \tag{2.5}$$
soit encore, en tenant compte de la valeur initiale $D_1 = s$,
$$s + \Delta D = \frac{\partial}{\partial u} D. \tag{2.6}$$

Ces relations montrent que les polynômes d'André ont les propriétés élémentaires suivantes :

Propriété 2.1. *Les polynômes D_n sont homogènes de degré total n en les variables s et $\sqrt{t}$. Ils sont divisibles par t pour $n \geqslant 2$ et leurs coefficients sont des entiers positifs.*

On a donc pour chaque $n \geqslant 2$,
$$D_n = \sum_{1 \leqslant k \leqslant n/2} s^{n-2k} t_k \, d_{n,k},$$
où les coefficients $d_{n,k}$ sont des entiers naturels et la relation (2.5) livre immédiatement la propriété suivante.

Propriété 2.2. *Pour $n \geqslant 2$, on a $d_{n,1} = 1$, et pour $2 \leqslant k \leqslant \frac{1}{2}n$, on a*
$$d_{n+1,k} = k \, d_{n,k} + (n+2-2k) \, d_{n,k-1}. \tag{2.7}$$
Par conséquent, tous les $d_{n,k}$ $(1 \leqslant k \leqslant \frac{1}{2}n)$ sont positifs. Posant maintenant
$$\bar{D}_n = \sum_{1 \leqslant k \leqslant n/2} \bar{t}^k \, d_{n,k},$$
la même relation (2.5) donne la formule de récurrence
$$\bar{D}_{n+1} = n\bar{t}\bar{D}_n + (\bar{t} - 2\bar{t}^2)\frac{\partial}{\partial \bar{t}} \bar{D}_n \qquad (n \geqslant 2) \tag{2.8a}$$
qui est équivalente à
$$\bar{\bar{D}}_{n+1} = \bar{t}(1 - 2\bar{t})^{\frac{1}{2}} \frac{\partial}{\partial \bar{t}} \bar{\bar{D}}_n, \tag{2.8b}$$
où
$$\bar{\bar{D}}_n = (1 - 2\bar{t})^{-n/2} \, \bar{D}_n.$$

On notera qu'en raison de (2.7) le polynôme D_n est divisible par s ssi n est impair.

2.2. Une relation différentielle

Nous établissons maintenant la généralisation naturelle de la deuxième égalité dans la relation (1.1).

Propriété 2.3. *On a l'identité*

$$2D'' = 2t - s^2 + D'^2. \tag{2.9}$$

Preuve. D'après (2.4) et (2.5), on a pour tout $n \in \mathbf{N}$

$$D_{n+3} = \Delta D_{n+2} = \sum_{0 \le j \le n} \left[{}^n_j\right](\Delta D_j)D_{n+2-j}.$$

Tenant compte de la symétrie des indices et de $\left[{}^n_j\right] = \left[{}_{n-j}^{\ n}\right]$, ceci donne

$$2\Delta D_{n+2} = \sum_{0 \le j \le n} \left[{}^n_j\right]\Delta(D_{j+1}D_{n+1-j}),$$

d'où

$$2D_{n+2} = \sum_{0 \le j \le n} \left[{}_j\right]D_{j+1}D_{n+1-j} + K_n, \tag{2.10}$$

où K_n est une fonction de s et t telle que $\Delta K_n = 0$. Comme les D_j sont des polynômes, K_n est un polynôme. D'autre part, comme

$$\Delta(t^p s^q) = pt^p s^{q+1} + qt^{p+1}s^{q-1} \qquad (p, q \in \mathbf{N}),$$

on voit que le terme de plus bas degré de K_n en t ne peut s'annuler que si ce degré est zéro. Comme, d'après la propriété 2.1, on a $D_{n+2}(s, 0) = 0$ pour tout $n \ge 0$, le polynôme K_n est nul pour $n \ge 1$. Enfin, on vérifie directement que $K_0 = 2t - s^2$. Ceci fait, la formule (2.9) s'obtient par sommation.

On notera que pour $n + 2 = 2m + 1$ impair, l'expression (2.10), avec $K_n = 0$, est symétrique et peut par conséquent s'écrire sous la forme (1.7) de l'introduction, soit de façon équivalente

$$D_{2m+1}/(2m-1)! = \sum_{0 \le j \le m-1} (D_{2j+1}/(2j)!)(D_{m-2j}/(2m-2j-1)!);$$

c'est-à-dire

$$D_{(1)''} = D_{(1)'}D_{(2)'}, \tag{2.11}$$

avec les notations déjà introduites dans le cas particulier de $s = t = 1$,

$$D_{(2)} = \sum_{0 \le m} (u^{2m}/(2m)!)D_{2m}(s, t),$$

$$D_{(1)} = D - D_{(2)}.$$

Nous en déduisons la formule suivante qui est la contre-partie polynomiale de (1.4).

Propriété 2.4. *On a*

$$D_{(1)'} = s \exp D_{(2)}. \tag{2.12}$$

7

Preuve. La formule (2.11) peut s'écrire

$$\frac{\partial}{\partial u} \log D_{(1)'} = \frac{\partial}{\partial u} D_{(2)'}.$$

D'où

$$D_{(1)'} = K(s, t) \exp D_{(2)},$$

où $K(s, t)$ est une fonction de s et t qui est déterminée en faisant $u = 0$ et en constatant que $D_2(u = 0) = 0$ et $D_{(1)'}(u = 0) = s$.

2.3. Fonction génératrice des polynômes d'André

Nous donnons maintenant des formules explicites pour D, D' et D''.

Propriété 2.5. *Posant* $r = (s^2 - 2t)^{\frac{1}{2}}$, $w = (s-r)/(s+r)$ *et* $E = \exp ru$, *on a les formules*

$$D = ru + 2 \log ((1-w)/(1-wE)); \tag{2.13}$$

$$D' = r(1+wE)/(1-wE); \tag{2.14}$$

$$D'' = wr^2 E/(1-wE)^2. \tag{2.15}$$

Preuve. L'équation (2.9) peut s'écrire

$$r = D''((D'-r)^{-1} - (D'+r)^{-1}),$$

d'où par intégration

$$ru = \log ((D'-r)/(D'+r)) + K(s, t),$$

où la fonction $K(s, t)$ est déterminée en faisant $u = 0$ et se trouve par conséquent égale à $-\log w$. Donc $(D'-r)/(D'+r) = w \exp ru$, ce qui est équivalent à (2.14). Maintenant le membre de droite de cette dernière équation peut s'écrire sous la forme

$$r(1 + (2w \exp ru)/(1 - w \exp ru)),$$

d'où par une nouvelle intégration

$$D = ru - 2 \log (1 - w \exp ru) + K(s, t).$$

Faisant de nouveau $u = 0$, on trouve

$$K(s, t) = 2 \log (1-w).$$

On obtient ainsi la formule (2.13). Enfin, la formule (2.15) s'obtient par simple dérivation.

Désignons par $D_{(0)'}$ la valeur du membre de droite de (2.14) pour $s = 0$. Posant $v = (2t)^{\frac{1}{2}} u$ et observant que $w = -1$ pour $s = 0$, on trouve

$$D_{(0)'} = (2t)^{\frac{1}{2}} \cdot i(1 - \exp iv)/(1 + \exp iv),$$

soit

$$D_{(0)'} = (2t)^{\frac{1}{2}} \operatorname{tg} \tfrac{1}{2} v. \tag{2.16}$$

Ce résultat a la conséquence très remarquable suivante:

Propriété 2.6. *Pour tout* k *positif, les coefficients* $d_{2k-1,k-1}$ *et* $d_{2k,k}$ *sont égaux a* $2^{1-k}[D_{2k}]_{s=t=1}$, *c'est-a-dire a* 2^{1-k} *fois le* k-*ème nombre d'Euler.*

Preuve. Prenant $n = 2k-1$, la récurrence (2.8) donne

$$d_{2k,k} = k\, d_{n,k} + (2k-1+2-2k)\, d_{n,k-1} = d_{2k-1,k-1}$$

puisque $d_{n,k} = 0$ en vertu de $n < 2k$. Les deux coefficients d mentionnés dans l'énoncé sont donc égaux.

Maintenant pour vérifier leur égalité avec le nombre $2^{1-k}[D_{2k}]_{s=t=1}$, il suffit d'observer que pour $s = 0$, tous les polynômes D_{2k-1} sont nuls et chacun des polynômes D_{2k} se réduit à $d_{2k,k}t^k$. Par conséquent,

$$D_{(0)'} = \sum_{1 \leqslant k} (u^{2k-1}/(2k-1)!)\, d_{2k,k}t^k.$$

On peut alors appliquer la formule (2.16) qui s'écrit

$$D_{(0)'} = (2t)^{\frac{1}{2}} \sum_{1 \leqslant k} (u^{2k-1}/(2k-1)!)(\tfrac{1}{2}t)^{(2k-1)/2} D_{2k},$$

soit

$$D_{(0)'} = \sum_{1 \leqslant k} (u^{2k-1}/(2k-1)!)2^{1-k} D_{2k}t^k.$$

Nous donnons enfin la formule binomiale

$$2d_{2n+2,n+1} = \sum_{0 \leqslant i \leqslant n-1} \left[\begin{smallmatrix} 2n \\ 2i+1 \end{smallmatrix}\right] d_{2i+2,i+1} d_{2n-2i,n-i} \qquad (n \geqslant 1), \quad (2.17)$$

qui se déduit immédiatement de la formule (2.9) lorsqu'on y fait $s = 0$ et $t = 1$, grâce à la propriété 2.6.

2.4. Relations avec les polynômes eulériens

Nous terminons cette section en établissant une relation entre les polynômes d'André et les polynômes eulériens. Pour la définition de ces derniers, nous renvoyons le lecteur à l'ouvrage de Riordan [1958], pp. 213–216, ou à notre précédent mémoire (Foata et Schützenberger [1970]).

Propriété 2.7. *Pour tout entier $n > 0$, le n-ème polynôme eulérien $A_n(x)$ est égal à*

$$\sum_{1 \leqslant k \leqslant (n+1)/2} d_{n+1,k}(2x)^{k-1}(1+x)^{n+1-2k},$$

où les $d_{n+1,k}$ sont les coefficients du $(n+1)$-ème polynôme d'André.

Preuve. Faisons la substitution

$$s = 1, \qquad t = 2x/(1+x)^2, \qquad u = (1+x)v$$

dans l'expression de $(D'-s)/t$ donnée par (2.14). Notant que la substitution envoie r sur $(1-x)/(1+x)$ et w sur x, on trouve

$$\frac{(1+x)\{\exp[(1-x)v]-1\}}{1-x\exp[(1-x)v]}.$$

Divisant par $1+x$, et ajoutant 1, on obtient

$$\frac{(1-x)\exp[(1-x)v]}{1-x\exp[(1-x)v]},$$

qui est l'expression classique de la fonction génératrice exponentielle des polynômes eulériens. Donc, pour $n > 0$, $A_n(x)$ est le polynôme obtenu en faisant la substitution $s = 1$, $t = 2x/(1+x)^2$ dans $(1+x)^{n-1}t^{-1}D_{n+1}$, ce qui est précisément le résultat annoncé.

On pourra noter que la relation de symétrie $x^n A_n(x^{-1}) = A_n(x)$ correspond à l'invariance $t = 2x^{-1}/(1+x^{-1})^2$.

3. Les permutations d'André

3.1. Quelques notions générales

Nous commençons par décrire en détail quelques notions de base.

Soit X un ensemble totalement ordonné ayant un nombre fini n d'éléments. Une *permutation* de X est une bijection $f : [n] \to X$ où $[n]$ désigne l'ensemble ordonné $\{1, 2, \ldots, n\}$ ($= \emptyset$ si $n = 0$). Nous l'identifierons au *mot* $1f. 2f. \ldots$ nf en les lettres de X. Puisque f est une bijection, chaque élément de X figurera exactement une fois dans ce mot. Pour abréger, nous écrirons $f \in X^1$ pour indiquer que f est une permutation de X ou son mot associé.

Soient maintenant $n \geqslant 2$ et $f \in X^1$; la *variation* de f est le mot $fV = v_1 v_2 \ldots v_{n-1}$ de longueur $n-1$ en les symboles $v_j = (+)$ et $(-)$ qui est défini pour chaque $j \leqslant n-1$ par

$$v_j = + \text{ si } jf < (j+1)f,$$
$$= - \text{ si } jf > (j+1)f.$$

Il est classique de dire que $[j, j+1]$ est une *montée* (resp. *descente*) ssi $v_j = +$ (resp. $= -$).

Soit maintenant $1 < j < n$:

$-[j-1, j+1]$ est une *double descente* ssi $[j-1, j]$ et $[j, j+1]$ sont deux descentes;

$-j$ est un *creux* ssi $[j-1, j]$ est une descente et $[j, j+1]$ une montée.

De façon analogue, la *variation circulaire* $f\mathring{V}$ est le mot de longueur n défini par $f\mathring{V} = fV . v_n$, où $v_n = +$ ou $-$ selon que $nf < 1f$ ou $nf > 1f$; autrement dit, v_n est défini pour $[n, 1]$ de la même manière que v_j était défini pour $[j, j+1]$.

D'une manière générale, une notion sera dite *circulaire* ssi dans sa définition il est convenu que "$n+1$" signifie "1". Par exemple pour $X = [8]$ et $f : [8] \to X$ identifié à $5\ 8\ 1\ 6\ 9\ 2\ 3\ 4\ 7$ on a $fV = +-++-+++$ ($\in \{+, -\}^8$), 1 et 2 sont les deux creux et f n'a pas de double descente. Comme $7 > 5$ on a $v_n = -$ et $f\mathring{V} = +-++-+++-$ ($\in \{+, -\}^9$); enfin comme $7 > 5$, mais $5 < 8$, la permutation f est sans *double descente circulaire*, donc aussi sans double descente.

Nous introduisons maintenant une notion plus spéciale et nous définissons la *variation réduite* de f comme le mot fU de longueur $\leqslant n-1$ en les symboles t et s qui est obtenu à partir de la variation fV en remplaçant d'abord toutes les paires $v_i v_{i+1}$ telles que $v_i = -$, $v_{i+1} = +$ par t, ensuite en remplaçant par

s les v_i restants. Par construction, $fU = s$ ssi $n = 2$. Dans notre exemple $fU = ststs$ puisque $fV = +(-+)+(-+)++$.

Rappelons la notation standard $|f|_x$ pour désigner le nombre d'occurrences d'une lettre x dans un mot f.

Propriété 3.1. *Le nombre des creux de f est $|fU|_t$, celui des montées est $\leqslant |fU|_t + |fU|_s$ avec égalité ssi f est sans double descente et se termine par une montée (c'est-à-dire $v_{n-1} = +$).*

La preuve est immédiate.

On définit de la même manière la *variation réduite circulaire* $f\mathring{U}$ en convenant d'écrire la lettre t à la fin du mot fU quand n est un creux circulaire (c'est-à-dire quand $v_{n-1} = -$ et $v_n = +$) et au début quand 1 est un *creux circulaire* (c'est-à-dire quand $v_n = -$ et $V_1 = +$). C'est ce second cas qui se produit dans notre exemple et l'on a donc

$$f\mathring{U} = t\,t\,s\,t\,s\,s,$$

puisque $f\mathring{V} = +)\,(-+)+(-+)++(-$.

On notera que si $n = 2$, $f\mathring{U}$ est toujours t.

On *conviendra* pour $n = 1$, $f\mathring{U} = s$ et $fU = e$ (c'est-à-dire le mot vide du monoide libre $\{s, t\}^*$).

3.2. Définition des permutations d'André

Nous appellerons *permutation d'André* sur X ($0 \leqslant \text{card } (X) = n < \infty$) toute permutation $f: [n] \to X$ *sans double descente* satisfaisant la condition caractéristique suivante:

(A) Soient $j, j' \in [n]$ tels que $1 < j < j'$ et
$$(j-1)f = \max \{(j-1)f, jf, (j'-1)f, j'f\},$$
$$j'f = \min \{(j-1)f, jf, (j'-1)f, j'f\}.$$
Il existe un j'' tel que $j < j'' < j'$ et que $j''f < j'f$.

De façon intuitive, en tenant compte de ce que f n'a pas de double descente, la condition peut être reformulée ainsi.

Si j et $j' > j$ sont deux creux tels que $jf > j'f$ et $(j-1)f > (j'-1)f$, il existe un creux j'' entre j et j' ($j < j'' < j'$) tel que $j''f < j'f$ et la même condition vaut quand $j' = n$ et que $[j'-1, j']$ est une descente.

Il résulte immédiatement de la définition que toute permutation ayant 0 ou 1 descente est une permutation d'André, car elle n'a pas de double descente et la deuxième condition est trivialement vérifiée.

Une permutation f ayant exactement deux descentes $[j, j+1]$ et $[j', j'+1]$ ($j < j'$) est une permutation d'André ssi les deux conditions suivantes sont réalisées

(i) $j+1$ et $j'+1$ sont des creux ou bien $j+1$ est un creux et $[j', j'+1]$ est une descente finale;

(ii) l'on a $jf < j'f$ ou bien $jf > j'f$ et $(j-1)f < (j'-1)f$.

Pour avoir une idée concrète de cette condition, le lecteur pourra vérifier que parmi les six permutations de $[6]$ qui sont de la forme $x\,2\,y\,3\,z\,1$ ($\{x, y, z\} = \{4, 5, 6\}$) et qui sont donc sans double descente puisqu'alternées, les permutations d'André sont les deux pour lesquelles $z = 6$.

En effet, puisque $2 = 2f < 3 = 4f$, la condition caractéristique ne s'applique qu'aux paires de creux $j = 2$ ou 4 et $j' = 6 > j$. Comme $jf = 2$ ou $4 > j'f = 1$ et comme il n'existe aucun creux j'' entre j et j' tel que $j''f < jf$ (puisque $4f = 3 > 6f = 1$), on doit avoir $(j-1)f < (j'-1)f$, c'est-à-dire $x < z$ et $y < z$.

Nous noterons D_n^* $(0 \leqslant n)$ l'ensemble des permutations d'André sur $[n]$ et $D^* = \bigcup_{0 \leqslant n} D_n^*$, en faisant comme d'usage la convention naturelle que pour $n = 0$, D_0^* est un singleton. Voici une table des D_n^* pour $n = 0, 1, 2, 3, 4$:

$$D_0^* = \{e\}, \qquad D_1^* = \{1\}, \qquad D_2^* = \{12, 21\},$$
$$D_3^* = \{123, 132, 213, 231, 312\},$$
$$D_4^* = \{1234, 1243, 1324, 1342, 1423,$$
$$2134, 2143, 2314, 2341, 2413,$$
$$3124, 3142, 3241, 3412,$$
$$4123, 4132\}.$$

On notera que $1 = \text{Card } D_0^* = \text{Card } D_1^*$; $2 = \text{Card } D_2^*$; $5 = \text{Card } D_3^*$; $16 = \text{Card } D_4^*$.

Par abus de notation, si $I = \{n'+1, \ldots, n'+m\}$ est un intervalle de $[n]$ et $f: [n] \to X$ une permutation, nous identifierons la restriction $f|I$ à la permutation $f': [m] \to If$ $(If \subset X)$ telle que $jf' = (n'+j)f$ identiquement.

Lemme 3.2. *Soit $f: [n] \to X$ une permutation d'André. Pour tout intervalle I de $[n]$, la restriction $f' = f | I$ de f à I est une permutation d'André.*

Preuve. Ceci découle de la structure des conditions "être sans double descente" et (A) qui ne font intervenir que les éléments d'un intervalle.

Nous introduisons maintenant deux familles spéciales de permutations d'André que nous appellerons respectivement (par abus de langage) *circulaires* et *augmentées*. Soit X un ensemble fini de cardinal n $(n \geqslant 0)$; une permutation d'André f sur X est dite *circulaire* (resp. *augmentée*) ssi son dernier élément nf est égal à min f (resp. max X). On note D (resp. A) l'ensemble des permutations d'André appartenant à D qui sont circulaires (resp. augmentées); on pose $D_n = D \cap D_n^*$ et $A_n = A \cap D_n^*$ $(n > 0)$ et l'on convient que D_0 est vide et que $A_0 = D_0^* = \{e\}$. On voit sur la liste ci-dessus que Card $D_j = \text{Card } A_j = 1$ pour $j = 1, 2$; Card $D_3 = 1$; Card $A_3 = 2$; Card $D_4 = 2$; Card $A_4 = 5$.

Propriété 3.3. *Soient $n \in \mathbb{N}$ et $f: [n+2] \to X$ une permutation quelconque telle que*

(i) $$(n+2)f = \min X.$$

Les trois conditions suivantes sont équivalentes:

(1) *La permutation f est une permutation d'André (qui est nécessairement circulaire).*

(2) *La restriction* $f' = f \,|\, [n+1]$ *est une permutation d'André augmentée.*

(3) *La restriction* $f'' = f \,|\, [n] = f' \,|\, [n]$ *est une permutation d'André et*

(ii) $$ j \in [n] \Rightarrow jf'' < (n+1)f'. $$

Preuve. Le lemme 3.2 donne immédiatement les implications

$$ f \in D^* \Rightarrow f' \in D^* \Rightarrow f'' \in D^*. $$

Supposons (1) et prenons $j' = n+2$. D'après (i), d'une part $[j'-1,\ j']$ est une descente, d'autre part on ne peut pas avoir $j''f < j'f$ pour $j'' < j'$. Donc d'après (A) on aura $(j-1)f < (j'-1)f$ pour tout $j < j'$ tel que $[(j-1),\ j]$ soit une descente.

Considérons $\bar{j}$ tel que $(\bar{j}-1)f = \max X$; le couple $[\bar{j}-1, \bar{j}]$ est une descente et par conséquent $\bar{j} = j'$, c'est-à-dire $(n+1)f = \max X$. La condition (1) implique donc (2).

Réciproquement supposons (3), c'est-à-dire que la restriction $f \,|\, [n]$ est une permutation d'André et que l'on a $(n+1)f = \max X$, $(n+2)f = \min X$. Il est clair que f n'a pas de double descente. D'autre part, prenant encore $j' = n+2$, la condition (A) est toujours satisfaite car il ne peut pas exister de creux $j < j'$ pour lequel $(j-1)f > (j'-1)f$. Donc (3) $\Rightarrow$ (1) et comme (2) $\Rightarrow$ (3) trivialement d'après $f' \in D^* \Rightarrow f'' \in D^*$, le résultat est établi.

Corollaire 3.4. *Pour tout* $n \geqslant 0$, *les ensembles* D_{n+2}, A_{n+1} *et* D_n^* *ont même cardinalité.*

3.3. Polynômes d'André en variables non commutatives

Pour simplifier, on appellera *polynômes d'André non commutatifs* les polynômes

$$ A_n U = \sum \{fU : f \in A_n\}, $$
$$ D_n \overset{\circ}{U} = \sum \{f\overset{\circ}{U} : f \in D_n\} \qquad (n \geqslant 0) $$

en les variables non commutatives s et t. Dans la propriété 3.10 ci-après, on trouvera deux relations de récurrence sur ces polynômes. Enfin, la liste des polynômes pour les premières valeurs de n est donnée à la fin de cette section.

Lemme 3.5. *Soit* $f: [n+1] \to X$ *une permutation d'André. Il existe exactement une valeur* $m \leqslant n$ *telle que*

(i) $f \,|\, [m] \in D$;

(ii) $m' \geqslant m,\ f \,|\, [m'] \in D \Rightarrow m' = m$.

Preuve. Il suffit de prendre $m = (\min X)f^{-1}$ et d'observer que $m = (\min ([m']f))f^{-1}$ pour tout $m' \geqslant m$.

On notera $f^{(1)}$ la restriction $f \mid [m]$ ($m = (\min X)f^{-1}$) et on appellera $f^{(1)}$ le *premier facteur* de f. La restriction $f \mid [n] \setminus [m]$ sera le *cofacteur de $f^{(1)}$* dans f et on utilisera souvent pour abréger la notation $f^{(1)-1}$ pour désigner $[m]$. L'importance de ce lemme est dans sa réciproque.

Propriété 3.6. *Une permutation $f \colon [n+1] \to X$ est une permutation d'André ssi posant $m = (\min X)f^{-1}$, les deux restrictions $f^{(1)} = f \mid [m]$ et $f' = f \mid [n] \setminus [m]$ sont des permutations d'André. Si ces hypothèses sont vérifiées et $n \geqslant 1$, f est augmentée si et seulement s'il en est de même de f'.*

Preuve. La partie directe résulte des lemmes 3.5 et 3.2. Supposons donc $f^{(1)}$, $f' \in D^*$ et sans perte de généralité $m < n$. Comme $mf = \min X$, $[m, m+1]$ est une montée. Donc f n'a pas de double descente puisque ni $f^{(1)}$ ni f' n'en ont.

Soit maintenant j et j' deux valeurs justiciables de la condition (A). Si $j, j' \in [m]$ ou $\in [n] \setminus [m]$, la condition (A) est satisfaite par f d'après l'hypothèse $f^{(1)}$, $f' \in D^*$. Si au contraire $j > m > j'$, la condition (A) est satisfaite par l'existence du creux $j'' = m$ entre j et j'.

Lemme 3.7. *Soit $f \colon [n+1] \to X$ une permutation d'André circulaire. Si $n = 0$, $f\mathring{U} = s$ et si $n > 0$, $f\mathring{U} = (f'U)t$ où $f' = f \mid [n]$. Par conséquent,*
$$D_{n+1}\mathring{U} = (A_n U)t \qquad \text{pour } n > 0.$$

Preuve. Le cas de $n = 0$ résulte de la définition même de $\mathring{U}$. Si $n \geqslant 1$, la variation de f se termine par une descente puisque $nf = \max X$, $(n+1)f = \min X$. Comme $(n+1)f < 1f$, la formule est encore une conséquence de la définition de $\mathring{U}$.

Lemme 3.8. *Soit $f \colon [n+3] \to X$ une permutation d'André circulaire. On a*
$$f\mathring{U} = g^{(1)}\mathring{U} . \bar{f}\mathring{U},$$
où $g^{(1)}$ est le premier facteur de $g = f \mid [n+1]$ et $\bar{f}$ le cofacteur de $g^{(1)}$ dans f.

Preuve. Le facteur $g^{(1)}$ est la restriction de f à $[m']$ où $m'f$ est le minimum de X privé de $\min X = (n+3)f$ et de $\max X = (n+2)f$. Donc $[m', m'+1]$ est toujours une montée de f.

Distinguons maintenant deux cas :

(i) $m' = 1$. On a $fV = +\bar{f}V$. Comme $\bar{f}\mathring{U}$ se termine par t puisque $n+3-m' \geqslant 2$, on a donc $f\mathring{U} = s . \bar{f}\mathring{U}$ et le résultat est établi.

(ii) $m' > 1$. Comme $g^{(1)} \in D$, $g^{(1)}$ se termine par la descente $[m'-1, m']$. Donc $fU = (g^{(1)}U)' t (\bar{f}U)$, où $(g^{(1)}U)'$ désigne le mot obtenu en supprimant le dernier s de $g^{(1)}U$. De façon équivalente, $fU = g^{(1)}\mathring{U} . \bar{f}U$, d'où encore $f\mathring{U} = g^{(1)}\mathring{U} . \bar{f}\mathring{U}$.

Corollaire 3.9. *Soit $f \colon [n+2] \to X$ une permutation d'André augmentée. On a*
$$fU = f^{(1)}\mathring{U} . f'U$$
où $f^{(1)}$ est le premier facteur de f et f' son cofacteur.

Preuve. Définissons la permutation $g : [n+3] \to X'$ par $g \mid [n+2] = f$ et $(n+3)g = \min X'$. Il est clair que g est une permutation d'André circulaire. Soient $g^{(1)}$ le premier facteur de $g \mid [n+1]$ et $\bar{g}$ le cofacteur de $g^{(1)}$ dans g. On a $g\mathring{U} = (fU)t$ (d'après le lemme 3.7), $f^{(1)} = g^{(1)}$, et enfin $\bar{g}\mathring{U} = (f'U)t$. Le lemme précédent donne d'autre part l'identité

$$g\mathring{U} = g^{(1)}\mathring{U} \cdot \bar{g}\mathring{U},$$

c'est-à-dire

$$(fU)t = f^{(1)}\mathring{U} \cdot (f'U)t.$$

Le corollaire est donc établi en supprimant la dernière lettre t de l'identité précédente.

Propriété 3.10. *Pour tout $n \geqslant 0$, on a les identités*

$$A_{n+2}U = \sum \begin{bmatrix} n \\ j \end{bmatrix} D_{j+1}\mathring{U} \cdot A_{n+1-j}U, \tag{3.1}$$

$$D_{n+3}\mathring{U} = \sum \begin{bmatrix} n \\ j \end{bmatrix} D_{j+1}\mathring{U} \cdot D_{n+2-j}\mathring{U}. \tag{3.2}$$

Preuve. La propriété 3.6 donne une bijection entre A_{n+2} et les triplés $(X' \cup X'', f^{(1)}, f')$, où $X' \cup X''$ est une partition de $X \setminus \{\min X, \max X\}, f^{(1)}$ une permutation circulaire d'André sur $X' \cup \{\min X\}$ et f' une permutation augmentée sur $X'' \cup \{\max X\}$. La première formule découle alors du corollaire 3.9 et la deuxième de la première et du lemme 3.7.

Tables 3.11. Pour terminer ce chapitre, nous donnons la liste des polynômes $A_n U$ et $D_n \mathring{U}$ pour les premières valeurs de n. Ces polynômes peuvent être évidemment calculés à partir des formules de récurrence (3.1) et (3.2):

$$A_1 U = 1,$$
$$A_2 U = s,$$
$$A_3 U = s^2 + t,$$
$$A_4 U = s^3 + 2st + 2ts,$$
$$A_5 U = s^4 + 3s^2 t + 5sts + 3ts^2 + 4t^2,$$
$$A_6 U = s^5 + 4s^3 t + 9s^2 ts + 9sts^2 + 4ts^3 + 12st^2 + 10tst + 12t^2 s;$$
$$D_1 \mathring{U} = s,$$
$$D_{n+1}\mathring{U} = (A_n U)t \text{ pour } n > 0.$$

4. Remarques

Comme il n'a pas été possible d'inclure dans le même article tous les résultats sur les polynômes d'André en variables non commutatives, nous renvoyons le lecteur à un prochain mémoire. Quelques ultimes remarques nous semblent cependant nécessaires.

Remarque 4.1. On a $D_1 \mathring{U} = s$ et $D_2 \mathring{U} = t$. D'autre part, la formule de récurrence (3.2) a la même structure formelle que la relation binomiale sur les polynômes *commutatifs* D_n qui s'écrivait en effet (voir formule (2.4))

$$D_{n+3} = \sum \begin{bmatrix} n \\ j \end{bmatrix} D_{j+1} D_{n+2-j} \qquad (n \geqslant 0). \tag{4.1}$$

Ceci montre que les polynômes $D_n\check{U}$ constituent bien une *version non commutative* des polynômes d'André $D_n(s, t)$.

Remarque 4.2. Lorsque les variables s et t commutent, on a aussi la *formule exponentielle*

$$\sum_{0 \leqslant n} (u^n/n!)D_{n+2} = t \exp\left[\sum_{0 \leqslant n} (u^n/n!)\, D_n\right]$$

(voir formule (2.1)). En fait, les formules (4.1) et (4.2) sont *équivalentes*. On peut s'en convaincre par l'argument suivant: La série formelle égale à t fois l'exponentielle de $\sum_{0 < n} (u^n/n!)D_n$ est unique. Ceci résulte du fait que l'exponentielle est une bijection de l'ensemble des séries formelles sans terme constant sur l'ensemble des séries formelles de terme constant égal à 1. Or par dérivation de (4.2) par rapport à u, et identification des termes de même puissance en u, on obtient justement les formules (4.1).

Cette équivalence n'est *plus* valable lorsqu'on suppose s et t non commutatifs. Plus exactement, on n'a *pas* de formule exponentielle ayant même structure formelle que (4.2) avec les polynômes $D_n\check{U}$. Seule subsiste la formule (3.2), qui doit donc être regardée comme la *généralisation non commutative de la formule exponentielle*.

Remarque 4.3. Une autre façon d'établir directement la formule exponentielle (4.2) sans recourir aux arguments analytiques de la section 2 est de faire appel aux techniques purement combinatoires du *composé partitionnel*, développées dans notre précédent mémoire (Foata et Schützenberger [1970]). L'ensemble D^* est, en effet, le composé partitionnel de l'ensemble D des permutations d'André circulaires. Indiquons rapidement comment on peut le démontrer. Soit $f = 1f.\, 2f.\, \ldots.\, nf$ $(n > 0)$ une permutation d'André. Elle admet une factorisation unique $(g^{(1)}, g^{(2)}, \ldots, g^{(k)})$ telle que

   (1) le produit de juxtaposition $g^{(1)}g^{(2)} \ldots g^{(k)}$ soit égal à f;

   (2) chaque $g^{(j)}$ est une permutation d'André circulaire;

   (3) la suite formée par les dernières lettres des mots $g^{(j)}$ est croissante.

Par exemple, la factorisation de

$$f = 8\ 6\ 9\ 7\ 12\ 13\ 1\ 2\ 4\ 11\ 14\ 15\ 3\ 10\ 5$$

est donnée par

$$(8\ 6\ 9\ 7\ 12\ 13\ 1,\quad 2,\quad 4\ 11\ 14\ 15\ 3,\quad 10\ 5).$$

L'existence et l'unicité de cette factorisation peuvent être démontrées en utilisant le lemme 3.8. Supposant s et t commutatifs, on pose pour tout $f \in D_n^*$ $(n > 0)$

$$f\mu \,.\, t = (f.\,\overline{n+1}\,.\,0)\check{U}.$$

Là encore, à l'aide du lemme 3.8, on peut vérifier que μ est *multiplicative*. D'après la proposition 3.12 de la référence citée plus haut, on en déduit l'identité

$$1 + \sum_{0 < n} (u^n/n!)D_n^*\mu = \exp\left[\sum_{0 < n} (u^n/n!)A_n\mu\right].$$

L'identité (4.2) en résulte en observant que

$$D_n^* \mu \cdot t = D_{n+2}(s, t) \quad \text{et} \quad A_n \mu = D_n(s, t) \quad \text{pour} \quad n > 0.$$

Remarque 4.4. Nous avons vu dans la propriété 2.3 que l'identité

$$2D_{n+2} = \sum_{0 \leqslant j \leqslant n} \begin{bmatrix} n \\ j \end{bmatrix} D_{j+1} D_{n+1-j} \qquad (n \geqslant 1) \tag{4.3}$$

sur les polynômes *commutatifs* $D_n(s, t)$ se déduisait facilement de l'identité

$$D_{n+3} = \sum_{0 \leqslant j \leqslant n} \begin{bmatrix} n \\ j \end{bmatrix} D_{j+1} D_{n+2-j} \qquad (n \geqslant 0).$$

Dans le cas des polynômes d'André *non commutatifs*, on peut établir égale-
ment l'identité

$$2D_{n+3}\mathring{U} = s \cdot D_{n+2}\mathring{U} + \sum_{1 \leqslant j \leqslant n} \begin{bmatrix} n+1 \\ j \end{bmatrix} D_{j+1}\mathring{U} \cdot D_{n+2-j}\mathring{U} + D_{n+2}\mathring{U} \cdot t^{-1}st$$

$$(n \geqslant 0) \tag{4.4}$$

qui est l'équivalent non commutatif de (4.3). Comparant (3.2) et (4.4),
on voit que pour obtenir (4.4), il suffit d'établir les formules

$$D_{n+3}\mathring{U} = \sum_{1 \leqslant j \leqslant n} \begin{bmatrix} n \\ j-1 \end{bmatrix} D_{j+1}\mathring{U} \cdot D_{n+2-j}\mathring{U} + D_{n+2}\mathring{U} \cdot t^{-1}st \qquad (n \geqslant 0). \tag{4.5}$$

Soit $w = u_1 \ldots u_k$ un mot en les lettres s et t; le mot *retourné* $\tilde{w}$ est défini par
$\tilde{w} = u_k \ldots u_1$. Les formules (4.5) se déduisent alors de (3.2) et de la *propriété
de symétrie* suivante: pour tout mot w en s et t, il y a dans l'ensemble A_n des
permutations d'André augmentées autant d'éléments f tels que $fU = w$ que
d'éléments g tels que $gU = \tilde{w}$. Cette propriété remarquable sera démontrée
dans un article ultérieur. Nous y introduirons la notion abstraite de *complexe
d'André*, qui nous permettra, en outre, de construire une bijection naturelle
entre l'ensemble des permutations d'André et celui des permutations
alternantes.

Références

D. André, 1879, Développements de sec x et de tang x, *C.R. Acad. Sci. Paris* **88,** 965–967.

D. André, 1881, Sur les permutations alternées, *J. Math. Pures Appl.* **7,** 167–184.

T. J. Buckholtz et D. E. Knuth, 1967, Computation of tangent, Euler and Bernoulli
numbers, *Math. Comp.* **21,** 663–688.

F. N. David et D. E. Barton, 1962, *Combinatorial Chance* (Griffin, London).

D. Foata and M.-P. Schützenberger, 1970, *Théorie géométrique des polynômes euleriéns*
(Springer, Berlin).

W. O. Kermack et A. G. McKendrick, 1938, Some properties of points arranged on a
Möbius surface, *Math. Gaz.* **22,** 66–72.

R. Mullin et G.-C. Rota, 1970, On the foundations of combinatorial theory, III: Theory
of binomial enumeration, *Graph Theory and its Applications* (B. Harris, ed.; Academic
Press, New York) 167–213.

N. Nielsen, 1923, *Traité élémentaire des nombres de Bernoulli* (Gauthier-Villars, Paris).

J. Riordan, 1958, *An Introduction to Combinatorial Analysis* (Wiley, New York).

J. N. Srivastava et al., eds., *A Survey of Combinatorial Theory*
© North-Holland Publishing Company, 1973

CHAPTER 17

Some Combinatorial Aspects of Coding Theory

J. M. GOETHALS

MBLE Research Laboratory, Brussels, Belgium

1. Introduction

Many aspects of coding theory are of a combinatorial, rather than of an algebraic nature. Typical problems, in this respect, are the construction of maximal codes, and of majority decodable codes, to which the present paper is devoted. Other interesting combinatorial aspects of coding, as for example the construction of perfect codes, have not been studied, neither the more algebraic aspects, for which the recent book by Berlekamp [1968] provides a good source of references.

The problem of constructing maximal codes, in the sense of Plotkin [1960], has been studied by Bose and Shrikhande [1959], who pointed out interesting connections with some combinatorial designs. Using their methods, we slightly extend their results. Some other good codes, recently constructed by Sloane and Seidel [1970], are also mentioned. This is the subject of Section 2.

The problem of constructing majority decodable linear block codes has been extensively studied in the recent literature. Almost all known majority decoding methods use the properties of block designs, mainly those arising from finite geometries over finite fields. Section 3 of the present paper is a tentative survey of the literature devoted to this problem, with special emphasis on geometric codes. Connections between geometric codes and polynomial codes, in the sense of Kasami, Lin and Peterson [1968b], are also mentioned.

This paper is primarily intended to stimulate the interest of specialists, of either coding or combinatorial theory, to the other field. No special prerequisites are required, in either field.

2. Plotkin's maximal binary codes

A set of m binary vectors of length n such that the Hamming distance[1] between any two of them is greater than or equal to d will be denoted by

---

[1] The Hamming distance between two vectors is the number of coordinates in which they differ.

$M(n, d; m)$. This set of vectors forms a binary code of length n and minimum distance at least d. We shall refer to an $M(n, d; m)$ *code*. According to Plotkin [1960], let us denote by $A(n, d)$ the maximum size of such a code. The following bound on $A(n, d)$,

$$A(n, d) \leqslant 2d/(2d-n), \quad \text{for} \quad 2d > n, \tag{2.1}$$

was obtained by Plotkin [1960], who also proved the following relations:

$$A(n, d) \leqslant 2A(n-1, d), \tag{2.2}$$

$$A(n, 2t-1) = A(n+1, 2t). \tag{2.3}$$

These relations permit us to derive a bound for the case $n = 2d$. For d odd, we obtain, from (2.3) and (2.1),

$$A(4t-2, 2t-1) = A(4t-1, 2t) \leqslant 4t, \tag{2.4}$$

and, for d even, (2.2) and (2.1) yield

$$A(4t, 2t) \leqslant 2A(4t-1, 2t) \leqslant 8t. \tag{2.5}$$

Among the $M(n, d; m)$ codes for which $2d \geqslant n$, those codes whose sizes m meet the bounds (2.1) for $2d > n$, or (2.4) or (2.5) for $2d = n$, will be called *maximal*, and we shall denote them by $M(n, d)$. Maximal codes $M(4t, 2t)$ have been constructed by Plotkin [1960] for $4t = 1+p$, p prime. It was also noticed by Plotkin [1960] that by using methods of Paley [1933], his construction could be extended to all orders $4t$ of the form

$$4t = 2^k(1+p^h), \quad p \text{ an odd prime, and } h, k \text{ integers.}$$

In addition, he conjectured that maximal codes $M(4t, 2t)$ exist for all integers t. In fact, as shown by Bose and Shrikhande [1959], this is equivalent to the famous conjecture that Hadamard matrices exist for all orders $n \equiv 0$ (mod 4).

A *Hadamard matrix* is a square matrix H of order n, all of whose elements are $+1$ or -1, such that

$$HH^{\mathrm{T}} = n\,I.$$

A necessary condition for their existence is $n = 2$, or $n \equiv 0$ (mod 4). A number of constructions are known, cf., for instance, Hall [1967], but it is still unknown whether Hadamard matrices exist for all orders $n \equiv 0$ (mod 4).

A *balanced incomplete block design*, with parameters v, b, r, k, λ, is a collection of v *points* and b subsets of points, called *blocks*, such that the following properties hold:

(i) any block contains exactly k points, with $0 < k < v$,

(ii) any point is contained in exactly r blocks,

(iii) any two distinct points are contained in exactly λ blocks, with $0 < \lambda < r$.

From the definition, the following relations between the parameters are obtained

$$bk = vr, \qquad r(k-1) = \lambda(v-1).$$

The point-block incidence matrix N of the design is defined by its elements $n_{i,l}$,

$$n_{i,l} = 1, \quad \text{if the i-th point belongs to the l-th block,}$$
$$n_{i,l} = 0, \quad \text{otherwise.}$$

The above conditions are equivalent to the following relations, for the $(0, 1)$-matrix N,

$$Nj = rj, \quad j^T N = kj^T, \quad NN^T = (r-\lambda)I + \lambda J,$$

where j is the all-one vector of appropriate length, and J is the all-one matrix.

For any balanced incomplete block design, we have $b \geqslant v$. Designs for which $b = v$ are called *symmetric*. For symmetric balanced incomplete block designs, we have $b = v$, $r = k$, $k(k-1) = \lambda(v-1)$, and the incidence matrix N satisfies

$$NN^T = N^T N = (k-\lambda)I + \lambda J.$$

For a symmetric balanced incomplete block design with parameters $v = 4t-1$, $k = 2t-1$, $\lambda = t-1$, and with incidence matrix N, the matrix

$$H = \begin{bmatrix} -1 & j^T \\ j & J-2N \end{bmatrix}$$

is a Hadamard matrix.

For further details concerning block designs and Hadamard matrices, we refer to any book on combinatorial mathematics, such as Hall [1967] or Ryser [1963].

Connections between maximal codes, block designs and Hadamard matrices were pointed out by Bose and Shrikhande [1959], who proved the following theorems.

Theorem 2.1. (Bose and Shrikhande [1959]). *There exist maximal codes $M(4t, 2t)$ and $M(4t-1, 2t)$ if and only if there exists a Hadamard matrix of order $4t$.*

Theorem 2.2. (Bose and Shrikhande [1959]). (i) *There exists a maximal code $M(4t-1, 2t)$ if and only if there exists a symmetric balanced incomplete block design with parameters $v = 4t-1$, $k = 2t$, $\lambda = t$.*

(ii) *There exists a maximal code $M(4t-2, 2t)$ if and only if there exists a balanced incomplete block design with parameters $v = 2t-1$, $b = 4t-2$, $r = 2t$, $k = t$, $\lambda = t$.*

Following the methods of Bose and Shrikhande [1959], we shall prove a slightly more general result, which includes Theorem 2.2. It can be stated as follows.

Theorem 2.3. *For even distance $d = ks$, there exists a maximal code $M((2k-1)s, ks)$ if and only if there exists a balanced incomplete block design with parameters $v = 2k-1$, $b = (2k-1)s$, $r = ks$, $k = k$, $\lambda = \frac{1}{2}ks$.*

Before proceeding with the proof, we make some further observations. There is a well-known isomorphism which maps the additive group of the binary field F_2 onto the multiplicative group $G = \langle \{+1, -1\}, \cdot \rangle$. This isomorphism maps the binary elements 0, 1 onto the real numbers $+1$, -1, respectively. It is extended in a natural way to the direct product $F^n = F \times F \times \ldots \times F$ (n times) mapping binary vectors onto vectors with real components $+1$ or -1. To the binary sum of two binary vectors corresponds the componentwise product of their $(+1, -1)$ image-vectors over the reals.[2] Since the Hamming distance d between two binary vectors is the weight (i.e., the number of non-zero components) of their modulo two sum, it follows that the inner product of their $(+1, -1)$ image-vectors in G^n is $n - 2d$. Therefore, $A(n, d)$ can be defined as the maximum number of vectors of length n, all of whose components are $+1$ or -1, such that the inner product of any two (distinct) of them is not greater than $n - 2d$. Let M be a matrix of order $A(n, d) \times n$, whose rows are the $(+1, -1)$-images of the vectors of a maximal code $M(n, d)$. Then, in the matrix product MM^T,

$$MM^T = nI + L, \qquad L = (\lambda_{i,l}), \tag{2.6}$$

we have $\lambda_{i,i} = 0$, $\lambda_{i,l} \leqslant n - 2d \leqslant 0$, for $i \neq l$. Defining the numbers k_i by

$$j^T M = (k_1, k_2, \ldots, k_n),$$

we get from (2.6)

$$j^T M M^T j = nm + \sum_{i \neq l} \lambda_{i,l} = \sum_{i=1}^{n} k_i^2 \geqslant 0, \tag{2.7}$$

with equality on the right if and only if $k_1 = k_2 = \ldots = k_n = 0$. Now, since for each $\lambda_{i,l}$, we have $\lambda_{i,l} \leqslant n - 2d$, we deduce from (2.7)

$$nm + m(m-1)(n - 2d) \geqslant 0, \tag{2.8}$$

with quality if and only if all k_i's are zero and all $\lambda_{i,l}$'s are $n - 2d$. For $n < 2d$, the inequality (2.8) yields the Plotkin bound (2.1), and the following lemma follows.

Lemma 2.4. *For any maximal code $M(n, d)$, the following properties hold:*

(i) *Each coordinate occurs as 0 in half the vectors, and as 1 in the remaining half. Hence, necessarily $A(n, d)$ is even.*

(ii) *The code is equidistant: any pair of distinct vectors has Hamming distance d, and this is only possible for d even.*

(iii) *For $2d - n = s$, n and d are of the form $n = (2k-1)s$, $d = ks$, where k and s are integers such that $ks \equiv 0 \pmod{2}$.*

Proof. We observed that equality in (2.8) is only possible for $k_1 = k_2 = \ldots = k_n = 0$, which yields part (i). We also observed that all $\lambda_{i,l}$ satisfy

---

[2] For a previous use of this mapping, which shows its usefulness, we refer to Goethals and Seidel [1970], Section 6.

$\lambda_{i,1} = n - 2d$ in this case. This proves part (ii). Finally, a necessary condition for the existence of a maximal code $M(n, d)$ is that $2d$ be divisible by $2d - n = s$, and since, from (i), $A(n, d) = 2d/(2d-n)$ is even, this implies $d \equiv 0$ (mod s), whence $d = ks$, and $n = (2k-1)s$, where from (ii), ks is even. This proves the lemma.

We now prove theorem 2.3.

Proof of Theorem 2.3. For $n = (2k-1)s$, $d = ks$, d even, we have, from (2.1), $A(n, d) \leqslant 2k$. Suppose a maximal code $M((2k-1)s, ks)$ exists with $2k$ vectors. There is no loss of generality in assuming that the first vector is all-zero. Then, the remaining $2k-1$ vectors have weight $d = ks$ and are exactly at a distance d apart, since the code necessarily is equidistant. It follows that any two distinct vectors have exactly $\lambda = \frac{1}{2}d$ ones in common. On the other hand, since any coordinate is one in half the vectors, exactly k vectors have a one in a given position. From these properties, it readily follows that the (0, 1)-matrix N, whose rows are the $2k-1$ vectors of constant weight $d = ks$ of the maximal code, satisfies

$$Nj = dj, \qquad j^{\mathrm{T}}N = kj^{\mathrm{T}}, \qquad NN^{\mathrm{T}} = \tfrac{1}{2}(dI + dJ),$$

where j is the all-one vector of appropriate length, and J is the all-one matrix. Hence N is the point-block incidence matrix of a balanced incomplete block design with parameters $v = 2k-1$, $b = n = (2k-1)s$, $r = d = ks$, $k = k$, and $\lambda = \frac{1}{2}ks$. Conversely, if such a block design exists, then its point-block incidence matrix together with the all-zero vector forms a maximal code $M(b, r)$. This proves the theorem.

Remark. Theorem 2.2 appears as a particular instance of Theorem 2.3. Part (i) is Theorem 2.3 for $s = 1$, $k = 2t$, and part (ii) is obtained with $s = 2$, $k = t$. It turns out that, if the block designs of Theorem 2.2 exist, then the maximal codes $M((2k-1)s, ks)$ exist for all integers s. Indeed, since $d = ks$ is even, then either k or s is even. If k is even, let $k = 2t$. Then s repetitions of the symmetric design $v = 4t-1$, $k = 2t$, $\lambda = t$ yield the design of Theorem 2.3. If, on the other hand, k is odd, then s is even and $\frac{1}{2}s$ repetitions of the design $v = 2k-1$, $b = 4k-2$, $k = \lambda = k$ yield the design of Theorem 2.3.

Symmetric designs $(4t-1, 2t, t)$ are usually called *Hadamard designs* since they simultaneously exist with a Hadamard matrix of order $4t$. The following theorem yields a construction for the designs $v = 2k-1$, $b = 4k-2$, $r = 2k$, $k = \lambda = k$, with k odd. It is a slight variation of a theorem of Van Lint and Seidel [1966].

Theorem 2.5. *If there exists a symmetric C-matrix of order $n = 4t+2$, then there exists a balanced incomplete block design with parameters $v = 4t+1$, $b = 8t+2$, $r = 4t+2$, $k = 2t+1$, $\lambda = 2t+1$.*

Before proceeding with the proof, we first give some details concerning

C-matrices. A C-*matrix* is a square matrix C of order n, with elements zero on the diagonal and $+1$ or -1 elsewhere, such that

$$CC^T = (n-1)I. \tag{2.9}$$

Necessary conditions for their existence are $n \equiv 0 \pmod 4$, or $n \equiv 2 \pmod 4$, $n-1 = a^2 + b^2$, a and b integers (Belevitch [1950, 1968], Raghavarao [1960]). The following operations generate an equivalence relation on the set of C-matrices of order n: multiplication by -1 of any row, multiplication by -1 of any column, simultaneous interchange of any two rows and of the corresponding columns. It can be shown (Delsarte, Goethals and Seidel [1971]) that any C-matrix is equivalent to a symmetric or to a skew-symmetric C-matrix according as its order n satisfies $n \equiv 2 \pmod 4$ or $n \equiv 0 \pmod 4$. For a skew C-matrix C, the matrix $H = C+I$ is a *skew-Hadamard matrix*. Symmetric C-matrices were called *conference matrices* by Belevitch [1950], who initiated their study in connection with the construction of networks for conference telephony. We quote, from constructions due to Paley [1933] and Goethals and Seidel [1967], that symmetric C-matrices exist for all orders $n = 1 + p^k$, p prime, k any integer, and for $n = 226$. The first undecided case is $n = 46$.

Let us now prove Theorem 2.5.

Proof of Theorem 2.5. Let there exist a C-matrix C of order $n = 4t+2$. Then, this matrix C is equivalent, under multiplication of some rows and columns by -1, to a matrix of the form

$$C = \begin{bmatrix} 0 & j^T \\ j & S \end{bmatrix}, \tag{2.10}$$

where j is the all-one vector of order $n-1$, and where the square matrix S of order $n-1$ satisfies

$$SS^T = (n-1)I - J, \qquad S^T = S, \qquad Sj = 0. \tag{2.11}$$

It readily follows that the $(0, 1)$-matrices $A = \frac{1}{2}(J - I - S)$ and $B = \frac{1}{2}(J - I + S)$ satisfy

$$A^2 + A = B^2 + B = tI + tJ, \qquad Aj = Bj = (2t)j,$$

whence the $(0, 1)$ matrix $N = (A+I, B+I)$ of order $(4t+1) \times (8t+2)$ satisfies

$$NN^T = (A+I)^2 + (B+I)^2 = (2t+1)I + (2t+1)J,$$

$$Nj = (4t+2)j, \qquad j^T N = (2t+1)j^T.$$

Hence, the matrix N is the point-block incidence matrix of a balanced incomplete block design with parameters $v = 4t+1$, $b = 8t+2$, $r = 4t+2$, $k = 2t+1$, $\lambda = 2t+1$. This proves the theorem.

Sloane and Seidel [1970] recently constructed a set of binary codes of length $n = 4t+1$ and minimum distance $d = 2t$, having $8t+2$ codewords. Their construction is quoted in the following theorem.

Theorem 2.6. (Sloane and Seidel [1970]). *If there exists a C-matrix of order* $4t+2$, *then there exists an* $M(4t+1, 2t; 8t+4)$ *code.*

Proof. Let the C-matrix C and the matrix S be defined as in the proof of Theorem 2.5, cf. (2.10) and (2.11). Then, using the properties of the matrix S, one easily verifies that in the matrix N of order $(8t+4) \times (4t+1)$, with N^T given by

$$N^T = [j, I+S, -j, I-S], \tag{2.12}$$

the inner product of any two distinct rows is less than or equal to $+1$. Since the elements of the matrix N are $+1$ or -1, its $8t+4$ rows are the $(+1, -1)$-images of a set of $8t+4$ binary vectors which are at least at a Hamming distance $d = \frac{1}{2}(4t+1-1) = 2t$ apart. This proves the theorem.

The codes of Theorem 2.6 have the property that $n-2d = 1$. For d odd, we get from (2.3) and (2.5)

$$A(4t-1, 2t-1) = A(4t, 2t) \leqslant 8t,$$

that is, $A(2d+1, d) \leqslant 4d+4$. On the other hand, for complementary codes (i.e., codes having the property that for any vector in the code, its binary complement is also in the code), Grey [1962], by use of a result of Rankin [1956], obtained the following bound

$$A(n, d) \leqslant \frac{8d(n-d)}{n-(n-2d)^2}, \quad \text{for} \quad n-\sqrt{n} < 2d \leqslant n,$$

which also yields

$$A(2d+1, d) \leqslant 4d+4. \tag{2.13}$$

Although Sloane and Seidel's codes are not complementary, they do meet the bound. Two cases are known for which better codes exist, namely:

(i) an $M(5, 2; 16)$ code obtained from the 16 even-weighted vectors of length 5;

(ii) an $M(13, 6; 32)$ code which can be obtained from Nadler's $M(12, 5; 32)$ code by adding an overall parity-check, cf. Nadler [1962].

For other values of n and d, with $n-2d = 1$ and d even, no better codes are known.

We conclude this section with a construction, due to Sloane and Whitehead [1970], which combines an $M(n, d; m_1)$ code with an $M(n, [\frac{1}{2}(d+1)]; m_2)$ code to form an $M(2n, d; m_1 m_2)$ code. This construction was already given by Plotkin [1960] for the case when d is even. It is quoted in the following theorem.

Theorem 2.7. (Plotkin [1960]; Sloane and Whitehead [1970]).

$$A(2n, d) \geqslant A(n, d)A(n, [\frac{1}{2}(d+1)]).$$

Proof. If there exists an $M(n, d; m_1)$ code and an $M(n, [\frac{1}{2}(d+1)]; m_2)$ code, then an $M(n, d; m_1 m_2)$ code can be constructed as follows: For any vector

x in the first code, and any vector y in the second code, we construct the vector of length $2n$,

$$(x+y, y),$$

obtained by juxtaposing y and the binary sum of x and y. There are $m_1 m_2$ distinct such vectors, and any two of them, with same x, are at least distance $2[\frac{1}{2}(d+1)] \geqslant d$ apart; while any two of them, with distinct x's, are at least distance d apart. This proves the theorem.

3. Majority decoding of linear block codes

3.1. Orthogonal check sets

A *linear code* (n, k) over a finite field $GF(q)$ is a k-dimensional subspace of the space of n-tuples of elements of $GF(q)$. Thus, a linear code (n, k) contains q^k code vectors, and the Hamming distance between any two of them is the weight of their difference, which is a vector of the code. Any linear code (n, k) can be defined as the set of q^k n-tuples which satisfy a set of $n-k$ linearly independent linear equations over $GF(q)$. In other words, any vector of the code is orthogonal to the $(n-k) \times n$ matrix of the coefficients of these $n-k$ linear equations. This matrix is usually called *parity-check matrix*, and its $n-k$ rows define the *parity-check rules*, which are the linear equations satisfied by any vector of the code. Quite obviously, any vector in the $(n-k)$-dimensional space generated by the rows of the parity-check matrix also defines a parity-check rule. We shall refer to the *parity-check space*, which consists of the $q^{n-k}-1$ parity check rules, defined by the non-zero vectors of the dual space of the code. This latter is usually called the *dual code*.

Codes are used for the correction (or detection) of errors occuring during transmission over a noisy channel. For a q-ary symmetric channel, with $q = p^m$, p prime, the effect of errors occuring during transmission can be described by the addition of an error-vector e to the transmitted codevector x, resulting in the reception of the vector $x+e$, where addition is performed in $GF(q)$. It follows that, for a linear code, application of the parity-check rules on the received vector will produce a result which only depends on the error-vector. The value resulting from the application of a parity-check rule on a received vector will be called a *parity check*. By a *symbol*, we mean a coordinate of a vector, at a given position. When a parity check rule has a nonzero coefficient at a position corresponding to a given symbol, we shall say that the symbol is *checked* by that parity check rule. For more details concerning linear codes, we refer to Berlekamp [1968].

Massey [1963] defined an *orthogonal check set of order r on a symbol* to consist of a collection of r parity check rules, each of which checks that particular symbol, and such that any other symbol is checked by at most one parity-check rule of the set. When the parity-check space contains orthogonal check sets of order r on each one of the n symbols, then a very simple decoding method, known as *majority decoding*, can be devised which permits to correct anyc ombination of $[r/2]$ or fewer errors. It proceeds as follows: Considering the r parity-check rules that are orthogonal on the first symbol, we observe that, for any pattern of t errors, there are at most t among the r parity check rules which fail to be satisfied, whenever the first symbol is not corrupted by error, and at least $r-(t-1)$ whenever one of the t errors affects the first symbol. Therefore, provided the number t of errors does not exceed $[r/2]$, it can be decided by a majority vote on the r parity checks whether or not the first symbol was corrupted by error, Moreover, the actual value of the error is given by a clear majority[3] of the r parity checks whenever the checking coefficient for the first symbol is 1 in each parity check rule, which is always possible since the parity check rules form a linear space. Now, applying the same procedure on each of the n symbols will correct any combination of $[/r2]$ or fewer errors. This procedure is known as *one-step majority decoding*, and has been somewhat generalized by Rudolph [1967]. Rudolph considered collections of r parity-check rules, each of which checks a given particular symbol, and such that any other symbol is checked by at most λ among the r rules. Such a set will be called a λ-*check set of order r*. Orthogonal sets are λ-check sets with $\lambda = 1$. It was shown by Rudolph [1967] that whenever λ-check sets of order r can be found, for each of the n symbols, in the parity-check space, and provided the number of errors does not exceed $[r/2\lambda]$, then these errors can be corrected by a majority decoding procedure with threshold, in one step. Very recently, Bastin [1970] and Ng [1970] independently improved Rudolph's decoding procedure by showing that it can be applied for a number of errors not exceeding $[(r+\lambda-1)/2\lambda]$. We briefly describe this decoding method, which will be called *one-step majority decoding with threshold*. Considering a λ-check set of order r on a given symbol, say the first, one easily verifies, using a similar argument as above, that for any pattern of t errors, at most $t\lambda$ among the r parity checks are nonzero when the first symbol is not affected by error, and at least $r-(t-1)\lambda$ are nonzero when one of the t errors affects the first symbol. Moreover, if the parity check rules are normalized so as to have their first checking coefficients equal to 1, then the actual value of the error on the first symbol is given by at least $r-(t-1)\lambda$ parity checks, when the first symbol is corrupted by error. Since we have

$$r-(t-1)\lambda \geqslant t, \quad \text{for} \quad t \geqslant [(r+\lambda-1)/2\lambda],$$

it follows that the actual value of the error is given by a majority of the r parity check rules, provided this majority is at least $[\frac{1}{2}(r+\lambda+1)]$, when the number of errors does not exceed $[(r+\lambda-1)/2\lambda]$. This suggests the following

---

[3] The clear majority is at least $[r/2]+1$, for a nonzero element. It is reduced to $[(r+1)/2]$ for the element 0 of $GF(q)$.

decoding procedure: For each received symbol, subtract the value given by the majority of the r parity checks which form a λ-check set on it, provided this majority is at least $[\frac{1}{2}(r+\lambda+1)]$; otherwise, leave the received symbol unchanged. This is Rudolph's majority decoding with threshold.

A well-known class of combinatorial designs can be used to construct λ-check sets of order r. These are the balanced incomplete block designs with parameters (v, b, r, k, λ). For each one of the v points, the r blocks containing that point form a λ-check set of order r on it. Therefore, when the b rows of the incidence matrix of the design are used as parity check rules for a linear code of length $n = v$, then this code is majority decodable for a number of errors not exceeding $[(r+\lambda-1)/2\lambda]$. However, this code might be trivial. Indeed, since for any block design we have $b \geqslant v$, the rank of the incidence matrix of the design can be as great as $n = v$ over a finite field $GF(q)$, in which case the code has dimension zero over $GF(q)$. In order to have a great dimension for the code, we must have a block design with low rank over a finite field. The problem of determining the rank of the incidence matrix of a design over a finite field is far from being solved in general; cf., for instance, MacWilliams and Mann [1968]. It is however completely solved now for the designs derived from finite geometries over finite fields; cf. Graham and MacWilliams [1966], Goethals and Delsarte [1968], Weldon [1968a, 1969], Smith [1967, 1969], Hamada [1968] and MacWilliams and Mann [1968], among others.

Rudolph [1967] first used incidence matrices of points and flats in finite geometries to construct one-step majority decodable linear codes. He observed that, usually, incidence matrices from finite geometries over $GF(q)$, $q = p^m$, p prime, have low rank over $GF(p)$, but was unable to derive an explicit formula for this rank. On the other hand, his decoding procedure was, in general, far from being able to correct the number of errors that the codes theoretically are able to correct. It turns out that another decoding procedure, known as *multistep majority decoding*, can be applied to these codes to correct a considerably greater number of errors. This decoding procedure, first introduced by Massey [1963], will now briefly be described. Suppose there exists, in the parity-check space of a code, a set of r parity-check rules, each of which checks a given subset of symbols with the same coefficients, and such that any other symbol is checked by at most one rule of the set. Then this set of parity check rules is said to form an *orthogonal check set of order r on the subset* of symbols, and a new parity check can be determined as follows: Let the coefficients (the same in each parity check rule of the set) on the given subset of symbols be $a_1, a_2, \ldots, a_S$. Then, with the same reasoning as above, provided the total number of errors does not exceed $[r/2]$, the value of

$$a_1 e_1 + a_2 e_2 + \ldots + a_S e_S, \tag{3.1}$$

where $e_1, e_2, \ldots, e_S$ are the errors on the subset, can be determined by a clear majority[4] of the r parity checks. Hence, the new parity check (3.1) can be determined. Now, suppose r new parity checks can be determined to form an orthogonal check set on a smaller subset. Then, in a second step, the same procedure will produce a new parity check with a smaller number of symbols involved. If this procedure can be applied several times until an orthogonal check set of order r has been constructed on a single symbol, then, at the last step, the actual value of the error on that symbol will be determined. If this can be done for each of the n symbols, in at most L steps, and if, at each step, the orthogonal check sets are of an order at least r, then the code is *L-step majority decodable* for a number of errors not exceeding $[r/2]$. Several authors have shown how this decoding procedure can be used to decode Rudolph's geometric codes; cf. Smith [1967], Chow [1967], Goethals and Delsarte [1968], Weldon [1968a, 1969]. The decoding algorithm devised by Reed [1954] for the class of codes discovered by Muller [1954], and known as Reed-Muller codes, is of this type; cf. Weldon [1968a,b] and Gore [1969b]. In fact, the Reed-Muller codes constitute the simplest class of geometric codes, which we shall now discuss with some details. For an elementary study of geometric codes, we refer to Delsarte and Goethals [1970].

3.2. Geometric codes

3.2.1. Euclidean geometry codes

An *Euclidean geometry* $EG(m, q)$ of dimension m over $GF(q)$ consists of q^m *points*, which can be identified with the q^m distinct m-tuples of elements of $GF(q)$, and a set of *t-flats*, which can recursively be defined as follows, for $t = 0, 1, \ldots, m-1$: The 0-flats are the points of the geometry. For any t-flat E_t and any point α of $EG(m, q)$, there is a unique $(t+1)$-flat E_{t+1} of $EG(m, q)$, which contains both E_t and α. This $(t+1)$-flat consists of the set of points

$$E_{t+1} = \{\beta + \omega(\alpha - \gamma) \mid \beta \in E_t, \ \omega \in GF(q)\},$$

where γ is any fixed point of E_t.

From this definition, the following properties are easily derived:

(i) *A t-flat contains q^t points; it has the structure of an Euclidean geometry $EG(t, q)$.*

(ii) *An $EG(m, q)$ contains $b_E(t, m, q)$ distinct t-flats, with $0 < t < m$, and*

$$b_E(t, m, q) = \frac{q^{m-t}(q^m - 1)(q^{m-1} - 1) \ldots (q^{m-t+1} - 1)}{(q^t - 1)(q^{t-1} - 1) \ldots (q - 1)}. \tag{3.2}$$

---

[4] The clear majority is at least $[r/2]+1$, for a nonzero element. It is reduced to $[(r+1)/2]$ for the element 0 of $GF(q)$.

(iii) *Each s-flat of EG(m, q) is contained in* $\lambda(s, t, m, q)$ *distinct t-flats of EG(m, q), with* $0 \leq s < t \leq m$, *and*

$$\lambda(s, t, m, q) = \frac{(q^{m-s}-1)(q^{m-s-1}-1)\dots(q^{m-t+1}-1)}{(q^{t-s}-1)(q^{t-s-1}-1)\dots(q-1)}. \tag{3.3}$$

(iv) *The incidence relations of points and t-flats in EG(m, q) define a balanced incomplete block design with parameters* $v = q^m$, $b = b_E(t, m, q)$, $r = \lambda(0, t, m, q)$, $k = q^t$, *and* $\lambda = \lambda(1, t, m, q)$ *for* $t > 1$, *or* $\lambda = 1$ *for* $t = 1$.

Rudolph [1967] considered the above block designs for the construction of codes that are one-step majority decodable with threshold. These codes have as parity check rules the incidence vectors of points and t-flats in an $EG(m, q)$. For $q = p^v$, p prime, he observed that the rank of the incidence matrix is relatively small over $GF(p)$. Hence, the codes have reasonable dimensions over $GF(p)$ (or over any extension field of $GF(p)$). However, the maximum number of errors that Rudolph's algorithm guarantees to correct is, in general, relatively small, even with the possible improvement of Bastin [1970] and Ng [1970]. Using methods of Kasami, Lin and Peterson [1968a], Weldon [1969] was able to show that a certain class of extended binary cyclic codes of length $n = q^m$, with $q = 2^v$, have the property that the incidence vector of any $(t+1)$-flat of an $EG(m, q)$ is contained in their dual space. He also showed that these codes are $(t+1)$-step majority decodable for a number of errors not exceeding $[r/2]$, with $r = (q^{m-t}-1)/(q-1)$. It turns out that these codes are the same as Rudolph's codes. In other words, the Euclidean geometry codes defined by Weldon [1969] are the largest codes which contain all $(t+1)$-flats in their dual space. This can be shown (not easily) by using some results of Delsarte [1969].

We shall give a slightly different and slightly more general definition of Euclidean geometry (= EG) codes than the one originally given by Weldon [1969].

Definition. The (t, v)-EG *code of length* q^m *over* $GF(p)$, *with* $q = p^v$, p *a prime or the power of a prime, is the linear code over* $GF(p)$ *having as parity-check rules the incidence vectors of points and* $(t+1)$-*flats in an* $EG(m, q)$.

We now briefly summarize the basic properties of EG codes.

Theorem 3.1. (Weldon [1969]). *The* (t, v)-EG *codes are* $(t+1)$-*step majority decodable, for a number of errors not exceeding* $[r/2]$, *with* $r = (q^{m-t}-1)/(q-1)$.

Proof. The $r = (q^{m-t}-1)/(q-1)$ incidence vectors of $(t+1)$-flats which contain a given t-flat form an orthogonal check set of order r on the given t-flat. This follows from the definition of t-flats since any point not contained in a t-flat determines with the t-flat a unique $(t+1)$-flat. Therefore, any parity check corresponding to the incidence vector of a t-flat can be determined from a clear majority of the r parity checks that are orthogonal on it, provided the number of errors does not exceed $[r/2]$. Now the theorem is easily proved by induction on t, since the orthogonal check sets constructed at each step are of an order at least equal to r.

For positive integers q and u, with $u < q^m$, let us write

$$u = u_0 + u_1 q + u_2 q^2 + \ldots + u_{m-1} q^{m-1},$$

with $0 \leqslant u_i \leqslant q-1$, for $i = 0, 1, \ldots, m-1$. Then the q-weight of u, $w_q(u)$, is defined by

$$w_q(u) = u_0 + u_1 + \ldots + u_{m-1},$$

that is, the sum of the "digits" of u expressed in radix-q form.

Theorem 3.2. *The rank over $GF(p)$ of the incidence matrix of t-flats and points in an $EG(m, q)$, with $q = p^\mu$, p prime, is equal to $R_p(t, m, q)$, where $R_p(t, m, q)$ is the number of integers u, $0 \leqslant u < q^m$, such that*

$$\max_{0 \leqslant i < \mu} w_q(p^i u) \leqslant (m-t)(q-1). \tag{3.4}$$

The proof of this theorem, which can be derived from Delsarte [1969], is rather long and will be omitted. It is based on a theorem of MacWilliams and Mann [1968], and follows a method, initiated by Graham and MacWilliams [1966], which was used by Goethals and Delsarte [1968] for projective geometry codes. A slightly different method was used by Smith [1967], which allowed Hamada [1968] to derive a (rather complicated) explicit formula for this rank. For the special case $t = m-1$, an explicit and simple formula is known.

Theorem 3.3. (MacWilliams and Mann [1968]). $R_p(m-1, m, p^\mu) = \binom{m+p-1}{p-1}^\mu$.

From the definition of EG codes, it follows that their dimension is given by the following corollary.

Corollary 3.4. *The (t, v)-EG code of length q^m, with $q = p_1^v$, $p_1 = p^\mu$, p prime, has dimension*

$$k = q^m - R_p(t+1, m, q)$$

over $GF(p_1)$.

Proof. From the definition, it follows that the dimension of the (t, v)-EG code is q^m minus the rank over $GF(p_1)$ of the incidence matrix of points and $(t+1)$-flats in $EG(m, q)$. But, obviously, this rank is the same as the rank over $GF(p)$, whence the corollary follows from Theorem 3.2.

From Theorem 3.1 it follows that the minimum distance of a (t, v)-EG code is at least

$$\frac{q^{m-t}-1}{q-1} + 1. \tag{3.5}$$

However, it can be shown that this minimum distance is, in general, greater than (3.5).

Theorem 3.5. (Chow [1968], Lin [1968]). *The minimum distance of a (t, v)-EG code of length q^m over $GF(p)$, with $q = p^v$, is at least*

$$(q+p)q^{m-t-1}. \tag{3.6}$$

The proof uses the fact that an *EG* code is an extended cyclic code (cf. Berlekamp [1968]) for which the Bose-Chaudhuri bound applies cf. (Bose and Chaudhuri [1960a,b]). It will not be reproduced here. For more details, we refer to Delsarte [1970a].

We conclude from Theorem 3.5 that the majority decoding algorithm does not use, in general, the full error-correcting ability of the code. It has, however, the great advantage of its simplicity. The complexity of this algorithm grows exponentially with the number of steps. It is therefore important to reduce, if possible, the number of steps of majority decoding. As shown by Weldon [1968b], this can be done for any *EG* code by combining the ordinary majority decoding with Rudolph's threshold decoding method.

Theorem 3.6. (Weldon [1968b]). *Any (t, v)-EG code is majority decodable in at most two steps, for a number of errors not exceeding $[r/2]$, with $r = (q^{m-t} - 1)/(q-1)$.*

Proof. For $t \leqslant 1$, the theorem follows from Theorem 3.1. Let us assume $t > 1$. Then, parity checks can be determined on all t-flats in a first step of majority decoding. In a second step, Rudolph's method is used with the incidence matrix of points and t-flats in $EG(m, q)$. Since the parameters r and λ of the design are such that

$$\frac{r}{\lambda} = \frac{q^m - 1}{q^t - 1} > \frac{q^{m-t} - 1}{q - 1},$$

it follows that this decoding algorithm has the same error-correcting ability as the ordinary $(t+1)$-step majority decoding algorithm of Theorem 3.1. This proves the theorem.

When m is a composite number, (t, v) *EG* codes of length q^m can usually be decoded in less than $(t+1)$ steps of majority decoding by using the parity check rules corresponding to sd-flats of $EG(m, q)$ which are s-flats in $EG(m/d, q^d)$, where d is a divisor of m. This was shown by Weldon [1968b] and Delsarte [1969]. By doing so, one only exploits a fraction of the totality of $(t+1)$-flats of $EG(m, q)$, which constitute a special kind of "geometries" as defined by Delsarte [1969]. It turns out that the p-rank of the incidence matrix of this fraction of the totality of $(t+1)$-flats is, in general, less than $R_p(t+1, m, q)$. Hence, as shown by Delsarte [1969], a code with higher dimension can be decoded with this modified algorithm.

3.2.2. Projective geometry codes

A *projective geometry* $PG(m, q)$ of dimension m over $GF(q)$ consists of $(q^{m+1} - 1)/(q-1)$ *points*, which can be identified with a maximal set of $(m+1)$-tuples no two of which are dependent over $GF(q)$, and a collection of *t-flats* which are recursively defined as follows, for $t = 0, 1, \ldots, m-1$: The 0-flats are the points of the geometry. For any *t*-flat P_t and any point α, there is a unique $(t+1)$-flat P_{t+1} in $PG(m, q)$ which contains both P_t and α. This $(t+1)$-flat consists of the set of points

$$P_{t+1} = \{\alpha\} \cup \{\beta + \omega\alpha \mid \beta \in P_t, \omega \in GF(q)\}.$$

From this definition, the following properties are easily derived:

(i) *A t-flat contains $(q^{t+1} - 1)/(q-1)$ points; it has the structure of a projective geometry $PG(t, q)$.*

(ii) *A $PG(m, q)$ contains $b_P(t, m, q)$ distinct t-flats, with $0 \leqslant t < m$, and*

$$b_P(t, m, q) = \frac{(q^{m+1} - 1)(q^m - 1) \ldots (q^{m-t+1} - 1)}{(q^{t+1} - 1)(q^t - 1) \ldots (q-1)}. \tag{3.7}$$

(iii) *Each s-flat of $PG(m, q)$ is contained in $\lambda(s, t, m, q)$ distinct t-flats of $PG(m, q)$ with $0 \leqslant s < t \leqslant m$, and $\lambda(s, t, m, q)$ given by (3.3).*

(iv) *The incidence relations of points and t-flats in $PG(m, q)$ define a balanced incomplete block design with parameters $v = (q^{m+1} - 1)/(q-1)$, $b = b_P(t, m, q)$, $r = \lambda(0, t, m, q)$, $k = (q^{t+1} - 1)/(q-1)$, and $\lambda = 1$ for $t = 1$, or $\lambda = \lambda(1, t, m, q)$ for $t > 1$.*

Rudolph [1967] considered the above block designs for the construction of codes that are one-step majority decodable with threshold. These codes have as parity-check rules the incidence vectors of points and *t*-flats in a $PG(m, q)$. The same remarks hold here as for Euclidean geometry codes. Goethals and Delsarte [1968], and independently, Smith [1967], Chow [1967], and Weldon [1968a], devised a multistep majority decoding procedure for these codes, which, except for $t = 1$, has a better error-correcting ability than Rudolph's decoding method. A particular class of these codes was introduced by Weldon [1966] under the name of *difference-set codes*. Following the method initiated by Graham and MacWilliams [1966] for difference-set codes, Goethals and Delsarte [1968] were able to calculate the dimension of these projective geometry $(= PG)$ codes, which can be defined as follows.

Definition. *The (t, v)-PG code of length $(q^{m+1} - 1)/(q-1)$ over $GF(p)$, with $q = p^v$, p a prime or the power of a prime, is the linear code over $GF(p)$ having as parity-check rules the incidence vectors of points and $(t+1)$-flats in a $PG(m, q)$.*

We briefly summarize some of their properties.

Theorem 3.7. (Goethals and Delsarte [1968], Smith [1967], Chow [1967], Weldon [1968a]). *The (t, v)-PG codes are $(t+1)$-step majority decodable, for a number of errors not exceeding $[r/2]$, with $r = (q^{m-t}-1)/(q-1)$.*

The proof is quite similar to the one of Theorem 3.1 and will be omitted.

Theorem 3.8. (Goethals and Delsarte [1968]). *The rank over $GF(p)$ of the incidence matrix of points and t-flats in a $PG(m, q)$, with $q = p^{\mu}$, p prime, is equal to $Q_p(t, m, q)$, where $Q_p(t, m, q)$ is the number of integers u, $0 \leqslant u < (q^{m+1}-1)/(q-1)$, such that*

$$\max_{0 \leqslant i < \mu} w_q[p^i u(q-1)] \leqslant (m-t)(q-1). \tag{3.8}$$

The proof of this theorem is rather long and will be omitted. An explicit, but rather complicated, formula for $Q_p(t, m, q)$ was obtained by Hamada [1963]. There exists between $Q_p(t, m, q)$ and $R_p(t, m, q)$, defined by (3.4), the following relation.

Theorem 3.9. $R_p(t, m, q) = Q_p(t, m, q) - Q_p(t, m-1, q).$

The proof will be omitted. It consists in showing that there is a one-to-one correspondence between the numbers u and u' satisfying, respectively,

$$(q^m - 1)/(q-1) \leqslant u < (q^{m+1}-1)/(q-1) \quad \text{and (3.8)},$$

and

$$0 \leqslant u' < q^m, \max_{0 \leqslant i < \mu} w_q(p^i u') \leqslant (m-t)(q-1):$$

Corollary 3.10.

$$Q_p(t, m, q) = \sum_{i=0}^{m-t} R_p(t, m-i, q).$$

In particular, we have, for the special case $t = m-1$:

Corollary 3.11. (Goethals and Delsarte [1968], MacWilliams and Mann [1968], Smith [1969]). *For $q = p^{\mu}$,*

$$Q_p(m-1, m, q) = 1 + \binom{m+p-1}{p-1}^{\mu}.$$

Corollary 3.12. *The dimension over $GF(p_1)$ of the (t, v)-PG code of length $(q^{m+1}-1)/(q-1)$, with $q = p_1^v$, $p_1 = p^{\mu}$, p prime, is given by*

$$k = (q^{m+1}-1)/(q-1) - Q_p(t+1, m, q).$$

Theorem 3.13. (Weldon [1968b]). *Any (t, v)-PG code is majority decodable in at most two steps, for a number of errors not exceeding $[r/2]$, with $r = (q^{m-t}-1)/(q-1)$.*

The proof is quite similar to the one of Theorem 3.6 and will be omitted.

3.3. Other majority decodable codes. Concluding remarks

3.3.1. Polynomial codes

An interesting class of codes has been recently investigated by many authors. These codes have the property that any vector of the code can be described by the set of values assumed by a polynomial at a set of points. The original definition of Reed-Muller codes was given by Reed [1954], by using this polynomial approach. It turns out that the geometric codes can also be described as polynomial codes. In his recent book, Berlekamp [1968] defined geometric codes as dual of subfield subcodes of particular polynomial codes in the class introduced by Kasami, Lin and Peterson [1968a,b] and Weldon [1968a] under the name of *Generalized Reed-Muller codes*. The connection between geometric codes and polynomial codes was pointed out by Lin [1968], and by Delsarte, Goethals and MacWilliams [1970]. Further results were obtained by Chen and Lin [1969]. Delsarte [1970a] obtained the *BCH* bounds for these codes. A more general class of polynomial codes, having a quite large automorphism group, was also investigated by Delsarte [1970b]. Finally, we want to mention the work of Gore [1970b], who described polynomial codes as a special kind of product codes, and obtained quite promising results for their decoding by threshold-majority methods, cf. Gore [1969a, 1970a]. Also, comments made by Gore and Cooper [1970], and by Cooper and Gore [1970] on the work of Kasami, Lin and Peterson [1968b] and of Chen and Lin [1969], are worth mentioning.

Other majority decodable codes are obtained by combining several majority decodable codes by the operation of direct product (= Kronecker product) of codes; cf. Gore [1970b], and Lin and Weldon [1970].

3.3.2. Majority decoding with block designs

As mentioned before, majority decoding with threshold can be used for the decoding of codes having in their parity-check space the incidence vectors of a block design. The number of errors that are correctable by this method is, in general, relatively small as compared with the actual error-correcting ability of the code. However, some codes are known for which a majority decoding method with threshold uses the full error-correcting ability of the code. For example, as shown by Assmus and Mattson [1969b] and Goethals [1971], the extended Golay (24, 12) binary code can be decoded in one or two steps by majority decoding methods, up to the minimum distance. Similar methods exist for the related (23, 12), (23, 11), (22, 12) and (22, 11) binary codes. It was also shown by Goethals [1970] that a similar method applies on the (48, 24) extended quadratic residue binary code. All these methods are based on the existence of t-designs in the parity-check spaces of the codes; cf. Assmus and Mattson [1967, 1969a].

Another class of majority decodable codes has been constructed by Smith [1968]. These codes have a parity check matrix of the form

$$H = [N^T, I],$$

where N is the point-block incidence matrix of a block design with v points and b blocks, and I is a unit matrix of order b. These codes have length $n = b + v$, and dimension $k = v$. For a balanced incomplete block design with parameters b, v, r, k, $\lambda = 1$, the code has minimum distance r, and a one-step majority decoding method using the full error-correcting ability of the code can be devised. These codes, however, have the disadvantage of their relatively low dimension. Designs with several association classes, or partially balanced incomplete block designs, cf. Bose [1963], can also be used, giving codes with higher dimension. For further details, we refer to Smith [1968].

Note added in proof

The author just received, when the paper was completed, several reports from the Coordinated Science Laboratory, University of Illinois (Chen [1970a, b], Chen and Chien [1970]). One of these reports (Chen and Chien [1970]) is a short survey on majority decoding, similar, although less complete, to the present one. The other two reports contain improved decoding methods for EG-codes (Chen [1970a]) and PG-codes (Chen [1970b]).

Acknowledgements

The author is gratefully indebted to P. Delsarte and J. J. Seidel, for reading the manuscript and making some useful suggestions.

References

E. F. Assmus, Jr. and H. F. Mattson, Jr., 1967, On tactical configuration and error-correcting codes, *J. Combin. Theory* **2**, 243–257.

E. F. Assmus, Jr. and H. F. Mattson, Jr., 1969a, New 5-designs, *J. Combin.Theory* **6**, 122–151.

E. F. Assmus, Jr. and H. F. Mattson, Jr., 1969b, Algebraic theory of codes, Ann. Rept. Sylvania Electronic Systems, Appl. Res. Lab., Waltham, Mass.

G. Bastin, 1970, A propos du décodage par décision majoritaire, *Revue H. F.*, **8**, 85–95.

V. Belevitch, 1950, Theory of $2n$-terminal networks with application to conference telephony, *Electr. Commun.* **27**, 231–244.

V. Belevitch, 1968, Conference networks and Hadamard matrices, *Ann. Soc. Sci. Bruxelles* **82**, 13–32.

E. R. Berlekamp, 1968, *Algebraic Coding Theory* (McGraw Hill, New York).

R. C. Bose, 1963, Strongly regular graphs, partial geometries and partially balanced designs, *Pacif. J. Math.* **13**, 389–419.

R. C. Bose and D. K. Ray-Chaudhuri, 1960a, On a class of error-correcting binary group codes, *Inf. Control* **3**, 68–79.

R. C. Bose and D. K. Ray-Chaudhuri, 1960b, Further results on error-correcting binary group codes, *Inf. Control* **3**, 279–290.

R. C. Bose and S. S. Shrikhande, 1959, A note on a result in the theory of code construction, *Inf. Control* **2**, 183–194.

C. L. Chen, 1970a, On decoding Euclidean geometry codes, Rept. R-479, Coord. Sci. Lab., Univ. of Illinois, Urbana, Ill.

C. L. Chen, 1970b, On decoding projective geometry codes, Rept. R-486, Coord. Sci. Lab., Univ. of Illinois, Urbana, Ill.

C. L. Chen and R. T. Chien, 1970, Recent developments in majority-logic decoding, Rept. R-487, Coord. Sci. Lab., Univ. of Illinois, Urbana, Ill.

C. L. Chen and S. Lin 1969, Further results on polynomial codes, *Inf. Control* **15**, 38–60.

D. K. Chow, 1967, A geometric approach to coding theory with applications to information retrieval, Rept. R-368, Coord. Sci. Lab., Univ. of Illinois, Urbana, Ill.

D. K. Chow, 1968, On threshold decoding of cyclic codes, *Inf. Control* **13**, 471–483.

A. B. Cooper and W. C. Gore, 1970, A recent result concerning the dual of polynomial codes, *IEEE Trans. on Information Theory* **IT-16**, pp. 638–640.

P. Delsarte, 1969, A geometric approach to a class of cyclic codes, *J. Combin. Theory* **6**, 340–358.

P. Delsarte, 1970a, BCH bounds for a class of cyclic codes, *SIAM Rev.* **19**, 420–429.

P. Delsarte, 1970b, On cyclic codes that are invariant under the general linear group, *IEEE Trans. on Information Theory* **IT-16**, pp. 760–769.

P. Delsarte and J. M. Goethals, 1970, Codes correcteurs d'erreurs et décodage par décision majoritaire, *Revue MBLE* **13**, 23–35.

P. Delsarte, J. M. Goethals and F. J. MacWilliams, 1970, On generalized Reed-Muller codes and their relatives, *Inf. Control* **16**, 403–442.

P. Delsarte, J. M. Goethals and J. J. Seidel, 1971, Orthogonal matrices with zero diagonal, Part II, *Canad. J. Math.* **23**, 816–832.

J. M. Goethals, 1970, On *t*-designs and threshold decoding, Univ. of North Carolina, Inst. of Statist., Mimeo Ser. No. 600.29.

J. M. Goethals, 1971, On the Golay perfect binary code, *J. Combin. Theory* **11**, 178–186.

J. M. Goethals and P. Delsarte, 1968, On a class of majority-logic decodable cyclic codes, *IEEE Trans. on Information Theory* **IT-14**, 182–188.

J. M. Goethals and J. J. Seidel, 1967, Orthogonal matrices with zero diagonal, *Canad. J. Math.* **19**, 1001–1010.

J. M. Goethals and J. J. Seidel, 1970, Strongly regular graphs derived from combinatorial designs, *Canad. J. Math.* **22**, 597–614.

W. C. Gore, 1969a, Generalized threshold decoding and the Reed-Solomon codes, *IEEE Trans. on Information Theory* **IT-15**, 81–87.

W. C. Gore, 1969b, The equivalence of L-step orthogonalization and a Reed decoding procedure, *IEEE Trans. on Information Theory* **IT-15**, 184–186.

W. C. Gore, 1970a, Threshold decoding of the generalized Reed-Muller codes, *Proc. Fourth Annual Princeton Conference on Information Sciences and Systems* (Princeton University, N.J.), p. 20.

W. C. Gore, 1970b, Further results on product codes, *IEEE Trans. on Information Theory* **IT-16**, 446–451.

W. C. Gore and A. B. Cooper, 1970, Comments on polynomial codes, *IEEE Trans. on Information Theory* **IT-16**, 635–638.

R. L. Graham and F. J. MacWilliams, 1966, On the number of information symbols in difference-set cyclic codes, *Bell Syst. Tech. J.* **45**, 1057–1070.

L. D. Grey, 1962, Some bounds for error-correcting codes, *IRE Trans. on Information Theory* **IT-8**, 200–202.

M. Hall, Jr., 1967, *Combinatorial Theory* (Blaisdell, Waltham, Mass.).

N. Hamada, 1968, The rank of the incidence matrix of points and *d*-flats in finite geometries, *J. Sci. Hiroshima Univ. Ser. A-I*, **32**, 381–396.

T. Kasami, S. Lin and W. W. Peterson, 1968a, New generalization of the Reed-Muller codes-Part I: primitive codes, *IEEE Trans. on Information Theory* **IT-14**, 189–199.

T. Kasami, S. Lin and W. W. Peterson, 1968b, Polynomial codes, *IEEE Trans. on Information Theory* **IT-14**, 807–814.

S. Lin, 1968, On a class of cyclic codes, *Error Correcting Codes* (H. B. Mann, ed.; Wiley, New York), pp. 131–148.

S. Lin, and E. J. Weldon, 1970, Further results on cyclic product codes, *IEEE Trans. on Information Theory* **IT-16,** 452–459.

F. J. MacWilliams and H. B. Mann, 1968, On the p-rank of the design matrix of a difference-set, *Inf. Control* **12,** 474–488.

J. L. Massey, 1963, *Threshold Decoding* (M.I.T. Press, Cambridge, Mass.).

D. E. Muller, 1954, Application of Boolean algebra to switching circuit design and to error detection, *IRE Trans. on Electronic Computers*, **EC-3,** 6–12.

M. Nadler, 1962, A 32-point $n = 12$, $d = 5$ code, *IRE Trans. on Information Theory* **IT-8,** 58.

S. W. Ng, 1970, On Rudolph's majority logic decoding algorithm, *IEEE Trans. on Information Theory* **IT-16,** 651–652.

R. E. A. C. Paley, 1933, On orthogonal matrices, *J. Math. Phys.* **12,** 311–320.

M. Plotkin, 1960, Binary codes with specified minimum distance, *IRE Trans. on Information Theory* **IT-6,** 445–450.

D. Raghavarao, 1960, Some aspects of weighing designs, *Ann. Math. Statist.* **31,** 878–884.

R. A. Rankin, 1956, On the minimal points of positive definite quadratic forms, *Mathematika* **3,** 15–24.

I. S. Reed, 1954, A class of multiple-error-correcting codes and the decoding scheme, *IRE Trans. on Information Theory* **PGIT-3,** 38–49.

L. D. Rudolph, 1967, A class of majority-logic decodable codes, *IEEE Trans. on Information Theory* **IT-13,** 305–307.

H. J. Ryser, 1963, *Combinatorial Mathematics*, Carus Math. Monographs (Wiley, New York).

N. J. A. Sloane and J. J. Seidel, 1970, A new family of nonlinear codes obtained from conference matrices, *Ann. N.Y. Acad. Sci.* **175,** 363–365.

N. J. A. Sloane and D. S. Whitehead, 1970, New family of single-error correcting codes, *IEEE Trans. on Information Theory* **IT-16,** 717–719.

K. J. C. Smith, 1967, Majority decodable codes derived from finite geometries, Univ. of North Carolina, Inst. of Statist., Mimeo Ser. No. 561.

K. J. C. Smith, 1968, An application of incomplete block designs to the construction of error-correcting codes, Univ. of North Carolina, Inst. of Statist., Mimeo Ser. No. 587.

K. J. C. Smith, 1969, On the p-rank of the incidence matrix of points and hyperplanes in a finite projective geometry, *J. Combin. Theory* **7,** 122–129.

J. H. van Lint and J. J. Seidel, 1966, Equilateral point sets in elliptic geometry, *Kon. Ned. Akad. Wetensch. Proc. Ser.* A **69** (= *Indag. Math.* **28**), 335–348.

E. J. Weldon, 1966, Difference-set cyclic codes, *Bell Syst. Tech. J.* **45,** 1045–1056.

E. J. Weldon, 1968a, New generalization of the Reed-Muller codes. Part II: Non-primitive codes, *IEEE Trans. on Information Theory* **IT-14,** 199–205.

E. J. Weldon, 1968b, Some results on majority-logic decoding, *Error Correcting Codes* (H. B. Mann, ed; Wiley, New York), pp. 149–162.

E. J. Weldon, 1969, Euclidean geometry cyclic codes, *Combinatorial Mathematics and its Applications, Proc. Chapel Hill Conf.* (R. C. Bose and T. A. Dowling, eds.; Univ. of North Carolina Press, Chapel Hill, N.Car.).

J. N. Srivastava et al., eds., *A Survey of Combinatorial Theory*
© North-Holland Publishing Company, 1973

CHAPTER 18

Some Augmentations of Bose-Chaudhuri Error Correcting Codes †

ALAN J. GROSS ‡ *

*California Center for Health Services Research, School of Public Health,
University of California, Los Angeles, Calif. 90024, U.S.A.*
and
*The Veterans Administration, Western Research Support Center,
Sepulveda, Calif., U.S.A.*

Summary

In this paper we investigate some of the properties of augmented Bose-Chaudhuri type parity check matrices. It is shown that some of the ensuing codes not only possess the property of correcting a predetermined number of independent errors, but also corrects certain bursts of errors that may occur in transmission of messages. Using the concept of a cycle it is shown how we may construct Bose-Chaudhuri type parity check matrices for which the number of message places is not of the form $(2^m - 1)/c$, where c is an integer greater than or equal to unity.

1. Introduction

Bose and Ray-Chaudhuri [1960] have defined a class of binary group codes with the property that if a message vector $\alpha \equiv (a_1, a_2, \ldots, a_n)$ of length $n = (2^m - 1)/c$, $c \geqslant 1$ an integer, is transmitted and if the received message vector $\beta \equiv (b_1, b_2, \ldots, b_n)$ contains t or fewer errors, where t is a predetermined number, $mt < n$, $a_i, b_i \in GF(2)$, $i = 1, 2, \ldots, n$, then the received message β is correctly decoded as α.

In this paper we show that augmenting the Bose-Chaudhuri parity check matrix by adjoining the column vector $j_{n \times 1}$ to it, where $j_{n \times 1} = (1, 1, \ldots, 1)$, the ensuing parity check matrix corrects not only t independent errors, but all error bursts of length $t + 1$.

Gross [1961] defines a (Galois) cycle as follows: Let p be an arbitrary prime. Suppose that y is a primitive element of $GF(p^m)$ and z is a primitive

† This research was supported in part by USPHS Grant No. 8 PO2–HS–00234, National Center for Health Services Research and Development.

‡ Alan J. Gross completed his Ph.D. dissertation, under the direction of Professor Bose, August 19, 1961. The title of his dissertation is "On the construction of burst-error-correcting codes."

* Now at the Department of Public Health, University of Massachusetts, Amherst, Mass., U.S.A.

element of $GF(p^{m-\nu})$, $m > \nu$. Further, suppose that W_1 and W_2 are integers dividing $p^m - 1$ and $p^{m-\nu} - 1$, respectively. Consider the vectors $(y^{W_1 i}, z^{W_2 i})$, $i = 0, 1, \ldots, l-1$. These vectors are said to form a Galois cycle of length l if:

(i) they are all distinct, and

(ii) $(y^{W_1 l}, Z^{W_2 l}) = (1, 1)$, where the first component "1" is the identity element of $GF(p^m)$ and the second component "1" is the identity element of $GF(p^{m-\nu})$.

In this paper we also consider the augmentation of Bose-Chaudhuri parity check matrices by examining the error-correcting properties of Bose-Chaudhuri parity check matrices whose elements are vectors forming Galois cycles in accordance with the above definition.

2. Augmented Bose-Chaudhuri codes that correct single and multiple bursts of errors

Definition 2.1. We say that an (n, k) binary group code is an *augmented Bose-Chaudhuri code* if the parity check matrix for this code is obtained from the Bose-Chaudhuri parity check matrix by adjoining to it one or more columns.

Lemma 2.1. *Let $D_{n,t}$ be the following Bose-Chaudhuri parity check matrix:*

$$D_{n,t} = \begin{bmatrix} \alpha_1 & \alpha_1^3 & \cdots & \alpha_1^{2t-1} \\ \alpha_2 & \alpha_2^3 & & \alpha_2^{2t-1} \\ \cdot & \cdot & & \cdot \\ \cdot & \cdot & & \cdot \\ \cdot & \cdot & & \cdot \\ \alpha_n & \alpha_n^3 & & \alpha_n^{2t-1} \end{bmatrix},$$

where $\alpha_i = y^{ci}$, $c \equiv c(n)$ is that integer such that $cn = 2^m - 1$, for a given pair (m, n), and y is a primitive element of $GF(2^m)$. Assume that $2t - 1 < n$. Then, given the rank of $D_{n,t}$ is $R(m, t)$, the rank of $D_{n,t}^*$ is $R(m, t) + 1$, where

$$D_{n,t}^* = \begin{bmatrix} \alpha_1 & \alpha_1^3 & \cdots & \alpha_1^{2t-1} & 1 \\ \alpha_2 & \alpha_2^3 & \cdots & \alpha_2^{2t-1} & 1 \\ \cdot & \cdot & & \cdot & \cdot \\ \cdot & \cdot & & \cdot & \cdot \\ \cdot & \cdot & & \cdot & \cdot \\ \alpha_n & \alpha_n^3 & & \alpha_n^{2t-1} & 1 \end{bmatrix}.$$

That is, $D_{n,t}^*$ is the augmentation of $D_{n,t}$ obtained by adjoining to it the column vector $j_{n \times 1}$.

The proof of Lemma 2.1 is found in Gross [1961].

Theorem 2.1. *The matrix $D_{n,t}^*$ is an appropriate parity check matrix for the binary group code (n, k) which corrects all single bursts of errors of length t*

or less as well as all sets of t independent errors, where $n = (2^m - 1)/c$ *is the number of message places and* $k = n - R(m, t) - 1$ *is the number of information places for the code.*

The proof of Theorem 2.1 is found in Gross [1963].

Example 2.1. *Suppose* $m = 5$, $c = 1$, *and* $t = 3$. *Then* $\alpha_i = y^i$, $i = 1, 2,$ ..., 31, *where* y *is a primitive element of* $GF(2^5)$ *whose characteristic polynomial is chosen to be* $y^5 + y + 1$. *The matrix* $D^*_{31,3}$ *which is given by*

$$D^*_{31,3} = \begin{bmatrix} \alpha_1 & \alpha_1^3 & \alpha_1^5 & 1 \\ \alpha_2 & \alpha_2^3 & \alpha_2^5 & 1 \\ \cdot & \cdot & \cdot & \cdot \\ \cdot & \cdot & \cdot & \cdot \\ \cdot & \cdot & \cdot & \cdot \\ \alpha_{30} & \alpha_{30}^3 & \alpha_{30}^5 & 1 \\ 1 & 1 & 1 & 1 \end{bmatrix} = \begin{bmatrix} 0\,1\,0\,0\,0 & 0\,0\,0\,1\,0 & 1\,0\,1\,0\,0 & 1 \\ 0\,0\,1\,0\,0 & 0\,1\,0\,1\,0 & 1\,0\,0\,0\,1 & 1 \\ \cdot & \cdot & \cdot & \cdot \\ \cdot & \cdot & \cdot & \cdot \\ \cdot & \cdot & \cdot & \cdot \\ 0\,1\,0\,0\,1 & 0\,1\,1\,0\,1 & 1\,1\,1\,0\,1 & 1 \\ 1\,0\,0\,0\,0 & 1\,0\,0\,0\,0 & 1\,0\,0\,0\,0 & 1 \end{bmatrix}$$

is an appropriate parity check matrix for the binary group code $n = 31$, $k = 31 - R(5, 3) - 1$, *which corrects all independent triple errors as well as all single error bursts of length four or less.*

Finally, we determine $R(5, 3)$ so that k can be evaluated.

Bose and Ray-Chaudhuri [1960] show that $R(m, t)$ is the number of distinct residue classes mod n among the integers

$$2^j u, \quad j = 0, 1, \ldots, m-1, \quad u = 1, 3, \ldots, 2t-1.$$

For $m = 5$, $n = 31$, and $t = 3$, the classes are:

1	2	4	8	16
3	6	12	24	17
5	10	20	9	18

Thus $k = 15$, and the code corresponding to $D^*_{31,3}$ has 31 message places and 15 information places.

Theorem 2.2. *Consider the following augmented Bose-Chaudhuri parity check matrix* $D^*_{n,2s+1}$ *given by*

$$D^*_{n,2s+1} = \begin{bmatrix} \alpha_1 & \alpha_1^3 & \cdots & \alpha_1^{4s+1} & 1 \\ \alpha_2 & \alpha_2^3 & \cdots & \alpha_2^{4s+1} & 1 \\ \cdot & \cdot & \cdots & \cdot & \cdot \\ \cdot & \cdot & \cdots & \cdot & \cdot \\ \cdot & \cdot & \cdots & \cdot & \cdot \\ \alpha_n & \alpha_n^3 & \cdots & \alpha_n^{4s+1} & 1 \end{bmatrix}$$

where $\alpha_i = y^{ci}$, $c \equiv c(n)$ *is that integer such that* $cn = 2^m - 1$ *for a given pair* (m, n) *and* y *is a primitive element of* $GF(2^m)$. *Then* $D^*_{n,2s+1}$ *is an appropriate parity check matrix for the binary group code* (n, k), $n = (2^m - 1)/c$, $k = n - R(m, 2s+1) - 1$, *which corrects a set of* $s+1$ *independent bursts of errors of* *length two or less, as well as all $2s+1$ independent errors, provided that $(2^m-1)/c > 4s+1$, $s \geqslant 1$. (N.B. all single bursts of length $2s+2$ are subsumed by all $s+1$ independent bursts of length two.)*

Proof. First of all, since $D_{n,2s+1}^*$ is an augmented Bose-Chaudhuri matrix, $D_{n,2s+1}^*$ possesses property P_{4s+2} — any set of $2s+1$ or less independent errors are corrected. Let δ_i denote the i-th row vector of $D_{n,2s+1}^*$, $i = 0, 1, \ldots, n$. We need show the following:

$$\sum_{v=1}^{2s+1} \delta_{i_v} \neq \sum_{v=1}^{s+1} (\delta_{i'_v} + \delta_{i'_{v+1}}), \tag{1}$$

where $\delta_{i_1}, \ldots, \delta_{i_{2s+1}}$; $\delta_{i'_1}, \ldots, \delta_{i'_{s+1}}$ are any $3s+2$ row vectors chosen from among δ_i, $i = 1, 2, \ldots, n$, and

$$\sum_{v=1}^{s+1} (\delta_{j_v} + \delta_{j_{v+1}}) \neq \sum_{v=1}^{s+1} (\delta_{j'_v} + \delta_{j'_{v+1}}), \tag{2}$$

where $\delta_{j_1}, \ldots, \delta_{j_{s+1}}$; $\delta_{j'_1}, \ldots, \delta_{j'_{s+1}}$ are any $2s+2$ vectors chosen from δ_i, $i = 0, 1, \ldots, n$. (1) is clearly satisfied since the ensuing vector on the left-hand side contains a one in its last position whereas the one on the right-hand side contains a zero in its last position. To show that (2) is satisfied, assume that for some set of vectors $\delta_{j_1}, \ldots, \delta_{j_{s+1}}$; $\delta_{j'_1}, \ldots, \delta_{j'_{s+1}}$ all distinct

$$\sum_{v=1}^{s+1} (\delta_{j_v} + \delta_{j_{v+1}}) = \sum_{v=1}^{s+1} (\delta_{j'_v} + \delta_{j'_{v+1}}). \tag{3}$$

Then

$$\sum_{v=1}^{s+1} \alpha_u^{j_v}(1 + \alpha_u) = \sum_{v=1}^{s+1} \alpha_u^{j'_v}(1 + \alpha_u) \tag{4}$$

holds for $u = 1, 3, 5, \ldots, 4s+1$. Since $n > 4s+1$, it follows that $1 + \alpha_u \neq 0$ for $u = 1, 3, 5, \ldots, 4s+1$. Thus

$$\sum_{v=1}^{s+1} \alpha_u^{j_v} = \sum_{v=1}^{s+1} \alpha_u^{j'_v} \tag{5}$$

holds for $u = 1, 3, 5, \ldots, 4s+1$, which implies

$$\sum_{v=1}^{s+1} \delta_{j_v} = \sum_{v=1}^{s+1} \delta_{j'}. \tag{6}$$

This, however, contradicts that $D_{n,2s+1}^*$ has property P_{4s+2}. This contradiction establishes the proof of the theorem.

Example 2.2. *Suppose $m = 5$, $s = 1$, and $c = 1$. Then, as in Example 2.1, we obtain the parity check matrix $D_{31,3}^*$. Thus, $D_{31,3}^*$ is an appropriate parity check matrix for the binary group code $(31, 15)$ which corrects not only all independent triple errors and all single error bursts of length four or less, but also corrects all independent double bursts of errors of length two or less.*

3. Bose-Chaudhuri codes augmented by means of Galois cycles

A Bose-Chaudhuri parity check matrix whose columns are formed by means of Galois cycles has the following form:

$$D_{n_1,n_2,t} = \begin{bmatrix} \gamma_1 & \gamma_1^3 & \cdots & \gamma_1^{2t-1} \\ \gamma_2 & \gamma_2^3 & \cdots & \gamma_2^{2t-1} \\ \cdot & \cdot & & \cdot \\ \cdot & \cdot & & \cdot \\ \cdot & \cdot & & \cdot \\ \gamma_{n_1 n_2} & \gamma_{n_1 n_2}^3 & \cdots & \gamma_{n_1 n_2}^{2t-1} \end{bmatrix}$$

where $\gamma_i^j = ((y^{c_1 i})^j, (z^{c_2 i})^j)$, $i = 1, 2, \ldots, n_1 n_2$, $c_i \equiv c(n_i)$ is that integer such that $c_i n_i = 2^{m_i} - 1$, for given pairs (m_i, n_i), $i = 1, 2$, and the vectors γ_i form a Galois cycle of length $n_1 n_2$. As a consequence of the construction of $D_{n_1,n_2,t}$, n_1 and n_2 are relatively prime.

The following lemma will be necessary in order to determine the rank of the matrix $D_{n_1,n_2,t}$.

Lemma 3.1. *Suppose $X^{n_1} - 1$ and $X^{n_2} - 1$ are polynomials whose coefficients belong to any field. If n_1 and n_2 are relatively prime integers, then the only common divisor of $X^{n_1} - 1$ and $X^{n_2} - 1$ is $X - 1$.*

Proof. Without loss of generality we assume $n_2 > n_1$. Since n_1 and n_2 are relatively prime, there exists pair of integers s_1 and s_2, $s_1 \leqslant 0$, $s_2 > 0$, such that $s_1 n_1 + s_2 n_2 = 1$. Assume that $P_v(X)$, the greatest common factor of $X^{n_1} - 1$ and $X^{n_2} - 1$, is of degree v. Then, since $P_v(X)|(X^{n_2} - 1)$, it follows that $P_v(X)|(X^{s_2 n_2} - 1)$, since $(X^{n_2} - 1)|(X^{s n_2} - 1)$ for all integers $s \geqslant 1$. Now $P_v(X)|[(X^{s_2 n_2} - 1) - (X^{n_1} - 1)]$ since $P_v(X)$ divides each polynomial. Thus, $P_v(X)|(X^{s_2 n_2 - n_1} - 1)$. Repeating this differencing procedure j times, where $j = -s_1$ we see that $P_v(X)$ must divide $X^{s_2 n_2 + s_1 n_1} - 1$, which implies that $P_v(X)$ divides $X - 1$, since $s_1 n_1 + s_2 n_2 = 1$. But clearly $X - 1$ divides $P_v(X)$ since $P_v(X)$ is the greatest common divisor of $X^{n_1} - 1$ and $X^{n_2} - 1$. Thus, $P_v(X) = X - 1$, which proves the lemma.

Recall now that the code words corresponding to the parity check matrix $D_{n_1,n_2,t}$ form a vector space orthogonal to $D_{n_1,n_2,t}$. Thus $(d_1, d_2, \ldots, d_{n_1 n_2})$ is a code word corresponding to $D_{n_1,n_2,t}$, $d_i \in \mathrm{GF}(2)$, $i = 1, 2, \ldots, n_1 n_2$, provided that

$$d_1 \gamma_1 + d_2 \gamma_2 + \cdots + d_{n_1 n_2} \gamma_{n_1 n_2} = 0,$$

$$d_1 \gamma_1^3 + d_2 \gamma_2^3 + \cdots + d_{n_1 n_2} \gamma_{n_1 n_2}^3 = 0,$$

$$\cdot \qquad \cdot \qquad \cdot$$
$$\cdot \qquad \cdot \qquad \cdot$$
$$\cdot \qquad \cdot \qquad \cdot$$

$$d_1 \gamma_1^{2t-1} + d_2 \gamma_2^{2t-1} + \cdots + d_{n_1 n_2} \gamma_{n_1 n_2}^{2t-1} = 0$$

all hold simultaneously. We now observe that $d_1, d_2, \ldots, d_{n_1 n_2}$ are chosen so that

$$d_1 + d_{n_1+1} \cdots + d_{n_1(n_2-1)+1} = e_1 \pmod{2},$$
$$d_2 + d_{n_1+2} \cdots + d_{n_1(n_2-1)+2} = e_2 \pmod{2},$$

$$d_{n_1} + d_{2n_1} \qquad + d_{n_1 n_2} \qquad = e_{n_1} \pmod{2},$$

and

$$d_1 + d_{n_2+1} + \cdots + d_{n_2(n_1-1)+1} = f_1 \pmod{2},$$
$$d_2 + d_{n_2+2} + \cdots + d_{n_2(n_1-1)+2} = f_2 \pmod{2},$$

$$d_{n_2} + d_{2n_2} + \cdots + d_{n_1 n_2} \qquad = f_{n_2} \pmod{2},$$

where $(e_1, \ldots, e_{n_1})$ and $(f_1, \ldots, f_{n_2})$ are code words corresponding to $D_{n_1,t}$ and $D_{n_2,t}$, respectively. $D_{n_i,t}$ is a Bose-Chaudhuri parity check matrix of length $n_i = (2^{m_i} - 1)/c_i$, having rank $R(m_i, t)$, $i = 1, 2$. The code word $(d_1, d_2, \ldots, d_{n_1 n_2})$ identified with the polynomial

$$S(X) = \{d_0 + d_1 X + \cdots + d_{n_1 n_2 - 1} X^{n_1 n_2 - 1}\} \bmod \{X^{n_1 n_2} - 1\},$$

where $d_{n_1 n_2} \equiv d_0$, which has as its roots $y^{c_1}, y^{3c_1}, \ldots, y^{(2t-1)c_1}, z^{c_2}, z^{3c_2}, \ldots,$ and $z^{(2t-1)c_2}$. For example,

$$S(y^{c_1}) = d_0 + d_1 y^{c_1} + \cdots + d_{n_1 n_2 - 1}(y^{c_1})^{n_1 n_2 - 1},$$
$$\equiv e_0 + e_1 y^{c_1} + \cdots + e_{n_1 - 1}(y^{c_1})^{n_1 - 1} = 0,$$

where $(e_0, \ldots, e_{n_1 - 1})$ is a code word corresponding to $D_{n_1,t}$ and $e_0 \equiv e_{n_1}$, and

$$S(z^{c_2}) = d_0 + d_1 z^{c_1} + \cdots + d_{n_1 n_2 - 1}(z^{c_1})^{n_1 n_2 - 1}$$
$$\equiv f_0 + f_1 z^{c_1} + \cdots + f_{n_2 - 1}(z^{c_1})^{n_2 - 1} = 0,$$

where $(f_0, \ldots, f_{n_2 - 1})$ is a code word corresponding to $D_{n_2,t}$ and $f_0 \equiv f_{n_2}$.

Lemma 3.2. *The rank of $D_{n_1,n_2,t}$ is $R(m_1, t) + R(m_2, t)$, where $R(m_i,t)$ is the number of distinct residue classes mod n_i among the integers $2^j u$, $u = 1, 3, \ldots,$ $2t-1$, $j = 0, 1, \ldots, m_i - 1$, $i = 1, 2$.*

Proof. Following Peterson's proof [1960], the rank of $D_{n_1, n_2, t}$ is the degree of the polynomial

$$g(X) = \text{L.C.M.}[P_1(X), \ldots, P_{2t-1}(X), Q_1(X), \ldots, Q_{2t-1}(X)],$$

where $P_w(X)$ is the minimum polynomial having y^{wc_1} as a root, and $Q_l(X)$ is the minimum polynomial having z^{lc_2} as a root, $w, l = 1, 3, \ldots, 2t-1$. It follows from Lemma 3.1 that $P_w(X)$ and $Q_l(X)$ are relatively prime, since

$P_w(X)$ divides $X^{n_1} - 1$ and $Q_w(X)$ divides $X^{n_2} - 1$ and neither $P_w(X)$ nor $Q_l(X)$ contains $X - 1$ as a factor. Thus

$$g(X) = g_1(X)g_2(X),$$

where

$$g_1(X) = \text{L.C.M.}[P_1(X), \ldots, P_{2t-1}(X)],$$

and

$$g_2(X) = \text{L.C.M.}[Q_1(X), \ldots, Q_{2t-1}(X)].$$

Since the rank of $D_{n_1,n_2,t}$ is the degree of $g(X)$ which is the sum of the degrees of $g_1(X)$ and $g_2(X)$, the proof is complete.

Summarizing these results we now have:

Theorem 3.1. *The matrix $D_{n_1,n_2,t}$ is an appropriate parity check matrix for the binary group (n', k') which corrects all sets of t independent errors, where $n' = n_1 n_2$ is the number of message places, and $k' = n' - R(m_1, t) - R(m_2, t)$ is the number of information places for the code provided $D_{n_1,n_2,t}$ has property P_{2t}.*

Example 3.1. *Suppose $m_1 = 5$, $c_1 = 1$, $m_2 = 2$, $c_2 = 1$, and $t = 3$. Then $\gamma_i = (y^i, z^i)$, $i = 1, 2, \ldots, 93$, where y is a primitive element of $\text{GF}(2^5)$ and z is a primitive element of $\text{GF}(2^2)$, where y and z have characteristic polynomials $y^5 + y + 1$ and $z^2 + z + 1$, respectively. The matrix $D_{31,3,3}$ which is given by*

$$D_{31,3,3} = \begin{bmatrix} \gamma_1 & \gamma_1^3 & \gamma_1^5 \\ \gamma_2 & \gamma_2^3 & \gamma_2^5 \\ \cdot & \cdot & \\ \cdot & \cdot & \\ \cdot & \cdot & \\ \gamma_{92} & \gamma_{92}^3 & \gamma_{92}^5 \\ \gamma_{93} & \gamma_{93}^3 & \gamma_{93}^5 \end{bmatrix} = \begin{bmatrix} 01000 & 01 & 00010 & 10 & 10100 & 11 \\ 00100 & 11 & 01010 & 10 & 10001 & 01 \\ \cdot & & \cdot & \cdot & & \cdot \\ \cdot & & \cdot & \cdot & & \cdot \\ \cdot & & \cdot & \cdot & & \cdot \\ 01001 & 11 & 01101 & 10 & 11101 & 01 \\ 10000 & 10 & 10000 & 10 & 10001 & 10 \end{bmatrix}$$

is an appropriate parity check matrix for the binary group code $n' = 93$, $k' = 93 - R(5, 3) - R(2, 3)$, which corrects all independent triple errors.

Note that since $D_{31,3,3}$ contains a column $j_{93 \times 1}$, where $j_{1 \times 93} = (1, 1, \ldots, 1)$ (column 13 of the matrix), it also corrects all single bursts of length four or less. $R(5, 3) = 16$ from Example 3.1. Since $R(2, 3)$ is the number of distinct residue classes mod 3 among $2^j u$ $j = 0, 1$, $u = 1, 3, 5$. The classes are:

$$\begin{array}{cc} 1 & 2 \\ 0 & 0 \\ 2 & 1 \end{array}$$

Thus, $R(2, 3) = 3$; $k' = 93 - 19 = 74$ and the code corresponding to $D_{31,3,3}$ has 93 message places and 74 information places provided $D_{31,3,3}$ has property P_6.

References

R. C. Bose and D. K. Ray-Chaudhuri, 1960, On a class of error correcting binary group codes, *Inf. Control* **3**, 69–79.

R. C. Bose and D. K. Ray-Chaudhuri, 1960, Further results on error correcting binary group codes, *Inf. Control* **3**, 279–290.

A. J. Gross, 1961, On the construction of burst-error-correcting codes, Ph.D. Dissertation, Univ. of North Carolina, Chapel Hill, N. Car., pp. 24–26.

A. J. Gross, 1963, Augmented Bose-Chaudhuri codes which correct single bursts of errors, *IRE Trans. Inf. Theory* **IT-9**, 121.

W. W. Peterson, 1960, Encoding and error-correcting procedures for the Bose-Chaudhuri codes, *IRE Trans. Inf. Theory* **IT-6**, 459–470.

J. N. Srivastava et al., eds., *A Survey of Combinatorial Theory*
© North-Holland Publishing Company, 1973

CHAPTER 19

On Order Statistics and Some Applications of Combinatorial Methods in Statistics †

SHANTI S. GUPTA

Purdue University, Lafayette, Ind. 47907, U.S.A.

and

S. PANCHAPAKESAN

Southern Illinois University, Carbondale, Ill., U.S.A.

1. Introduction

Research in the area of order statistics has been steadily and rapidly growing especially during the last two decades. The extensive role of order statistics in several areas of statistical inference has made it imperative and useful to gather these results and present them in varied manner to suit diverse interests. The present paper is an instance of such an attempt.

Historically, formal investigation in the sampling theory of order statistics dates back to 1902 when Karl Pearson solved the problem of finding the mean of the difference between the rth and the $(r+1)$th order statistics in a sample of n observations from a continuous population. Tippett [1925] found the mean of the sample range and tabulated for certain sample sizes ranging from 3 to 1000, the cumulative distribution function (cdf) of the largest order statistic in a sample from a standard normal population. Asymptotic results were first obtained by Fisher and Tippett [1928], who determined under certain regularity conditions the limiting distributions of the largest and the smallest order statistics as the sample size increases indefinitely by a method of functional equations. These early developments and subsequent research over a period of nearly a quarter of a century have been nicely summarized by Wilks [1948] in a survey paper. Since then, a huge volume of research has been accomplished in this field dealing with several aspects of the problems involving order statistics. Besides the basic distribution theory and limit laws, attention has been focused by several authors on problems involving order statistics in the theory of estimation and testing of hypotheses and in multiple decision and multiple comparison procedures. Many of these results are embodied in books and monographs;

† Research supported in part by the Office of Naval Research Contract N00014–67–A–0226–00014 at Purdue University. Reproduction in whole or in part is permitted for any purpose of the United States Government.

mention should be made of Gumbel [1958], Sarhan and Greenberg [1962], Miller [1966] and David [1970].

The modest objective of the present authors is to state some of the basic results in the theory of order statistics, describe the trend of the work done in certain areas by referring to what might be called "landmark" papers and indicate some of the recent results. In doing so, certain areas where order statistics play an important role have not been considered with no reflection on the nature of their importance in applications; for example, multiple comparison problems and slippage tests. A few topics have been treated to a very limited extent. The basic theory (Section 2), results concerning moments and inequalities (Section 3) and problems concerning estimation and hypotheses testing (Section 7) come under this category. Section 4 discusses some important asymptotic results relating to the papers of Gnedenko [1943], Smirnov [1949], Rényi [1953], Berman [1962], Pyke [1965] and Kiefer [1970a]. Applications of combinatorial methods in the general distribution theory and fluctuation theory have been described in Sections 5 and 6. These results are mainly concerned with the applications of the ballot lemma and its generalizations and the use of the equivalence principle proved by Andersen [1953]. The last section discusses the role of order statistics in the subset selection problems and the algebraic structure involved in identification problems.

2. Basic distribution theory

Let $X_1, X_2, \ldots, X_n$ be independent and identically distributed random variables each having an absolutely continuous distribution function $F(x)$ and the corresponding density function $f(x)$. Let the ordered variables be denoted by

$$X_{(1)} < X_{(2)} < \cdots < X_{(n)}. \tag{2.1}$$

If the situation demands more clarity, $X_{(r)}$ will be denoted by $X_{r,n}$. It is well-known that $f_r(x)$ and $F_r(x)$, namely, the density and the cdf of $X_{(r)}$ are given by

$$f_r(x) = r\binom{n}{r}F^{r-1}(x)[1-F(x)]^{n-r}f(x) \tag{2.2}$$

and

$$F_r(x) = \sum_{i=r}^{n} \binom{n}{i}F^i(x)[1-F(x)]^{n-i} \tag{2.3}$$

$$= I_{F(x)}(r, n-r+1),$$

where $I_p(a, b)$ is the incomplete beta function defined by

$$I_p(a, b) = \frac{\Gamma(a+b)}{\Gamma(a)\Gamma(b)} \int_0^p t^{a-1}(1-t)^{b-1}\,\mathrm{d}t, \tag{2.4}$$

$$a, b > 0, \qquad 0 \leqslant p \leqslant 1.$$

The joint density function $f_{r,s}(x, y)$ of $X_{(r)}$ and $X_{(s)}$ $(1 \leqslant r < s \leqslant n)$ is given by

$$f_{r,s}(x, y) = \begin{cases} \dfrac{n!}{(r-1)!(s-r-1)!(n-s)!} F^{r-1}(x)[F(y)-F(x)]^{s-r-1} \\ \qquad \times [1-F(y)]^{n-s}f(x)f(y), \qquad \text{if } x \leqslant y, \\ 0 \qquad\qquad\qquad\qquad\qquad\qquad \text{otherwise.} \end{cases} \qquad (2.5)$$

If ξ_p is the unique quantile of order p of the parent distribution $F(x)$, then it is easy to see that

$$P\{X_{(r)} \leqslant \xi_p \leqslant X_{(s)}\} = \sum_{i=r}^{s-1} \binom{n}{i}p^i(1-p)^{n-i}, \qquad (2.6)$$

a result which enables us to obtain distribution-free confidence interval for ξ_p.

Let us define for $r = 1, 2, \ldots, n$,

$$Y_r = F(X_r), \qquad Z_r = -\log Y_r \qquad (2.7)$$

and

$$Y_{(r)} = F(X_{(r)}), \qquad Z_{(r)} = -\log Y_{(n+1-r)}. \qquad (2.8)$$

It is known and it can easily be verified that $Z_1, \ldots, Z_n$ are independent and have the exponential distribution $F(z) = 1-\mathrm{e}^{-z} (z > 0)$. By defining $X_{(0)} \equiv 0$ and

$$W_r = (n-r+1)(Z_{(r)}-Z_{(r-1)}), \qquad r = 1, \ldots, n, \qquad (2.9)$$

we have

$$X_{(r)} = F^{-1}\exp(-Z_{(n+1-r)}) \qquad (2.10)$$

$$= F^{-1}\exp\left(-\left(\frac{W_1}{n} + \cdots + \frac{W_{n+1-r}}{r}\right)\right).$$

A well-known result due to Sukhatme [1937] states that $W_1, \ldots, W_n$ are independent and exponentially distributed with cdf $= 1-\mathrm{e}^{-x} (x > 0)$. An important and very useful fact when dealing with order statistics is that they form a Markov process. To be precise, $\{X_{(r)}: 1 \leqslant r \leqslant n\}$ is a non-homogeneous discrete-parameter, real-valued Markov process whose initial measure is $F_1(x) = 1-[1-F(x)]^n$ and whose transition distribution function $P\{X_{(r+1)} \leqslant x| X_{(r)} = y\}$ is the distribution of the minimum of $(n-r)$ independent observations on the distribution F truncated at y, that is,

$$P\{X_{(r+1)} \leqslant x| X_{(r)} = y\} = 1-[1-F(x)]^{n-r}[1-F(y)]^{-n+r}, \quad x > y. \quad (2.11)$$

This Markov property was first pointed out by Kolmogorov [1933]. Further it is clear from (2.10) that $Y_{(r)}/Y_{(r+1)} = \mathrm{e}^{-W_{n+1-r}/r}$, $r = 1, \ldots, n$, are all independent $[Y_{(n+1)} \equiv 1]$. Hence $[Y_{(r)}/Y_{(r+1)}]^r$, $r = 1, \ldots, n$, are independent and uniformly distributed on $(0, 1)$.

Now, the joint density of $X_{(1)}, \ldots, X_{(n)}$ is given by

$$f(x_1, \ldots, x_n) = \begin{cases} n! f(x_1) \cdots f(x_n), & x_1 < x_2 < \cdots < x_n, \\ 0 & \text{otherwise.} \end{cases} \qquad (2.12)$$

Define

$$D_r = X_{(r)} - X_{(r-1)}, \; r = 2, \ldots, n. \tag{2.13}$$

Then $D_2, \ldots, D_n$ are called the spacings. For some distributions we may define $X_{(0)}$ and $X_{(n+1)}$ suitably depending on F, and let $D_1 = X_{(1)} - X_{(0)}$ and $D_{n+1} = X_{(n+1)} - X_{(n)}$. For example, if F is the uniform distribution on $(0, 1)$, then $X_{(0)} = 0$ and $X_{(n+1)} = 1$. If $F(x) = \lambda e^{-\lambda x}$, $x > 0$, we will just define $X_{(0)} = 0$. Thus, depending on the particular F, the number of spacings considered could be different. For the general spacing we see that the joint density of $D_2, \ldots, D_n$ is given by

$$f_D(d_2, \ldots, d_n) = \begin{cases} n! \displaystyle\int_{-\infty}^{\infty} \prod_{r=2}^{n} f(x + d_2 + \cdots + d_r) \; dx, \; d_2, \ldots, d_n > 0, \\ 0 \qquad\qquad\qquad\qquad\qquad\qquad\qquad\qquad \text{otherwise,} \end{cases} \tag{2.14}$$

and the density of D_r is

$$f_{D_r}(y) = \frac{n!}{(r-2)!(n-r)!} \int_{-\infty}^{\infty} F^{r-2}(x)[1 - F(x+y)]^{n-r} f(x) f(x+y) \, dx. \tag{2.15}$$

In the case of the exponential spacings, $D_1, \ldots, D_n$ are independent exponential random variables with parameters $\lambda n, \lambda(n-1), \ldots, \lambda$. Equivalently, the normalized spacings $\lambda(n-r+1)D_r$, $1 \leqslant r \leqslant n$, are independent and identically distributed exponentially with mean unity. The exponential spacings can be looked upon as holding times of a continuous parameter Markov process. The first unified approach to the distribution theory of uniform spacings is given by Darling [1953]. A good discussion of spacings can be found in Pyke [1965]. The use of spacings in tests of hypotheses is discussed in a subsequent section.

3. Moments of order statistics and bounds

Some important results are concerned with moments of order statistics from specific distributions, particularly the normal distribution, and bounds for the moments under certain assumptions on the parent distribution. When $f(x)$ is symmetric about the origin, we have

$$E(X_{r,n}) = -E(X_{n-r+1,n}) \tag{3.1}$$

and

$$\mathrm{Cov}\,(X_{r,n}, X_{s,n}) = \mathrm{Cov}\,(X_{n-s+1,n}, X_{n-r+1,n}). \tag{3.2}$$

In the case of the standard normal distribution, the means, variances and covariances have been calculated for different ranges of values of n by Sarhan and Greenberg [1956], Teichroew [1956] and Harter [1961a]. Bose and

Gupta [1959] have discussed the evaluation of the exact moments of order statistics in the normal case. By defining

$$I_n(a) = \int_{-\infty}^{\infty} [\Phi(ax)]^n \, e^{-x^2} \, dx, \tag{3.3}$$

they have obtained the recurrence relation

$$I_{2m+1}(a) = \sum_{r=1}^{2m+1} \frac{(-1)^{r+1}\binom{2m+1}{r}I_{2m-r+1}(a)}{2^r} \tag{3.4}$$

which is used to obtain the moments up to $n = 5$.

Moments of order statistics from other continuous distributions have been considered by several authors and tables are available to varying extents. Some of the distributions considered are uniform (Hastings et al. [1947]), gamma (Gupta [1960, 1962], Breiter and Krishnaiah [1968]), double exponential (Govindarajulu [1966]), logistic (Gupta and Shah [1965], Shah [1966], Gupta et al. [1967]) and Cauchy (Barnett [1966]). As for the discrete distributions, the mean and variance of the smaller of two binomial variates are considered by Craig [1962] and Shah [1966a]. Gupta and Panchapakesan [1967] have discussed order statistics arising out of binomial population and have tabulated the first two moments of the largest and the smallest of M independent and identical binomial random variables, each denoting the number of successes in N independent trials with p as the probability of a success, for $N = 1(1)20$, $M = 1(1)10$ and $p = 0.05(0.05)0.50$.

In some cases we are interested in inequalities concerning the moments of order statistics from distributions F and G which are partially ordered in a certain sense in the space of probability distributions. Van Zwet [1964] considers convex ordered and s-ordered distributions. Some special cases of partial ordering and some properties of order statistics from partially ordered distributions are of interest in selection problems. Suppose that $X_1, \ldots, X_n$ and $Y_1, \ldots, Y_n$ are two independent random samples from continuous distributions F and G, respectively. Let $X_{r,n}$ and $Y_{r,n}$ $(r = 1, \ldots, n)$ denote the order statistics based on each of the two sets of observations. If F is star-shaped with respect to G, that is, $F(0) = G(0) = 0$ and $G^{-1}F(x)/x$ is increasing in $x > 0$, then it has been shown by Barlow and Gupta [1969] that the distribution of $X_{r,n}$ is star-shaped with respect to that of $Y_{r,n}$. Further, for $0 < c < 1$,

$$P\{\max(X_n/X_1, \ldots, X_n/X_{n-1}) \geq c\} \geq P\{\max(Y_n/Y_1, \ldots, Y_n/Y_{n-1}) \geq c\}, \tag{3.5}$$

a result which is used to obtain a lower bound on the probability of a correct selection. Comparisons between linear combinations of order statistics from F and G have been studied by Barlow and Proschan [1966], where (a) F is star-shaped w.r.t. G and (b) F is convex-ordered w.r.t. G. It is known that (b) implies (a). These results have applications in life testing where the underlying distribution has monotone failure rate or monotone failure rate on the average. For illustrating the nature of the results, we state the following theorem proved by Barlow and Proschan.

Theorem 3.1. *Let F be star-shaped w.r.t. G. Then $EX_{r,n}/EY_{r,n}$ is* (i) *decreasing in r,* (ii) *increasing in n, and* (iii) *$EX_{n-r,n}/EY_{n-r,n}$ is decreasing in n.*

If $G(x) = 1 - e^{-x}$, $x \geqslant 0$, then F is an IFRA (increasing failure rate on the average) distribution and from the above theorem it follows that

$$EX_{r,n} \bigg/ \sum_{j=1}^{r} \frac{1}{n-j+1}$$

is decreasing in r and increasing in n.

Theorem 3.1 can also be used to obtain bounds on $EX_{r,n}$. If we assume that F and G have the same mean θ, we obtain

$$\theta EY_{r,n}/EY_{r,r} \leqslant EX_{r,n} \leqslant \theta EY_{r,n}/EY_{1,n-r+1}. \tag{3.6}$$

Barlow and Proschan have also obtained a number of interesting special results when G is exponential and $G^{-1}F(x)$ is convex.

4. Asymptotic theory

We first state some basic results using the notations of Section 2.

Theorem 4.1. (Rényi [1953]). *If $r \geqslant 1$ is a fixed integer, $nZ_{(r)}$, $nY_{(r)}$ and $n(1 - Y_{(n-r+1)})$ all have in the limit $(n \to \infty)$ the gamma distribution with density $e^{-x} x^{r-1}/\Gamma(r)$, $x \geqslant 0$.*

Let $0 < \lambda_1 < \lambda_2 < \cdots < \lambda_k < 1$ and define $n_j = [n\lambda_j] + 1$, where $[\cdot]$ denotes the integral part. Let $\xi_{\lambda_j}, j = 1, \ldots, k$, be the population quantiles. Then we have the following theorem.

Theorem 4.2. (Mosteller [1946]). *If $f(x)$ is differentiable in the neighborhoods of ξ_{λ_j} and $f(\xi_{\lambda_j}) \neq 0$ $(j = 1, \ldots, k)$, then the joint distribution of the sample quantiles $X_{(n_1)}, \ldots, X_{(n_k)}$ tends to a k-dimensional normal distribution $N_k(\mu, \Sigma)$, where $\mu' = (\xi_{\lambda_1}, \xi_{\lambda_2}, \ldots, \xi_{\lambda_k})$ and $\Sigma = (\sigma_{ij})$ where*

$$\sigma_{ij} = \frac{\lambda_i(1 - \lambda_j)}{nf(\xi_{\lambda_i})f(\xi_{\lambda_j})}, \quad i \leqslant j.$$

A result concerning the moments of the order statistics is given in the following theorem due to Bickel [1967].

Theorem 4.3. *Suppose $\lim_{x \to \infty} |x|^\varepsilon [1 - F(x) + F(-x)] = 0$ for some $\varepsilon > 0$. Then*

(a) *for any natural number $k \geqslant 0$, $0 < \alpha < 1$, there exists $N(k, \alpha, \varepsilon)$ such that $E(X_{r,n}^k)$ exists for $\alpha n \leqslant r \leqslant (1-\alpha)n$ and $n \geqslant N(k, \alpha, \varepsilon)$. Conversely, if $E|X_{r,n}^k| < \infty$ for some k, n, then for some $\varepsilon > 0$, $\lim_{x \to \infty} |x|^\varepsilon [1 - F(x) + F(-x)] = 0$.*

(b) $E(X_{r,n} - F^{-1}(r/(n+1)))^k = n^{-k/2} \sigma^k(p_n)\mu_k + o(n^{-k/2})$ *uniformly for* $\alpha n \leqslant r \leqslant (1-\alpha)n$, n *sufficiently large, where* $p_n = r/n$, $\sigma^2(p_n) = [(r/n)(1-r/n) f(F^{-1}(r/n))]^{-2}$ *and* μ_k *is the k th central moment of the standard normal distribution.*

Theorem 4.3(b) has been proved in the literature under various regularity conditions by several authors including Chu and Hotelling [1955], Blom [1958] and Sen [1959].

Van Zwet [1964] establishes asymptotic expressions for $E(X_{r,n})$, $\mu_k(X_{r,n})$ and $m(X_{r,n})$, where the latter two denote the kth central moment and median of $X_{r,n}$. We assume that (i) F is twice differentiable on its support I with continuous second derivative F'' on I, (ii) $F'(x) > 0$ on I and (iii) there exist integers r and n, $1 \leqslant r \leqslant n$, such that $E(X_{r,n})$ exists. Let G be the inverse function of F. The following results have been obtained by Van Zwet.

Lemma 4.1. *If* $\lim r/n = q$, $0 < q < 1$, *as* n *tends to infinity, then*
(a) *for sufficiently large* n, $E(X_{r,n})$ *exists and*

$$E(X_{r,n}) = G(r/(n+1)) + \tfrac{1}{2}G''(r/(n+1))\mu_2(r, n) + o(n^{-1}), \qquad (4.1)$$

(b) *for sufficiently large* n, $\mu_k(X_{r,n})$ $(k = 2, 3, \ldots)$ *exists and*

$$\mu_k(X_{r,n}) \begin{cases} = (G'(r/(n+1)))^k \, \mu_k(r, n) + o(n^{-(k+1)/2}), & k \text{ even} \\ = (G'(r/(n+1)))^k \, \mu_k(r, n) + \tfrac{1}{2}k(G'(r/(n+1)))^{k-1} \, G''(r/(n+1)) \\ \quad \times [\mu_{k+1}(r, n) - \mu_2(r, n)\mu_{k-1}(r, n)] + o(n^{-(k+1)/2}), & k \text{ odd,} \end{cases}$$

(c) *for all* $\varepsilon > 0$,

$$m(X_{r,n}) = G\left(\frac{r}{n+1}\right) + \tfrac{1}{3}\frac{2r-n-1}{(n+1)^2}G'\left(\frac{r}{n+1}\right) + o(n^{-\frac{3}{2}+\varepsilon}), \qquad (4.3)$$

where $\mu_j(r, n)$ *is the jth central moment of the beta distribution with density* $r\binom{n}{r}y^{r-1}(1-y)^{n-r}$, $0 < y < 1$.

As a consequence of the above results, the following theorem has been obtained.

Theorem 4.4. *If* $\lim r/n = q$, $0 < q < 1$, *as* n *tends to infinity, then, for sufficiently large* n,

(a) $F(E(X_{r,n}))$ *exists and*

$$F(E(X_{r,n})) = \frac{r}{n+1} + \frac{r(n+1-r)}{2(n+1)^3}\frac{G''(r/(n+1))}{G'(r/(n+1))} + o(n^{-1}), \qquad (4.4)$$

(b) $\mu_{2k+1}(X_{r,n})/\sigma^{2k+1}(X_{r,n})$ *exists* $(k = 1, 2, \ldots)$ *and*

$$\mu_{2k+1}(X_{r,n})/\sigma^{2k+1}(X_{r,n}) = \mu_{2k+1}(r, n)(\mu_2(r, n))^{-k-\frac{1}{2}}$$
$$+ 2^{-k}\frac{(2k+1)!}{(k-1)!}\left[\frac{r(n+1-r)}{(n+1)^3}\right]^{\frac{1}{2}}\frac{G''(r/(n+1))}{G'(r/(n+1))} + o(n^{-\frac{1}{2}}), \quad (4.5)$$

(c) $\{E(X_{r,n}) - m(X_{r,n})\}/\sigma(X_{r,n})$ exists and

$$\frac{E(X_{r,n}) - m(X_{r,n})}{\sigma(X_{r,n})} = -\frac{2r - n - 1}{3\sqrt{r(n+1-r)(n+1)}}$$

$$+\tfrac{1}{2}\left[\frac{r(n+1-r)}{(n+1)^3}\right]^{\frac{1}{2}} \frac{G''(r/(n+1))}{G'(r/(n+1))} + o(n^{-\frac{1}{2}}). \qquad (4.6)$$

The result for $E(X_{r,n})$ is well-known (see David and Johnson [1954]); the result for $F(E(X_{r,n}))$ derived from it closely resembles the corresponding expression given by Blom [1958], who obtains his result under slightly different conditions.

One of the important areas where fruitful research has been accomplished is the theory of extreme order statistics. Contributions have been made in this area nearly over a period of five decades by several authors among whom notably are Fisher and Tippett [1928], Gumbel [1958, 1962] and Gnedenko [1943]. The important problem is to find $\mathscr{L}_k$, the family of all possible (nondegenerate) limit distributions for sequences of the form $b_n^{-1}(X_{k,n} - a_n)$, where a_n and b_n ($b_n > 0$) are constants. For $k = n$, a complete solution with specification of domains of attraction was given by Gnedenko [1943]. His results were generalized by Smirnov (1949) who obtained the following theorem.

Theorem 4.5. *The family $\mathscr{L}_k$ is given by*

$$\Lambda_1^{(k)}(x) = \begin{cases} 0, & x \leqslant 0, \alpha > 0, \\ \dfrac{1}{(n-k)!}\displaystyle\int_{x^{-\alpha}}^{\infty} e^{-t}\, t^{n-k}\, dt, & x > 0, \alpha > 0, \end{cases}$$

$$\Lambda_2^{(k)}(x) = \begin{cases} \dfrac{1}{(n-k)!}\displaystyle\int_{(-x)^{\alpha}}^{\infty} e^{-t}\, t^{n-k}\, dt, & x \leqslant 0, \alpha > 0, \\ 0, & x > 0, \alpha > 0, \end{cases} \qquad (4.7)$$

$$\Lambda_3^{(k)}(x) = \frac{1}{(n-k)!}\int_{e^{-x}}^{\infty} e^{-t}\, t^{n-k}\, dt, \qquad -\infty < x < \infty.$$

Berman [1962] shows that the limiting distribution for the maximal order statistic of a random number of independent indentically distributed random variables under certain general conditions is a mixture of distributions of $\mathscr{L}_n$.

Consider a sequence consisting of the sets of random variables $X_{n,1}, \ldots, X_{n,N_n}$; $n = 1, 2, \ldots$. Assume that $E(X_{n,k}) = 0$, $E(X_{n,k}^2) < \infty$ and the random variables in any set are independent.

Let

$$F_{n,k}(x) = P\{X_{n,k} < x\},$$

$$S_{n,k}(x) = \sum_{v=1}^{k} X_{n,v}, \tag{4.8}$$

$$B_n^2 = V(S_{n,N_n}) = \sum_{k=1}^{N_n} V(X_{n,k}).$$

Suppose that

$$\lim_{n\to\infty} \frac{1}{B_n^2} \sum_{k=1}^{N_n} \int_{|x|>\varepsilon B_n} x^2 \, dF_{n,k}(x) = 0, \quad \varepsilon > 0. \tag{4.9}$$

Under the above conditions the following results have been obtained by Rényi [1953].

Theorem 4.6.

(a) $\displaystyle \lim_{n\to\infty} P\{\max_{1\leqslant k\leqslant N_n} S_{n,k} < xB_n\} = \begin{cases} \sqrt{2/\pi} \displaystyle\int_0^x e^{-t^2/2} \, dt, & x > 0, \\ 0, & x \leqslant 0. \end{cases}$

(b) $\displaystyle \lim_{n\to\infty} P\{\max_{1\leqslant k\leqslant N_n} |S_{n,k}| < xB_n\} = \begin{cases} (4/\pi) \displaystyle\sum_{k=0}^{\infty} (-1)^k \frac{e^{-(2k+1)^2\pi^2/8x^2}}{(2k+1)}, & x > 0, \\ 0, & x \leqslant 0. \end{cases}$

(c) $\displaystyle \lim_{n\to\infty} P\{-yB_n \leqslant \min_{1\leqslant k\leqslant N_n} S_{n,k} \leqslant \max_{1\leqslant k\leqslant N_n} S_{n,k} < xB_n\}$

$$= \begin{cases} (4/\pi) \displaystyle\sum_{k=0}^{\infty} e^{-(2k+1)^2\pi^2/2(x+y)^2} \frac{\sin\left[(2k+1)\pi x/(x+y)\right]}{2k+1}, & x > 0 \text{ and } y \geqslant 0, \\ 0, & x \leqslant 0 \text{ or } y < 0. \end{cases}$$

(d) *Let* $A_{n}^2 = V(S_{n,M_n})$ *with* $1 \leqslant M_n < N_n$ *and* $\displaystyle\lim_{n\to\infty} A_n/B_n = \lambda \ (0 \leqslant \lambda < 1).$

Then

$$\lim_{n\to\infty} P\{\max_{M_n<k\leqslant N_n} |S_{n,k}| < yB_n\}$$

$$= \begin{cases} (4/\pi) \displaystyle\sum_{k=0}^{\infty} (-1)^k \frac{e^{-(2k+1)^2\pi^2/8y^2}}{2k+1} \left(1 - \sqrt{2/\pi} \int_{y/\lambda}^{\infty} e^{-u^2/2} \, du + \rho_k\right), & y > 0, \\ 0, & y \leqslant 0, \end{cases}$$

where

$$\rho_k = \frac{2\lambda \, e^{-y^2/2\lambda^2}}{y\sqrt{2\pi}} \int_0^{(2k+1)\pi/2} e^{\lambda^2 u^2/2y^2} \sin u \, du.$$

If $y = x$, (c) reduces to (b). In the special case $M_n = 1$ (i.e. for $\lambda = 0$), (d) is identical with (b). For the case where all the variables $X_{n,k}$ have the same distribution, the parts (a) and (b) were proved by Erdös and Kac [1946].

The classical theory of the limiting distribution of the maximum in sequences of independent random variables has been generalized in two directions, namely, (1) when the random variables are exchangeable and (2) when the number of random variables considered in the determination of the maximum is itself a random variable N_n, depending on a non-negative integer-valued parameter n. Let $\{X_n : n \geqslant 1\}$ be a sequence of exchangeable random variables defined on $(\Omega, \mathscr{A}, P)$, i.e. the joint df denoted by $G_m(x_1, \ldots, x_m)$ for each m is given according to the fundamental theorem of de Finetti (see Loéve [1960], p. 365) by

$$G_m(x_1, \ldots, x_m) = \int_\Omega G_\omega(x_1) \cdots G_\omega(x_m)\, dP(\omega), \tag{4.10}$$

where for fixed x, $G_\omega(x)$ is a random variable and for each $\omega \in \Omega$, $G_\omega(x)$ is a df in x. For any sequence $X_1, \ldots, X_n, P\{X_{(n)} \leqslant x\} = EG_\omega^n(x)$. The problem is to find sequences $\{a_n\}$ and $\{b_n\}$ and a df $L(x)$ such that $a_n > 0$ and

$$\lim_{n \to \infty} P\{a_n^{-1}(X_{(n)} - b_n) \leqslant x\} = \lim_{n \to \infty} EG_\omega^n(a_n x + b_n) = L(x) \tag{4.11}$$

for all $x \in C_L$, the set of continuity points of L. Let $\Lambda_i(x) = \Lambda_i^{(n)}(x)$, $i = 1, 2, 3$. The following results are due to Berman [1962].

Theorem 4.7. *Suppose that there exists a sequence of positive numbers* $\{a_n\}$ *and a df* $F(x)$ *in the domain of attraction of* $\Lambda_1(x)$ *such that* $\lim_{n \to \infty} F^n(a_n x) = \Lambda_1(x)$. *Then*

(a) *there exists a nondegenerate df* $L(x)$ *such that for all* $x \in C_L$,

$$\lim_{n \to \infty} P\{a_n^{-1} X_{(n)} \leqslant x\} = \lim_{n \to \infty} EG_\omega^n(a_n x) = L(x)$$

iff there exists a df $A(y)$ *such that for all* $y \in C_A$,

$$\lim_{u \to \infty} P\left\{\frac{\log G_\omega(u)}{\log F(u)} \leqslant y\right\} = A(y),$$

where $A(y)$ *satisfies the conditions*

$$A(\infty) - A(0-) = 1; \quad A(0+) - A(0-) < 1.$$

(b) $L(x)$ *is necessarily of the form*

$$L(x) = \begin{cases} 0, & x < 0, \\ \displaystyle\int_0^\infty [\Lambda_1(x)]^y\, dA(y), & x \geqslant 0. \end{cases}$$

Berman has obtained similar results for the case where we have a sequence of positive numbers $\{a_n\}$ and a real number x_0 and a df $F(x)$ in the domain of attraction of $\Lambda_2(x)$ such that $\lim_{n \to \infty} F^n(a_n x + x_0) = \Lambda_2(x)$ and for the case where there exist sequences $\{a_n\}$ and $\{b_n\}$ ($a_n > 0$) and a df $F(x)$ in the domain of attraction of $\Lambda_3(x)$ such that $\lim_{n\to\infty} F^n(a_nx+b_n) = \Lambda_3(x)$. The limiting distribution $L(x)$ is a mixture of $\Lambda_2(x)$ and $\Lambda_3(x)$, in each case, respectively.

Berman has also investigated the case of random number of random variables. Let $\{X_n : n \geqslant 1\}$ be a sequence of independent random variables with common df $F(x)$ which is in the domain of attraction of $\Lambda(x)$, one of the three extreme-value df's $\Lambda_i(x)$, $i = 1, 2, 3$. Let $\{N_n : n \geqslant 1\}$ be a sequence of nonnegative, integer-valued random variables distributed independently of the sequence $\{X_n\}$. Let N_n have the distribution given by $P\{N_n = k\} = p_n(k)$, $k \geqslant 0$, where for fixed n, $p_n(k) \geqslant 0$, $\sum_{k=0}^{\infty} p_n(k) = 1$. Define a sequence of random variables W_n as follows:

$$W_n = \begin{cases} -\infty, & N_n = 0, \\ X_{(N_n)}, & N_n > 0. \end{cases} \tag{4.12}$$

Then the df of W_n is

$$P\{W_n \leqslant x\} = \sum_{k=0}^{\infty} p_n(k) F^k(x). \tag{4.13}$$

This df is not necessarily proper:

$$\lim_{x\to-\infty} P\{W_n \leqslant x\} = p_n(0) \geqslant 0. \tag{4.14}$$

Suppose $N_n \to \infty$ in probability as $n \to \infty$. Then we have the following theorem.

Theorem 4.8. *There exists a df $L(x)$ such that for all $x \in C_L$,*

$$\lim_{n\to\infty} P\left\{a_n^{-1}(W_n-b_n) \leqslant x\right\} = \lim_{n\to\infty} \sum_{k=0}^{\infty} p_n(k) F^k(a_nx+b_n)$$
$$= L(x),$$

iff there exists a df $A(y)$ such that for all $y \in C_L$,

$$\lim_{n\to\infty} P\{n^{-1}N_n \leqslant y\} = A(y),$$

where $A(y)$ satisfies the conditions

   (i) $A(\infty) - A(0-) = 1$, $A(0+) - A(0-) < 1$, *or*

   (ii) $A(0+) - A(0-) = 0$, $0 < A(\infty) - A(0-) \leqslant 1$, *or*

   (iii) $A(0+) - A(0-) = 0$, $A(\infty) - A(0-) = 1$,

according as $\Lambda(x)$ is $\Lambda_1(x)$ or $\Lambda_2(x)$ or $\Lambda_3(x)$. Further $L(x)$ is a mixture of the appropriate $\Lambda(x)$ in each case.

If $\{N_n\}$ and $\{X_n\}$ are not necessarily independent of each other and if there exists a positive number c such that $n^{-1}N_n \to c$ in probability, then it has been shown that, for every x,

$$\lim_{n\to\infty} P\{a_n^{-1}(W_n-b_n) \leqslant x\} = \Lambda^c(x), \tag{4.15}$$

where $\Lambda^c(x)$ is of the same type as $\Lambda(x)$.

In the above set-up, let $N_n/n \to N$ in probability, where N is a random variable satisfying $P\{N \leqslant 0\} = 0$. It is known that if $E(X_1) = 0$, $V(X_1) = 1$, then as $n \to \infty$,

$$P\left\{\frac{S_{N_n}}{\sqrt{N_n}} \leqslant x\right\} \to \frac{1}{\sqrt{2\pi}} \int_{-\infty}^{x} e^{-t^2/2} \, dt \qquad \text{for all } x. \tag{4.16}$$

This result was proved by Anscombe [1952] in the case where N is constant with probability 1. Rényi [1960] extended Anscombe's result to discrete random variables N. Finally, Blum, Hanson and Rosenblatt [1963] and Mogyoródi [1962], independently obtained a proof for arbitrary positive N. Barndorff-Neilson [1964] proves the following result for $X_{(n)}$.

Theorem 4.9. *Let $\{a_n\}$ and $\{b_n\}$ be sequences of constants with $a_n > 0$ for all n and Λ be a nondegenerate df. The following three statements are equivalent:*

(i) $P\{X_{(n)} \leqslant a_n x + b_n\} \to \Lambda(x) \qquad$ *for all $x \in C_\Lambda$,*

(ii) $P\{X_{(N_n)} \leqslant a_{N_n} x + b_{N_n}\} \to \Lambda(x) \qquad$ *for all $x \in C_\Lambda$,*

(iii) $P\{X_{(N_n)} \leqslant a_n x + b_n\} \to \displaystyle\int_0^\infty [\Lambda(x)]^s \, dP[N \leqslant s], \qquad$ *for all $x \in C_\Lambda$,*

as $n \to \infty$.

The above theorem was independently discovered about the same time by Lamperti. As is well-known, if (i) holds, then Λ is one of the three extreme value distributions Λ_i, $i = 1, 2, 3$. Equivalence of (i) and (iii) when N is constant with probability 1 is the result of Berman [1964].

For a study of $X_{(k)}$ as a stochastic process with emphasis on limit theorems, the reader is referred to Dwass [1964] and Lamperti [1964]. Dwass discusses the three possible extremal processes and Lamperti studies the joint limiting behavior of $X_{(n)}$ and $X_{(n-1)}$ considered as a two-dimensional process. Limit laws for maxima of a sequence of random variables defined on a Markov Chain have been studied by Fabens and Neuts [1970], and Resnick and Neuts [1970].

Another area of research under asymptotic results is the theory of spacings. Let us first consider n independent observations from a continuous distribution $F(x)$. Define

$$U_{r,n} = F(X_{r,n}). \tag{4.17}$$

Then $U_{r,n}(r = 1, \ldots, n)$ are order statistics from the uniform distribution on $(0, 1)$. For the purpose of notational convenience, let us define slightly modified spacings

$$D'_{nr} = (n+1)(U_{r,n} - U_{r-1,n}). \tag{4.18}$$

Let $\{g_n : n \geqslant 1\}$ be a sequence of real Borel-measurable functions and consider the random variable

$$G_n = \sum_{r=1}^{n+1} g_n(D'_{nr}). \tag{4.19}$$

Many of the tests based on spacings considered in the literature are of this form. As pointed out by Pyke [1965], prior to 1953 there was no unified approach to the problem of finding the limiting df of a statistic of the form (4.19). Earlier, the asymptotic normality of G_n was obtained for special forms of $g_n(x)$ by Moran [1947], Sherman [1950] and Kimball [1950] using different methods. It was Darling [1953] who provided the first general method of deriving limit theorems for G_n by applying the method of steepest descent to a simple formula for the characteristic function of G_n. LeCam [1958] gave a more easily applied general approach to this problem. Suppose g_n is defined on $[0, \infty)$ and $\{Y_r: r \geqslant 1\}$ is a sequence of independent exponential random variables with mean 1. Set $S_n = n^{-\frac{1}{2}} \sum_{r=1}^{n+1} (Y_r - 1)$. Then the df of G_n is the same as the conditional df of $J_n = \sum_{r=1}^{n+1} g_n(Y_r)$ given that $S_n = 0$. The approach of LeCam is to use information about the joint limiting behavior of (J_n, S_n) to derive the desired conditional limiting distribution of J_n, given $S_n = 0$. For the details of this approach and other results concerning the weak convergence of the empirical distribution function for uniform spacings and limit theorems for functions of general spacings, one may refer to Pyke [1965].

In conclusion of this section we briefly state some recent large sample results concerning quantiles and the deviation between the sample quantile process and the sample df. It is fitting here to mention some of the remarks made by Weiss [1970] and Kiefer [1970b]. In deriving the asymptotic distribution of a set of sample quantiles, the usual approach is to study the joint probability density function as the sample size increases. This technique gets complicated enough when each element of the sample is itself a k-dimensional random variable and we seek the joint asymptotic distribution of a quantile of the first co-ordinates in the sample, a quantile of the second co-ordinates in the sample, . . ., a quantile of the kth co-ordinates in the sample. The simple approach used by Weiss studies limit probabilities of events concerning sample quantiles by rewriting them as events concerning multinomial random variables. Weiss [1970] illustrates the use of this method in some nonstandard cases.

Let $X_1, X_2, \ldots$ be independent and identically distributed with common twice differentiable univariate df F on the unit interval I. Assume that $\inf_{x \in I} F'(x) > 0$ and $\sup_{x \in I} F''(x) < \infty$ and let $\xi_p = F^{-1}(p)$. Also let S_n and $Y_{p,n}$ denote the sample df and the sample quantile of order p, respectively, both based on $(X_1, \ldots, X_n)$; i.e.

$$nS_n(x) = [\text{number of } X_i \leqslant x, 1 \leqslant i \leqslant n] \qquad (4.20)$$

and

$$Y_{p,n} = \inf \{x: S_n(x) = p\}. \qquad (4.21)$$

Define

$$R_n(p) = Y_{p,n} - \xi_p + [S_n(\xi_p) - p]/F'(\xi_p). \qquad (4.22)$$

The study of $R_n(p)$ was initiated by Bahadur [1966]. Later Kiefer [1967] showed that, for $u > 0$,

$$\lim_{n \to \infty} P\{n^{\frac{3}{4}}F'(\xi_p)R_n(p) \leqslant u\} = 2 \int_0^\infty \Phi(k^{-\frac{1}{2}}u) \, d_k\Phi(k/\sigma_p) \qquad (4.23)$$

and that

$$\limsup_{n \to \infty}\{\pm F'(\xi_p)R_n(p)[2^5 3^{-3}\sigma_p^2 n^{-3}(\log\log n)^3]^{-\frac{1}{4}}\} = 1 \qquad (4.24)$$

with probability 1, where Φ is the standard normal df and $\sigma_p = [p(1-p)]^{\frac{1}{2}}$. Let

$$R_n^\pm = \sup_{p\in I} + F'(\xi_p)R_n(p)$$
$$R_n^* = \max(R_n^+, R_n^-). \qquad (4.25)$$

Kiefer [1970a] proves the following results.

Theorem 4.10. *For $Q_n = R_n^+$ or R_n^- or R_n^*,*
(a) $n^{\frac{3}{4}}Q_n(D_n \log n)^{-\frac{1}{2}} \to 1$ *in probability as $n \to \infty$ where*

$$D_n = n^{\frac{1}{2}} \sup_x|S_n(x) - F(x)|.$$

(b) $\limsup_{n \to \infty} n^{\frac{3}{4}}(\log n)^{-\frac{1}{2}}(\log\log n)^{-\frac{1}{4}}Q_n = 2^{-\frac{1}{4}}$ *with probability 1.*

The consequence of part (a) is that, for $t > 0$,

$$\lim_{n \to \infty} P\{n^{\frac{3}{4}}(\log n)^{-\frac{1}{2}}Q_n > t\} = 2 \sum_{m=1}^\infty (-1)^{m+1} e^{-2m^2t^4}. \qquad (4.26)$$

For some of the consequences of part (b) and a list of open problems, the reader is referred to Kiefer [1970a].

5. Combinatorial methods in order statistics

One may see that some elementary combinatorial arguments are always involved in the study of order statistics, for example, in writing the cdf of the rth order statistic based on n observations. But in order to throw light on applications of combinatorial methods of deeper significance we interpret order statistics in a broad sense to include Kolmogorov-Smirnov statistic which requires knowledge of the actual ordered observations in the sample only up to a monotonic increasing transformation. Many combinatorial problems arise when we want to compare theoretical and empirical distribution functions. A fundamental theorem of much application in this area is a generalized version of the classical ballot theorem. A brief but interesting summary of the historical development of the classical ballot theorem and some of its generalizations is given in Takács [1970] to whom the following theorem is due.

Theorem 5.1. *Let $k_1, k_2, \ldots, k_n$ be nonnegative integers with sum $k_1 + k_2 + \cdots + k_n = k \leqslant n$. Among the $n!$ permutations of $(k_1, k_2, \ldots, k_n)$ there are exactly $(n-1)!(n-k)$ for which the rth partial sum $\sum_1^r k_i$ is less than r for all $r = 1, 2, \ldots, n$.*

The above theorem was first obtained by Takács in 1960 and the proofs first given by him [1961, 1962] were based on mathematical induction. Later in 1967 he gave a direct combinatorial proof of the theorem. Recently this theorem has been formulated by Takács [1970] in the following slightly more general form.

Theorem 5.2. *Let* $X_1, X_2, \ldots, X_n$ *be exchangeable random variables taking on nonnegative integer values. Set* $S_r = X_1 + X_2 + \cdots + X_r$ *for* $r = 1, 2, \ldots, n$. *Then*

$$P\{S_r < r \text{ for } r = 1, \ldots, n \mid S_n = k\} = \begin{cases} 1 - \dfrac{k}{n} & \text{for } k = 0, 1, \ldots, n, \\ 0 & \text{otherwise,} \end{cases} \tag{5.1}$$

where the conditional probability is defined up to an equivalence.

As an example of a useful application of Theorem 5.1, consider n random points which are distributed independently and uniformly on the interval $(0, t)$. Let $\chi(u)$ $(0 \leqslant u \leqslant t)$ be c times the number of points in the interval $(0, u]$ where c is a positive constant. Then, by using Theorem 5.1 (see Takács [1970]), we can show that

$$P\{\chi(u) \leqslant u \text{ for } 0 \leqslant u \leqslant t\} = \begin{cases} 1 - \dfrac{nc}{t} & \text{for } 0 \leqslant nc \leqslant t, \\ 0 & \text{otherwise.} \end{cases} \tag{5.2}$$

Another important combinatorial theorem which together with Theorem 5.2 leads to many applications is the following theorem which is due to Andersen [1953] and Feller [1959].

Theorem 5.3. *Let* $X_1, X_2, \ldots, X_n$ *be interchangeable random variables taking on real values. Define* $S_r = X_1 + \cdots + X_r$ *for* $r = 1, 2, \ldots, n$ *and* $S_0 = 0$. *Denote by* N_n *and* N_n^*, *respectively, the number of positive and nonnegative members in the sequence* $S_1, S_2, \ldots, S_n$. *Denote by* L_n *and* L_n^*, *the subscripts of the first and the last maximal members in the sequence* $S_0, S_1, \ldots, S_n$. *We have*

$$P\{N_n = j\} = P\{L_n = j\} \tag{5.3}$$

and

$$P\{N_n^* = j\} = P\{L_n^* = j\} \tag{5.4}$$

for $j = 0, 1, \ldots, n$.

Theorems 5.2 and 5.3 can be combined to yield the following interesting result.

Theorem 5.4. (Takács [1970]). *Let* $X_1, X_2, \ldots, X_n$ *be interchangeable random variables taking on nonnegative integers. Set* $S_r = X_1 + \cdots + X_r$ *for*

$r = 1, 2, \ldots, n$ and $S_0 = 0$. *Denote by* Δ_n *the number of subscripts* $r = 1, 2,$
$\ldots, n$ *for which* $S_r < r$ *holds. If* $P\{S_n = n-1\} > 0$ *then we have*

$$P\{\Delta_n = j | S_n = n-1\} = 1/n \tag{5.5}$$

for $j = 1, 2, \ldots, n$.

Takács [1970] has proved a number of auxiliary theorems which can be
used in the theory of order statistics. All these theorems are consequences of
Theorems 5.2 and 5.3 and are concerned with the distributions of $\Delta_n^{(c)}$, the
number of subscripts $r = 1, 2, \ldots, n$ for which $S_r < r+c$ where $c = 0, \pm 1,$
$\pm 2, \ldots$. In particular, $\Delta_n = \Delta_n^{(0)}$.

We shall now indicate the applications of these results to the problem of
comparing a theoretical and an empirical distribution function. Let $X_1, X_2,$
$\ldots, X_n$ be mutually independent random variables having common df $F(x)$.
Let $F_n(x)$ be the empirical df, i.e., $nF_n(x) =$ the number of variables $\leqslant x$.
Consider

$$D_n^+ = \sup_{-\infty < x < \infty} [F_n(x) - F(x)] \tag{5.6}$$

$$= \max_{1 \leqslant r \leqslant n} [F_n(X_{(r)}) - F(X_{(r)})].$$

If we assume that $F(x)$ is a continuous df, then the joint distribution of $\delta_n(r) =$
$F_n(X_{(r)}) - F(X_{(r)})$ $(r = 1, 2, \ldots, n)$ does not depend on $F(x)$ and consequently
the distributions of D_n^+, ρ_n and ρ_n^* are also independent of $F(x)$, where ρ_n
denotes the number of non-negative elements among $\delta_n(r)$ $(r = 1, 2, \ldots, n)$
and ρ_n^* denotes the largest r for which $\delta_n(r)$ attains its maximum. By using
the auxiliary theorems we can obtain the distributions of D_n^+, ρ_n and ρ_n^*.
These distributions have been obtained earlier by several authors (see
Takács [1970]) and are given below.

Theorem 5.5.
(a) *If* $0 < x \leqslant 1$, *then*

$$P\{D_n^+ \leqslant x\} = 1 - \sum_{nx \leqslant j \leqslant n} \frac{nx}{n+nx-j} \binom{n}{j} \left(\frac{j}{n}-x\right)^j \left(1+x-\frac{j}{n}\right)^{n-j}. \tag{5.7}$$

(b) *For* $j = 1, 2, \ldots, n$,

$$P\{\rho_n = j\} = P\{\rho_n^* = j\} = n^{-1} \sum_{i=1}^{j} i^{-1} \binom{n}{i-1} \left(\frac{i}{n}\right)^{i-1} \left(1-\frac{i}{n}\right)^{n-i}. \tag{5.8}$$

Also of interest is the problem of comparing two empirical distribution
functions. Let $X_1, X_2, \ldots, X_m$ and $Y_1, Y_2, \ldots, Y_n$ be independent random
samples from the distributions $F(x)$ and $G(x)$, respectively. Denote by
$F_m(x)$ and $G_n(x)$ the empirical distribution functions of the two samples.

Define

$$D^+(m, n) = \sup_{-\infty < x < \infty} [F_m(x) - G_n(x)] \tag{5.9}$$

$$= \max_{1 \leqslant r \leqslant n} [F_m(Y_{(r)}) - G_n(Y_{(r)} - 0)].$$

Let $\gamma_c(m, n)$ denote the number of subscripts $r = 1, 2, \ldots, n$ for which $F_m(Y_{(r)}) < G_n(Y_{(n)}) - c/n$, where $c = 0, \pm 1, \ldots, \pm(n-1)$ and let $\tau(m, n)$ denote the smallest $r = 1, 2, \ldots, n$ for which $F_m(Y_{(r)}) - G_n(Y_{(r)} - 0)$ attains its maximum. If F and G are identical continuous distributions, then the distributions of $D^+(m, n)$, $\gamma_c(m, n)$ and $\tau(m, n)$ do not depend on F. When $n = mp$, these distributions can be derived easily by appealing to the auxiliary theorems discussed earlier and a simple probability result relating to the drawing of the ith white ball at the $(i+s)$th draw when the balls are drawn without replacement from a box containing m black and n white balls. We state below the results relating to $D^+(m, n)$.

Theorem 5.6.

(a) *If $n = mp$, where p is a positive integer, and $c = 0, 1, \ldots, n$, then*

$$P\{D^+(m, n) \leqslant c/n\} = 1 - \frac{1}{\binom{m+n}{n}} \sum_{(c+1)/p \leqslant s \leqslant m} \frac{c+1}{n+c+1-sp}$$

$$\times \binom{sp+s-c-1}{s}\binom{m+n+c-sp-s}{m-s}. \tag{5.10}$$

(b) *For $0 < x \leqslant 1$ and $n = mp$,*

$$\lim_{p \to \infty} P\{D^+(m, n) \leqslant c/n\} = P\{D_m^+ \leqslant x\} \tag{5.11}$$

where $c = [nx]$.

The null distribution of $D^+(m, n)$ was first obtained for the case $m = n$ by Gnedenko and Korolyuk [1951] and independently by Drion [1952]. Later Korolyuk [1955] obtained the distribution when $n = mp$ (p is an integer), and Hodges [1957] derived the distribution when the sample sizes differ by unity. All these results were obtained by using a random walk model. Recently, Steck [1969] has obtained the non-null distribution of $D^+(m, n)$ when $G^k = F(0 < k \leqslant 1)$ by expressing $D^+(m, n)$ explicitly in terms of the ranks of one sample in the ordered combined sample and deriving its distribution in terms of the joint distributions of these ranks.

Another result of Steck [1969] which is of interest here relates to the null case. Let $b_1 \leqslant b_2 \leqslant \cdots \leqslant b_m$ and $c_1 \leqslant c_2 \leqslant \cdots \leqslant c_m$ be increasing sequences of integers such that $i-1 \leqslant b_i < c_i \leqslant n+i+1$. Define $N(b; c)$ to be the number of ways the event $\{b_i < R_i < c_i; i = 1, \ldots, m\}$ can occur where R_i denotes the rank of X_i in the ordered combined sample obtained by pooling the random samples $(X_1, \ldots, X_m)$ and $(Y_1, \ldots, Y_n)$. It is well recognized that $N(b; c)$ determines the null distribution of $D^+(m, n)$. Steck has established that

$$N(b; c) = \det (d_{ij})_{m \times m} \tag{5.12}$$

where

$$d_{ij} = \begin{cases} 0 & \text{if } i-j > 1 \text{ or if } c_i - b_j \leqslant 1, \\ \begin{pmatrix} c_i - b_j + j - i - 1 \\ j - i + 1 \end{pmatrix} & \text{otherwise.} \end{cases} \tag{5.13}$$

The above result, for which Mohanty [1971] has given a direct elementary proof, yields the distribution of $D^+(m, n)$ for proper choices of the sequences $\{b_i\}$ and $\{c_i\}$.

Vincze [1970] observes that proofs of Kolmogorov-Smirnov type distribution theorems can be simplified by using a certain generalization of the ballot lemma by G. Tusnády stated below.

Theorem 5.7. *Let $A_0, A_1, \ldots, A_n$ be a complete system of events, for which $P(A_0) = q$ and $P(A_j) = p$, $j = 1, 2, \ldots, n$ holds. Making n independent observations, let v_i be the number of cases in which A_i occurred. Then*

$$P\left\{ \sum_{i=1}^{j} v_i < j : \; j = 1, 2, \ldots, n \right\} = q. \tag{5.14}$$

The above theorem is equivalent to the following theorem of Daniels [1945].

Theorem 5.8. *If $F_n(x)$ denotes the empirical distribution function corresponding to a sample of size n taken on a random variable with distribution $F(x)$ which is uniform in $(0, 1)$, then*

$$P\left\{ \frac{F_n(x) - F(x)}{F(x)} < y, 0 \leqslant x \leqslant 1 \right\} = \frac{y}{y+1} \; (0 \leqslant y \leqslant \infty). \tag{5.15}$$

The equivalence of Theorems 5.7 and 5.8 was utilized by K. Sarkadi to give an independent proof of Theorem 5.7. Vincze [1970] also refers to a generalized ballot lemma of E. Csáki which is closely related to results of Nef [1964] and gives the following formulation of the generalized ballot lemma in terms of the empirical distribution function.

Let $F_n(x)$ be the empirical distribution function belonging to a sample of size n taken on a random variable distributed uniformly in $(0, 1)$. Let λ denote the number of (horizontal) intersections of the graph of $F_n(x)$ with the straight line $y = x/np$ $(np \leqslant 1)$. Then

$$P\{\lambda \geqslant l\} = l! \binom{n}{l} p^l \; (l = 0, 1, \; , \ldots, n). \tag{5.16}$$

6. Some combinatorial methods in fluctuation theory and the distribution of the maxima

In this section we describe some results concerning the partial sums of a sequence of random variables. Although fluctuations of partial sums of random variables have been investigated in special cases for a long time, and even in more general cases for the purpose of finding limit theorems, the idea of using combinatorial methods for analyzing the partial sums of a fixed finite set of more general random variables goes to the credit of E. Sparre Andersen who made a fundamental contribution in this area.

Let $\{X_k\}$, $k = 1, 2, \ldots, n$, be a sequence of independent and identically distributed random variables, with partial sums $S_0 = 0$, $S_1 = X_1, \ldots,$ $S_n = X_1 + \cdots + X_n$. Let

$$N_n = \text{the number of positive } S_n \text{ among } S_1, \ldots, S_n.$$
$$L_n = \text{the smallest index } k(= 0, 1, \ldots, n) \text{ with } S_k = \max_{0 \leqslant m \leqslant n} S_m \quad (6.1)$$

The variable N_n serves in a way as a "measure" of the ups and downs of the sequence $S_0, S_1, \ldots, S_n$. For any permutation $\sigma: i_1, \ldots, i_n$ of the integers $1, 2, \ldots, n$, define $N_n(\sigma)$ and $L_n(\sigma)$ as in (6.1) in terms of the partial sums $S_k(\sigma) = X_{i_1} + \cdots + X_{i_k}$ of the permuted variables $\sigma(X_1, \ldots, X_n) = (X_{i_1}, \ldots, X_{i_n})$.

By the basic assumption it is implied that $N_n(\sigma)$ and $L_n(\sigma)$ have the same distributions as N_n and L_n. As a matter of fact, if we consider the whole class $\{N_n(\sigma)\}_{(\sigma)}$ of $n!$ variables, each element of the class has the same distribution as N_n. By successfully seeking properties of the whole class which do not depend upon the particular values of the variables $X_1, \ldots, X_n$, we can as well carry out the analysis for a set of numbers $x_1, \ldots, x_n$ instead of the variables $X_1, \ldots, X_n$. The following theorem due to Andersen [1953] gives the equivalence principle.

Theorem 6.1. $\{N_n(\sigma)\}_{(\sigma)} \equiv \{L_n(\sigma)\}_{(\sigma)}$.

There are two essential facts connected with the above theorem. First, $N_n(\sigma)$ and $L_n(\sigma)$ are integers between 0 and n, so that there will be multiplicities among the integers which are assumed by the $n!$ terms in each set. The theorem asserts that these multiplicities are exactly the same in the two sets $\{N_n(\sigma)\}_{(\sigma)}$ and $\{L_n(\sigma)\}_{(\sigma)}$. Secondly, the identity holds for all sets of numbers $x_1, \ldots, x_n$ and, therefore, it is not directly concerned with probability theory.

Theorem 6.1, restated in terms of the sequence of random variables, gives

$$P\{N_n = j\} = P\{L_n = j\}. \quad (6.2)$$

This is exactly (5.3) of Theorem 5.3. Thus, the equivalence principle permits us to translate statements concerning the position of maximal terms into statements concerning the number of positive terms: usually the statements of the first kind are more readily proved whereas those of the second kind are more important. Further, Andersen [1953] has shown that, if $X_1, \ldots, X_n$ are independent and identically distributed with continuous and symmetric distributions,

$$P\{L_n = m\} = \binom{2m}{m}\binom{2n-2m}{n-m}2^{-2n}, \qquad 0 \leqslant m \leqslant n. \tag{6.3}$$

It should be pointed out that the joint distribution of N_n and L_n is *not* distribution-free.

Let $R_{n0} \geqslant R_{n1} \geqslant \cdots \geqslant R_{nn}$ be an ordering of the partial sums $S_0, S_1, \ldots, S_n$. Since the distribution of X_1 is continuous, there is a unique index m such that $R_{nk} = S_m$, with probability one. We define $L_{nk} = m$ if $R_{nk} = S_m$. Darling [1951] found the distribution of L_{nk} in terms of products of binomial coefficients, but he gave no results for joint distributions. Baxter [1962] has proved the following theorem.

Theorem 6.2. *For all* $0 \leqslant m,k \leqslant n$ $(n \geqslant 1)$,

$$P\{L_{nm} = 0, L_{nk} = n\} = \begin{cases} \dfrac{1}{2n}\dbinom{2m}{m}\dbinom{2n-2k}{n-k}2^{-2n-2m+2k}, & m < k, \\[2ex] 0, & m = k, \\[2ex] \dfrac{1}{2n}\dbinom{2k}{k}\dbinom{2n-2m}{n-m}2^{-2n-2k+2m}, & m > k. \end{cases} \tag{6.4}$$

We note that $L_{nm} = 0$ is equivalent to $N_n = m$. Also, $L_{nk} = n$ means that there are exactly k partial sums greater than S_n. Thus Theorem 6.2 provides the joint distribution of the number of partial sums less than $S_0(= 0)$ and the number of partial sums greater than S_n. In particular, for $k = 0$,

$$P\{N_n = m, L_n = n\} = \frac{1}{2n}\binom{2n-2m}{n-m}2^{-2n+2m}, \qquad 1 \leqslant m \leqslant n. \tag{6.5}$$

Now, we consider again a sequence of mutually independent random variables with a common distribution function and define

$$\begin{aligned} a_n &= P\{S_n > 0\}, \qquad a_n^* = P\{S_n \geqslant 0\}, \\ \mu_n &= P\{S_n > S_j, j = 0, \ldots, n-1\}, \qquad n \geqslant 1, \\ \mu_n^* &= P\{S_n \geqslant S_j, j = 0, \ldots, n-1\}, \qquad n \geqslant 1. \end{aligned} \tag{6.5}$$

Then, the generating functions $\mu(t) = \sum \mu_n t^n$ and $\mu^*(t) = \sum \mu_n^* t^n$ have been obtained by Andersen [1953, 1954] and also by Spitzer [1956] in the form:

$$\mu(t) = \exp \sum_1^\infty \frac{a_n}{n} t^n \tag{6.6}$$

and

$$\mu^*(t) = \exp \sum_1^\infty \frac{a_n^*}{n} t^n. \tag{6.7}$$

The above result shows that the knowledge of the sequences $\{a_n\}$ and $\{a_n^*\}$ suffices for the calculation of the distribution of the position of the maximal term and of the number of positive terms in $\{S_0, \ldots, S_n\}$.

The probability

$$v_n^* = P\{S_1 \leqslant 0, \ldots, S_n \leqslant 0\} \tag{6.8}$$

has the associated generating function

$$v^*(t) = \exp\left[\sum_1^\infty \frac{1-a_n}{n} t^n\right] = [(1-t)\mu(t)]^{-1}. \tag{6.9}$$

Let

$$p_{k,n} = P\{S_k > S_j \text{ for } j < k, \quad S_k \geqslant S_j \text{ for } k < j \leqslant n\}. \tag{6.10}$$

Then Feller [1959] gives the following theorem.

Theorem 6.3. $p_{k,n} = \mu_k v_{n-k}^*.$

As we can see, $p_{k,n}$ is the probability that the first maximum in $(S_0, S_1, \ldots, S_n)$ occurs at the place numbered k. If $p_{k,n}^*$ denotes the probability that the last maximum occurs at the place numbered k, then

$$p_{k,n}^* = \mu_k^* v_{n-k}. \tag{6.11}$$

Instead of $a_n = P\{S_n > 0\}$, let us consider more generally the truncated distribution function

$$F_n(x) = P\{0 < S_n \leqslant x\}, \quad x > 0 \tag{6.12}$$

with the Laplace transform

$$\phi_n(\lambda) = \int_{0+}^\infty e^{-\lambda x} \, dF_n(x). \tag{6.13}$$

Define

$$H_n(x) = P\{S_j < S_n \leqslant x, j = 0, \ldots, n-1\}, \quad x > 0,$$

and

$$h_n(\lambda) = \int_0^\infty e^{-\lambda x} \, dH_n(x), \quad h(\lambda, t) = 1 + \sum_1^\infty h_n(\lambda) t^n, \tag{6.15}$$

The following theorem is due to Spitzer [1956].

Theorem 6.4. $h(\lambda, t) = \exp\left[\sum_1^\infty \frac{t^n}{n} \phi_n(\lambda)\right].$

For $\lambda = 0$, this theorem gives (6.6).

Again considering a sequence $X_1, X_2, \ldots, X_n, \ldots$ of independent and identically distributed random variables, having continuous distribution function, let us define X_n to be *outstanding* if it is larger than all previous observations, that is, $X_n > \max_{1 \leqslant k \leqslant n-1} X_k$. Let A_n be the event that X_n is an outstanding observation ($n = 1, 2, \ldots$). Rényi [1962] has obtained some results concerning the outstanding observations based on the simple but surprising fact that the events $A_1, \ldots, A_n, \ldots$ are independent and $P(A_n) = 1/n$. The results of Rényi are contained in the following theorem.

Theorem 6.5. *Let* $X_{v_1}, X_{v_2}, \ldots, X_{v_k}, \ldots$ *be all the outstanding observations of* $\{X_k\}, k = 1, 2, \ldots,$ *Then*

(a) $\lim_{k \to \infty} v_k^{1/k} = e$ *with probability* 1 *and*

(b) $(\log v_k - k)/k^{\frac{1}{2}}$ *is asymptotically* $(k \to \infty)$ *normal with mean* 0 *and variance* 1.

If we define α_N as the number of outstanding observations among $X_1, \ldots, X_N$, then Theorem 6.5 says that $\lim_{N \to \infty} \alpha_N/\log N = 1$ with probability 1 and the distribution of $(\alpha_N - \log N)/(\log N)^{\frac{1}{2}}$ is asymptotically standard normal.

7. Some estimation and hypothesis testing problems based on order statistics

Order statistics have been employed in many problems of estimation and testing of hypotheses. The usual methods, in some cases, lead to estimators involving order statistics. An example of practical interest where the observations arise in an ordered sequence is a life test experiment where a certain number of units are put on test and their failure times are observed. The literature has grown so enormously in this area that any attempt to survey all the results will be beyond the aim of the present paper. We will be content with a brief outline of some of the problems investigated.

An important paper is that of Lloyd [1952] in which he considers the least-squares estimates of location and scale parameters using order statistics. Suppose that $X_1, X_2, \ldots, X_n$ are independent observations on X having a continuous distribution $F(x - \mu/\sigma)$, $\sigma > 0$, where μ and σ are not necessarily the mean and standard deviation, respectively. Define $U = (X - \mu)/\sigma$. Then $U_{(r)} = (X_{(r)} - \mu)/\sigma$ $(r = 1, 2, \ldots, n)$ can be regarded as ordered observations on U. Let $E(U_{(r)}) = \alpha_r$, $V(U_{(r)}) = v_{rr}$, $\text{Cov}(U_{(r)}, U_{(s)}) = v_{rs}$ and V be the matrix (v_{rr}). Under the generalized Gauss-Markov linear model, one can obtain μ^* and σ^*, the least-squares (l.s.) estimates of the parameters μ and σ. The formulas for the estimates and their dispersion matrix simplify considerably when X has a symmetric distribution, in which case we can take μ to be the center of the distribution and σ a symmetric measure of dispersion. Since the l.s. estimates are linear compounds of the ordered observations with minimal variance, $V(\mu^*) \leqslant \sigma^2/n$, where $\sigma^2 = V(X)$. Lloyd has obtained conditions to determine when $V(\mu^*) < \sigma^2/n$, i.e., the l.s. estimate is more efficient than the sample mean. It turns out that $V(\mu^*) < \sigma^2/n$ unless μ^* is the sample mean, a result due to Downton [1953].

Blom [1956, 1958, 1962] addressed himself to the problem of unbiased nearly best linear estimates where one settles for an estimate with nearly minimum variance. He investigated how such an approximation to the best linear estimate can be found. He has also dealt briefly with relaxing the unbiasedness, seeking nearly unbiased, nearly best estimates.

Bennett [1952], in his unpublished thesis, studied the asymptotic properties of estimates which are linear functions of the order statistics with continuous weight functions. Following Bennett [1952] and Jung [1955, 1962], let $T_n = n^{-1}\sum_{j=1}^{n} J(j/(n+1))X_{j,n}$ where $J(\cdot)$ is a well-behaved function. Bennett obtained asymptotically optimal J's for both the uncensored and multi-censored cases, but did not derive the asymptotic normality of the estimates. Some of his results were independently obtained by Jung [1955] under rather restrictive conditions. Plackett [1958] and Weiss [1963] independently considered the case where all observations below the pth and above the qth sample quantiles $(0 < p < q < 1)$ are censored and obtained asymptotic normality for suitable linear combinations of the available order statistics. Chernoff, Gastwirth and Johns [1967] obtain a quite general theorem concerning the conditions under which the statistics of the form $T_n = n^{-1}\sum c_{j,n} h(X_{j,n})$ are asymptotically normally distributed. They specialize their results to the case where $c_{j,n} = J(j/(n+1))$. These theorems involve the decomposition $T_n = \mu_n + Q_n + R_n$, where μ_n is non-random, $Q_n = n^{-1}\sum \alpha_{j,n}(Z_j - 1)$ where Z_j's are independent and identically distributed exponential random variables, $n^{\frac{1}{2}}Q_n$ is asymptotically normal and R_n is asymptotically negligible. Results overlapping with those of Chernoff $et\ al.$ have been obtained by Govindarajulu [1965] whose technique is based on some unpublished results of Le Cam and whose main result requires bounds on $J(u)$ and $J'(u)$ as $u \to 0$ or 1, which is not necessary for the results of Chernoff $et\ al.$

Problems of estimation of parameters using censored data from normal as well as non-normal distributions have been studied by several authors. Among the non-normal distributions considered are Gamma (Harter and Moore [1967]), Log normal (Harter and Moore [1966]), Double Exponential (Govindarajulu [1966]), Weibull (Cohen [1965], Gumbel [1958]) and Logistic (Gupta, Shah and Qureishi [1967]) just to mention a few. The published literature on life testing and reliability problems is quite vast and the reader is referred to the bibliographies of Mendenhall [1958], Govindarajulu [1964] and a short classified list of David [1970], p. 124.

In the problems of estimation using only some of the order statistics, an interesting question is how to choose or "space" the order statistics to obtain good estimates. Let us choose $0 < \lambda_1 < \lambda_2 < \cdots < \lambda_k < 1$. The sample quantiles are $X_{(n_j)}, j = 1, \ldots, k$, where $n_j = [n\lambda_j] + 1$. Ogawa [1951] considered estimation of the location (μ) and scale (σ) parameters based on sample quantiles in large samples. In these cases, the relative efficiency of an estimate which is a function of the chosen order statistics as compared to those which are based on the whole sample is defined by the ratio of the amounts of information in Fisher's sense in the two cases. The best linear unbiased estimator in each case is found to be efficient for a given spacing $\lambda_1, \ldots, \lambda_k$. However, the efficiency can be raised by suitably choosing the values of $\lambda_1, \ldots, \lambda_k$ for which the relative efficiency of an estimator attains its maximum. Such a set of $\lambda_1, \ldots, \lambda_k$ is called an optimum spacing. Ogawa [1962a] has shown that, in the case of normal distribution, the optimum spacing for the location parameter μ is necessarily a symmetric one. Ogawa [1962c] has also considered optimum spacing for the scale parameter of the exponential distribution. The problem of optimum spacing for the asymptotically best linear estimate (ABLUE) of μ when σ is known has so far been considered for three symmetric distributions with support $(-\infty, \infty)$. For the normal and logistic distributions it has been proved, respectively, by Higuchi [1954] and Gupta and Gnanadesikan [1966] that the optimum spacing is symmetric. The question of whether the optimum spacing for the ABLUE of μ when σ is known is symmetric for any distribution which is symmetric and has the support $(-\infty, \infty)$ has been raised and answered in the negative by Kulldorff [1971b] who gives a counter-example. Optimum spacings for the ABLUE of the location parameter μ of an extreme value distribution of Type I ($\Lambda_3^{(n)}(x-\mu)$ given by (4.7)) and the scale parameter σ of an extreme value distribution of Type II or III ($\Lambda_1^{(n)}(x/\sigma)$ or $\Lambda_2^{(n)}(x/\sigma)$ given by (4.7)) have been considered by Kulldorff [1971a] by making use of the previous results for the scale parameter of an exponential distribution. The problem of determining the optimum choice of the ranks $n_1 < n_2 < \cdots < n_k$ of order statistics in a small sample of size n for estimating the parameters of the exponential distribution $F(x) = 1 - e^{-(x-\alpha)/\sigma}$ ($x \geqslant \alpha, \sigma > 0$) has been dealt with by Harter [1961b] and Siddiqui [1963] for the case $k > 1$, and by Ukita [1955], Harter [1961b] Sarhan, Greenberg and Ogawa [1963] and Siddiqui [1963] for $k = 2$. The case of general k has been investigated by Kulldorff [1963].

The problems of testing of hypotheses in life test models illustrate the use of order statistics. Some quick tests based on order statistics have been used in several situations; see David and Johnson [1956]. Tests for outliers and slippage are further specific problems where test statistics are based on ordered observations.

As regards the use of spacings in testing of hypotheses, Ogawa [1962b] considered the test for the hypothesis H: $\mu = \mu_0$ for the normal mean and the test of the homogeneity of several means. He also discusses selection of the optimum spacing for testing purposes. In another paper [1962d] he discusses the test for H: $\sigma = \sigma_0$ where σ is the scale parameter of the exponential. Pyke [1965] gives limit theorems for general spacings which are useful in obtaining asymptotic results on the power of goodness-of-fit tests based on spacings. Proschan and Pyke [1967] have discussed tests for monotone failure rate using test statistics based on spacings. Recently, Sethuraman and Rao [1970] have discussed the Pitman efficiencies of tests based on spacings in the goodness-of-fit problem.

8. Multiple decision (selection and ranking) problems

The goal in selection and ranking problems can be roughly described as follows. Suppose there are k populations $\pi_1, \ldots, \pi_k$ which are ranked in a certain sense. There are two basic approaches. We may either want to select one of them as the "best" or select a subset of the given populations so that the selected subset contains the "best". Obviously, basing our inference on a sample, we will be content if we can say that the probability of our selection being a correct selection (CS) is at least P^* ($1/k < P^* < 1$). In the case of selecting a subset, we can achieve this regardless of the true states of the distributions. In the case of selecting one of them as the best, we shall require that the condition on the minimum probability of a correct selection be met whenever the true best population is sufficiently apart from the second best. This is the indifference zone formulation, of Bechhofer [1954] in its simplest form. The former, known as the subset selection formulation, is due to Gupta [1956].

For a detailed account of subset selection formulation, one can refer to Gupta [1965], and Gupta and Panchapakesan [1969, 1971]. The monograph of Bechhofer, Kiefer and Sobel [1968] describes the basic formulation of selection and ranking problems using indifference zone approach.

Many of the multiple decision problems encountered in practice have a common algebraic structure and these problems are called identification problems. Let $\{X_{ij}\}$, $i = 1, \ldots, k$, be k independent sequences of random variables. For $i = 1, \ldots, k$, the X_{ij} have a common distribution $F^0_{j_i}$, where $(F^0_{j_1}, \ldots, F^0_{j_k})$ is a permutation of k known probability laws $F^0_1, \ldots, F^0_k$. The space $\Omega = \{(F^0_{j_1}, \ldots, F^0_{j_k})\}$ can be viewed as the permutation group S_k on k elements. We denote a typical element of S_k by $(\alpha(1), \ldots, \alpha(k)) = (\alpha_1, \ldots, \alpha_k)$, which is the result of the permutation α on $(1, \ldots, k)$. We may briefly use α to denote an element of S_k. Now, S_k can be regarded either as the space of all possible states of nature or as a group of transformations (permutations) operating on Ω. As a result of this dual interpretation of S_k, if $\alpha, \beta \in S_k$, then $\beta\alpha$ can be considered as the element of Ω arising from the permutation β operating on element α of Ω.

We say that α is the true element of Ω if the sequences $\{X_{\alpha,j}\}$ have the distributions F^0_i, $i = 1, \ldots, k$. If α^{-1} denotes the inverse permutation of α, then we can also say (when α is true) that X_{ij} has the distribution $F^0_{\alpha^{-1}(i)}$ ($i = 1, 2, \ldots, k$).

It is also convenient to think in terms of k numbered populations $\pi_1, \pi_2, \ldots, \pi_k$ with the sequence $\{X_{ij}\}$ coming from π_i. By saying that α is the true

9

element of Ω we mean that the population π_i has the distribution $F^0_{\alpha^{-1}(i)}$, or more briefly, $\alpha^{-1}(i)$, $i = 1, 2, \ldots, k$. Thus the correct pairing of the populations with the distribution functions can be written in two equivalent ways,

$$\begin{pmatrix} F^0_1 & F^0_2 & \cdots & F^0_k \\ \pi_{\alpha_1} & \pi_{\alpha_2} & \cdots & \pi_{\alpha_k} \end{pmatrix} = \begin{pmatrix} F^0_{\alpha^{-1}(1)} & F^0_{\alpha^{-1}(2)} & \cdots & F^0_{\alpha^{-1}(k)} \\ \pi_1 & \pi_2 & \cdots & \pi_k \end{pmatrix}. \tag{8.1}$$

An identification problem must satisfy certain requirements. Let D^t denote the space of possible decisions and d a typical element of D^t. Then we require that there exists a group Γ of transformations homomorphic to S_k and operating on D^t. In other words, if g_α is the element of Γ corresponding under the homomorphism $\mathscr{H}$ to α, then $g_\alpha g_\beta d = g_{\alpha\beta} d$ for $\alpha, \beta \in S_k$ and $d \in D^t$. Further it is also required that the loss function $W(\alpha, d)$ has the invariance property, that is, $W(\alpha, d) = W(\beta\alpha, g_\beta d)$ for $\alpha, \beta \in S_k$ and $d \in D^t$.

Finally, many decision procedures which are used in conjunction with an identification problem have a corresponding invariant structure under a group of transformations g_α isomorphic to S_k. Let $\alpha x_j = \alpha(x_{1j}, \ldots, x_{kj}) = (x_{\alpha_1 j}, \ldots, x_{\alpha_k j})$, where x_{ij} is a realization of X_{ij}. Let Δ denote a subset of D^t and $P_m\{\Delta | x\}$, the probability of arriving at one of the decisions belonging to Δ on the basis of the observations $x = (x_1, \ldots, x_m)$. Then for an invariant procedure, $P_m\{\Delta | x\} = P_m\{g_\alpha(\Delta) | \alpha x\}$, where $\alpha x = (\alpha x_1, \ldots, \alpha x_m)$.

The description of the basic structure of identification problems given above is on the line of Bechhofer, Kiefer and Sobel. They have provided several examples to illustrate the basic structure and the properties of minimal invariant sets.

Before we pass on to discuss the role of order statistics in the context of subset selection procedures, we will briefly explain an identification problem and its connection with a ranking problem. Let $\pi_1, \pi_2, \ldots, \pi_k$ be k populations and F_{θ_i} be the distribution associated with π_i ($i = 1, 2, \ldots, k$). For the identification problem, we assume that ranked values of the θ_i, denoted by $\theta^0_{[1]} \leqslant \theta^0_{[2]} \leqslant \cdots \leqslant \theta^0_{[k]}$, are known *a priori*. However, the true pairing of the π_i with the $\theta^0_{[j]}$ is unknown to the experimenter and he has no *a priori* knowledge relevant to the true pairing of the π_i with the $\theta^0_{[j]}$ ($i, j = 1, 2, \ldots, k$). Suppose that $\theta^0_{[k-1]} < \theta^0_{[k]}$. Then an identification goal would be "to identify the population π_i associated with $\theta^0_{[k]}$". For ranking problem, we assume that the ordered values of the θ_i, denoted now by $\theta_{[1]} \leqslant \theta_{[2]} \leqslant \cdots \leqslant \theta_{[k]}$, are unknown *a priori* and that the true pairing or any knowledge relevant to the true pairing of the π_i with the $\theta_{[j]}$ ($i, j = 1, 2, \ldots, k$) is not available to the experimenter. The ranking goal corresponding to the identification goal stated above would be "to select a population π_i associated with $\theta_{[k]}$." But the formulation of the ranking problem will be complete only if the experimenter specifies certain constants, and then states an associated probability requirement involving these constants which must be guaranteed. Several ranking procedures have been investigated for specific cases under this formulation by several authors, notably, Bechhofer and Sobel among them. Generally these procedures have been proposed on heuristic grounds. As one can intuitively see, the decisions in all these cases depend on the sample observations through the ordered values of statistics T_i ($i = 1, 2, \ldots, k$).

We now discuss the role of order statistics in subset selection problems. Let $\pi_1, \pi_2, \ldots, \pi_k$ be k independent populations with continuous distributions F_{θ_i} ($i = 1, \ldots, k$), $\theta_i \in \Theta$, an interval on the real line. We assume that $\{F_\theta\}$ is a stochastically increasing family. The ordered values of the unknown parameters θ_i are denoted by $\theta_{[1]} \leqslant \theta_{[2]} \leqslant \cdots \leqslant \theta_{[k]}$. The population associated with $\theta_{[k]}$ (or $\theta_{[1]}$) is called the best. In the case of a tie, we assume that one of the contenders is tagged as the best. The objective is to select a subset of the given populations and claim that the probability that the best population is included in the selected subset is at least P^* ($1/k < P^* < 1$) regardless of the true configuration of the parameters. Let T_i be a suitable statistic based on n independent observations from π_i ($i = 1, 2, \ldots, k$). The selection rule proposed in most of the specific situations for selecting a subset containing the population associated with $\theta_{[k]}(\theta_{[1]})$ is of the form:

$\quad$ R: Select π_i iff $T_i \geqslant T_{\max} - d_1$ ($T_i \leqslant T_{\min} + d_2$) or

$\quad$ R: Select π_i iff $c_1 T_i \geqslant T_{\max}$ ($T_i \leqslant c_2 T_{\min}$)

where $T_{\max} = \max(T_1, \ldots, T_k)$, $T_{\min} = \min(T_1, \ldots, T_k)$ and the constants $c_1, c_2 \geqslant 1$ and $d_1, d_2 \geqslant 0$ are to be determined so that the basic probability requirement is satisfied. The above rules are particular cases of a general class of rules discussed by Gupta and Panchapakesan [1970]. Usually, T_i is a sufficient statistic for θ_i and preserves the stochastic ordering. In all these cases, the standard technique is to obtain the expression for $P\{CS|R\}$, the probability of a correct selection using the rule R, evaluate its infimum over all parametric configurations and determine the constant of the procedure by equating this infimum to P^*. Because of the stochastic ordering of the T_i, the infimum of $P\{CS|R\}$ is to be found over only the equal parameter configuration, namely, $\theta_1 = \theta_2 = \cdots = \theta_k = \theta$ (say). Exceptional situations arise in certain procedures for multinomial cells and in some rules using rank sums. In most of the problems we can establish the monotonic behavior of $P\{CS|R\}$ in θ by verifying certain sufficient condition (see Gupta and Panchapakesan [1970]). Thus we obtain the infimum of $P\{CS|R\}$ and depending upon the type of the procedure used, we get one of the following relations:

$$
\begin{aligned}
P\{T_k \geqslant T_{\max} - d_1\} &= P^*, \\
P\{T_1 \leqslant T_{\min} + d_2\} &= P^*, \\
P\{c_1 T_k \geqslant T_{\max}\} &= P^*, \\
P\{T_1 \leqslant c_2 T_{\min}\} &= P^*,
\end{aligned}
\tag{8.2}
$$

where $T_1, \ldots, T_k$ are independent and identically distributed random variables with a common distribution G (say). Define

$$
\begin{aligned}
U_1 &= \max \{T_1 - T_k, T_2 - T_k, \ldots, T_{k-1} - T_k\}, \\
U_2 &= \min \{T_2 - T_1, T_3 - T_1, \ldots, T_k - T_1\}, \\
V_1 &= \max \{T_1/T_k, T_2/T_k, \ldots, T_{k-1}/T_k\}, \\
V_2 &= \min \{T_2/T_1, T_3/T_1, \ldots, T_k/T_1\}.
\end{aligned}
\tag{8.3}
$$

It should be pointed out that V_1 and V_2 arise when the T_i are non-negative random variables. We see that the constants in (8.2) are either appropriate percentage points of the distribution of the random variables in (8.3) or related to these percentage points. The constants are given by the appropriate equation of the following set:

$$
A(k, d_1) = \int_{-\infty}^{\infty} G^{k-1}(x + d_1) \, dG(x) = P^*,
$$

$$
B(k, d_2) = \int_{-\infty}^{\infty} [1 - G(x - d_2)]^{k-1} \, dG(x) = P^*,
$$

$$
I(k, c_1) = \int_{0}^{\infty} G^{k-1}(c_1 x) \, dG(x) = P^*,
\tag{8.4}
$$

$$
J(k, c_2) = \int_{0}^{\infty} [1 - G(x/c_2)]^{k-1} \, dG(x) = P^*.
$$

In each problem we know the specific form of G. Tables of constants are available in the literature in several special cases of G for selected values of k and P^*.

Gupta [1963] discusses among other things, the integral

$$
F_N(H; \rho) = \int_{-\infty}^{\infty} F^N((x\rho^{\frac{1}{2}} + H)(1 - \rho)^{-\frac{1}{2}}) \, dF(x),
\tag{8.5}
$$

where F denotes the cdf of a standard normal variable. It can be seen that $F_N(H; \rho)$ is the probability that the maximum of a set of N equally correlated standardized normal random variables does not exceed H or the probability that the minimum of this set exceeds $-H$.

Consider, for example, the rule

$$
R: \text{Select } \pi_i \text{ iff } T_i \geq T_{\max} - d_1,
$$

in the case of k normal populations with unknown means $\theta_1, \ldots, \theta_k$ and a common known variance σ^2, where the T_i are the sample means. One can see clearly the possibilities of using procedures which involve more order statistics than just $T_{\max}$. If $T_{[1]} \leq T_{[2]} \leq \cdots \leq T_{[k]}$ are the ordered means, then Seal [1955] considered a class of procedures D_c defined below.

D_c: Select the population corresponding to $T_{[i]}$ if and only if

$$
T_{[i]} \geq c_1 T_{[1]} + \cdots + c_{i-1} T_{[i-1]} + c_i T_{[i+1]} + \cdots + c_{k-1} T_{[k]} - \sigma t(P^*, c)/\sqrt{n},
$$

where $c = (c_1, \ldots, c_{k-1})$ is a vector whose components are arbitrary real numbers such that $c_i \geqslant 0$ and $\sum_{i=1}^{k-1} c_i = 1$ and, $t(P^*, c)$ is chosen so as to satisfy the basic probability requirement. It is possible to propose procedures involving other functions of all the order statistics $T_{[1]}, \ldots, T_{[k]}$, but these essentially present difficult distribution problems in terms of evaluating the infimum of $P\{CS|R\}$ and explains to an extent the absence of complete investigations of such procedures in the literature.

Acknowledgement

The authors wish to thank Mr. T. Santner for a critical reading of the paper and some helpful comments.

References

E. S. Andersen, 1953, On sums of symmetrically dependent random variables, *Skand. Aktuarietidskr.* **36**, 123–138.

E. S. Andersen, 1954, On the fluctuations of sums of random variables, II, *Math. Scand.* **2**, 195–223.

F. J. Anscombe, 1952, Large sample theory of sequential estimation, *Proc. Cambridge Philos. Soc.* **48**, 600–607.

R. R. Bahadur, 1966, A note on quantiles in large samples, *Ann. Math. Statist.* **37**, 577–580.

R. E. Barlow and S. S. Gupta, 1969, Selection procedures for restricted families of probability distributions, *Ann. Math. Statist.* **40**, 905–917.

R. E. Barlow and F. Proschan, 1966, Inequalities for linear combinations of order statistics from restricted families, *Ann. Math. Statist.* **37**, 1574–1592.

O. Barndorff-Neilson, 1964, On the limit distribution of the maximum of a random number of independent random variables, *Acta Math. Acad. Sci. Hungar.* **15**, 399–403.

V. D. Barnett, 1966, Order statistics estimators of the location of the Cauchy distribution, *J. Am. Statist. Assoc.* **61**, 1205–1218. Correction: *Ibid.* **63**, 383–385.

G. Baxter, 1962, On a generalization of the finite arcsine law, *Ann. Math. Statist.* **33**, 909–915.

R. E. Bechhofer, 1954, A single-sample multiple decision procedure for ranking means of normal populations with known variances, *Ann. Math. Statist.* **25**, 16–39.

R. E. Bechhofer, J. Kiefer and M. Sobel, 1968, *Sequential Identification and Ranking Procedures, with Special Reference to Koopman-Darmois Populations*, Statistical Research Monographs, vol. 3 (The University of Chicago Press, Chicago).

C. A. Bennett, 1952, Asymptotic properties of ideal linear estimators, Ph.D. Thesis, University of Michigan, Ann Arbor, Mich.

S. M. Berman, 1962, Limiting distribution of the maximum term in sequences of dependent random variables, *Ann. Math. Statist.* **33**, 894–908.

S. M. Berman, 1964, Limiting distribution of the maximum of a diffusion process, *Ann. Math. Statist.* **35**, 319–329.

P. J. Bickel 1967, Some contributions to the theory of order statistics, *Proc. Fifth Berkeley Symp. on Mathematical Statistics and Probability* **1** (L. Le Cam and J. Neyman, eds.), pp. 575–591 (University of California Press, Berkeley, California).

G. Blom, 1956, On linear estimates with nearly minimum variance, *Ark. Mat.* **3**, 365–369.

G. Blom, 1958, *Statistical Estimates and Transformed Beta-Variables* (Almqvist and Wiksell, Uppsala, Sweden; Wiley, New York).

G. Blom, 1962, Nearly best linear estimates of location and scale parameters. In *Contributions to Order Statistics* (A. E. Sarhan and B. G. Greenberg, eds.), pp. 34–46. (Wiley, New York).

J. R. Blum, D. L. Hanson and J. I. Rosenblatt, 1963, On the central limit theorem for the sum of a random number of independent random variables, *Z. Wahrscheinlichkeitstheorie und Verw. Gebiete* **1**, 389–393.

R. C. Bose and S. S. Gupta, 1959, Moments of order statistics from a normal population, *Biometrika* **46**, 433–440.

M. C. Breiter and P. R. Krishnaiah, 1968, Tables for the moments of gamma order statistics, *Sankhya Ser.* B **30**, 59–72.

H. Chernoff, J. L. Gastwirth and M. V. Johns, Jr., 1967, Asymptotic distribution of linear combinations of functions of order statistics with applications to estimation, *Ann. Math. Statist.* **38**, 52–72.

J. T. Chu and H. Hotelling, 1955, The moments of the sample median, *Ann. Math. Statist.* **26**, 593–606.

A. C. Cohen, Jr., 1965, Maximum likelihood estimation in the Weibull distribution based on complete and on censored samples, *Technometrics* **7**, 579–588.

C. C. Craig, 1962, On the mean and variance of the smaller of two drawings from a binomial population, *Biometrika*, **49**, 566–569.

H. E. Daniels, 1945, The statistical theory of the strengths of bundles of threads, I, *Proc. Roy. Soc. London* A, **183**, 405–435.

D. A. Darling, 1951, Sums of symmetrical random variables, *Proc. Am. Math. Soc.* **2**, 511–517.

D. A. Darling, 1953, On a class of problems relating to the random division of an interval, *Ann. Math. Statist.* **24**, 239–253.

F. N. David and N. L. Johnson, 1954, Statistical treatment of censored data, I: fundamental formulae, *Biometrika* **41**, 228–240.

F. N. David and N. L. Johnson, 1956, Some tests of significance with ordered variables (with discussion), *J. Roy. Statist. Soc.* B **78**, 1–31.

H. A. David, 1970, *Order Statistics* (Wiley, New York).

F. Downton, 1953, A note on ordered least-squares estimation, *Biometrika* **40**, 457–458.

E. F. Drion, 1952, Some distribution-free tests for the difference between two empirical cumulative distribution functions, *Ann. Math. Statist.* **23**, 563–574.

M. Dwass, 1964, Extremal processes, *Ann. Math. Statist.* **35**, 1718–1725.

P. Erdös and M. Kac, 1946, On certain limit theorems on the theory of probability, *Bull. Am. Math. Soc.* **52**, 292–302.

A. J. Fabens and M. F. Neuts, 1970, The limiting distribution of the maximum term in a sequence of random variables defined on a Markov chain, *J. Appl. Probability*, **7**, 754–760.

W. Feller, 1959, On combinatorial methods in fluctuation theory. In *Probability and Statistics* (U. Grenander, ed.), pp. 75–91 (Almqvist and Wiksell, Stockholm; Wiley, New York).

R. A. Fisher and L. H. C. Tippett, 1928, Limiting forms of the frequency distribution of the largest or smallest member of a sample, *Proc. Cambridge Philos. Soc.* **24**, 180–190.

B. V. Gnedenko, 1943, Sur la distribution limite du terme maximum d'une série aléatoire, *Ann. Math.* **44**, 423–453.

B. V. Gnedenko and V. S. Korolyuk, 1951, On the maximum discrepancy between two empirical distributions, *Dokl. Akad. Nauk SSSR* **80**, 525–528 (in Russian; English Transl.: *Selected Translations in Mathematical Statistics and Probability* (IMS and AMS) **1** (1961) 13–16).

Z. Govindarajulu, 1964, A supplement to Mendenhall's bibliography on life testing and related topics, *J. Am. Statist. Assoc.* **59**, 1231–1291.

Z. Govindarajulu, 1965, Asymptotic normality of linear functions of order statistics in one and multi-samples, Tech. Rept. U.S. Air Force Grant No. AF–AFOSR–741–65, pp. 1–26.

Z. Govindarajulu, 1966, Best linear estimates under symmetric censoring of the parameters of a double exponential population, *J. Am. Statist. Assoc.* **61**, 248–258.

E. J. Gumbel, 1958, *Statistics of Extremes* (Columbia University Press, New York).

E. J. Gumbel, 1962, Statistical theory of extreme values (main results). In *Contributions to Order Statistics* (A. E. Sarhan and B. G. Greenberg, eds.), pp. 56–93 (Wiley, New York).

S. S. Gupta, 1956, On a decision rule for a problem in ranking means, Mimeo Ser. No. 150, Institute of Statistics, University of North Carolina, Chapel Hill, N.Car.

S. S. Gupta, 1960, Order statistics from the gamma distribution, *Technometrics* **2**, 243–262.

S. S. Gupta, 1962, Gamma distributions. In *Contributions to Order Statistics* (A. E. Sarhan and B. G. Greenberg, eds.), pp. 431–450 (Wiley, New York).

S. S. Gupta, 1963, Probability integrals of multivariate normal and multivariate *t*, *Ann. Math. Statist.* **34**, 792–828.

S. S. Gupta, 1965, On some multiple decision (selection and ranking) rules, *Technometrics* **7**, 225–245.

S. S. Gupta and M. Gnanadesikan, 1966, Estimation of the parameters of the logistic distribution, *Biometrika* **53**, 565–570.

S. S. Gupta and S. Panchapakesan, 1967, Order statistics arising from independent binomial populations, Mimeo Ser. No. 120, Department of Statistics, Purdue University, Lafayette, Ind.

S. S. Gupta and S. Panchapakesan, 1969, Selection and ranking procedures. In *The Design of Computer Simulation Experiments* (T. H. Naylor, ed.), pp. 132–160 (Duke University Press, Durham, N.Car.).

S. S. Gupta and S. Panchapakesan, 1970, On a class of subset selection procedures, Mimeo. Ser. No. 225, Department of Statistics, Purdue University, Lafayette, Ind.

S. S. Gupta and S. Panchapakesan, 1971, Contributions to multiple decision (subset selection) rules, multivariate distribution theory and order statistics, Mimeo. Ser. No. 260, Department of Statistics, Purdue University, Lafayette, Ind.

S. S. Gupta, A. S. Qureishi and B. K. Shah, 1967, Best linear unbiased estimators of the parameters of the logistic distribution using order statistics, *Technometrics* **9**, 43–56.

S. S. Gupta and B. K. Shah, 1965, Exact moments and percentage points of the order statistics and the distribution of the range from the logistic distribution, *Ann. Math. Statist.* **36**, 907–920.

H. L. Harter, 1961a, Expected values of normal order statistics, *Biometrika* **48**, 151–165. Correction: *Ibid.* **48**, 476.

H. L. Harter, 1961b, Estimating the parameters of negative exponential populations from one or two order statistics, *Ann. Math. Statist.* **32**, 1078–1090.

H. L. Harter and A. H. Moore, 1966, Local-maximum likelihood estimation of the parameters of three-parameter lognormal populations from complete and censored samples, *J. Am. Statist. Assoc.* **61**, 842–855.

H. L. Harter and A. H. Moore, 1967, Asymptotic variances and covariances of maximum likelihood estimators, from censored samples, of the parameters of Weibull and gamma populations, *Ann. Math. Statist.* **38**, 557–570.

C. Hastings, Jr., F. Mosteller, J. W. Tukey and C. P. Winsor, 1947, Low moments for small samples: a comparative study of order statistics, *Ann. Math. Statist.* **18**, 413–426.

J. L. Hodges, Jr. 1957, The significance probability of the Smirnov two-sample test, *Ark. Mat.* **3**, 469–486.

I. Higuchi, 1954, On the solutions of certain simultaneous equations in the theory of systematic statistics, *Ann. Inst. Statist. Math.* **5**, 77–90.

J. Jung, 1955, On linear estimates defined by a continuous weight function, *Ark. Mat.* **3**, 199–209.

J. Jung, 1962, Approximation of least-squares estimates of location and scale parameters. In *Contributions to Order Statistics* (A. E. Sarhan and B. G. Greenberg, eds.), pp. 28–33 (Wiley, New York).

J. Kiefer, 1967, On Bahadur's representation of sample quantiles, *Ann. Math. Statist.* **38**, 1323–1342.

J. Kiefer, 1970, (a) Deviations between the sample quantile process and the sample df. (b) Old and new methods for studying order statistics and sample quantiles. In *Nonparametric Techniques in Statistical Inference* (M. L. Puri, ed.), pp. 299–319 and 349–357 (Cambridge University Press, London).

B. F. Kimball, 1950, On the asymptotic distribution of the sum of powers of unit frequency differences, *Ann. Math. Statist.* **21**, 263–271.

A. N. Kolmogorov, 1933, Sulla determinazionne empirica di une legge di distribuzione, *Giorn. Ist. Ital. Attuari* **4**, 83–91.

V. S. Korolyuk, 1955, On the discrepancy of empirical distribution functions for the case of two independent samples, *Izv. Akad. Nauk. SSSR Ser. Mat.* **19**, 81–96 (in Russian; English Transl.: *Selected Translations in Mathematical Statistics and Probability* (IMS and AMS) **4** (1963) 105–121).

G. Kulldorff, 1963, Estimation of one or two parameters of the exponential distribution on the basis of suitably chosen order statistics, *Ann. Math. Statist.* **34**, 1419–1431.

G. Kulldorff, 1971a, A note on the optimum spacing of sample quantiles from the six extreme value distributions, Tech. Rept. No. 26, Texas Agr. and Mech. University, College Station, Texas.

G. Kulldorff, 1971b, Private communication.

J. Lamperti, 1964, On extreme order statistics, *Ann. Math. Statist.* **35**, 1726–1737.

L. LeCam, 1958, Une théoreme sur la division d'une intervalle par des points pres au hasard, *Publ. Inst. Statist. Univ. Paris* **7**, 7–16.

E. H. Lloyd, 1952, Least-squares estimation of location and scale parameters using order statistics, *Biometrika* **39**, 88–95.

M. Loéve, 1960, *Probability Theory* (2nd edition) (Van Nostrand, New York).

W. Mendenhall, 1958, A bibliography on life testing and related topics, *Biometrika* **45**, 521–543.

R. G. Miller, Jr., 1966, *Simultaneous Statistical Inference* (McGraw-Hill, New York).

J. Mogyoródi, 1962, A central limit theorem for the sum of a random number of independent random variables, *Magyar Tud. Akad. Mat. Kutató Int. Közl.* **7**, 409–424.

S. G. Mohanty, 1971, A short proof of Steck's result on two-sample Smirnov statistics, *Ann. Math. Statist.* **42**, 413–414.

P. A. P. Moran, 1947, The random division of an interval, *J. Roy. Statist. Soc. B* **9**, 92–98.

F. Mosteller, 1946, On some useful "inefficient" statistics, *Ann. Math. Statist.* **17**, 377–408.

W. Nef. 1964, Über die Differenz zwischen theoretischer und empirischer Verteilungsfunktion, *Z. Wahrscheinlichkeitstheorie und Verw. Gebiete* **3**, 154–162.

J. Ogawa, 1951, Contributions to the theory of systematic statistics, I, *Osaka J. Math.* **3**, 175–213.

J. Ogawa, 1962, (a) Determinations of optimum spacings in the case of normal distribution, (b) Tests of significance using sample quantiles, (c) Optimum spacing and grouping for the exponential distribution, (d) Tests of significance and confidence intervals. In

Contributions to Order Statistics (A. E. Sarhan and B. G. Greenberg, eds.), pp. 272–283, 291–299, 371–380 and 380–382 (Wiley, New York).

K. Pearson, 1902, Note on Francis Galton's problem, *Biometrika* **1**, 390–399.

R. L. Plackett, 1958, Linear estimation from censored data, *Ann. Math. Statist.* **29**, 131–142.

F. Proschan and R. Pyke, 1967, Tests for monotone failure rate, *Proceedings of the Fifth Berkeley Symposium on Mathematical Statistics and Probability* **3** (L. LeCam and J. Neyman, eds.), pp. 293–312 (University of California Press, Berkeley).

R. Pyke, 1965, Spacings (with discussion), *J. Roy. Statist. Soc.* B **27**, 395–449.

A. Rényi, 1953, On the theory of order statistics, *Acta Math. Acad. Sci. Hungar.* **4**, 191–231.

A. Rényi, 1960, On the central limit theorem for the sum of a random number of random variables, *Acta Math. Acad. Sci. Hungar.* **11**, 97–102.

A. Rényi, 1962, Théorie des éléments sailants d'une suite d'observations. In *Combinatorial Methods in Probability Theory*, pp. 104–115 (Mathematical Institute, Aarhus University, Denmark).

S. I. Resnick and M. F. Neuts, 1970, Limit laws for maxima of a sequence of random variables defined on a Markov chain, *Adv. Appl. Probability* **2**, 323–343.

A. E. Sarhan and B. G. Greenberg, 1956, Estimation of location and scale parameters by order statistics from singly and doubly censored samples. Part I: The normal distribution up to samples of size 10, *Ann. Math. Statist.* **27**, 427–451. Correction: *Ibid.* **40**, 325.

A. E. Sarhan and B. G. Greenberg, ed., 1962, *Contributions to Order Statistics* (Wiley, New York).

A. E. Sarhan, B. G. Greenberg and J. Ogawa, 1963, Simplified estimates for the exponential distribution, *Ann. Math. Statist.* **34**, 102–116.

K. C. Seal, 1955, On a class of decision procedures for ranking means of normal populations, *Ann. Math. Statist.* **26**, 387–398.

P. K. Sen, 1959, On the moments of the sample quantiles, *Calcutta Statist. Assoc. Bull.* **9**, 1–20.

J. Sethuraman and J. S. Rao, 1970, Pitman efficiencies of tests based on spacings. In *Nonparametric Techniques in Statistical Inference* (M. L. Puri, ed.), pp. 405–415 (Cambridge University Press. London).

B. K. Shah, 1966a, A note on Craig's paper on the minimum of binomial variates, *Biometrika* **53**, 614–615.

B. K. Shah, 1966b, On the bivariate moments of order statistics from a logistic distribution, *Ann. Math. Statist.* **37**, 1002–1010.

B. Sherman, 1950, A random variable related to the spacing of sample values, *Ann. Math. Statist.* **21**, 339–361.

M. M. Siddiqui, 1963, Optimum estimators of the parameters of negative exponential distributions from one or two order statistics, *Ann. Math. Statist.* **34**, 117–121.

N. V. Smirnov, 1949, Limit distributions for the terms of a variational series (English Transl.: *Am. Math. Soc. Transl. Ser.* 1, **67** (1952)).

F. Spitzer, 1956, A combinatorial lemma and its application to probability theory, *Trans. Am. Math. Soc.* **82**, 323–339.

G. P. Steck, 1969, The Smirnov two sample tests as rank tests, *Ann. Math. Statist.* **40**, 1449–1466.

P. V. Sukhatme, 1937, Tests of significance for samples of the χ^2 population with two degrees of freedom, *Ann. Eugen.* **8**, 52–56.

L. Takács, 1961, The probability law of the busy period for two types of queueing process, *Operations Res.* **9**, 402–407.

L. Takács, 1962, A generalization of the ballot problem and its application in the theory of queues, *J. Am. Statist. Assoc.* **57**, 327–337.

L. Takács, 1967, *Combinatorial Methods in the Theory of Stochastic Processes* (Wiley, New York).

L. Takács, 1970, Combinatorial methods in order statistics. In *Nonparametric Techniques in Statistical Inference* (M. L. Puri, ed.), pp. 359–384 (Cambridge University Press, London).

D. Teichroew, 1956, Tables of expected values of order statistics and products of order statistics for samples of size twenty and less from the normal distribution, *Ann. Math. Statist.* **27**, 410–426.

L. H. C. Tippett, 1925, On the extreme individuals and the range of samples from a normal population, *Biometrika* **17**, 264–387.

Y. Ukita, 1955, On the efficiency of order statistics, *J. Hokkaido College Sci. and Art* **6**, 54–65.

W. R. van Zwet, 1964, *Convex Transformation of Random Variables*, Mathematical Center Tracts 7 (Mathematisch Centrum, Amsterdam).

I. Vincze, 1970, On Kolmogorov-Smirnov type distribution theorems. In *Nonparametric Techniques in Statistical Inference* (M. L. Puri, ed.), pp. 385–401 (Cambridge University Press, London).

L. Weiss, 1963, On the asymptotic distribution of an estimate of a scale parameter, *Naval Res. Logist. Quart.* **10**, 1–11.

L. Weiss, 1970, Asymptotic distributions of quantiles in some nonstandard cases. In *Nonparametric Techniques in Statistical Inference* (M. L. Puri, ed.), pp. 343–348 (Cambridge University Press, London).

S. S. Wilks, 1948, Order Statistics, *Bull. Am. Math. Soc.* **54**, 6–50.

J. N. Srivastava et al., eds., *A Survey of Combinatorial Theory*
© North-Holland Publishing Company, 1973

CHAPTER 20

Construction of Block Designs†

MARSHALL HALL, Jr.

California Institute of Technology, Pasadena, Calif., 91109, U.S.A.

1. Introduction

The paper by R. C. Bose [1939] is the outstanding classic on the construction of block designs. For this reason it seems appropriate to discuss the present status of this very large subject. A large section of the author's book (Hall, [1967]) is devoted to this subject, and the theorems of Bruck-Ryser-Chowla and other results now classical will be mentioned only briefly here.

Recent applications of group theory and coding theory will be discussed, the recent constructions of designs with $v = 56$, $k = 11$, $\lambda = 2$ and with $v = 79$, $k = 13$, $\lambda = 2$ will be given and some current attacks will also be discussed.

2. Block designs with $r \leqslant 15$

In Hall [1967], Table I lists as unknown those designs with the parameters as shown in the following table:

Number	v	b	r	k	λ	Present status
35	46	69	9	6	1	Unknown.
44	51	85	10	6	1	Unknown.
52	45	55	11	9	2	Residual of 53. Constructed.
53	56	56	11	11	2	Constructed.
54	100	110	11	10	1	Residual of 55. Unknown.
55	111	111	11	11	1	Unknown.
59	22	33	12	8	4	Unknown.
61	33	44	12	9	3	Residual of 64. Constructed.
64	45	45	12	12	3	Constructed.
66	61	122	12	6	1	Unknown.
74	40	52	13	10	3	Unknown.

† This research was supported in part by ONR contract N00014-67-A0094-0010.

Number	v	b	r	k	λ	Present status
77	66	143	13	6	1	Unknown.
78	66	78	13	11	2	Residual of 79. Constructed.
79	79	79	13	13	2	Constructed.
80	144	156	13	12	1	Residual of 81. Unknown.
81	157	157	13	13	1	Unknown.
85	22	44	14	7	4	Unknown.
90	85	170	14	7	1	Unknown.
100	28	42	15	10	5	Unknown.
104	36	36	15	15	6	Constructed.
105	43	43	15	15	5	Does not exist.
106	46	69	15	10	3	Unknown.
107	56	70	15	12	3	Residual of 109. Unknown.
109	71	71	15	15	3	Unknown.
110	76	190	15	6	1	Unknown.
114	136	204	15	10	1	Unknown.

Some of these listings were in error. The symmetric designs 64 and 104 were already known and so also their residual designs. A solution for the residual design of 104 was listed. An elegant solution for 104 is the following taking as a base block modulo (6, 6) the following:

$$(1, 1), (2, 2), (3, 3), (4, 4), (5, 5)$$
$$(0, 1), (0, 2), (0, 3), (0, 4), (0, 5) \qquad (2.1)$$
$$(1, 0), (2, 0), (3, 0), (4, 0), (5, 0)$$

The design number 64 may be obtained from a representation of the simple group G of order 25920 as a permutation group on 45 letters. Here G is generated by t and v where

$$t = (1, 19, 7, 20, 37)(2, 3, 4, 17, 29, 33)(3, 9, 14, 38, 15)$$
$$(4, 21, 25, 22, 28)(5, 18, 27, 31, 11)(6, 35, 26, 40, 42)$$
$$(8, 36, 39, 12, 24)(10, 32, 13, 45, 16)(23, 43, 44, 30, 41), \qquad (2.2)$$
$$v = (1, 5, 7)(2, 4, 8)(3, 6, 9)(10, 18, 15, 12, 17, 14, 11, 16, 13)$$
$$(21, 28, 40, 44, 27, 30, 31, 41, 26)(22, 38, 25, 39, 34, 42, 35, 24, 29)$$
$$(19, 37, 43, 45, 36, 33, 32, 23, 20).$$

If we take as an initial block

$$B_1 = \{16, 17, 18, 19, 23, 27, 28, 32, 36, 37, 41, 45\}, \qquad (2.3)$$

then the 45 images of B_1 under the action of G are the blocks of a symmetric design with $v = 45$, $k = 12$, $\lambda = 3$. Design number 105 does not exist, as a consequence of the Bruck-Ryser-Chowla theorem, since the equation $x^2 = (k-\lambda)y^2 + (-1)^{\frac{1}{2}(v-1)}\lambda z^2$ has no solutions in non-zero integers for these parameters.

3. Some recent constructions of block designs

A permutation group G on v letters is said to be a rank m group if G is transitive on the v letters and if the subgroup G_α fixing the letter α has exactly m orbits (including $\{\alpha\}$ itself as an orbit). Thus "rank 2" and "doubly transitive" are the same property. Recently, rank 3 groups have been studied at some length. The basic theory of such groups has been given by Higman [1964].

A rank three group G is a permutation group transitive on a set of v letters, $\Omega = \{a_1, a_2, \ldots, a_v\}$ such that for $a \in \Omega$, G_a has orbits

$$\{a\}, \Delta(a), \Gamma(a), \tag{3.1}$$

where $\Delta(a)$ has length k (for all a) and $\Gamma(a)$ has length l (for all a) and $v = 1+k+l$. We choose our notation so that for $g \in G$, $\Delta(a)^g = \Delta(a^g)$, where for $X \in \Omega, g \in G$, X^g is the image of X under g. We also choose the notation so that $k \leqslant l$. Intersection numbers λ and μ are defined by

$$| \Delta(a) \cap \Delta(b) | = \begin{cases} \lambda \text{ for } b \in \Delta(a) \\ \mu \text{ for } b \in \Gamma(a). \end{cases} \tag{3.2}$$

Here λ and μ do not depend on the particular choice of a and b. The parameters satisfy

$$\mu l = k(k-\lambda-1). \tag{3.3}$$

If the order of G is odd, then $k = l$, and v must be a prime power. In the more interesting case when the order of G is even then either

(I) $k = l,$ and $\mu = \lambda+1 = k/2$

or

(II) $d = (\lambda-\mu)^2+4(k-\mu)$ is a square.

In case II, the permutation representation of G considered as a matrix representation is the direct sum of three irreducible representations of degrees $f_1 = 1, f_2$ and f_3, where

$$\begin{Bmatrix} f_2 \\ f_3 \end{Bmatrix} = \frac{2k+(\lambda-\mu)(k+l)\mp(k+l)\sqrt{d}}{\mp2\sqrt{d}}. \tag{3.3}$$

These are strong conditions on the parameters and with the help of computers, listings have been made of parameters satisfying the conditions. But there may be no group with parameters satisfying the conditions. A particular case which had aroused interest was that with $v = 3250, k = 57, l = 3192$, $\lambda = 0, \mu = 1$, and it has recently been shown by Michael Aschbacher [1] that there is no rank 3 group with these parameters.

For G of even order, the relations $a \in \Delta(b)$ and $b \in \Delta(a)$ imply each other. Then the blocks $\Delta(a_1), \ldots, \Delta(a_v)$ form a partially balanced design D with two associate classes in the sense of Bose and Shimamoto [1952]. Here we say that a and b are first associates if $b \in \Delta(a)$ and second associates if $b \in \Gamma(a)$.

If $\lambda = \mu$, the blocks $\Delta(a)$ form a symmetric block design. If we put $\Delta^*(a) = \{a, \Delta(a)\}$ then

$$| \Delta^*(a) \cap \Delta^*(b) | = \begin{cases} \lambda+2 & \text{if} \quad b \in \Delta(a) \\ \mu & \text{if} \quad b \in \Gamma(a). \end{cases} \tag{3.4}$$

Thus if $\lambda+2 = \mu$, the blocks $\Delta^*(a)$ form a symmetric block design.

The simple group of order 25920 is a rank 3 group in its representation on 45 letters, and here $k = 12$, $l = 32$, $\lambda = \mu = 3$. Here the blocks $\Delta(a)$ give the symmetric block design with $v = 45$, $k = 12$, $\lambda = 3$ given in the preceding section. This group also has a rank 3 representation on 36 letters which yields a block design $v = 36$, $k = 15$, $\lambda = 6$, different from the one given in section 2. It also has two different rank 3 representations on 40 letters with $k = 12$, $l = 27$, $\lambda = 2$, $\mu = 4$. In both cases, the blocks $\Delta^*(a)$ yield symmetric designs with $v = 40$, $k = 13$, $\lambda = 4$, but in one case these are the planes in the projective space PG(3, 3) and in the other they are not.

The simple group $G = L_3(4)$ of order 20,160 is a collineation group of the plane of order 4. In this plane, an oval is a set of 6 points no three on a line. There are 168 ovals in the plane moved in three orbits of 56 each by $L_3(4)$. These three orbits are interchanged by an outer automorphism of $L_3(4)$ of order 3. Representing G on one of these orbits, G is of rank 3 on 56 letters with $k = 10$, $l = 45$, $\lambda = 0$, $\mu = 2$. Here the blocks $\Delta^*(a)$ form a symmetric block design [7] with $v = 56$, $k = 11$, $\lambda = 2$. Here $G = \langle a, c \rangle$, where

$$\begin{aligned} a = &(1, 2, 3, 4, 5, 6, 7)(8, 9, 10, 11, 12, 13, 14)(15, 16, 17, 18, 19, 20, 21) \\ &(22, 23, 24, 25, 26, 27, 28)(29, 30, 31, 32, 33, 34, 35) \\ &(36, 37, 38, 39, 40, 41, 42)(43, 44, 45, 46, 47, 48, 49) \\ &(50, 51, 52, 53, 54, 55, 56), \end{aligned} \tag{3.5}$$

$$\begin{aligned} c = &(1)(2, 8, 41)(3, 27, 28)(4, 36, 31)(5, 20, 53)(6, 14, 22)(7, 42, 54) \\ &(9, 29, 34)(10, 52, 17)(11, 24, 46)(12, 30, 48)(13, 55, 33) \\ &(15, 26, 32)(16, 21, 56)(18, 40, 35)(19, 23, 49)(25, 50, 44) \\ &(37, 51, 47)(38, 39, 43)(45). \end{aligned}$$

Here $\Delta^*(1)$ is the set

$$B_1 = \{1, 12, 19, 23, 30, 37, 45, 48, 49, 51\}, \tag{3.6}$$

and its images under G form the symmetric design number 53 with parameters $v = 56$, $k = 11$, $\lambda = 2$.

Michael Aschbacher [1971] set out to find the largest possible group of automorphisms which a symmetric design with $v = 79$, $k = 13$, $\lambda = 2$ might possibly have. He was able to show that the largest possible group was of order 110, $G = \langle x, w \rangle$ where $x^{11} = 1$, $w^{10} = 1$, $w^{-1}xw = x^7$ and

$$\begin{aligned} x = &(1)(2)(3, 4, 5, 6, 7, 8, 9, 10, 11, 12, 13)(14, 15, 16, 17, 18, 19, 20, 21, 22, 23, 24) \\ &(25, 26, 27, 28, 29, 30, 31, 32, 33, 34, 35)(36, 37, 38, 39, 40, 41, 42, 43, 44, 45, 46) \\ &(47, 48, 49, 50, 51, 52, 53, 54, 55, 56, 57)(58, 59, 60, 61, 62, 63, 64, 65, 66, 67, 68) \\ &(69, 70, 71, 72, 73, 74, 75, 76, 77, 78, 79), \end{aligned} \tag{3.7a}$$

$$w = (1,2)(3)(4,10,8,5,6,13,7,9,12,11)(14)(15,21,19,16,17,24,18,20,23,22)$$
$$(25,36,47,58,69)(26,43,52,60,72,35,40,53,67,77)$$
$$(27,39,57,62,75,34,44,48,65,74)(28,46,51,64,78,33,37,54,63,71)$$
$$(29,42,56,66,70,32,41,49,61,79)(30,38,50,68,73,31,45,55,59,76). \quad (3.7b)$$

Here G has 4 orbits on the points, namely

$$\{1,2\}, \{3,\ldots,13\}, \{14,\ldots,24\} \quad \text{and} \quad \{25,\ldots,79\}.$$

There are also 4 orbits on the blocks and representatives of blocks are

$$\begin{aligned}
&B_1: 1, 2, 3, 4, 5, 6, 7, 8, 9, 10, 11, 12, 13; \\
&B_2: 1, 2, 14, 15, 16, 17, 18, 19, 20, 21, 22, 23, 24; \\
&B_3: 1, 3, 14, 26, 29, 40, 41, 52, 56, 61, 67, 70, 72; \\
&B_4: 5, 12, 19, 20, 25, 38, 40, 43, 45, 52, 53, 73, 76.
\end{aligned} \qquad (3.8)$$

Here B_1 and B_2 are each fixed by G, B_3 has 22 images and B_4 has 55 images under G.

If a group G is doubly transitive on v points then the images under G of any set of k points will form a block design, though possibly one with an extremely large number of blocks. If a set of k points can be found such that the subgroup H of G taking these into themselves is of index v in G, then their images form a symmetric block design. But it is exceedingly obscure as to when this happens. If the subgroup G_{ab} of G fixing two points a and b fixes $k \geqslant 3$ points, then the images of such a set of points form a block design with $\lambda = 1$. G is called a Jordan group. Most of the known Jordan groups are collineation groups of geometries in which the k points are points of a subspace. A different kind of example has been found by Graham Higman [1971] in representing the Higman-Sims simple group of order 44,352,000 as a doubly transitive group on 176 points. In this case a symmetric design with $v = 176$, $k = 50$, $\lambda = 14$ is given by the representation. This design is as unusual as the group itself.

4. Various approaches to construction of designs

The Bruck-Ryser-Chowla theorem, essentially based on the Hasse-Minkowski theorems on rational equivalence of quadratic forms, shows that certain symmetric block designs do not exist. The Connor-Hall theorem asserts that if a residual design exists when $\lambda = 1$ or 2, then the symmetric design also exists. The combination of these results shows that designs with certain parameters do not exist. Up to the present, no other set of parameters has been shown impossible in a design. On the other side of the picture, if the elementary relations $bk = vr$ and $r(k-1) = \lambda(v-1)$ are satisfied, then Hanani has shown that for $k = 3, 4, 5$ the designs exist (naturally excluding $v = 15$, $b = 21$, $r = 7$, $k = 5$, $\lambda = 2$, the only case where the Connor-Hall theorem applies with $k = 3, 4$ or 5). His constructions are recursive. But for

$k \geqslant 6$ these methods have not been effective, and the special arguments used for $k = 3, 4$ or 5 do not appear to generalize. Richard Wilson has investigated cases with $k = 6$, $\lambda = 1$ and developed an extensive theory of compositions.

In the absence of further general results on the existence or non-existence of designs, it is desirable to examine more closely those designs with small parameters which are still unknown. Investigations have been made of some of these. For number 35, $v = 46, b = 69, r = 9, k = 6, \lambda = 1$, it is possible to make a complete listing of the 17 blocks which are the full set of blocks containing one or both of two distinct points. Up to isomorphism, there are less than 50 such starts. In general, for any block design let a_{ij} be the incidence number such that $a_{ij} = 1$ if the ith point is in the jth block and $a_{ij} = 0$ if not. Writing $L_j = \sum_i a_{ij}x_i$, we have

$$
\begin{aligned}
L_1^2 + \cdots + L_b^2 &= (r - \lambda)(x_1^2 + \cdots + x_v^2) + \lambda(x_1 + \cdots + x_v)^2 = Q, \\
L_1 + \cdots + L_b &= r(x_1 + \cdots + x_v).
\end{aligned}
\tag{4.1}
$$

If 17 blocks are known, the completion depends on the quadratic form Q^*, where

$$
Q^* = Q - L_1^2 \cdots - L_{17}^2.
\tag{4.2}
$$

Study of the eigenvalues and eigenvectors of Q^* was carried out by the author and Leonard Baumert. One of these starts was completely eliminated and several others were strongly restricted, but no further progress has been made. These parameters have also been investigated by John Brown and Ernest Parker, and probably by others.

John Thompson has begun a serious attack on the plane of order 10, number 55. He takes the 111 points as the basis of a vector space over GF(2). Then each line L is associated with the vector which is one for points of L and zero for other points. The 111 lines L span a vector space V of dimension 56 and the orthogonal space $V^\perp$ is the subspace of dimension 55 spanned by all combinations $L_i + L_j$. If A_n is the number of vectors in V with exactly n ones, it is easily shown that $A_{111} = 1$, and that $A_{111-n} = A_n$, and that $A_{4t+2} = A_{4t+1} = 0$. The set of numbers $A_0, A_1 \ldots, A_{111}$ are called the weight numbers and a formula of MacWilliams (Berlekamp [1968], p. 400) relates the weight numbers for $V^\perp$ to those of V. Using this formula and the easily proved fact that $A_i = 0$ for $i = 1, \ldots, 10$, and that $A_{11} = 111$ (only a line has weight 11), it is possible to reduce the determination of all the weight numbers to the determination of A_{12}, A_{15} and A_{16}. If $A_{12} = A_{15} = A_{16} = 0$ then A_{19} is positive. A vector with weight 12 yields 12 points which form a complete oval, i.e., any line goes through two of these points or none. A vector of weight 15, 16 or 19 corresponds to a certain kind of configuration of that many points. A computer search at the Bell Telephone Laboratories has recently shown that $A_{15} = 0$. It can be shown that there exist configurations of 16 or 19 points whether or not there are any ovals. But this information has still not been enough to determine whether or not the plane exists.

A plane of order 10 cannot have an involutory collineation, and in fact can conceivably have collineations of order 3 or order 5 but of no other prime order. But a plane of order 12, number 81, might have a fairly large collineation group. An involutory collineation must be an elation. The author is investigating whether or not there might be a plane of order 12 with a group of elations of order 12 with a fixed axis and fixed center. This calculation is within range of testing by computer.

Certain other designs will exist if a kind of composition is possible. Thus if there is a design number 85 with $v = 22$, $b = 44$, $r = 14$, $k = 7$, $\lambda = 4$ which contains a repeated block

$$\begin{aligned} B_1&: 1, 2, 3, 4, 5, 6, 7 \\ B_2&: 1, 2, 3, 4, 5, 6, 7 \end{aligned} \qquad (4.2)$$

then it is easily shown that each of the remaining 42 blocks contains exactly two of 1, 2, 3, 4, 5, 6, 7 and as $\lambda = 4$, each of the 21 pairs i, j, where $1 \leqslant i < j = 7$ occurs exactly two more times. The remaining 15 points 8, . . ., 22 on the remaining 42 blocks will form a design with $v = 15$, $b = 42$, $r = 14$, $k = 5$, $\lambda = 4$. This design exists, and it may be possible to combine this with points 1, . . ., 7 to form a design number 85. This would be of particular interest because of the work of Van Lint and Ryser [1972] on designs with repeated blocks.

Similarly for design number 59 with $v = 22, b = 33, r = 12, k = 8, \lambda = 4$, if we have three initial blocks of the shape

$$\begin{aligned} & 1, 2, 3, 4, 5, 6, 7, 8 \\ & 1, 2, 3, 4, 9, 10, 11, 12 \\ & 5, 6, 7, 8, 9, 10, 11, 12 \end{aligned} \qquad (4.3)$$

then each of the remaining 30 blocks contains exactly 4 of 1, 2, . . ., 12 and the remaining 10 points form a design with $v = 10, b = 30, r = 12, k = 4, \lambda = 4$. This design exists and it may be possible to expand it to design 59.

Note added in proof

Solutions to number 85 have been sent to me by R. Stanton and S. S. Shrikhande. One solution has 2 base blocks modulo 22, namely 0, 6, 11, 15, 18, 20, 21 and 0, 5, 7, 8, 9, 13, 19.

References

M. Aschbacher, 1971, The nonexistence of rank three permutation groups of degree 3250 and subdegree 57, *J. Algebra*.

M. Aschbacher, 1971, On collineations of symmetric block designs, *J. Combin. Theory* **19**, 538–540.

E. R. Berlekamp, 1968, *Algebraic Coding Theory* (McGraw-Hill, New York).

R. C. Bose, 1939, On the construction of balanced incomplete block designs, *Ann. Eugenics* **9**, 353–399.

R. C. Bose and T. Shimamoto, 1952, Classification and analysis of partially balanced incomplete block designs with two associate classes, *J. Am. Statist. Assoc.* **47**, 151–184.

M. Hall, Jr., 1967, *Combinatorial Theory* (Blaisdell, Waltham, Mass.).

M. Hall, Jr., R. Lane and D. Wales, 1970, Designs derived from permutation groups, *J. Combin. Theory* **8**, 12–22.

D. G. Higman, 1964, Finite permutation groups of rank 3, *Math. Z.* **86**, 145–156.

G. Higman, 1969, On the simple group of D. G. Higman and C. W. Sims, *Illinois J. Math.* **13**, 74–80.

F. J. MacWilliams, N. S. A. Sloan and J. G. Thompson, On the existence of a projective plane of order 10, *J. Combin. Math.*, to appear.

J. H. van Lint and H. J. Ryser, 1972, Block designs with repeated blocks, *Discrete Math.* **3**, to appear.

R. Wilson, 1972, An existence theorem for pairwise balanced block designs, I: Composition theorems and morphisms; II: The structure of PBD-closed sets and the existence conjectures, *J. Combin. Theory* (to appear).

J. N. Srivastava et al., eds., *A Survey of Combinatorial Theory*
© North-Holland Publishing Company, 1973

CHAPTER 21

A Survey of Graphical Enumeration Problems †

FRANK HARARY and EDGAR M. PALMER

University of Michigan, Ann Arbor, Mich. 48104, U.S.A.

Abstract

We present a wide range of graphical enumeration problems, far more comprehensive than any which has appeared previously. Although various sophisticated, recondite, and specialized terminology may confuse the situation, the fact is that very many pattern and configuration problems become graphical in nature when properly reformulated. Furthermore, the conceptual difficulty of the problem is more easily identified when recast in terms of graphs or variations on graphs. Within the hundreds of problems indicated, there is adequate material to occupy research scholars for generations.

1. Setting the stage

There exist several earlier lists of exactly twenty-seven unsolved problems in graphical enumeration. That restriction is now abandoned and myriads of open questions are exposed. Our terminology will faithfully follow the books Harary [1969] and Harary and Palmer [to appear]. In the latter book, all known results and methods on graphical enumeration are presented, with just one exception: the counting of rooted planar maps. Fortunately, the definitive work on this topic appears in this volume and the contribution of Tutte (Tutte [1972]).

There are still relatively few graph theorists who regard enumeration as one of their principal areas of interest. However, activity in this field is gradually accelerating. More scholars are attempting the solution of existing problems and frequently rediscover some of the classical counting methods involving enumeration under group action, while also devising clever and powerful new methods. Very often, the first task to be accomplished in counting graphs with property "P" is the discovery of a structure theorem which says that a graph has this abstract property if and only if it has a certain appearance, which then gives a clue for counting.

† Work supported in part by grants from the Air Force Office of Scientific Research and the National Science Foundation.

In this and the remaining sections, we list the problem areas as P1.1, P1.2, ... Each such area in turn may contain several individual problems which will be indicated. In some cases, data in the form of the first few terms of the generating function will be included.

We include no explicit problems on counting trees of various types, for the existing methods appear always adequate. For counting labeled trees, see Moon [1970], and for unlabeled trees, Harary and Palmer [to appear], Chapter 3.

There are unsolved problems in counting both labeled and unlabeled graphs. In general, the succeeding sections include questions open in both cases, except where otherwise specified. Usually, the labeled case is more manageable than the unlabeled case because there is less symmetry to cope with. However, there are two notable exceptions for which unlabeled configurations have been counted but not the labeled ones:

P1.1. Labeled self-converse digraphs

The unlabeled self-converse digraphs were counted in Harary and Palmer [1966a] using the Power Group Enumeration Theorem.

P1.2. Labeled self-complementary graphs

The number of self-complementary graphs and digraphs was determined by Read [1963]. The labeled cases are untouched.

2. Digraphs

There are many unsolved problems involving digraphs which are better stated later along with the corresponding problems for graphs in the succeeding sections. Nevertheless, there are some which merit separate mention here because they involve structural properties exclusive to digraphs.

P2.1. Strong digraphs

A digraph is *strongly connected* or *strong* (Harary et al. [1965], p. 51) if each pair of points are mutually reachable by directed paths. Our good friend R. W. Robinson succeeded in enumerating both labeled and unlabeled strong digraphs several years ago, but has not yet found the time to write up these interesting, important, and difficult results. It is to be hoped that he will do so within the present decade. His methods involve the condensation D^* of an arbitrary digraph D in which the points of D^* are the strong components

Fig. 2.1. The strong digraphs of order 3.

of D, together with techniques of cycle index sums developed in Chapter 8 of Harary and Palmer [to appear]. All the digraphs of order 4 are listed (Harary [1969], pp. 227–330). There are 83 strong digraphs of order 4 as indicated in Harary and Palmer [1966c]. Hence their counting series begins:

$$x + x^2 + 5x^3 + 83x^4 + \cdots$$

P2.2. Unilateral digraphs

A digraph is *unilaterally connected* or *unilateral* if for any two points, at least one is reachable from the other. Since its unilateral components do not partition a digraph, the method of Robinson mentioned above for counting strong digraphs is not readily specialized to count unilateral digraphs, whose counting series begins

$$x + 2x^2 + 11x^3 + 171x^4 + \cdots.$$

P2.3. Digraphs with a source

A point in a digraph is called a *source* if all other points can be reached from it. The directional dual is a *sink*. Of course, there are the same number of digraphs with a source as with a sink, as these are converse collections. The series for these starts

$$x + 2x^2 + 12x^3 + 184x^4 + \cdots.$$

In a strong digraph, every point is both a source and a sink. The counting of digraphs with both a source and a sink is also open, except for the claim of R. W. Robinson that his approach to counting strong digraphs applies to this problem, as well as to the determination of the number of digraphs with a specified number of sources.

P2.4. Transitive digraphs

A digraph is *transitive* if the presence of arcs uv and vw implies that of arc uw. It is very easy to see that transitive digraphs of order p correspond precisely to finite topologies on a set of p elements. These have only been enumerated in the labeled case using Stirling numbers of the second kind.

The first four terms in the series for unlabeled transitive digraphs are

$$x + 3x^2 + 9x^3 + 32x^4 + \cdots.$$

Even the next term is unknown. For the labeled case, the first nine terms have been computed, and each coefficient takes exponentially more computer time.

Acyclic transitive digraphs correspond to partial orders, and remain uncounted, although Robinson [1970] has counted acyclic digraphs.

P2.5. Digraphs both self-complementary and self-converse

The only digraphs of order 3 which are both self-complementary and self-converse are the cyclic and transitive triples. The counting series starts

$$x + x^2 + 2x^3 + 4x^4 + \cdots.$$

This is an interesting new type of problem which seems to call for a "double Burnside Lemma".

P2.6. Eulerian digraphs

Eulerian graphs have been counted also by Robinson [1969], but the techniques are not adaptable to the corresponding problem for digraphs. The counting series for eulerian digraphs begins
$$x + x^2 + 3x^3 + 12x^4 + 68x^5 + \cdots.$$
A. Kotzig raised the special case of this question for eulerian tournaments, whose series begins
$$x + x^3 + x^5 + 3x^7 + \cdots.$$

3. Graphs with given structural properties

Some graphs with certain structural properties have been counted. Examples include trees, unicyclic graphs, functional digraphs, connected graphs, blocks, and block-graphs. But other such types of graphs remain uncounted. We consider in this section eight categories of graphs:

(a) hamiltonian,
(b) eulerian,
(c) graphs with local subgraphs,
(d) identity graphs,
(e) symmetric graphs,
(f) graphs with a square root,
(g) line and total graphs,
(h) clique and interval graphs.

Each of these contains in turn several individual counting problems, so that there are literally dozens of open questions in this section. Furthermore, whenever a new structural property is introduced into the graphical literature, we immediately take this as a challenge to find how many such graphs exist.

P3.1. Hamiltonian graphs

A graph or digraph is *hamiltonian* if it contains a spanning cycle. These have not been counted for labeled or unlabeled graphs or digraphs. The series for unlabeled graphs begins
$$x^3 + 3x^4 + 8x^5 + 48x^6 + \cdots,$$
while that for unlabeled digraphs starts
$$x^2 + 4x^3 + 60x^4 + \cdots.$$
This is the most publicized special case of graphs containing a specified subgraph, namely a spanning cycle. One can also stipulate graphs with other kinds of subgraphs such as a 1-factor, a 1-basis, etc.

P3.2. Eulerian graphs

Robinson [1969] counted eulerian graphs of order p. It is most difficult to count eulerian (p, q) graphs, where the number of lines is also an enumeration parameter. For eulerian graphs of order 6, the counting polynomial, where the parameter gives the number of lines, is

$$x^6 + x^7 + 2x^8 + x^9 + x^{10} + x^{11} + x^{12}.$$

Another interesting parameter for eulerian graphs would involve the minimum number of cycles whose union is the entire graph. This stems from the theorem, see Harary [1969] p. 64, that a connected graph is eulerian if and only if its set of lines can be partitioned into cycles.

P3.3. Local subgraphs

Given a graph H, the problem is to find the number of graphs of order p such that each point lies in a subgraph isomorphic to H. In the easiest example, H is a triangle and the series begins

$$x^3 + 2x^4 + 7x^5 + 37x^6 + \cdots.$$

We also ask for the number of graphs each line of which lies in a triangle. Similar questions can be raised for digraphs.

P3.4. Identity graphs

There are no nontrivial graphs of order less than 6 which have the identity group. For $p = 6$, there are eight identity graphs. Asymptotically, most graphs have the identity group, but there is no exact formula for order p. The same problem is raised for digraphs, where the series begins

$$x + x^2 + 7x^3 + 137x^4 + \cdots.$$

Identity trees have been counted (Harary and Prins [1959]), and also identity unicyclic graphs and identity functional digraphs by Stockmeyer [1971]. In principle, Stockmeyer obtained a formula for the number of graphs with given automorphism group. But its use entails the knowledge of the entire lattice of subgroups of the symmetric group S_p. Thus while this includes theoretically the determination of the number of identity graphs and digraphs of given order, it cannot be properly regarded as a solution to this problem.

P3.5. Symmetric graphs

In a *point-symmetric* graph, the automorphism group is transitive on the points. (This is why others sometimes call them vertex-transitive graphs). A *line-symmetric* graph is then defined as expected. A *symmetric* graph is both point symmetric and line-symmetric.

Turner [1967] counted point-symmetric graphs with a prime number p of points only. Even this number when p is a prime power has not been found. Chao [1971] proved that there exists a symmetric graph of prime order p, regular of degree n, if and only if n is even and divides $p-1$, and furthermore that such a graph is unique up to isomorphism. No other results are known about the number of symmetric graphs.

P3.6. Graphs with a square root

The *square* G^2 of a graph G has the same points as G, with u and v adjacent in G^2 whenever their distance in G is either 1 or 2. Other powers G^3, G^4, ... are defined similarly. Graphs which have a square root have been characterized (Harary [1969], p. 24), and so have other powers. Similar results are also known for digraphs. The counting of graphs and digraphs which have an nth root may not be an impossible problem.

P3.7. Line graphs and total graphs

The *line graph* $L(G)$ has the lines of graph G as its points with adjacency of lines as in G. A *line graph* is the line graph of some graph. This concept was introduced by Whitney [1932], who showed that a line graph H is the line graph of only one graph unless $H = K_3$. It follows trivially that the number of connected line graphs of order $p > 3$ is exactly the number of connected graphs with p lines. From the table in Cadogan [1966] of connected graphs, the generating function for connected line graphs with a given number of points begins

$$x + x^2 + 2x^3 + 5x^4 + 12x^5 + 30x^6 + 79x^7 + 227x^8 + \cdots.$$

The problem here is to find a more direct method, as well as to count them with a given number of points *and* lines. Such an approach would probably use one of the structure theorems for line graphs given in Harary [1969], Chapter 8.

Connected digraphs with q arcs are not the same in number as connected line digraphs of order q, as shown in Harary and Norman [1961]. Hence we do not even have this roundabout approach to the counting of line digraphs which works for line graphs of given order.

The *total graph* $T(G)$ has the points *and* lines of G as its point set, with adjacency defined more or less as expected; see Harary [1969], p. 82. It is known that the total graph $T(G)$ is the square of the subdivision graph $S(G)$ obtained by inserting a new point of degree 2 into each line of G. But this does not seem to facilitate a formula for the number of total graphs.

P3.8. Clique graphs and interval graphs

We define a few special kinds of intersection graphs.

A *clique* of a graph G is a maximal complete subgraph. The *clique graph* of G has the cliques of G as its points, with adjacency determined by non-empty intersection of two cliques. An *interval graph* has intervals on the real line as its points, with adjacency again determined by intersection. In a *rigid circuit graph*, there are no induced cycles of length greater than three.

These classes of graphs are known to be related to each other by their characterization theorems, and also to other similar structures such as "comparability graphs". For example, clique graphs were characterized structurally by Roberts and Spencer [1971], as reported in Harary [1969], p. 20.

The counting of these types of graphs represents a fiendish problem area. Recently, a young Harvard biologist named Joel Cohen asked us to count interval graphs, which he encountered in the study of genealogical chains.

4. Graphs with given parameter

All definitions not given here can be found in Harary [1969]. Counting problems for graphs with given parameters partition themselves naturally into sets of parameters which are closely related. If these questions were not so divided, we would have many more problems, one for each parameter. Our ten categories are:

(a) radius and diameter,
(b) girth and circumference,
(c) minimum and maximum degrees,
(d) connectivity,
(e) independence and covering numbers,
(f) clique numbers,
(g) intersection number,
(h) arboricity,
(i) genus and thickness,
(j) chromatic numbers.

For each of these categories, we define the related invariants, describe the partial progress which has been attained, and indicate if the problem is intractable.

The counting of graphs having a combination of parameters is much more difficult as the constraints become quite binding.

It is much easier to propose enumeration problems than to solve them. For example, each time a new parameter is introduced into graph theory (such as the *Ramsey number* $r(G)$ of a graph with no isolated points defined in Harary and Chvátal [1972], our distorted viewpoint immediately formulates an associated counting question. For another class of examples of such new problems, take a polynomial associated with a graph, such as the chromatic polynomial or the characteristic polynomial of the adjacency matrix studied in Harary *et al.* [1971], and ask for the number of graphs with the same polynomial.

P4.1. Radius and diameter

Not research and development, but radius and diameter are r and d here. The *eccentricity* $e(v)$ of a point v of a graph G is the maximum distance between v and any other point u. The *radius* $r(G)$ is the minimum eccentricity in G and the *diameter* $d(G)$ is the maximum.

Thus a graph G has radius 1 if and only if it has a point v_0 adjacent to all other points. It is easy to tell the number of graphs of order p with radius 1 because this number is precisely G_{p-1}, the total number of graphs of order $p-1$.

The proof is obvious: $G - v_0$ is an arbitrary graph with $p - 1$ points. Even for graphs with radius 2, there are no immediate neat solutions.

For trees, the diameter is approximately double the radius. Trees with prescribed diameter d were counted in Harary and Prins [1959], but for graphs this is a different story. Of course, the only p-point graph with $d = 1$ is K_p.

For digraphs, there are corresponding problems. The inradius and out-radius of a digraph D were defined in Harary *et al.* [1965], p. 162. These invariants definitely exist when D is strong. (Norman liked to pronounce the latter invariant as a homonym to "outrageous".) The counting of strong digraphs with prescribed inradius, outradius, and diameter is mentioned for completeness.

P4.2. Girth and circumference

The *girth* of a graph is the minimum length of a cycle in it; the *circumference* the maximum. Let c_n be the number of cycles of length n in a graph G, $n \geqslant 3$. Then $(c_3, c_4, c_5, \ldots, c_p)$ may be called the cycle length distribution sequence. How many graphs are there with such a prescribed sequence? The answer to this one general question would count the following structures with only a little more trouble:

(a) graphs with given girth, and *a fortiori*,

(b) graphs containing a triangle,

(c) graphs with given circumference,

(d) hamiltonian graphs.

The counting of unicyclic graphs is a very minor step in this direction.

P4.3. Minimum and maximum degrees

The minimum degree δ and the maximum degree Δ are natural parameters to consider for counting problems. Although there are theoretical formulations for counting graphs with a given partition (Parthasarathy [1968]) and digraphs with given partitions (Harary and Palmer [1966b]), these do not bear directly on δ and Δ.

The following initial progress has been made on counting graphs with $\delta \geqslant n > 0$. When $n = 1$, these are all the graphs with no isolated points, and are readily reckoned. The case $n = 2$ comprises all graphs with no endpoints as well as no isolates. These were counted by Robinson [1970a] using the method of cycle index sums developed for counting blocks. For $n = 3$, these are the homeomorphically irreducible graphs with no end points. R. W. Robinson also counted those but has not yet written up the formulas.

P4.4. Connectivity

The *connectivity* κ (*line-connectivity* λ) of a graph G is the minimum number of points (lines) whose removal from G results in a graph which is either disconnected or trivial. Then G is *n-connected* if $\kappa \geqslant n$; G is *n-line-*

connected if $\lambda \geqslant n$. Thus the number of graphs with connectivity n is the number of $(n+1)$-connected graphs minus the number of n-connected graphs. So we need only mention n-connected graphs henceforth.

Both 0-connected and 1-connected graphs have been counted as these are the disconnected and connected graphs, respectively. The 2-connected graphs are the same as blocks (except for the graph K_2) and these were counted by Robinson [1970a].

The enumeration of n-connected graphs for $n \geqslant 3$ await more powerful methods than now exist.

Since (as is well-known) $\kappa \leqslant \lambda$, one can also ask for the number of graphs with given κ and λ, and as a special case for those graphs having $\kappa = \lambda$.

P4.5. Independence and covering numbers

A set of points (lines) is *independent* if no two are adjacent. The *point-independence number* is the maximum number β_0 of independent points of G, and the *line-independence number* β_1 is defined similarly.

A point v and a line x *cover each other* if v is on x. The *point-covering number* α_0 of G is the minimum number of points which cover all the lines, and the *line-covering number* α_1 switches "points" and "lines".

Since it is a classic equation in graph theory that $\alpha_0 + \beta_0 = p = \alpha_1 + \beta_1$, we need only mention the problems of counting graphs with given point-independence number β_0 and with given β_1. Intuitively these seem easier than counting graphs with given α_0 and with given α_1. In practice, it makes no difference: by the above equation, the questions are interchangeable, and besides, nobody is likely to solve either of them.

A few other covering numbers, equally intractable for enumeration, are mentioned. Two points (lines) *cover each other* if they are adjacent. Let

α_{00} = the minimum number of points needed to cover all the other points.

α'_{00} = the minimum number (if any) of independent points which cover all other points. α_{11} and α'_{11} are defined similarly for lines.

α'_0 = the minimum number (if any) of independent points which cover all all the lines.

α'_1 then switches points and lines.

It is not even known for which graphs these primed invariants exist.

P4.6. Clique numbers

There are several invariants associated with cliques, see P3.8. One of these is the greatest clique order of G which we have just encountered in the form $\beta_0(\bar{G})$, the maximum number of independent points in the complement of G. Some other clique numbers are:

(a) the number of cliques,

(b) the minimum order of a clique,

(c) the minimum number of cliques which cover all the points of G,

(d) similarly for covering the lines of G,

(e) the maximum number of point-disjoint cliques.

The invariant (e) is of course the point-independence number of the clique graph of G, while the first, (a), is the number of points in the clique graph. In general, any invariant of the clique graph of G becomes in this way an invariant of G itself. None of these problems seem possible.

P4.7. Intersection number

The intersection number $\omega = \omega(G)$ of a given graph G is the minimum number of elements in a set S such that there is a family $S_1, \ldots, S_p$ of distinct nonempty subsets of S whose union is S, and V_i and V_j are adjacent in G if and only if $S_i \cap S_j \neq \emptyset$. Another variation on this invariant is ω_0 which differs from ω only in that the sets S_i need not be distinct, so that for example $\omega_0(K_p) = 1$.

There does not seem to exist any method for counting graphs with given intersection number or any of its possible variations.

P4.8. Arboricity

The *arboricity* of a graph G is the minimum number of line-disjoint acyclic subgraphs whose union is G. Perpetrating a glorious groaning pun, our friend R. C. Read called the maximum number of line-disjoint non-acyclic subgraphs whose union is G the *anarboricity* of G. These provide a typical example of a covering (arboricity) and packing (anarboricity) pair of invariants, as noted in Harary [1970]. Another such pair is given by the (covering) path number, also known as the *pathos*, which is the smallest number of paths whose union is G, and its packing counterpart, the *apathy*.

Such invariants appear hopeless for use as enumeration parameters, as does also the number of spanning trees in a graph, called its *complexity* by Brooks *et al.* [1940].

P4.9. Genus, thickness, coarseness, crossing number

These and other topological invariants appear to be the most intractable of all as far as enumeration is concerned. In the easiest case, they all reduce to planar graphs. The *genus* γ of G is the minimum genus n of an orientable surface on which G can be embedded with no pair of edges intersecting. The *thickness* θ is the smallest number of planar subgraphs whose union is G. The *coarseness* ξ is the greatest number of line-disjoint nonplanar subgraphs whose union is G. And the *crossing number* v is the smallest number of pairs of edges which intersect when G is drawn in the plane. When G is planar, $\gamma = v = 0$, $\theta = 1$, and ξ is not defined.

There are many other related topological invariants of a graph, each at least as impossible to count as the above four. These will not be defined here and include the "maximum genus", the "toroidal thickness", "toroidal coarseness", "toroidal crossing number", and the "rectilinear crossing number".

P4.10. Chromatic number

The *chromatic number* χ of a graph G is the minimum number of colors needed for its points so that no two adjacent points have the same color. Then G is *n-chromatic* if $\chi = n$, *n-colorable* if $\chi \leqslant n$ and *n-colored* if the points are colored using exactly n colors. The *n*-colored graphs were counted by Robinson [1968] and the 2-colorable graphs by Harary and Prins [1963]. Even the 3-colorable graphs appear impossible to enumerate at present.

The *line-chromatic number* χ' (*total-chromatic number* χ'') of G is quickly described as the chromatic number of the line graph (total graph) of G. The corresponding counting problems will probably have to await the enumeration of *n*-colorable graphs.

5. Subgraphs of a given graph

Most of the problems in this section ask for the number of dissimilar subgraphs of a given graph G which are isomorphic to a certain graph H. Thus the group of G determines whether or not two occurrences of subgraph H are regarded as equivalent. Analogous questions can also be posed for digraphs.

P5.1. Hamiltonian cycles

The number of spanning cycles in a given graph or digraph can be theoretically expressed in the labeled case using the method of Cartwright and Gleason [1966] in terms of the adjacency matrix. But the calculation of such numbers is forbidding, and becomes even more tricky when the number of similarity classes is wanted. Only for special graphs such as K_p can the answer be written effortlessly from first combinatorial principles. There is of course just one type of spanning cycle in K_p and the labeled number is $(p-1)!/2$.

The most interesting special case of this problem arises when the given graph is the *n*-cube Q_n because of applications to coding theory. Two spanning cycles of an *n*-cube are *similar* if there is an automorphism of Q_n that maps one onto the other. It is easily observed that there is only one similarity class of spanning cycles in Q_2 and Q_3. Gilbert [1956] has shown that the series for this unlabeled problem begins

$$x^2 + x^3 + 9x^4 + \cdots$$

but the numbers are not known for $n > 4$. The number of labeled hamiltonian cycles in Q_n has also been found only for $n \leqslant 4$. An undetermined amount of computer time is required to calculate just the next coefficient.

P5.2. Cycles of a given length

The problem asks for the number of dissimilar cycles of length k in a graph of order p, generalizing P5.1. It is easily solved in special cases. For example, there is only one similarity class of cycles of length $2k$ in the complete bipartite graph $K_{m,n}$ when $m, n \geqslant k$. The answer to the labeled version of the same question is $\binom{m}{k}\binom{n}{k}k!/(2k)$. A solution of the labeled problem in terms of the adjacency matrix for cycles of length 3, 4 and 5 in a given graph or digraph was found in Harary and Manvel [1971].

P5.3. Complete graphs

As mentioned above, the number of triangles K_3 in a given labeled graph can be determined. The problem of determining the number of occurrences of K_n, $n > 3$, in a given labeled graph is open. For the unlabeled case, the knowledge of not only the number of labeled triangles, but their behavior with respect to the group of the given graph, tells the number of dissimilar triangles.

P5.4. Spanning trees

The number of labeled spanning trees of a given labeled graph can be determined from the Matrix-Tree Theorem. But there is no corresponding method for determining the number of nonisomorphic spanning trees of a given graph. The difficulty of the problem is indicated by the example in which the given graph is the complete graph of order p. Then the number of dissimilar spanning trees is the number of trees of order p, a problem first solved by Cayley.

P5.5. Factors

Let G be a labeled graph which does have a 1-factor, in accordance with the criterion of Tutte; see Harary [1969], p. 86. The number of different 1-factors of G is not known except in very special cases. For example, K_{2n} has $(2n)!/(2^n n!)$ labeled 1-factors and $K_{n,n}$ has just $n!$ of them.

A *factorization* of G, if any, is a partition of the lines of G into 1-factors. It is known that K_{2n} has a 1-factorization but the number of these has not been settled except for small n. Recently, W. Wallis established that K_8 has exactly six 1-factorizations. By an older theorem of König (see Harary [1969], p. 85), every regular bigraph such as $K_{n,n}$ has a 1-factorization, so that the same question can be asked for these.

P5.6. Eulerian trails in a given eulerian graph

There is an explicit formula for the number of eulerian trails in a given digraph; see Harary [1969], p. 204. For graphs, however, no progress has been made. One possible approach would begin with a given eulerian graph G and consider all its orientations which result in eulerian digraphs $D_1, D_2, \ldots$. If e_i is the number of eulerian trails in D_i, then $\sum e_i$ is the total number of oriented eulerian trails G. But this is easier said than done. For example, the special case $G = K_{2n+1}$ requires the availability of the adjacency matrices of the eulerian tournaments of order $2n+1$.

6. Supergraphs of a given graph

If H is a subgraph of G, then G is a *supergraph* of H. The problems in this section ask for the number of graphs of order p which are supergraphs of a given graph H. Extremal graph theory may be useful in counting such graphs. For example, take $H = K_3$. Turán's theorem shows that if G has at least $[p^2/4]$ lines then G contains K_3. Therefore the counting problem need only be handled for graphs with less than $[p^2/4]$ lines. On the other hand a solution to such a counting problem may solve the corresponding extremal problem. Versions of all of these problems can also be raised for digraphs.

P6.1. Cycles

Counting the supergraphs of a triangle is the same as counting graphs of girth 3. But supergraphs of the cycle C_n of order $n \geqslant 4$ do not correspond to graphs of girth n. The series for $n = 4$ begins

$$3x^4 + 16x^5 + 111x^6 + \cdots.$$

P6.2. Complete graphs

The problem of determining the number of supergraphs of K_n has already arisen in different guises for $n = 3$. Therefore our interest here focuses on $n \geqslant 4$. Turán's general theorem solves the associated extremal problem; see Harary [1969], p. 18.

P6.3. Complete bipartite graphs

The problem is to count supergraphs of $K_{m,n}$. Since $K_{2,2}$ is a cycle of order 4, we have already encountered a difficult special case. Note that the supergraphs of $K_{1,n}$ are those graphs whose maximum degree is at least n, already mentioned in problem P4.3.

P6.4. Paths

Counting supergraphs of the path P_n of order n is easy for small n. For example if $n = 4$, the only connected graphs of order $\geqslant 4$ which are *not* supergraphs of P_n are the stars $K_{1,m}$. A solution for each n would involve knowledge of those connected graphs of order $\geqslant n$ whose spanning trees all have diameter less than n.

7. Enumeration equivalent to the four color conjecture

There exists a *bona fide* method for settling the 4CC (Four Color Conjecture) if only certain graphical enumeration problems could be solved.

4CC: Every planar graph is 4-colorable.

EE4CC: The number of planar graphs equals the number of 4-colorable planar graphs.

The counting series for these two classes of graphs are known to have the same first thirty-nine coefficients; see Harary [1969], p. 5. There is some latitude with regard to two different degrees of freedom:

(1) The parameter can vary. One can use either p points, or p points and

q lines, or q lines, or diameter d, or any other parameter for which there is some hope.

(2) The type of graph can vary quite a bit provided only that both the planar graphs and the 4-colorable planar graphs share the same properties, which may include:

  (a) planar graphs (as in EE4CC),
  (b) rooted planar graphs,
  (c) line-rooted planar graphs,
  (d) planar graphs rooted at a triangle,
  (e) labeled planar graphs,
  (f)–(j) the preceding five properties with plane graphs in place of planar graphs,
  (k) plane graphs rooted à la Tutte.

The latter rooting of plane graphs, or as Tutte prefers, plane maps, is accomplished in three stages:

  (i) Select an arbitrary edge x of a plane graph G.
  (ii) Orient x arbitrarily.
  (iii) Arbitrarily designate one of the two regions incident with x as the exterior.

There are q possibilities for step (i) as G has q lines, and 2 possibilities for each of the other two steps. Multiplying these together, we see that the total number of Tutte-orientations of a plane graph is $4q$, which is reflected in the celebrated paper Harary and Tutte [1966], on the automorphism group of a planar graph.

Fortunately, Tutte [1972] has provided the definitive comprehensive survey of the art of enumerating Tutte-oriented plane maps, and we refer the reader to his work for all information related to this approach.

P7.1. Planar and plane graphs

All trees are planar, so the number of planar trees equals the number of trees. Plane unicyclic graphs are easily counted by using rooted plane trees for the figure counting series and taking the dihedral group as the configuration group in the Pólya Enumeration Theorem.

For both plane and planar graphs, all ten variations (a)–(j) above are open problems.

P7.2. n-colorable graphs

This has been solved only for $n = 2$, and is the same as problem P4.10.

P7.3. n-colorable planar graphs

This problem has not been attempted for any $n > 1$, and may well be possible for $n = 2$, we hope to solve this simple special case and publish it before anyone else does.

P7.4. Self dual plane graphs

Given a plane graph G, its *dual* G^* is constructed as follows: place a point in each region of G including the exterior region and if two regions have a line x in common, join the corresponding points by a line x^* crossing only x. The result is always a plane *pseudograph* in which loops and multiple lines are allowed. The tetrahedron is self dual, while the cube and the octahedron are duals, as are the dodecahedron and the icosahedron.

Conclusion

The confrontation of manageable solutions to P7.3 for $n = 4$ and to P7.1 would settle the EE4CC and hence the 4CC itself. Of course this would also be accomplished by comparing any other pair of classes of planar graphs which are obtained in the same way, as for example by the same kind of rooting. It is safe to predict that the 4CC will not be settled for the first time, if any, by means of the EE4CC.

8. Summary of problem areas

Labeled

    P1.1. Labeled self-converse digraphs
    P1.2. Labeled self-complementary graphs

Digraphs

    P2.1. Strong digraphs
    P2.2. Unilateral digraphs
    P2.3. Digraphs with a source
    P2.4. Transitive digraphs
    P2.5. Digraphs both self-complementary and self-converse
    P2.6. Eulerian digraphs

Graphs with given structural properties

    P3.1. Hamiltonian graphs
    P3.2. Eulerian graphs
    P3.3. Local subgraphs
    P3.4. Identity graphs
    P3.5. Symmetric graphs
    P3.6. Graphs with a square root
    P3.7. Line graphs and total graphs
    P3.8. Clique graphs and interval graphs

Graphs with given parameter

    P4.1. Radius and diameter
    P4.2. Girth and circumference
    P4.3. Minimum and maximum degrees

10

This particular listing of graphical enumeration problems which are unsolved contains forty items. But there are many more variations mentioned within most of the problems. Other open questions which are not transparently graphical in nature have been deliberately omitted. It appears that in general new counting methods will be required to solve the problems in this list.

References

R. L. Brooks, C. A. B. Smith, A. H. Stone and W. T. Tutte, 1940, The dissection of rectangles into squares, *Duke Math. J.* **7**, 312–340.

C. C. Cadogan, 1966, The Möbius function and connected graphs, Sci. Rept. UWI-CC4, Univ. of the West Indies, Kingston, Jamaica.

C. Y. Chao, 1971, On the classification of symmetric graphs with a prime number of vertices, *Trans. Am. Math. Soc.* **158**, 247–256.

D. Cartwright and T. Gleason, 1966, The number of paths and cycles in a digraph, *Psychometrika* **31**, 179–199.

E. N. Gilbert, 1956, Enumeration of labeled graphs, *Canad. J. Math.* **8**, 405–411.

F. Harary, 1969, *Graph Theory* (Addison-Wesley, Reading, Mass.).

F. Harary, 1970, Covering and packing in graphs, I, *Ann. N.Y. Acad. Sci.* **175**, 198–205.

F. Harary and V. Chvátal, 1972, Generalized Ramsey theory for graphs, II; small diagonal numbers, *Proc. Ann. Math. Soc.* **32**, 389–394.

F. Harary and B. Manvel, 1971, On the number of cycles in a graph, *Mat. Časopis* **21**, 55–63.

F. Harary and R. Z. Norman, 1961, Some properties of line digraphs, *Rend. Circ. Mat. Palermo*, **9**, 161–168.

F. Harary and E. M. Palmer, 1966a, Enumeration of self-converse digraphs, *Mathematika* **13**, 151–157.

F. Harary and E. M. Palmer, 1966b, The groups of the small digraphs, *J. Indian Statist. Assoc.* **4**, 155–169.

F. Harary and E. M. Palmer, 1966c, Enumeration of locally restricted digraphs, *Canad. J. Math.* **18**, 853–860.

F. Harary and E. M. Palmer, to appear, *Graphical Enumeration* (Academic Press, New York).

F. Harary and G. Prins, 1959, The number of homeomorphically irreducible trees, and other species, *Acta Math.* **101**, 141–162.

F. Harary and G. Prins, 1963, Enumeration of bicolourable graphs. *Canad. J. Math.* **15**, 237–248.

F. Harary and W. T. Tutte, 1966, On the order of the group of a planar map, *J. Combin. Theory* **1**, 394–395.

F. Harary, R. Z. Norman and D. Cartwright, 1965, *Structural Models: an introduction to the theory of directed graphs* (Wiley, New York).

F. Harary, C. King, A. Mowshowitz and R. C. Read, 1971, Cospectral graphs and digraphs, *Bull. London Math. Soc.* **3**, 321–328.

J. W. Moon, 1970, *Counting Labelled Trees* (Canad. Math. Congress, Montreal).

O. Ore, 1967, *The Four Color Problem* (Academic Press, New York).

K. R. Parthasarathy, 1968, Enumeration of ordinary graphs with given partition, *Canad. J. Math.* **20**, 40–47.

R. C. Read, 1963, On the number of self-complementary graphs and digraphs, *J. London Math. Soc.* **38**, 99–104.

R. W. Robinson, 1968, Enumeration of colored graphs, *J. Combin. Theory* **4**, 181–190.

R. W. Robinson, 1969, Enumeration of Euler graphs, *Proof Techniques in Graph Theory* (Academic Press, New York), pp. 147–153.

R.W. Robinson, 1970a, Enumeration of non-separable graphs, *J. Combin.Theory* **9**, 327–356.

R. W. Robinson, 1970b, Enumeration of acyclic digraphs, *Proc. Second Chapel Hill Conf. on Combinatorial Mathematics and its Applications* (University of North Carolina, Chapel Hill, N. Car.), pp. 391–399.

R. S. Roberts and J. H. Spencer, 1971, A characterization of clique graphs, *J. Combin. Theory* **10B**, 102–108.

P. K. Stockmeyer, 1971, The enumeration of graphs with prescribed automorphism group, Doctoral Dissertation, Univ. of Michigan, Ann Arbor, Mich.

J. Turner, 1967, Point-symmetric graphs with a prime number of points, *J. Combin. Theory* **3**, 136–145.

W. T. Tutte, 1972, The enumerative theory of planar maps, *A Survey of Combinatorial Theory* (J. N. Srivastava *et al.*, eds.; North-Holland, Amsterdam), pp. 437–448 (this volume).

H. Whitney, 1932, Congruent graphs and the connectivity of graphs, *Am. J. Math.* **54**, 150–168.

N. Srivastava et al., eds., *A Survey of Combinatorial Theory*
© North-Holland Publishing Company, 1973

CHAPTER 22

On Spectrally Bounded Graphs

A. J. HOFFMAN

IBM Inc., New York, N.Y., U.S.A.

1. Introduction

Let G be a graph (undirected, on a finite number of vertices, with at most one edge joining a pair of vertices, and no edge joining a vertex to itself). Two vertices are said to be adjacent if they are joined by an edge. The valence of a vertex is the number of vertices adjacent to it. The set of vertices of G is denoted by $V(G)$, the set of edges by $E(G)$.

If G is a graph, the adjacency matrix of $G \equiv A(G)$ is defined by

$$A(G) = A = (a_{ij}) = \begin{cases} 1 & \text{if vertex } i \text{ and vertex } j \text{ are adjacent,} \\ 0 & \text{otherwise.} \end{cases}$$

Thus, $A(G)$ is a real symmetric $(0, 1)$ matrix of order $|V(G)|$, with diagonal all 0. If A is any real symmetric matrix, we denote its eigenvalues in descending order by $\lambda_1(A) \geq \lambda_2(A) \geq \ldots$, in ascending order by $\lambda^1(A) \leq \lambda^2(A) \leq \ldots$. If $A = A(G)$, we will write $\lambda_i(G)$ and $\lambda^i(G)$ for $\lambda_i(A(G))$, $\lambda^i(A(G))$.

Let $\mathscr{G}$ be an infinite set of graphs. In the investigations given in the bibliography, a lower bound on $\lambda^1(G)$, as G varies in $\mathscr{G}$, played a key role; sometimes a specific lower bound (most especially -2), sometimes the existence of some unspecified bound. It is this latter situation we will explore in detail. For given $\mathscr{G}$, it may be true or false that there exists some λ such that $\lambda^1(G) \geq \lambda$ for all $G \in \mathscr{G}$. We shall give two graph theoretic characterizations of those families $\mathscr{G}$ for which a uniform lower bound exists, one "local" in terms of excluded subgraphs, one "global" in describing how each graph in $\mathscr{G}$ is constructed.

2. Statement of characterizations

We shall need further definitions. If G is a graph, $\bar{G}$ is the graph with $V(G) = V(\bar{G})$, and two distinct vertices are adjacent in $\bar{G}$ if and only if they are not adjacent in G. If H and G are graphs, we say $H \subset G$ (H is a subgraph of G) if $V(H) \subset V(G)$, and two vertices in $V(H)$ are adjacent in H if and only if they were adjacent in G. A clique on t vertices, denoted by K_t, is a graph in which each pair of distinct vertices is adjacent. $\bar{K}_t$ is called an independent set on t

vertices. The symbol C_t denotes a claw on t vertices (i.e., a set of $t+1$ vertices, one of which is adjacent to all the others, of which no two are adjacent; this graph is also sometimes denoted $K_{1,t}$). The symbol H_t denotes a graph on $2t+1$ vertices, $2t$ of which form a K_{2t}, while the remaining vertex is adjacent to exactly t of these $2t$ vertices.

If G and H are graphs with $V(G) = V(H)$, we shall define a distance $d(G, H)$. Write

$$A(G) + A(\tilde{G}) = A(H) + A(\tilde{H}),$$

where

$$(A(\tilde{G}))_{ij} = 1 \quad \text{if and only if } (A(G))_{ij} = 0, \ (A(H))_{ij} = 1$$

$$(A(\tilde{H}))_{ij} = 1 \quad \text{if and only if } (A(G))_{ij} = 1, \ (A(H))_{ij} = 0.$$

Then $d(G, H)$ = the largest of the valences of the vertices in $\tilde{G}$ and $\tilde{H}$.

Theorem. *Let $\mathcal{G}$ be an infinite set of graphs. Then the following statements about $\mathcal{G}$ are all true or all false:*

(i) *there exists a number λ such that, for all $G \in \mathcal{G}$, $\lambda^1(G) \geq \lambda$;*

(ii) *there exists a positive integer l such that, for all $G \in \mathcal{G}$, $C_l \nsubseteq G$ and $H_l \nsubseteq G$;*

(iii) *there exists a positive integer L such that, for each $G \in \mathcal{G}$, there exists a graph H with $d(G, H) \leq L$, and H contains a distinguished family of cliques $K^1, K^2, \ldots$ such that*

(iiia) *each edge of H is in at least one K^i,*

(iiib) *each vertex of H is in at most L of the cliques $K^1, K^2, \ldots$,*

(iiic) *$|V(K^i) \cap V(K^j)| \leq L$ for $i \neq j$.*

This theorem was first announced by Hoffman [1970a]. Some consequences of the theorem are reported in Hoffman [1970a,b, 1971]. In addition, a portion of these results have been used by Howes [1970] to characterize the families $\mathcal{G}$ for which there is a uniform upper bound on $\lambda_2(G)$ for all $G \in \mathcal{G}$.

Before proceeding to the proof, which occupies the remainder of the paper, some remarks are in order. Let $\mathcal{G} = \{G_1, G_2, \ldots\}$. If $\mathcal{H} = \{H_1, H_2, \ldots\}$ is an infinite sequence of graphs such that there exist two infinite sequences of indices $i_1 < i_2 \ldots$ and $j_1 < j_2 \ldots$ with $G_{i_k} \subset H_{j_k}$, $k = 1, 2, \ldots$, then we will say $\mathcal{G} \subset \mathcal{H}$. The significance of (i) $\Leftrightarrow$ (ii) in the theorem can now be restated: if $\mathcal{G} = \{G_1, G_2, \ldots\}$ is any sequence of graphs such that $\lambda^1(G_i) \to -\infty$, then $\{C_1, H_1, C_2, H_2, C_3, H_3, \ldots\} \subset \mathcal{G}$.

The second remark concerns (iii). It would be desirable, if true, to avoid the intervention of H in (iii) and assert (iiia,b,c) for G rather than H. But the intervention of H cannot be avoided. Let G_n be the cocktail party graph of order n (i.e., G_n is a graph on $2n$ vertices, in which each vertex is adjacent to all but one of the remaining vertices). Note $\lambda^1(G_n) = -2$ for all $n \geq 2$.

Let k be the largest number of vertices in a clique $\{K^i\}$, say $k = |V(K^1)|$. Let v be a vertex not adjacent to a vertex, say w, in K^1. Then v and each of the $k-1$ vertices in K^1 other than w is contained in at least one distinguished clique by (iiia). There are at most L such cliques, by (iiib), and each contains less than L vertices other than v, by (iiic). It follows that $k-1 \leq L^2$. Since $k = \max |V(K^i)|$, it follows that w is adjacent to at most $L(k-1) \leq L^3$ vertices of G_n. But this cannot be true for $2n-1 \geq L^3$.

3. Proof of Theorem

In this section, we show that (i) $\Rightarrow$ (ii) and (iii) $\Rightarrow$ (i). In the next section, we will show (ii) $\Rightarrow$ (iii).

Throughout the proof, we shall use the fact that $G \subset H$ implies $\lambda^1(G) \geq \lambda^1(H)$, which follows from the fact that $G \subset H$ means that $A(G)$ is a principal submatrix of $A(H)$.

To prove (i) $\Rightarrow$ (ii), it is sufficient to record that $\lambda^1(C_t)$, $\lambda^1(H_t)$ both tend to $-\infty$ for t large, which we have proved elsewhere (Hoffmann [1971]), or which the reader can easily establish himself. This and the preceding paragraph complete the proof.

To prove that (iii) implies (i), assume a graph G satisfies (iii) for some L. We will prove

$$\lambda^1(G) \geq -3L - \binom{L}{2}(L-1), \tag{3.1}$$

which will prove (i).

Let M be the (0, 1) incidence matrix of vertices of H versus distinguished cliques $K^1, K^2, \ldots$. Thus $m_{ij} = 1$ if $v_i \in K^j$, and 0 otherwise.

Lemma 3.1. $MM^T = A(H) + S$, where

$$S = (s_{ij}) \tag{3.2}$$

is a matrix with all entries nonnegative, and

$$\lambda_1(S) \leq L + \binom{L}{2}(L-1). \tag{3.3}$$

Proof. To prove (3.2), note first that MM^T has all entries nonnegative, $A(H)_{ii} = 0$ for all i, hence $s_{ii} \geq 0$ for all i. If two vertices in H are adjacent, then there is at least one distinguished clique containing both by (iiia), so $(MM^T)_{ij} \geq 1$, $(A(H))_{ij} = 1$, so $s_{ij} \geq 0$. If two different vertices i and j of H are not adjacent, $(MM^T)_{ij} = 0$, $A(H)_{ij} = 0$, $s_{ij} = 0$. Thus, in all cases $s_{ij} \geq 0$.

To prove (3.3), we note that, by the Perron-Frobenius theorem, $\lambda_1(S) \leq \max_i \sum_j s_{ij}$. Also, by (iiib), $s_{ii} \leq L$. Hence, to prove (3.3), it is sufficient to prove that, for each i,

$$\sum_{j \neq i}' s_{ij} \leq (L-1)\binom{L}{2}. \tag{3.4}$$

But, the left side of (3.4) is the number of 2×2 matrices in M which have one row i and consist entirely of 1's. But the number of 1's in row i is at most L by (iiib), so the number of pairs of columns which are candidates for a 2×2 "box" is at most $\binom{L}{2}$. Two such columns can produce at most $L-1$ boxes, by (iiic). Thus (3.4) is proved, and hence the lemma.

To complete the proof of (3.1), we invoke the theorem that, if A and B are real symmetric matrices, and $C = A+B$, then

$$\lambda^1(A)+\lambda^1(B) \leqq \lambda^1(C) \leqq \lambda^1(A)+\lambda_1(B). \tag{3.5}$$

By the Perron-Frobenius theorem, if G is a graph in which each vertex has valence at most L, then

$$-L \leqq \lambda^1(G) \leqq \lambda_1(G) \leqq L. \tag{3.6}$$

From $A(G)+A(\tilde{G}) = A(H)+A(\tilde{H})$, (3.5) and (3.6), we conclude

$$\lambda^1(G) \geqq \lambda^1(H)-2L. \tag{3.7}$$

From Lemma 3.1 and (3.5), we conclude

$$0 \leqq \lambda^1(MM^{\mathrm{T}}) \leqq \lambda^1(H)+\lambda_1(S) \leqq \lambda^1(H)+L+(L-1)\binom{L}{2}. \tag{3.8}$$

But (3.7) and (3.8) imply (3.1).

4. Proof of the Theorem (continued)

In this section, we prove (ii) $\Rightarrow$ (iii). Our reasoning here is entirely graph theoretical, since the concept of eigenvalue plays no role in the statement of (ii) or (iii). We shall prove that there exists a function $L(l)$ such that, if G satisfies (ii) for some l, then G satisfies (iii) for $L = L(l)$.

The strategy of the proof is as follows: We shall first look for large cliques in G ("large" depends on l), and define an equivalence relation on large cliques. The equivalence classes of large cliques will be shown to have properties (iiib) and (iiic) and, if additional edges (forming a graph $\tilde{G}$ in which each vertex has "small" valence) are added, the equivalence classes will be cliques. It will also turn out that the edges in G not contained in any large clique form a graph $\tilde{H}$ in which each vertex has small valence. Thus the distinguished cliques in H will be the equivalence classes of large cliques in G.

To define large cliques, we need first the Ramsey function $R(l)$ which satisfies: if $|V(G)| \geqq R(l)$, then $K_l \subset G$ or $\bar{K}_l \subset G$. We also need a function $f(m, r, l)$ defined recursively on triples of positive integers:

$$f(1, r, l) = r+1,$$
$$f(m+1, r, l) = \max \{r+mr(l-2)+1, f(m, r+l-1, l)\}.$$

Let

$$N = N(l) = \max \{l^2+l+2, l+R(l), f(l, 1, l)\}.$$

Define $\mathscr{W}$ to be the set of all cliques $K \subset G$ such that $|V(K)| \geqq N$.

Lemma 4.1. *If K, $K' \in \mathscr{W}$, define*

$$K \sim K'$$

if each vertex of K is adjacent to all but at most $l-1$ vertices of K'. Then $\sim$ is an equivalence relation.

Proof. Reflexivity is clear, since $|V(K)| \geqq l$. To prove symmetry, assume there is a vertex v in K' not adjacent to at least l vertices in K, and let A denote that set of l vertices in K. Each vertex in A is not adjacent to at most $l-2$ vertices in K' other than v, since $K \sim K'$. Hence, the set of vertices in K' each not adjacent to at least one vertex in A consists of v and at most $l(l-2)$ other vertices. Since $N > l+l(l-2)+1$, it follows that K' contains at least l vertices each of which is adjacent to each vertex in A. Call that set of l vertices B. Then v, A, B generate an H_l, contrary to hypothesis. This contradiction proves that $\sim$ is symmetric.

To prove transitivity, assume $K^1 \sim K^2$, $K^2 \sim K^3$, $K^1 \nsim K^3$. Then K^3 contains a vertex v not adjacent to a set C of l vertices in K^1. Since $N > 2l+l(l-1)-1$, and $K^1 \sim K^2$, it follows that K^2 contains a subset D of $2l-1$ vertices each of which is adjacent to all vertices in C. But since $K^3 \sim K^2$, D contains some subset F of l vertices adjacent to v. Then C, F, v generate an $H_l \subset G$, which is a contradiction.

Henceforth, the letter E will denote any equivalence class of cliques in $\mathscr{W}$, and $V(E)$ will be the union of all vertices of all cliques in E.

Lemma 4.2. *Let E be an equivalence class, $v \in V(E)$. Then v is adjacent to all but at most $R(l)-1$ other vertices in $V(E)$.*

Proof. Let $K^v \in E$ be a clique containing v. By Ramsey's theorem, if $F \subset V(E)$, $|F| \geqslant R(l)$, and every vertex in F not adjacent to v, then F contains a K_l or a $\overline{K}_l$. If $\overline{K}_l \subset F$, then since $|V(K^v)| > l^2 - 2l + 1$, there exists a vertex $w \in K^v$ adjacent to all vertices in $\overline{K}_l$. Thus $C_l \subset G$, a contradiction.

If $K_l \subset F$, then $|V(K^v)| > l+l(l-2)+1$ implies that K^v contains a set of l vertices each adjacent to all the vertices in K_l, thus generating an H_l.

Lemma 4.3. *If $K, K' \in \mathscr{W}$, $K \in E$, and $V(K') \subset V(E)$, then $K \sim K'$.*

Proof. Assume the contrary. Then there exists a vertex $v \in K'$ not adjacent to as many as l vertices in K, thus adjacent to at most $l-1$ vertices in K, thus not adjacent to more than $N-l$ vertices in K. But since $N \geqq l+R(l)$, v is not adjacent to more than $R(l)$ vertices of K, contradicting Lemma 4.2.

Lemma 4.4. *If E_1 and E_2 are different equivalence classes,*

$$|V(E_1) \cap V(E_2)| \leqq R(N)-1.$$

Proof. If $|V(E_1) \cap V(E_2)| \geqslant R(N)$, then by Ramsey's theorem, $V(E_1) \cap V(E_2)$ contains K_N or $\overline{K}_N$ as a subgraph. It is impossible for $\overline{K}_N$ to be a subgraph, by Lemma 4.2. If K_N were a subgraph, let $K \in E_1$. By Lemma 4.3,

$K_N \sim K$, so $K_N \in E_1$. Similarly, $K_N \in E_2$. Therefore, $E_1 = E_2$, contrary to hypothesis.

Lemma 4.5. *If $K, K' \in \mathscr{W}$, $Z \subset V(K')$, $|Z| = l$, each vertex of Z adjacent to all but at most $l-1$ vertices of K, then $K \sim K'$.*

Proof. Assume the contrary; so there is a vertex $v \in V(K')$ adjacent to fewer than l vertices in K. Since $|V(K)| \geq N$, there are at least $(2l-1)$ vertices in K each of which is adjacent to all vertices in Z. Vertex v is not adjacent to at least l of them, thus an H_l would be generated.

Lemma 4.6. *Let $f(m, r, l)$ be the function defined at the beginning of Section 4. Let $n \geq f(m, r, l)$, and let $K^1, \ldots, K^m \in \mathscr{W}$ be inequivalent cliques, v a vertex in each of $V(K^1), \ldots, V(K^m)$, $|V(K^i)| \geq n$, $i = 1, \ldots, m$. Then there exist sets $S_i \subset V(K^i)-v$, $i = 1, \ldots, m$, such that $|S_i| = r$, and $i \neq j$ implies that each vertex in S_i is adjacent to no vertices in S_j.*

Proof (by induction on m). If $m = 1$, then $n \geq r+1$; the lemma holds. Assume the lemma to be true for some m and all r; we shall show that it holds for $m+1$ and all r.

Since $n \geq f(m+1, r, l) \geq f(m, r+l-1, l)$, it follows from the induction hypothesis that there exist subsets $S_i' \subset V(K')$, $i = 1, \ldots, m$, $|S_i'| = r+l-1$, and each vertex in S_i' adjacent to no vertices in S_j' for $i \neq j$. By Lemma 4.5, at most $l-1$ vertices in S_i' ($i = 1, \ldots, m$) are each adjacent to at least l vertices in K^{m+1}. Consequently, S_i' contains a subset S_i, $|S_i| = r$, such that each vertex in S_i is adjacent to at most $l-2$ vertices in K^{m+1} other than v. Since

$$|V(K^{m+1})| \geq r+mr(l-2)+1,$$

there exists a subset $S_{m+1} \subset V(K^{m+1})$, $|S_{m+1}| = r$, such that each vertex in S_{m+1} is adjacent to no vertex of any S_i, $i = 1, \ldots, m$. This completes the induction.

Lemma 4.7. *Each vertex is contained in fewer than l equivalence classes.*

Proof. Since $N \geq f(l, 1, l)$, a contradiction of Lemma 4.7 would produce, by Lemma 4.6, $aC_l \subset G$.

Lemma 4.8. *Let $\tilde{H}$ be the graph formed by edges of G not in any clique in W. Then every vertice in $\tilde{H}$ has valence at most $R(\max(N-1, l))$.*

Proof. If not, then by Ramsey's theorem we would have $C_l \subset G$, or the edges in $\tilde{H}$ adjacent to v would be in a clique in $\mathscr{W}$, contradicting the definition of $\tilde{H}$.

It is clear that, following the strategy outlined at the beginning of this section, we have proved (ii) $\Rightarrow$ (iii), with

$$L(l) = \max\{R(l)-1, R(N)-1, R(\max(N-1, l), l-1\} = \max\{R(N)-1\},$$

where N is defined at the beginning of this section.

We are very grateful to Leonard Howes for his valuable help in the development and exposition of this material.

References

M. Doob, 1970, On characterizing certain graphs with few eigenvalues by their spectra, *Linear Algebra Appl.* **3,** 461–482.

A. J. Hoffman, 1960a, On the uniqueness of the triangular association scheme, *Ann. Math. Statist.* **31,** 492–497.

A. J. Hoffman, 1960b, On the exceptional case in a characterization of the arcs of a complete graph, *IBM J. Res. Develop.* **4,** 497–504.

A. J. Hoffman, 1965, On the line graph of a projective plane, *Proc. Am. Math. Soc.* **16,** 297–302.

A. J. Hoffman, 1968, Some recent results on spectral properties of graphs, *Beiträge zur Graphentheorie* (H. Sachs, H. J. Voss and H. Walther, eds.; B. G. Teubner Verlagsgesellschaft, Leipzig), pp. 75–80.

A. J. Hoffman, 1969a, The eigenvalues of the adjacency matrix of a graph, *Combinatorial Mathematics and its Applications* (R. C. Bose and T. C. Dowling, eds.; Univ. of North Carolina Press, Chapel Hill, N. Car.), pp. 578–584.

A. J. Hoffman, 1969b, The change in the least eigenvalue of the adjacency matrix of a graph under imbedding, *SIAM J. Appl. Math.,* 664–671.

A. J. Hoffman, 1970a, $-1-\sqrt{2}$?, *Combinatorial Structures and Their Applications* (R. Guy, H. Hanani, N. Taner, J. Schonheim, eds.; Gordon and Breach, New York), pp. 173–176.

A. J. Hoffman, 1970b, On eigenvalues and colorings of graphs, *Graph Theory and its Applications* (B. Harris, ed.; Academic Press, New York), pp. 79–91.

A. J. Hoffman, 1971, On vertices near a given vertex of a graph, *Studies in Pure Mathematics, papers presented to Richard Rado* (L. Mirsky, ed.; Academic Press, London), pp. 131–136.

A. J. Hoffman and Leonard Howes, 1970, On eigenvalues and colorings of graphs, II, *Intern. Conf. on Combinatorial Mathematics* (A. Gewirtz and L. Quintas, eds.), *Ann. N.Y. Acad. Sci.* **175,** 238–242.

A. J. Hoffman and A. M. Ostrowski, On the least eigenvalue of a graph of large minimum valence containing a given graph, *Linear Algebra Appl.* (to appear).

A. J. Hoffman and D. K. Ray-Chaudhuri, 1965a, On the line graph of a finite affine plane, *Canad. J. Math.* **17,** 687–694.

A. J. Hoffman and D. K. Ray-Chaudhuri, 1965b, On the line graph of a symmetric balanced incomplete block design, *Trans. Am. Math. Soc.* **116,** 238–252.

A. J. Hoffman and D. K. Ray-Chaudhuri, On a spectral characterization of regular line graphs (unpublished).

Leonard Howes, 1970, On subdominantly bounded graphs, Doctoral Dissertation, City Univ. of New York.

D. K. Ray-Chaudhuri, 1967, Characterization of line graphs, *J. Combin. Theory* **3,** 461–482.

J. J. Seidel, 1968, Strongly regular graphs with $(-1, 1, 0)$ adjacency matrix having eigenvalue 3, *Linear Algebra Appl.* **1,** 281–298.

J. N. Srivastava et al., eds., *A Survey of Combinatorial Theory*
© North-Holland Publishing Company, 1973

CHAPTER 23

Combinatorial Search Problems

G. O. H. KATONA

Mathematical Institute, Hungarian Academy of Sciences, Budapest, Hungary

1. Introduction

The basic problem is the following: *We have a finite set $X = \{x_1, \ldots, x_n\}$ and we want to identify an unknown element x_i of X testing some subsets A of X whether A contains x_i or not.*

There are many practical problems of this type. The first one (known from mathematical problems) is the following (see Dorfmann [1943] and Sterrett [1953]):

1. *"Wasserman-type" blood test of a large population.* X is the set of some men. The test may be divided conveniently into two parts: (1) A sample of blood is drawn from every man. (2) The blood sample is subjected to a laboratory analysis which reveals the presence or absence of "syphilitic antigen". The presence of syphilitic antigen is a good indication of infection; for the second step, instead of carrying out the test individually we can pour together some samples. Carrying out the second step on the mixture we may determine whether the given subset of men contains an infected man or not.

2. *Diagnosis of a sick TV set.* X is the set of parts of the TV set. First we see that there is a good picture. The trouble must be in the "sound-channel", which is a subset of the set of parts of the TV set. Similarly, by different tests we can determine whether certain subsets contain the ill part or not.

3. *Chemical analysis.* Assume we have an unknown chemical element and we want to identify it. X is the set of chemical elements. We pour some chemical to the unknown one; if its colour turns red, we know that it belongs to a subset of the set of chemical elements; in the contrary case, it does not. After carrying out some such tests, we can identify the unknown element.

4. *Defective coin problem.* X consists of 27 coins, one of them is defective. The defective coin is heavier than the good ones. We have an equal arm balance, and we want to identify the defective coin by weighings. If we put on the balance two sets of coins of equal size, then we can see which one contains the defective coin, and if they are equally heavy then the remaining set must contain it. In the previous examples we divided the set X into two subsets (A and its complement $X - A$). However, in this case we divide X into three disjoint subsets, and after the weighing we know which one

contains the unknown defective coin. Thus, this problem is generalization of the original problem.

The above examples differ in many things.

(A) (α) In the 3rd and 4th (and probably in the 2nd) example there is exactly one unknown defective† element.

(β) In the 1st example the elements may be infected independently with equal probability. It may occur that all the persons are infected or that all of them are healthy.

(B) In these examples the next subset may be (α) dependent or (β) independent on the answers of the previous tests. If the person or the machine performing the tests has a sufficiently large memory, then it may depend on the answers; in the contrary case it may not.

(C) (α) In the 1st example we may choose any subsets for test. (β) However, in the cases of the 2nd, 3rd and 4th examples, the electrical construction, the chemical properties and the condition that two subsets of three parts are equally sized produce restrictions for choosing sets.

(D) (α) In the 1st, 2nd and 3rd example we test a subset of X; in other words, we divide X into two subsets (A and $X-A$). The answer says which one contains the (or an) unknown element.

(β) In the 4th example we divide into three parts. Practically, in the 3rd example we always divide into many parts; pouring the testing chemical we can get many different colours. From the colour we may determine to which subset the unknown element belongs. The number of subsets may change from step to step.

(E) Our aim (in all the cases) is to minimize either (α) the average number of tests, or (β) the maximal number of tests.

There are many other different questions. We do not want to list all of them in order not to frighten away the reader. We shall investigate some of them later. There is one more reassuring fact: We do not know the solutions of all the problems obtained by combination of the cases (A), (B), (C) and (D).

We shall not investigate three kinds of problems: (1) the method "element by element"; p_i is the probability of x_i being wrong, c_i is the cost of testing x_i, determine the optimal order of the tests; (2) the case in which X is infinite; (3) sequential decoding of information theory. Problem 1 has no combinatorial aspects, problem 2 has some, but its methods are rather analytical. Finally, problem 3 has some connections with problems treated in this survey paper; however, these problems are very involved and the connections are not clear yet.

Let us first examine (for warming up) a trivial case: (Aα), (Bβ), (Cα), (Dα), (Eα) = (Eβ). We have a finite set $X = \{x_1, \ldots, x_n\}$ and exactly one

---

† Sometimes we say briefly "unknown element" or "unknown".

unknown element x_i. We have to determine a family $A_1, \ldots, A_m$ of subsets in such a way, that

after knowing whether $A_1, \ldots, A_m$ contains x_i or not we can determine x_i.

$$(1.1)$$

$(B\beta)$ means that we test all A_j's independently of the answers; $(C\alpha)$ means that we can use any subset of X for A_j's. The number of tests does not depend on the unknown element x_i; it is m. Thus, we have to minimize m under the condition (1.1), where the A_j's run over all the subsets of X.

Put $B_j^1 = A_j$ and $B_j^2 = X - A_j$ $(1 \leqslant j \leqslant m)$. If we know whether A_j $(1 \leqslant j \leqslant m)$ contains x_i or not, we also know whether $B_1^{i_1} B_2^{i_2} \cdots B_m^{i_m}$ $(i_1, \ldots, i_m = 1$ or $2)$ contains x_i or not. These sets are disjoint for different sequences $i_1, \ldots, i_m$. Conversely, if we know which $B_1^{i_1} \cdots B_m^{i_m}$ contains x_i, then we know whether A_j $(1 \leqslant j \leqslant m)$ contains x_i or not (depending on i_j). Thus, (1.1) is equivalent with the condition that

$$B_1^{i_1} B_2^{i_2} \cdots B_m^{i_m} \text{ contains at most 1 element for each } i_1, \ldots, i_m, \quad (1.2)$$

and if we write $i_j = 1$ if $x_i \in A_j$ and $i_j = 2$ if $x_i \notin A_j$ then $B_1^{i_1} \cdots B_m^{i_m}$ is the unknown element.

Moreover, (1.2) is equivalent to the following condition:

For each pair x_j, x_k $(j \neq k)$ there is an A_l such that

$$x_j \in A_l \quad \text{and} \quad x_k \notin A_l$$

or

$$(1.3)$$

$$x_j \notin A_l \quad \text{and} \quad x_k \in A_l.$$

Indeed, if (1.3) does not hold, then $x_j \in B_l^i$ and $x_k \in B_l^i$ are satisfied at the same time $(i = 1$ or $2)$. Choosing $i_1, \ldots, i_m$ in such a way that $x_j \in B_1^{i_1} \cdots B_m^{i_m}$, it has another element x_k, in contradiction with (1.2). Conversely, if (1.2) does not hold, then for some x_j, x_k $(j \neq k)$ and $i_1, \ldots, i_m$ we have $x_j, x_k \in B_1^{i_1} \cdots B_m^{i_m}$. In this case, $x_j \in B_l^{i_l}$ and $x_k \in B_l^{i_l}$, that is, $x_j \in A_l$ and $x_k \in A_l$ hold at the same time $(1 \leqslant l \leqslant m)$ in contradiction with (1.3).

We call a family of subsets $A_1, \ldots, A_m$ a *separating system* if they satisfy either (1.1) or (1.2) or (1.3).

There is a 4th characterization of separating systems. Define the 0, 1 matrix $M = (a_{ij})$ in the following way:

$$a_{ij} = 1 \quad \text{iff} \quad x_j \in A_i \quad (1 \leqslant i \leqslant m, \quad 1 \leqslant j \leqslant n).$$

Then (1.3) is equivalent to:

$$M \text{ has different columns.} \quad (1.4)$$

After these preliminary remarks, our first mathematical problem becomes very easy: *Given n, determine the minimal m such that there exists an $m \times n$ matrix with different columns.* The number of different columns is 2^m, thus $2^m \geqslant n$ necessarily holds. In other words $m \geqslant \log n$ (we shall always use logarithms with basis 2) or $m \geqslant \{\log n\}$, where $\{x\}$ denotes the least integer $\geqslant x$. This estimation is best possible: choosing n columns arbitrarily from the different $2^{\{\log n\}}$ 0, 1 sequences, we obtain a good matrix M.

Theorem 1.1. *If X is a finite set of n elements, then the minimal separating system consists of $\{\log n\}$ elements.*

2. Connections with noiseless encoding

Let us restrict ourselves now to the case $(A\alpha)$, $(B\alpha)$, $(C\alpha)$, $(D\alpha)$, $(E\alpha)$.

We have again a finite set $X = \{x_1, \ldots, x_n\}$; exactly one element x_i of X is defective (wrong, unknown) with probabilities $p_1, \ldots, p_n$. Further, there are subsets A_1, $A_j(e_1, \ldots, e_{j-1})$ where A_1 is the first test, and $A_j(e_1, \ldots, e_{j-1})$ $(1 < j \leqslant m; e_1, \ldots, e_{j-1} = 0$ or $1)$ is the j-th test when the answer of the previous tests were $e_1, \ldots, e_{j-1}$ ($e_k = 1$ means: it contains x_i; $e_k = 0$ means: it does not contain x_i).

If $A_{l+1}(e_1, \ldots, e_l)$ is not defined, but $A_l(e_1, \ldots, e_{l-1})$ is
then the answers $e_1, \ldots, e_{l-1}, e_l$ (together with the subsets   (2.1)
$A_1, A_2(e_1), \ldots, A_k(e_1, \ldots, e_{l-1})$)) uniquely determine x_i.

We call such a family of subsets a *strategy*.

If we fix x_i for a moment, then the sequence $e_1(i), \ldots, e_{l_i}(i)$ of answers is uniquely determined $((A_{l_i+1}(e_1(i), \ldots, e_{l_i}(i)))$ is not defined). The number of necessary tests is l_i. The average number of tests is

$$\sum_{i=1}^{n} p_i l_i. \tag{2.2}$$

We have to minimize (2.2) over the strategies, where the A's run over all the subsets of X.

Observe that in this way we corresponded a 0, 1 sequence $e_1(i), \ldots, e_{l_i}(i)$ with every x_i. This is a *code* in the language of information theory. The sequences are called codewords. This code has a simple property: There are no two different i and j such that $l_j \geqslant l_i$ and

$$e_1(i) = e_1(j), \ldots, e_{l_i}(i) = e_{l_i}(j).$$

In other words, no codeword is a *segment* of another one. We say that it is a *prefix code*.

This definition is adopted for the case when we use codewords formed from r different symbols $y_1, \ldots, y_r$ instead of 0 and 1.

Conversely, if we have a prefix code $x_i \rightarrow e_1(i), \ldots, e_{l_i}(i)$ formed from 0's and 1's, then we can define a strategy in the following way

$$A_1 = \{x_i : e_1(i) = 1\} \tag{2.3}$$
$$A_j(e_1, \ldots, e_{j-1}) = \{x_i : e_1(i) = e_1, \ldots, e_{j-1}(i) = e_{j-1}, e_j(i) = 1\}$$
$$(j > 1)$$

where $e_1, \ldots, e_{j-1}$ is a fixed sequence of 0's and 1's. If the set on the right hand side is empty, we do not define $A_j(e_1, \ldots, e_{l-1})$. For any fixed x_i we get the results $e_1(i), \ldots, e_{l_i}(i)$, writing 1 if the testing subset contains x_i and 0 if not. (It is easy to see by induction.) After these l_i tests, x_i is uniquely defined by the prefix property of the code. Thus (2.3) is a strategy and we found a correspondence between the prefix codes and the strategies; moreover, this correspondence is length-preserving: the length of the codeword of x_i is equal to the number of tests necessary to identify x_i.

This correspondence was described by Sobel [1960] (cf. Section 12), by Picard [1965] and by Campbell [1968], and it may also have been known to earlier authors. However, the optimal prefix codes and optimal strategies do not coincide, as Sobel [1967] has noticed in his Appendix. He also investigated this connection in another paper (Sobel [1970]).

This correspondence allows us to use the following well known theorem of information theory:

Noiseless Coding Theorem. *If the symbols $x_i, \ldots, x_n$ are encoded by the symbols $y_1, \ldots, y_m$ in a prefix way, then*

$$L = \sum_{i=1}^{n} p_i l_i \geqslant \frac{-\sum_{i=1}^{n} p_i \log p_i}{\log m} = \frac{H(P)}{\log m} \tag{2.4}$$

where $P = (p_1, \ldots, p_n)$ $(p_i > 0, \sum p_i = 1)$ is the vector of probabilities of the symbols $x_1, \ldots, x_n$ and l_i is the length of the codeword of x_i.

On the other hand, we can always find a prefix code satisfying the inequality

$$L \leqslant \frac{H(P)}{\log m} + 1. \tag{2.5}$$

We do not prove it here. The reader can find it in any information-theoretical book (e.g. Feinstein [1958]).

Substituting $m = 2$, this theorem gives us good estimates for the minimum of the average test-number:

$$H(P) \leqslant L \leqslant H(P) + 1. \tag{2.6}$$

However, it remains an open question what is the exact minimum. To answer this question let us examine some simple properties of the (in average sense) shortest code. Assume $p_1 \geqslant \cdots \geqslant p_n$.

Lemma 2.1. *For the optimal prefix code, $l_1 \leqslant \cdots \leqslant l_n$.*

Proof. If there is a pair i, j such that $p_i > p_j$ and $l_i > l_j$, then changing the code words of x_i and x_j the average increases by $p_i l_j + p_j l_i - p_i l_i - p_j l_j = (p_i - p_j)(l_j - l_i)$ and this is negative. The lemma is proved.

Lemma 2.2. *If $l_i = l_n$, then $e_1(i), \ldots, e_{l_i-1}(i), 1 - e_{l_i}(i)$ is also a code word together with $e_1(i), \ldots, e_{l_i-1}(i), e_{l_i}(i)$.*

Proof. In the contrary case change the code word $e_1(i), \ldots, e_{l_i}(i)$ for $e_1(i), \ldots, e_{l_i-1}(i)$. The new word can not be a segment of another one (the only possibilities $e_1(i), \ldots, e_{l_i}(i)$ and $e_1(i), \ldots, 1 - e_{l_i}(i)$ are excluded). Conversely, any segment of the new code word is a segment of $e_1(i), \ldots, e_{l_i}(i)$

and this is impossible by the prefix property. Thus, the new code is prefix, too. The average code length is smaller; this is a contradiction. The proof is completed.

Denote by $L(p_1, \ldots, p_n)$ the average code length $\sum p_i l_i$ for a given code and by $L_{\min}(p_1, \ldots, p_n)$ its minimum for prefix codes. Let us consider a code with average code length $L_{\min}(p_1, \ldots, p_n)$. By Lemma 2.1, x_n has a code of maximal length: $e_1(n), \ldots, e_{l_n}(n)$. If we change its last element, then the new sequence is also a code word:

$$(e_1(n), \ldots, 1 - e_{l_n}(n)) = (e_1(i), \ldots, e_{l_n}(i)) \qquad (i \neq n).$$

Here $l_i = l_n$, thus, again by Lemma 2.1, $l_i = l_{n-1} = l_n$. Changing the code words of x_i and x_{n-1}, the average code length does not change; we may assume $i = n-1$. Let us omit the code words of x_{n-1} and x_n and take a new one for both of them: $e_1(n), \ldots, e_{l_n-1}(n)$. This code is prefix again, and its average code length is smaller by $p_{n-1} + p_n$.

$$L_{\min}(p_1, \ldots, p_n) = L(p_1, \ldots, p_{n-1} + p_n) + p_{n-1} + p_n.$$

Hence

$$L_{\min}(p_1, \ldots, p_n) \geq L_{\min}(p_1, \ldots, p_{n-1} + p_n) + p_{n-1} + p_n \qquad (2.7)$$

follows. On the other hand, given a code with average code length $L_{\min}(p_1, \ldots, p_{n-1} + p_n)$ then we can form a new prefix code writing 0 and 1 at the end of the code word with probability $p_{n-1} + p_n$. The average code length is enlarged by $p_{n-1} + p_n$:

$$L_{\min}(p_1, \ldots, p_n) \leq L_{\min}(p_1, \ldots, p_{n-1} + p_n) + p_{n-1} + p_n. \qquad (2.8)$$

(2.7) and (2.8) result in

$$L_{\min}(p_1, \ldots, p_n) = L_{\min}(p_1, \ldots, p_{n-1} + p_n) + p_{n-1} + p_n. \qquad (2.9)$$

We have the following important result:

Theorem 2.1. *We reach the optimal code with the following Huffman procedure: Assume that a code with average code length $L_{\min}(p_1, \ldots, p_{n-1} + p_n)$ is determined, where p_{n-1} and p_n are the two smallest probabilities. Write 0 and 1 at the end of the code word with probability $p_{n-1} + p_n$. This is the optimal code for $P = (p_1, \ldots, p_n)$. The optimal code for $P = (1)$ is the void sequence.*

This theorem was first proved by Huffman [1952], but it was independently found by Zimmerman [1959] in the language of search.

A simple example: $P = (0.4; 0.25; 0.2; 0.15)$

The code for $(0.6; 0.4)$ is 0, 1.
The code for $(0.4; 0.35; 0.25)$ is 1, 00, 1.

The code for $(0.4; 0.25; 0.2; 0.15)$ is $1, 01, 000, 001$.

Theorem 2.1 gives us the answer to our question. The next question arises: Is there any difference between search theory and noiseless code theory? The answer is clear: there are many differences.

1. Code theory does not give solutions for the problems of type $(A\beta)$ or, in general, for the problems where there are two unknown elements with positive probability.

2. In case $(C\beta)$ the possible restrictions for the testing subsets give restrictions for the corresponding codes. However, these restrictions are different from the usual restrictions of code theory.

3. Perhaps the most important difference is that at a noiseless channel we have many symbols to transmit. Thus, we consider the sequences of length N formed from $x_1, \ldots, x_n$ and we transmit these sequences as new symbols. By this method we may approximate the lower bound of (2.6) arbitrarily good. The Huffman procedure has less interest in this case. However, in the case of search we have usually only one set and one unknown. Here the Huffman procedure has a great interest.

In any case, if the code theorems do not give the exact solution of a search problem, they give (sometimes good) estimates.

We have to mention that in the case when we can divide the set by one test into m subsets, then we can also use the noiseless coding theorem and a modified form of the Huffman procedure.

3. Results

3.1. *Case* $(A\alpha)$, $(B\alpha)$, $(C\alpha)$, $(D\alpha)$, $(E\alpha)$

After these long preliminaries we start the real survey of results.

First consider the following problem: Just one of the elements $x_1, \ldots, x_n$ is defective with equal probability; what is the minimum of the average number of tests necessary to identify the defective element? This problem is obviously a particular case of the problem treated in the previous section. Theorem 2.1 gives an algorithm to determine $L_{\min}(1/n, \ldots, 1/n)$; however, in this special case we may determine the exact value.

Lemma 3.1. *The code words of the code having average length $L_{\min}(1/n, \ldots, 1/n)$ can have just two different lengths, which are consecutive integers.*

Proof. Assume $l_1 \leqslant \cdots \leqslant l_n$. If $l_1 \leqslant l_n - 2$ then consider the code word $e_1(n), \ldots, e_{l_n}(n)$. By Lemma 2.1 there exists a code word of the form $e_1(n), \ldots, 1 - e_{l_n}(n)$. Change the code words

$$\left.\begin{array}{l} e_1(n), \ldots, e_{l_n-1}(n), e_{l_n}(n) \\ e_1(n), \ldots, e_{l_n-1}(n), 1 - e_{l_n}(n) \\ e_1(1), \ldots, e_{l_1}(1) \end{array}\right\} \quad \text{for} \quad \left\{\begin{array}{l} e_1(n), \ldots, e_{l_n-1}(n) \\ e_1(1), \ldots, e_{l_1}(1), 0 \\ e_1(1), \ldots, e_{l_1}(1), 1. \end{array}\right.$$

It is easy to see that the new code is prefix. However, the average code length is increased by $(2(l_1+1)+l_n-1)/n-(2l_n+l_1)/n = (l_1-l_n+1)/n$, which is negative by the assumption $l_1 = l_n-2$. The new code has a smaller average length. This is a contradiction. We proved $l_1 \geqslant l_n-1$. The proof is completed.

Choosing an arbitrary $0, 1$ sequence $c_1, \ldots, c_{l_n-1}$ of length l_n-1, either it is a code word or one of the sequences

$$
\begin{aligned}
e_1, \ldots, e_{l_n-1}, 0, \\
e_1, \ldots, e_{l_n-1}, 1
\end{aligned}
\tag{3.1}
$$

is a code word. In the contrary case we would change a code word of length l_n for $c_1, \ldots, c_{l_n-1}$ preserving the prefix property and decreasing the average length. This is a contradiction. However, by Lemma 2.1 if one of the sequences (3.1) is a code word then the second one is also a code word. Thus either $e_1, \ldots, e_{l_n-1}$ or both of (3.2) are code words. Denoting by s the number of code words of length l_n-1 we have

$$
s+\tfrac{1}{2}(n-s) = 2^{l_n-1}. \tag{3.2}
$$

Here $0 \leqslant s < n$, and $\tfrac{1}{2}n \leqslant \tfrac{1}{2}(n+s)$. Using (3.2), we obtain the inequality

$$
\tfrac{1}{2}n \leqslant 2^{l_n-1} < n,
$$

or $\log n \leqslant l_n < \log n+1$. It results in $l_n = \{\log n\}$, where $\{x\}$ denotes the least integer $\geqslant x$. On the other hand, we may count s from (3.2):

$$
s = 2^{\{\log n\}}-n.
$$

The average is

$$
\frac{s(\{\log n\}-1)+(n-s)\{\log n\}}{n} = \{\log n\}-\left(\frac{2^{\{\log n\}}}{n}-1\right).
$$

Theorem 3.1.

$$
L_{\min}\left(\frac{1}{n}, \ldots, \frac{1}{n}\right) = \{\log n\}-\left(\frac{2^{\{\log n\}}}{n}-1\right).
$$

This theorem was first proved by Sandelius [1961]. Sobel [1968b] has it also as a by-product. The proof published here is different from both that of Sandelius and that of Sobel.

By this method it is easy to see the following generalizations (Katona and Lee):

Theorem 3.2. *Let* $p_1 \geqslant \cdots \geqslant p_n$ *be given probabilities, where* $p_{n-1}+p_n \geqslant p_1$. *Then*

$$
L_{\min}(p_1, \ldots, p_n) = \{\log n\}-\sum_{i=1}^{s} p_i,
$$

where $s = 2^{\{\log n\}}-n$.

Theorem 3.3. *If* $p_1 \geqslant \cdots \geqslant p_n$ *and* $p_n + kp_{n-1} > p_1$ *then*

$$l_n - l_1 \leqslant k;$$

that is, the number of different code lengths is at most $k+1$.

Sobel [1968b] determined another special case:

Theorem 3.4. *If*

$$p_i = \frac{i}{\frac{1}{2}n(n+1)} \quad (1 \leqslant i \leqslant n)$$

then† *for* $2^{[\log n]} \leqslant n < 3 \cdot 2^{[\log n]-1}$,

$$L_{\min}(p_1, \ldots, p_n) = ([\log n]+2) + \tfrac{1}{2}n(n+1)(3 \cdot 2^{2[\log n]-3} - 3 \cdot 2^{[\log n]-2}(2n+1))$$

and for $3 \cdot 2^{[\log n]-1} \leqslant n < 2^{[\log n]+1}$,

$$L_{\min}(p_1, \ldots, p_n) = ([\log n]+2) + \tfrac{1}{2}n(n+1)(3 \cdot 2^{2[\log n]-1} - 3 \cdot 2^{[\log n]-1}(2n+1)).$$

3.2. *Case* (Aα), (Bα), (Cα), (Dα), (Eβ)

In this case the probabilities do not play any role. We may choose them $p_1 = 1/n$ $(1 = i = n)$, and use Theorem 3.1:

$$L_{\min}\left(\frac{1}{n}, \ldots, \frac{1}{n}\right) = \{\log n\} - \left(\frac{2^{\{\log n\}}}{n} - 1\right).$$

However, the maximum $\geqslant$ the average. Denote by l the minimum of the maximal test number. Thus

$$l \geqslant \{\log n\} - \left(\frac{2^{\{\log n\}}}{n} - 1\right). \tag{3.3}$$

Here

$$1 \leqslant \frac{2^{\{\log n\}}}{n} < 2$$

and

$$\{\log n\} - \left(\frac{2^{\{\log n\}}}{n} - 1\right) > \{\log n\} - 1. \tag{3.4}$$

From (3.3) and (3.4) we obtain

$$l \geqslant \{\log n\}.$$

However, by Theorem 1.1 we can construct a strategy even by independent tests (case (Bβ)) with test length $\{\log n\}$.

Theorem 3.5. *Assume we have strategies for* $X = \{x_1, \ldots, x_n\}$. *The minimum (runs over strategies) of the maximal number of tests is*

$$l = \{\log n\}.$$

† [x] denotes the largest integer $\leqslant x$.

It is quite interesting that in the case $(A\alpha)(C\alpha)(E\beta)$ the optimality problems are equivalent for $(B\alpha)$ and $(B\beta)$; we do not obtain anything if we choose the next test depending on the answers of the previous tests.

3.3. *Case* $(A\alpha)$, $(B\beta)$, $(C\alpha)$, $(D\alpha)$, $(E\alpha) = (E\beta)$.

Theorem 1.1 gives answer for this case.

3.4. *Case* $(A\alpha)$, $(B\alpha)$, $(C\beta)$, $(D\alpha)$, $(E\alpha)$.

It is a very natural assumption that a linear order $x_1 < \cdots < x_n$ is given in X, and the admissible subsets are of type $\{x_1, \ldots, x_j\}$ or $\{x_j, \ldots, x_n\}$ $(1 \leqslant j \leqslant n)$.

For example, we want to classify apples according to their sizes; $x_1, \ldots, x_n$ are the classes, and the unknown x_i is the class of the given apple. We carry out the tests by holes. If the given apple falls through the hole then its class x_i belongs to $\{x_1, \ldots, x_h\}$ for some j depending on the hole. Conversely, if it does not fall through, then x_i belongs to $\{x_{j+1}, \ldots, x_n\}$.

What does this restriction mean on the language of codes? The tests $\{x_1, \ldots, x_j\}$ and $\{x_{j+1}, \ldots, x_n\}$ are equivalent; assume we use always the type $\{x_{j+1}, \ldots, x_n\}$. If $A_1 = \{x_{j+1}, \ldots, x_n\}$, then for the corresponding code we have

$$e_1(1) = \cdots = e_1(j) = 0, \qquad e_1(j+1) = \cdots = e_1(n) = 1.$$

Similarly, if $A_k(e_1, \ldots, e_{k-1}) = \{x_{l+1}, \ldots, x_n\}$ then considering the set $T = \{t : e_1(t) = e_1, \ldots, e_{k-1}(t) = e_{k-1}\}$ we have again

$$e_k(t) = 0 \qquad \text{if} \quad t \leqslant l, \quad t \in T$$

and

$$e_k(t) = 1 \qquad \text{if} \quad t > l, \quad t \in T.$$

Reformulating, it means that for any pair (t, u),

$$e_1(t) = e_1(u), \ldots, e_{k-1}(t) = e_{k-1}(u), e_k(t) = 0, e_k(u) = 1$$

for some k; that is, the code words are in *lexicographic* order. We say that the code is *alphabetical* if it possesses this property.

Our problem is to determine the prefix alphabetical code with minimal average length. Denote this average by $A_{\min}(p_1, \ldots, p_n)$.

Gilbert and Moore [1959] gave an efficient construction for alphabetical codes, which ensures the following estimation:

Theorem 3.6.

$$H(P) \leqslant A_{\min}(p_1, \ldots, p_n) \leqslant H(P) + 2, \tag{3.5}$$

where $P = (p_1, \ldots, p_n)$.

Proof. The left hand side is a consequence of the left hand side of (2.6) and of the trivial inequality

$$L_{\min}(p_1, \ldots, p_n) \leqslant A_{\min}(p_1, \ldots, p_n). \tag{3.6}$$

We prove the right hand side of (3.5) by a construction. Define the numbers $q_1, \ldots, q_n$ and $l_1, \ldots, l_n$ as follows:

$$q_j = \sum_{i=1}^{j-1} p_i + \tfrac{1}{2}p_j, \tag{3.7}$$

$$l_j = \{-\log p_j\} + 1.$$

Let the first l_j digits of the binary expansion of the number q_j be the code of x_j. If the prefix property does not hold, then the code of some x_i is a segment of the code of another x_j. It means that q_i and q_j have the same binary digits on the first l_i places. In other words,

$$|q_i - q_j| \leqslant \frac{1}{2^{l_i}} \leqslant \frac{1}{2^{-\log p_i + 1}} = \tfrac{1}{2}p_i,$$

and this contradicts (3.7), since

$$|q_i - q_j| \geqslant \tfrac{1}{2}p_i + \tfrac{1}{2}p_j > \tfrac{1}{2}p_i.$$

The constructed code is prefix, indeed. The alphabetical property is trivially satisfied. The average length is

$$\sum_{j=1}^{n} p_j l_j \leqslant \sum_{j=1}^{n} p_j(\{-\log p_j\} + 1) \leqslant \sum_{j=1}^{n} p_j(-\log p_j + 2) = H(P) + 2.$$

The proof is completed.

Knuth [1971] and further Hu and Tucker [1970] worked out algorithms to determine a good alphabetical code.

In the paper of Hu and Tucker the *tentative-connecting* algorithm is written down. This need not be directly associated with an alphabetical code, but it is proved that there exists an alphabetical code with the same code word lengths as the code generated by the tentative-connecting algorithm.

A code is equivalent to the following tree: The nodes are the different possible segments of the code words (including the void sequence, which is called *root*), and two nodes are connected if one of them is a segment of the other and their lengths differ one. The *terminal* nodes are the code words. The tentative-connecting algorithm determines the tree rather than the code.

We start the algorithm with the subtree consisting of the terminal nodes $c_1, \ldots, c_n$ with the given order (no edges). Every terminal node has weight p_j. We take the minimal sum of the form $p_j + p_{j+1}$ $(1 \leqslant j < n)$, we draw a new node d with weight $p_j + p_{j+1}$ and we connect d with c_j and c_{j+1}. We have a new subtree and a new *construction sequence*: $c_1, \ldots, c_{i-1}, d, c_{i+2}, \ldots, c_n$. In general, assume we have a subtree and its roots and terminal nodes form a construction sequence $d_1, \ldots, d_k$ (some of the d's are c's); they have weights $q_1, \ldots, q_k$. d_i and d_j $(i < j)$ are *tentative connected* if there is no d_k $(i < k < j)$ such that $d_k = c_l$ for some l. We form the minimal sum $q_i + q_j$ where d_i and d_j are tentative connected $(i < j)$. We connect the new code e with d_i and d_j.

The new construction sequence is $d_1, \ldots, d_{i-1}, e, d_{i+1}, \ldots, d_{j-1}, d_{j+1}, \ldots, d_k$. The corresponding weights are $q_1, \ldots, q_{i-1}, q_i + q_j, q_{i+1}, \ldots, q_{j-1}, q_{j+1}, \ldots, q_k$. We continue this procedure until the construction sequence consists of one element.

Observe that in this language, the Huffman algorithm means that we choose the minimal sum $q_i + q_j$ without any restriction.

Notice that

$$A_{\min}\left(\frac{1}{n}, \ldots, \frac{1}{n}\right) = L_{\min}\left(\frac{1}{n}, \ldots, \frac{1}{n}\right).$$

This is a consequence of the fact that in this case the average does not depend on the order.

By the methods of Lemma 3.1, the following theorem is easy to verify (Katona and Lee).

Theorem 3.7. *If $p_i + p_{i+1} > p_j$ $(1 \leqslant i < n, 1 \leqslant j \leqslant n)$ then the minimal alphabetic code can have only two different code word lengths, which are consecutive integers.*

It is an interesting question how the Huffman algorithm is modified if we have a prescribed bound b for the code lengths $(b > \{\log n\})$

$$l_i \leqslant b \qquad (1 \leqslant i \leqslant n);$$

but we are interested in the minimal average length. Cesari [1968] has a partial solution for the problem.

3.5. *Case* (Aα), (Bα), (Cβ), (Dα), (Eβ).

In this case the solutions of the problems of the preceding section are trivial and identical to Theorem 3.5.

However, there are other problems which are too difficult in the case of (Eα).

Suppose we have n coins $x_1, \ldots, x_n$, one of them being defective (say x_i). The weight w_i of a non-defective coin is

$$1 \leqslant w_i \leqslant 1 + \delta$$

and $w_i = 1 + \varepsilon$ $(\varepsilon > \delta)$. We can use scales (not equal arm balance), thus by one test we may determine the weight of a subset A of $\{x_1, \ldots, x_n\}$. If the number $|A|$ of elements of A is less than $[\varepsilon/\delta]$, then $x_i \in A$ if and only if the weight of the subset A is $\geqslant |A| + \varepsilon$, because in the contrary case its weight is less than $|A| + \delta[\varepsilon/\delta] \leqslant |A| + \varepsilon$.

This example raises the following problem. Suppose a finite set $X = \{x_1, \ldots, x_n\}$ is given. Determine the minimum of the maximal test number for strategies consisting of subsets of at most k elements (k is fixed $\leqslant n$). Denote this minimum by $f_k(n)$. We know from Theorem 3.5 that

$$f_k(n) \geqslant \{\log n\}.$$

If $k \geqslant \frac{1}{2}n$, we do not have an essential restriction by $|A| \leqslant k$, for instead of

$|A| > k$ we can use the complement $X-A$, where $|X-A| < n-k \leqslant k$. Thus, by Theorem 3.5

$$f_k(n) = \{\log n\} \qquad \text{if} \quad [\tfrac{1}{2}n] \leqslant k. \tag{3.8}$$

Assume now $\tfrac{1}{2}n > k$. It is clear that $f_k(n)$ is a monotonically increasing function of n. Suppose for the optimal strategy $|A_1| = l \ (1 \leqslant l \leqslant k)$ holds. If the subsets $A_j(e_1, \ldots, e_{j-1})$ form a strategy, then $A_1 \cap A_j(e_1, \ldots, e_{j-1})$ and $(X-A_1) \cap A_j(e_1, \ldots, e_{j-1})$ form a strategy on A_1 and $X-A_1$, respectively. Similarly, $|A_j(e_1, \ldots, e_{i-1})| \leqslant k$ results in $|A_1 \cap A_j(e_1, \ldots, e_{j-1})| \leqslant k$ and

$$|(X-A_1) \cap A_j(e_1, \ldots, e_{j-1})| \leqslant k.$$

For these strategies the maximal number of test is at least $f_k(l)$ and $f_k(n-l)$, respectively. We have the following inequality

$$f_k(n) \geqslant 1 + \max (f_k(l) + f_k(n-l)). \tag{3.9}$$

Here $l \leqslant n-l$ by $l \leqslant k$ and $k < \tfrac{1}{2}n$. Applying the monotonicity of $f_k(n)$ we have

$$\max (f_k(l), f_k(n-l)) = f_k(n-l) \tag{3.10}$$

and

$$f_k(n-l) \geqslant f_k(n-k). \tag{3.11}$$

Substitute (3.10) and (3.11) into (3.9):

$$f_k(n) \geqslant 1 + f_k(n-k). \tag{3.12}$$

Applying $v = \{n/k\} - 2$ times (3.12),

$$f_k(n) \geqslant v + f_k(n-kv)$$

follows. Here $n-kv \leqslant 2k$ is trivial; for the last term we can apply (3.8)

$$f_k(n) \geqslant v + \{\log (n-kv)\}.$$

However, it is easy to construct a strategy with this maximal test number; $A_1 = \{x_1, \ldots, x_k\}$, $A_2 = \{x_{k+1}, \ldots, x_{2k}\}, \ldots, A_v = \{x_{(v-k)k-1}, \ldots, x_{vk}\}$ are the first v tests. They are independent from the previous answers. After these tests we know that either $x_i \in A_j$ for some $j \, (1 \leqslant j \leqslant v)$ or $x_i \in \{x_{dk+1}, \ldots, x_n\}$ In the first case we have a strategy with maximal length $\{\log k\}$ to identify x_i by Theorem 3.5. In the second case we have a strategy with $\{\log (n-kv)\}$. Here $n-kv > k$, and the maximal length is $v + \{\log (n-kv)\}$. The conjecture of Vigassy is proved:

Theorem 3.8. *The minimum of the maximal test number of a strategy given to identify one of the n elements is $v + \{\log (n-kv)\}$ if the subsets used on the strategy can have at most k elements $(k < n)$.*

The next problem is a typical problem of computers: *Suppose there are given n numbers $y_1, \ldots, y_n$ whose values are unknown and pairwise unequal. We wish to order them using only binary comparisons.*

In other words we have an unknown permutation x_i from all the permutations $x_1, \ldots, x_{n!}$ of $y_1, \ldots, y_n$. The subsets we can use for tests consist of the permutations where y_i precedes y_j (for some fixed i and j ($i \neq j$). There are $n!$ permutations, thus by Theorem 3.5 the minimum of the maximal test number is

$$l \geqslant \log (n!). \tag{3.13}$$

Steinhaus [1950] proposed the following algorithm: Assume we have already ordered $y_1, \ldots, y_t$. We compare y_{t+1} first with $y_{\{t/2\}}$, secondly with $y_{\{t/4\}}$ or $y_{\{3t/4\}}$ depending on the answer of the first test, and so on . . . . The number of tests is maximally

$$l \leqslant \{\log 2\} + \{\log 3\} + \cdots + \{\log (n-1)\} < \log ((n-1)!) + n - 3. \tag{3.14}$$

Steinhaus conjectured in [1950] this procedure to be optimal, however, in [1958] he disproved the conjecture. Asymptotically, the lower (3.13) and the upper (3.14) bounds are equivalent, but we do not know the best algorithm up to now. Ford and Johnson [1959] determined an algorithm better than Steinhaus' one. (See also Wells [1965], and Cesari [1968].)

A generalization of the above problem is to find and order the t largest y's. This generalization does not belong to the general search problem treated here. But we can generalize it toward this direction: *The n objects $x_1, \ldots, x_n$ are divided into disjoint classes. We wish to determine just the class to which the unknown x_i belongs.*

In our case: $x_1, \ldots, x_n$ are the permutations of $y_1, \ldots, y_n$. The classes consist of the permutations where the last t elements are fixed. The number of classes is $n(n-1) \cdots (n-t+1)$.

If $t = 1$, it is easy to see that

$$l = n-1.$$

The case $t = 2$ has been solved by Schreier [1932], Slupecki [1949–51] and Sobel [1968a]. The case of general t is obviously unsolved. For estimations see Hadian and Sobel [1970]. A further considered but unsolved problem is to determine the minimax of binary comparisons sufficient to identify the tth largest element from $y_1, \ldots, y_n$. Kislicyn [1964], Hadian and Sobel [1969], and Hadian [1969] worked out algorithms.

R. C. Bose and Nelson [1961] modified Steinhaus' problem: We wish to determine the natural order of the given (pairwise different) numbers $y_1, \ldots, y_n$ by binary changes instead of binary comparisons. That is, if $y_i < y_j$ ($i \neq j$) there is no change, if $y_i > y_j$, we use the order $y_1, \ldots, y_{i-1}$, $y_j, y_{i+1}, \ldots, y_{j-1}, y_i, y_{j+1}, \ldots, y_n$. What is the minimum of the maximal number of steps needed to determine the natural order?

The minimum is not known, but a good algorithm is given by R. C. Bose and Nelson [1961]. About the ordering problems see also David [1959] and Moon [1968].

3.6. *Case* (Aα), (Bβ), (Cβ), (Dα), (Eα) = (Eβ)

We have to determine the minimal m for which there exist subsets $A_1, \ldots,$ A_m of $X = \{x_1, \ldots, x_n\}$ constituting a strategy and satisfying $|A_i| \leqslant k$ ($k < \frac{1}{2}n$). If the subsets of a strategy do not depend on the previous answers, then they form simply a separating system (see the Introduction). It is proved by Katona [1966] that this minimal m is equal to the minimal m such that there exist non-negative integers $s_0, \ldots, s_m$ satisfying

$$mk = \sum_{j=0}^{m} j s_j,$$

$$n = \sum_{j=0}^{m} s_j, \tag{3.15}$$

$$s_j \leqslant \binom{m}{j} \qquad (0 \leqslant j \leqslant m).$$

By this fact, the next theorem was proved.

Theorem 3.9. *Suppose that* $A_1, \ldots, A_m \subset X = \{x_1, \ldots, x_n\}$ *satisfy the condition* $|A_j| \leqslant k$ $(1 \leqslant j \leqslant m)$ (k *is given* $< \frac{1}{2}n$) *and constitute a separating system. Under this condition, for the minimum of m the inequalities*

$$\frac{\log n}{\log (en/k)} \frac{n}{k} \leqslant \min m \leqslant \left\{ \begin{matrix} \log 2n \\ \log (n/k) \end{matrix} \right\} \begin{matrix} n \\ k \end{matrix}$$

hold.

Dickson [1969] introduced the concept of the completely separating system. (It does not have, probably, a nice interpretation in search theory, but it is interesting in itself): $A_1, \ldots, A_m$ is a *completely separating system* if for any pair x_i, x_j ($i \neq j$) there is a k such that $x_i \in A_k$, $x_j \notin A_k$.

What is the minimum of m such that there exists a completely separating system $A_1, \ldots, A_m$ for $\{x_1, \ldots, x_n\}$? This is solved asymptotically by Dickson [1969] and Spencer [1970] proved

Theorem 3.10. *The minimal m for which a completely separating system* $A_1, \ldots, A_m$ *exists is*

$$\min \left\{ m : \binom{m}{[\frac{1}{2}m]} \geqslant n \right\}.$$

Two subsets A_1 and A_2 of X are said to be *qualitative independent* if none of the sets $A_1, A_2, A_1\bar{A}_2, \bar{A}_1A_2, \bar{A}_1\bar{A}_2$ is empty, where $\bar{A}$ denotes the complement $X-A$. In other words, testing first by A_1, we obtain some information by testing A_2, independently of the answer of the first test. For instance if $A_1A_2 = \emptyset$ then after the answer $x_i \in A_1$ the test A_2 does not give any information. Rényi [1971] asked what is the maximum of the pairwise qualitative independent sets. He solved the problem for even n in the following way: If $A_1, \ldots, A_m$ are qualitative independent, then it is easy to see that $A_1, \bar{A}_1, \ldots,$

A_m, $\bar{A}_m$ form such a system that none of them is contained in another one. By Sperner's theorem [1928] we obtain

$$2m \leqslant \binom{n}{[\frac{1}{2}n]} \quad \text{and} \quad m \leqslant \frac{1}{2}\binom{n}{\frac{1}{2}n}.$$

This estimation is the best possible because we can choose $\frac{1}{2}\binom{n}{\frac{1}{2}n}$ pairwise disjoint subsets of $\frac{1}{2}n$ elements. For odd n this estimation is not the best possible. The right value is

$$\binom{n-1}{\frac{1}{2}(n-1)}$$

(see Rényi [1971]). The maximal number of r-wise qualitative independent sets is not yet determined. An estimation is given in Rényi's book [1971].

3.7. *Case* (Aβ), (Bα), (Cα), (Dα), (Eα)

Each of the elements $x_1, \ldots, x_n$ can be defective independently with probability p. We can not use the results of encoding type for this model, but it can be done for a transformed variant: Let $x'_1, \ldots, x'_{2^n}$ be the subsets of $X = \{x_1, \ldots, x_n\}$. Exactly one x'_i of the elements of $X' = \{x'_1, \ldots, x'_{2^n}\}$ is "defective" (it is the subset of all defective elements). However, the testing subsets $A \subset X$ are also transformed. A has a defective element if and only if $A = x'_j$ has a common element with the set x'_i of defective elements. It is equivalent to $x'_i \in A'$ where A' is the set of x_k's non-disjoint to x'_j. However, since such subsets A' are very special, we reduced our problem to a problem of type (Aα), (Bα), (Cβ), (Dα), (Eα).

The restrictions for the testing subsets are very particular. We can not solve the problem, but an easy lower bound for the average number of tests follows from (2.6):

$$L_{\min}(P) \geqslant H(P),$$

where $P = (p^n, p^{n-1}q, p^{n-1}q, \ldots, q^n)$. It is well known (see e.g. Feinstein [1958]) that $H(P) = n(-p \log p - (1-p) \log (1-p))$ holds in this case. We obtain for the average test number

$$L \geqslant n(p \log p - (1-p) \log (1-p)). \tag{3.16}$$

However, it is not the best possible lower bound. For example, Ungar [1960] proved the following

Theorem 3.11. *If $p \geqslant \frac{1}{2}(3-\sqrt{5})$ then*

$$L \geqslant n$$

and for $0 \leqslant p < \frac{1}{2}(3-\sqrt{5})$ there is a strategy with

$$L < n.$$

For large P, $L \geqslant n$ is obviously a better estimate than (3.16), and it is exact in this case, since for the strategy "element by element" $L = n$. In this case the combinatorial search fails. However, Ungar's theorem ensures that for small P it is a good method. Sobel [1960] and Sobel and Groll [1959] worked out good strategies for searching. These procedures give upper estimates for the optimal average test number $L_{min}(n)$. For example, Sobel [1960] (partly personal communication) has proved

$$\lim_{n \to \infty} \frac{L_{min}(n)}{n} \leqslant -p \log p - q \log q + \frac{p}{1-q^x},$$

where x is the smallest integer such that $1 - q^x - q^{x+1} \geqslant 0$. The right hand side is $\leqslant p \log p - q \log q + 1$, or, if p is small, then it is $\leqslant -p \log p - q \log q + 2p$.

The next problem does not belong formally to this section, but it is a very closed generalization of the problem treated here. We have three types of elements in X: good, mediocre and defective ones. Testing a subset A of X it shows the "minimum" of its elements: The test says "good" if all the elements of A are good; it says "mediocre" if there is at least one mediocre element in A, but none of them is defective; it says "defective" if there is at least one defective element in A. The elements of X are good, mediocre and defective independently with probabilities q_1, q_2 and q_3 $(q_1+q_2+q_3 = 1)$, respectively. Kumar [1970] has a result analogous to Ungar's theorem: If $q_1 \geqslant \frac{1}{2}(q_2 - 1 + (5q_2^2 - 6q_2 + 5)^{\frac{1}{2}})$ then $L \geqslant n$; that is, the test "element by element" is the best possible. On the other hand, if $q_1 < \frac{1}{2}(q_2 - 1 + (5q_2^2 - 6q_2 + 5)^{\frac{1}{2}})$ then there is a better strategy satisfying $L < n$. Similarly, Kumar [1970] gives a good strategy, which is a generalization of Sobel [1960] and Sobel-Groll [1959].

3.8. Cases $(A\alpha)$, $(B\alpha)$, $(C\alpha)$, $(D\alpha)$, $(E\beta)$ and $(A\beta)$, $(B\beta)$, $(C\alpha)$, $(D\alpha)$, $(E\alpha) = (E\beta)$

These cases are uninteresting, because for $p = \frac{1}{2}$, (3.16) gives $L \geqslant n$, and this is a lower bound for these cases. The strategy "element by element" is the best one.

3.9. Case $(A\beta)$, ..., $(C\beta)$, $(D\alpha)$, ...

These problems are not considered in the literature. Sobel [1960] is the only author that points out that his strategy is *alphabetical* in the sense that the testing subsets are "intervals" in the ordered set $\{x_1, \ldots, x_n\}$.

3.10. Case $(A\gamma)$

We did not introduce this case in Section 1. The common generalization of the cases $(A\alpha)$ and $(A\beta)$ is the case when the probabilities $p(A)$ of A $(\subset X)$ being the set of defective elements are given for all A.

In this generality the problem is too hard to solve. A very particular case is when $p(A) = 1/\binom{n}{2}$ for $|A| = 2$ and $p(A) = 0$ otherwise. (Assume $(B\alpha)$, $(C\alpha)$, $(D\alpha)$, $(E\alpha)$). It is easy to transform it into a problem of type $(A\alpha)$, $(B\alpha)$,

(Cβ), (Dα), (Eα) considering the set of unordered pairs (x_i, x_j) $(i \neq j)$. For this modified problem we may apply Theorem 3. More exactly, the formula of Theorem 3 gives a lower bound for the minimum of the average test number. Moreover, Sobel (1968b) proved that we can reach this lower bound for infinitely many n's.

Theorem 3.12. *There are exactly two defective elements in the set* $\{x_1, \ldots, x_n\}$, *all possibilities with equal probabilities. For the optimal strategy the average test length is denoted by* $L_2(n)$. *Then*

$$L_2(n) = \{\log \tbinom{n}{2}\} - (2^{\{\log \binom{n}{2}\}}/\tbinom{n}{2} - 1),$$

if

$$n = 2^{\frac{1}{2}(m+1)} + [\tfrac{1}{3}(2^{\frac{1}{2}(m-1)} - 4)] \text{ for some odd } m \geqslant 1,$$

or

$$n = 2^{\frac{1}{2}m} + [\tfrac{1}{3}(2^{\frac{1}{2}(m+2)} - 4)] \text{ for some even } m \geqslant 0.$$

For the remaining n's there is a small difference between the lower bound and the average of the strategy worked out by Sobel [1968].

Sobel and Groll [1966] investigated the problem (Aβ) in the case when we do not know the exact value of p and we use an *a priori* distribution by the test as well as tests to get a Bayes solution of the problem. This problem is more statistical than combinatorial.

3.11. *Case* (Dβ)

In this case at each test we divide X into disjoint subsets and the result of the test shows us which one (or which ones) includes the unknown element(s). If the number of disjoint subsets is at each test a constant (say r), then many problems can be solved (and they are) in the same way as for (Dα). We do not want to repeat them.

It may occur that the number r of subsets depends on the situation, that is, on the previous tests and previous results. Picard [1965] generalized Theorem 2.1 (Huffman algorithm) toward this case.

There is one classical problem which belongs typically to this case: the so called "defective coin problem". The basic situation is the following: *We have n coins, and one of them is defective, with probability* 1. *The good coins weigh* 1 *and the defective one weighs* $1 + \varepsilon$ $(0 < \varepsilon < 1)$. *We wish to find the defective coin using an equal arm balance.* Let X be the set of coins. Taking a subset A and a subset B $(A \cap B = \emptyset)$ on the right and left hand side of the balance, respectively, we may obtain three results: balance and unbalance in two ways. In the first case we know that the defective coin x_i is in $X - A - B$ and in the second case we know which one of A and B includes x_i. One test divides X into three parts, and says which one includes x_i. However, there is a restriction: $|A| = |B|$. If we try to weigh subsets with different cardinalities, no information is obtained. (The problem belongs to case (Aα), (Bα), (Cβ), (Dβ), (Eα) or (Eβ).) Let us generalize our Theorem 3.1:

$$L^3_{min}\left(\frac{1}{n}, \ldots, \frac{1}{n}\right) = \{\log_3 n\} - [\tfrac{1}{2} \cdot 3^{\{\log_3 n\}} - \tfrac{1}{2}n]/n. \tag{3.17}$$

It gives a lower bound for the average test number. This lower bound is attainable even under the condition $|A| = |B|$ (see Cairns [1963] and Baranyai (a)) except for $n = 6$, when the minimum of the average is 2. On the other hand, the last term in (3.17) is less than 1 (if $\{log_3 n\} = \log_3 n$ then 0), thus $\{L^3_{min}(1/n, \ldots, 1/n)\} = \{\log_3 n\}$ gives a lower bound for the maximal number of tests sufficient to identify the defective coin.

A different problem is if in a test we can use only the elements of the subset containing the defective one according to the last test (that is we cannot weigh the coins proved to be good). Equation (3.17) is again a lower bound.

However, because of the restrictions we can not reach (3.17) for every n (for we can not reach $3^{\{\log_3 n\}} - n \equiv 3 \pmod 4$). Cairns [1963] (for a new simpler proof see Baranyai (a)) determined the optimal strategy which is optimal for both cases (Eα) and (Eβ):

Theorem 3.13. *The optimal strategy is the following: If we know that the defective coin is an element of an n' element subset, then let us weigh m coins against m other ones from this subset, where m is the odd one of the numbers $[\tfrac{1}{3}(n'+1)]$ and $[\tfrac{1}{3}(n'+4)]$. The maximum test number is $\{\log_3 n\}$ for this strategy, and the average test number is also optimal.†*

The case when there are more (but fixed number h) defective coins is more complicated, if we assume that we are not able to determine the number of defectives in a subset by one test because the weights w_i of the defectives are different (but $1 < w_i < 1 + 1/h$). For particular results see Cairns ($h = 2$) [1963], Bellmann and Gluss [1961] and Smith [1947].

A different problem is proposed by Shapiro [1960] and Fine [1960]. Again, we have n coins, some of them being defective. The weights of good and defective coins are a and b, respectively. We may use for tests scales (not equal arm balance). Thus, by one test we are able to determine the number of defective coins in the tested subset. Determine $l(n)$, the minimum of the maximal test number needed to determine all the defectives. Many authors (Cantor [1964], Shapiro and Söderberg [1963], Erdős and Rényi [1963]) have asymptotical results for $l(n)$. Finally, Lindström [1964, 1965, 1966] proved

$$\lim_{n \to \infty} \frac{l(n) \log n}{n} = 2.$$

† Baranyai noticed that this is not true if $n' = 3^a + 2$, and the right $m = 3^{a-1}$.

4. Random search

Let us go back to the simplest case: exactly one of the elements $x_1, \ldots, x_n$ is defective, x_i is defective with probability p_i. Again, subsets are used to test (any subset). Rényi proposed to choose the subsets randomly, with probability $1/2^n$. Is the number of tests much larger than in the traditional case? The answer is definitely "no" (Rényi [1962a, 1961a]):

Theorem 4.1. *If the subsets* $A_1, \ldots, A_m$ *of* $X = \{x_1, \ldots, x_n\}$ *are chosen independently with probability* $1/2^n$, *and* $P(n, m)$ *denotes the probability of the event that* $A_1, \ldots, A_m$ *constitute a separating system then*

$$\lim_{n \to \infty} P(n, 2 \log n + c) = \begin{cases} 1 & \text{if } c = \infty \\ e^{-1/2^{c+1}} & \text{if } c \text{ is finite} \\ 0 & \text{if } c = -\infty. \end{cases}$$

It means that if we choose e.g. $m = 2 \log n + 6$, then for sufficiently large n the system $A_1, \ldots, A_m$ is a separating system with probability $e^{-1/2^7} \sim 0.99$. Comparing with Theorem 1, choosing randomly the subsets, we have to test roughly twice as many as the minimal number of systematically selected subsets which determine uniquely the unknown element. If the costs of the tests are small and the costs of working out a systematical plan are large, then it is better to use the random search.

However, we do not need a separating system to identify the unknown x_i. It is sufficient if $A_1, \ldots, A_m$ *separates* x_i from the other elements, that is, if they satisfy (1.3) for x_i and for an arbitrary x_k ($1 \leqslant k \leqslant m$, $i \neq k$). Denote by $S(n, m)$ the probability of the event that $A_1, \ldots, A_m$ separates x_i (it does not depend on i).

Theorem 4.2. (Rényi [1962b]). *If* $c > 0$, *then*
$$S(n, \log n + c) \geqslant e^{-1/2^c}.$$

It means that if we use seven more questions than at the systematical search, then we find the unknown element with probability $e^{-1/2^7} \sim 0.99$. This is very surprising.

The random choice of the subsets with probability means that we choose subsets with sizes about $n/2$. If the probability of choice of a subset $|A|$ is $p^{|A|} q^{n-|A|}$ then the chosen subsets have about p^n elements. A generalization (Rényi [1961b]) of Theorem 4.2 says that in this case we need about $m = \log n/[H(p, 1-p)]$ tests. (Compare with Theorem 3.9.)

Again, the next problem was proposed and solved by Rényi [1961b]: It may occur that our tests are not reliable. The result of a test is right and false with probabilities β and $1-\beta$, respectively (obviously these cases are independent from the results of the other tests). In this case we need about $\log n/[1 - H(\beta, 1-\beta)]$ tests to identify the unknown element. (There are strong connections with Gallager's random coding [1968].)

For further generalizations see Rényi [1961b] and [1965].

Finally, we wish to mention a result of Rényi [1970] which appeared after his tragic death. A *q-regular strategy* is a strategy which divides into exactly q parts the subset which is known to include the unknown element (case (Aα), (Bα), (Cα), (Dα)). It is easy to see that in this case $n = 1 \pmod{q-1}$. Rényi determined the number $Cq(n)$ of different (they are not different if they differ only in the permutations of the elements $x_1, \ldots, x_n$) q-regular strategies:

$$C_q(n) = \frac{(kq)!}{k!\,n!} \quad \text{where} \quad n = k(q-1)+1.$$

Similarly, the total number $D(n)$ of different strategies for $X = \{x_1, \ldots, x_n\}$ is

$$D(n) = \frac{1}{n}\sum_{k=1}^{n-1}\binom{n-2}{k-1}\binom{n+k-1}{k} \sim \frac{\sqrt{3-2\sqrt{2}}}{4\sqrt{\pi}}\frac{(3+2\sqrt{2})^n}{n^{\frac{3}{2}}}$$

(see also Rényi [1969]). Recently, Chorneyko and Mohanty have a generalization of these results.

5. Open problems

Comparing the several sections and combining their conditions it is easy to obtain a large number of open problems. We want to emphasize some of them (it does not mean that they are the most important ones, they are the most interesting only to the author):

1. Generalize the Huffman algorithm for the case (Aγ). More exactly: Probabilities $p(A)$ are given for any $A \subset X = \{x_1, \ldots, x_n\}$ of the event that A is the set of defective elements.

A *general strategy* is a strategy which is able to determine all the defective elements. Find an algorithm which determines the general strategy with the minimal average number. (See Theorem 2.1, case (Aγ) and the beginning of (Aβ).)

2. Find the conditions under which it is possible to determine the optimal average length (see Theorems 3.1, 3.2, 3.3 and 3.4).

3. Generalize the results for alphabetical codes (see Theorems 3.7, 3.1, 3.2, 3.3 and 3.4).

4. Generalize the Huffman algorithm for the case if we can use only subsets with size $\leqslant k$ ($k < \frac{1}{2}n$) (see Theorems 3.8 and 3.9).

5. Determine the best strategy for Steinhaus' problem, or at least give a better lower bound (see (3.13)).

6. Determine the minimal number m for which there exists a separating system $A_1, \ldots, A_m$ satisfying $|A_1| \leqslant k$ ($i = 1, \ldots, m$, k fixed $< \frac{1}{2}n$). Theorem 3.9 gives good estimates for this minimum. Generalize (3.15) for the case (Dβ) when $A_1, \ldots, A_m$ are partitions into r parts and the sizes of the first $r-1$ parts are bounded.†

† Very recently it is solved by Zs. Baranyai (b). (Added in December, 1971.)

11

7. Determine the minimal number m for which there exists a completely separating system $A_1, \ldots, A_m$ satisfying $|A_i| \leqslant k$ $(i = 1, \ldots, m,$ k fixed $< \frac{1}{2}n)$. (See Theorem 3.10.)

8. Determine the minimal number m for which there exists a system $A_1,$ $\ldots, A_m$ $(\subset X)$ such that for any x_i, x_j $(i \neq j)$ there are disjoint A_k and A_l $(A_k \cap A_l = \emptyset)$ with $x_i \in A_k, x_j \in A_l$.

9. Determine the maximal m for which there are subsets $A_1, \ldots, A_m$ such that any r different of them are qualitative independent (none of the sets of type $A_{i_1} \bar{A}_{i_2} \cdots A_{i_r}$ are empty). (See the end of section 3.6.)

10. Find a better estimate than (3.16) (see also problem 1).

11. Generalize Theorem 3.13 for a *"three-arm balance"* which has three equally sized arms (with angles $\frac{2}{3}\pi$), and which is balanced only if three equal weights are weighed.

12. $X = \{x_1, \ldots, x_n\}$. There is exactly one defective element. It is x_i with probability p_i. We can test any subset $A \subset X$ whether $x_i \in A$ or not. The next test may depend on the results of the previous tests (case (Aα),(Bα),(Cα),(Dα)). However, the tests are noisy, that is, we received the contrary results with probability q. Find an algorithm which determines the strategy which has the minimal average length, but discovers the defective element with given probability $1 - \varepsilon$. (In the language of codes: variable length (not black) code with minimal code length with error probability ε.)

13. There are exactly one defective element and one mediocre element in the set X, with probabilities $p_1, \ldots, p_n$ and $q_1, \ldots, q_n$. Which strategy minimize the maximal number of tests needed to identify both elements (see the end of section 3.1 and Theorem 3.12).

References

Zs. Baranyai, (a), to be published later.

Zs. Baranyai, (b), to be published later.

R. Bellman and B. Gluss, 1961, On various versions of the defective coin problem, *Inf. Control* **4**, 118–131.

R. C. Bose and R. I. Nelson, 1961, A sorting problem, Case Inst. Technol., Computing Center, Rept. No. 1043, pp. 1–22.

S. S. Cairns, 1963, Balance scale sorting, *Am. Math. Monthly* **70**, 136–148.

L. L. Campbell, 1968, Note on the connection between search theory and coding theory, *Proc. Colloq. on Information Theory* (A. Rényi, ed.; János Bolyai Math. Soc., Budapest).

D. G. Cantor, 1964, Determining a set from the cardinalities of its intersections with other sets, *Canad. J. Math.* **16**, 94–97.

Y. Cesari, 1968, Questionnaire, codage et tris, Institute Blaise Pascal, Paris.

Y. Cesari, 1970, Optimisation des questionnaires avec contrainte de rang.

I. Z. Chorneyko and S. G. Mohanti, On the enumeration of pseudo-search codes, submitted to *Studia Sci. Math. Hungar.*

H. A. David, 1959, *The Method of Paired Comparisons* (Hafner Publ. Co., New York).

T. I. Dickson, 1969, On a problem concerning separating systems of a finite set, *J. Combin. Theory* **7**, 191–196.

R. Dorfman, 1943, The detection of defective members of large populations, *Ann. Math. Statist.* **14**, 436–440.

P. Erdös and A. Rényi, 1963, On two problems of information theory, *Publ. Math. Inst. Hungar. Acad. Sci.* **8**, 241–254.

A. Feinstein, 1958, *Foundations of Information Theory* (McGraw-Hill, New York).

N. J. Fine, 1960, Solution EI 399, *Am. Math. Monthly* **67**, 697.

L. R. Ford and S. M. Johnson, 1959, A tournament problem, *Am. Math. Monthly* **66**, 387–389.

R. G. Gallager, 1968, *Information Theory and Reliable Communication* (Wiley, New York).

E. N. Gilbert and E. F. Moore, 1959, Variable-length binary encodings, *Bell Syst. Tech. J.* **38**, 933–967.

A. Hadrian, 1969, Optimality properties of various procedures for ranking *n* different numbers using only binary comparisons, Tech. Rept. No. 117, Dept. of Statistics, Univ. of Minnesota.

A. Hadrian and M. Sobel, 1969, Selecting the *t*th largest of *n* items using binary comparisons, Tech. Rept. No. 121, Dept. of Statistics, Univ. of Minnesota.

A. Hadrian and M. Sobel, 1970, Ordering the *t* largest items using binary comparisons, *Combinatorial Math. and its Appl.*, Univ. of North Carolina, Chapel Hill, N.C.

T. C. Hu and A. C. Tucker, 1970, Optimum binary search trees, *Combinatorial Mathematics and its Applications* (Univ. of North Carolina, Chapel Hill, N. Car.).

D. A. Huffman, 1952, A method for the construction of minimum redundancy codes, *Proc. I.R.E.* **40**, 1098.

G. Katona, 1966, On separating systems of a finite set, *J. Combin. Theory* **1**, 174–194.

G. Katona and M. A. Lee, Some remarks on the construction of optimal codes, submitted to *Acta Math. Acad. Sci. Hungar.*

S. S. Kislicyn, 1964, On the selection of the *k*th element of an ordered set of pairwise comparison, *Sibirsk. Mat. Zh.* **5**, 557–564 (in Russian).

D. E. Knuth, 1971, Optimum binary search trees, *Acta Inform.* **1**, 14–25.

S. Kumar, 1970, Group-testing to classify all units in a trinomial sample, *Studia Sci. Math. Hungar.* **5**, 229–247.

B. Lindström, 1964, On a combinatorial detection problem I, *Publ. Math. Inst. Hungar. Acad. Sci.* **9**, 195–206.

B. Lindström, 1965, On a combinatorial problem in number theory, *Canad. Math. Bull.* **8**, 477–490.

B. Lindström, 1966, On a combinatorial detection problem II, *Studia Sci. Math. Hungar.* **1**, 353–361.

J. W. Moon, 1968, *Topics on Tournaments* (Holt, Rinehart and Winston, New York), p. 48.

C. Picard, 1965, *Théorie des Questionnaires* (Gauthier-Villars, Paris).

A. Rényi, 1961a, On random generating elements of a finite boolean algebra, *Acta. Sci. Math. (Szeged)* **22**, 75–81.

A. Rényi, 1961b, On a problem of information theory, *Publ. Math. Inst. Hungar. Acad. Sci.* **6**, 505–516.

A. Rényi, 1962a, Statistical laws of accumulation of information, *Bull. Inst. Intern. Statist.* **39**(2), 311–316.

A. Rényi, 1962b, Az információ-akkumuláció statisztikus törvényszerüségéről, *Magyar Tud. Akad. III Oszt. Közl.* **12**, 15–33.

A. Rényi, 1965, On the theory of random search, *Bull. Am. Math. Soc.* **71**, 809–828.

A. Rényi, 1969, *Lectures on the Theory of Search*, Mimeo Series No. 600.7, Dept. of Statistics, Univ. of North Carolina, Chapel Hill, N. Car.

A. Rényi, 1970, On the enumeration of search-codes, *Acta Math. Acad. Sci. Hungar.* **21,** 27–33.

A. Rényi, 1971, *Foundations of Probability* (Holden-Day, San Francisco).

M. Sandelius, 1961, On an optimal search procedure, *Am. Math. Monthly* **68,** 138–154.

J. Schreier, 1932, On a tournament elimination system, *Mathesis Polska,* **7** 154–160 (in Polish).

H. S. Shapiro, 1960, Problem E 1399, *Am. Math. Monthly* **67,** 82.

J. Slupecki, 1949–51, On the system *S* of tournaments, *Colloq. Math.* **2,** 286–290.

C. A. B. Smith, 1947, The counterfeit coin problem, *Math. Gaz.* **31,** 31–39.

M. Sobel, 1960, Group testing to classify efficiently all defectives in a binomial sample, *Information and Decision Processes* (R. E. Machol, ed.; McGraw-Hill, New York), pp. 127–161.

M. Sobel, 1967, Optimal group testing, *Proc. Colloq. on Information Theory*, Bolyai Math. Society, Debrecen, Hungary.

M. Sobel, 1968a, On the ordering of the *t* best of *n* items using binary comparisons, Tech. Rept. No. 113, Dept. of Statistics, Univ. of Minnesota (submitted for publication).

M. Sobel, 1968a, Binomial and hypergeometric group-testing, *Studia Sci. Math. Hungar.* **3,** 19–42.

M. Sobel, 1970, A characterization of binary codes that correspond to a class of group-testing procedures, Tech. Rept. No. 148, Dept. of Statistics, Univ. of Minnesota.

M. Sobel and P. A. Groll, 1959, Group testing to eliminate efficiently all defectives in a binomial sample, *Bell System Tech. J.* **38,** 1179–1252.

M. Sobel and P. A. Groll, 1966, Binomial group-testing with an unknown proportion of defectives, *Technometrics* **8,** 631–656.

M. Sobel and S. Kumar, 1971a, Finding a single defective in binomial group-testing, *J. Am. Statist. Assoc.* (accepted for publication).

M. Sobel and S. Kumar, 1971b, Group-testing with at most *c* tests for finite *c* and *c*, Tech. Rept. No. 146, Dept. of Statistics, Univ. of Minnesota.

E. Sperner, 1928, Ein Satz über Untermengen einer endlichen Menge, *Math. Z.* **27,** 544–548.

J. Spencer, 1970, Minimal completely separating systems, *J. Combin. Theory* **8,** 446–447.

H. Steinhaus, 1950, *Mathematical Snapskost* (Oxford Univ. Press, New York).

H. Steinhaus, 1958, *One Hundred Problems in Elementary Mathematics* (Pergamon Press, London), Problems 52, 85.

A. Sterrett, 1957, On the detection of defective members of large populations, *Ann. Math. Statist.* **28,** 1033.

P. Ungar, 1960, The cut-off point for group testing, *Commun. Pure Appl. Math.* **13,** 49–54.

J. Vigassy, personal communication.

M. B. Wells, 1965, Application of a language for computing in combinatorics, IFIP Congress.

S. Zimmerman, 1959, An optimal search procedure, *Am. Math. Monthly* **66,** 690–693.

J. N. Srivastava et al., eds., *A Survey of Combinatorial Theory*
© North-Holland Publishing Company, 1973

CHAPTER 24

Some Problems in Combinatorial Geometry

DOUGLAS KELLY and GIAN-CARLO ROTA

Massachusetts Institute of Technology, Cambridge, Mass. 02139, U.S.A.

Years ago, when we first had the privilege of meeting R. C. Bose at North Carolina and of exchanging our callow views on the subject of combinatorial geometry with a wealth of insight and experience resulting from a lifetime of dedication and lasting achievements, we were struck by one remark, which to our minds was later to prove prophetic: "We combinatorialists have much to gain from the study of algebraic geometry, if not by its direct applications to our field, at least by the analogies between the two subjects." In the ensuing years, we repeatedly found instances when this insight was to lead to developments which reveal a horizon of new possibilities, and, we should like to propose, offer at present the best hope yet for a successful attack along new and, what is more important, algebraic paths to the important problems of the subject. We shall not discuss the solid body of results that have sprung so far from an application of this guiding analogy: suffice it to mention the bracket ring, defined by one of the present authors but first studied in depth by Neil White, where for the first time an algebraic object is functorially associated with *every* geometry, thus giving a universal solution to the problem of representations, Brylawski's theory of the Tutte-Grothendieck ring, reminiscent of the enormously successful Gröthendieck ring of algebraic geometry, Thorkell Helgason's study of the flats of a combinatorial geometry as generalizations of Grassmanians, and finally, the crowning achievements of Crapo in problems of extensions of geometries, which revive long-forgotten ideas of synthetic projective geometry by giving them a definite purpose and wholly novel impetus.

It is our present purpose to outline some further problems of a very general nature, which can perhaps be attacked by adopting the very general and systematic approach which has been for some time the mark of commutative algebra. We hasten to add that none of the problems we are about to state would have even been conceivable without the penetrating pioneering work of such men as Birkhoff, Dilworth, Edmonds, Rado, Tutte and Whitney, which transformed what began as a narrow and eccentric subject, into what many now consider to be one of the first thriving and firm settlements in the boundless steppe of combinatorics.

For the sake of the uninitiated readers it would be remiss not to recall at least one definition. A *geometry* on a set S, which we shall probably assume to be finite, is a closure relation on the subsets of S (that is, if $A \subseteq S$, a function $A \to \bar{A}$ such that $\bar{A} \supseteq A$, $\bar{\bar{A}} = \bar{A}$ and $\bar{A} \supseteq \bar{B}$ if $A \supseteq B$) enjoying the *MacLane-Steinitz exchange property*: given two elements $p,q \in \underline{S}$, and a subset $A \subseteq S$, if $p \in \overline{A \cup q}$ but $p \notin \bar{A}$ then $q \in \overline{A \cup p}$. A *combinatorial geometry* (c.g.) enjoys the further property that the closure of the null set is the null set, and every set consisting of a single point is closed. Without going into further detail, suffice it to say that the theory of combinatorial geometries can be considered as a combinatorial analog of linear algebra: the prime example of a c.g. is a subset of projective space, with the ordinary linear span as the closure, but some of the most interesting examples have no representation whatsoever as points in projective space. For further definitions and esults quoted in this paper, we refer to the book quoted at the end. Several other terms have been used in place of geometry, by the successive discoverers of the notion; stylistically, these range from the pathetic to the grotesque. The only surviving one is "matroid", still used in pockets of the tradition-bound British Commonwealth.

The most interesting problem of the field, and to some extent its raison d'être, is the *critical problem*, which we shall state for sets S in a projective space P_n over a finite field (although more general formulations are available). Find the linear subspaces $V \subseteq P_n$ of maximum dimension d, such that $V \cap S$ is the empty set. The number $c = n - d$ is called the critical exponent. We conjecture that the size of c depends on the nonexistence of certain minors, called obstructions, in the geometry: a classical instance of this conjecture is Hadwiger's conjecture. Of course the number of obstructions should be finite, and the list of obstructions should depend on the field. How the absence of obstructions acts to prevent the critical exponent from being high is a mystery which is understood only in the simplest cases. We nevertheless believe that one day the mechanism of obstructions will be unraveled. This would of course pave the way for the solution of several of the deepest extremal combinatorial problems of today: Hadwiger's conjecture and with it the four-color conjecture, the far more fascinating and as yet untouched Duffin-Tutte flow problem, the capacity of the 5-graph (White), the problem of linear codes (Dowling), the extremal problems of Beniamino Segre's profound theory of "points in independent position", and many others can be recast as critical problems.

How to go about relating obstructions to the critical exponent? We can take a hint from a vaguely similar problem in algebraic geometry. The critical problem, as stated above for subsets S of P_n, can be easily reformulated in the following equivalent form: find the minimum number c of hyperplanes $H_1, \ldots, H_c$ in P_n which *distinguish* the set S, that is, having the property that for every point p in S there exists at least one hyperplane H_i not containing the point p. Stated in this form, the critical problem reminds us of an old problem of algebraic geometry: speaking in classical terms, find the minimum number of "conditions" on a hypersurface in order that it shall contain a given algebraic variety V. It is well-known that this number is given by a polynomial $H(V, \lambda)$ in the variable λ, the Hilbert polynomial, which for large—and often all—positive integer values of λ gives the number of "conditions" to be imposed on a form of degree λ in order that it contain the variety (see Zariski-Samuel for a precise algebraic statement). Strikingly enough, the critical problem is also related to a polynomial: the *Birkhoff polynomial* (or characteristic, or chromatic, or Poincaré polynomial), defined as $p(S, \lambda) = \Sigma\mu(0, x)\lambda^{n-r(x)}$, where x ranges over all flats, or closed sets, of the geometry, μ is the Möbius function of the lattice of flats, and $r(x)$ is the rank, or dimension, of the flat x. For positive integer $x = q^i$, when q is the size of the field, the Birkhoff polynomial gives the number of sequences of i hyperplanes which distinguish the points of S.

The analogy between the Birkhoff polynomial and the Hilbert polynomial is too striking to be dismissed. The coefficients of both are positive and alternate in sign, they are in both cases often unimodal (increasing to a maximum and then decreasing), the lowest coefficient is the arithmetic genus in one case and the value $\mu(0, S)$ of the Möbius function in the other. The coefficients of the Hilbert polynomial are also known to be Betti numbers for a resolution of the surface V; this leads us to the conjecture that a similar resolution should exist for any c.g. where Betti numbers would be the coefficients of the Birkhoff polynomial. The discovery of such a resolution would very probably shed much light on the mystery of the critical problem. In the case of the geometry of contractions of the complete graph, such a resolution has been constructed by Peter H. Sellers, but no one has yet picked up the ball from him.

The analogy between algebraic surfaces and combinatorial geometries could be carried a great deal farther. A c.g. over a field can be viewed as a set S, together with a vector space V of functions with values in the field. The closed sets of the geometry are then obtained as intersections of zero-sets of functions in V, by the classical hull-kernel construction adapted from ideal theory. In this context, the critical problem becomes the problem of studying the minimum number of functions having a given zero-set, which will be a flat of the geometry. This bears a certain resemblance to the problem of finding functions on a Riemann surface having a given set of zeros and poles.

We shall now summarize the preceding discussion, independent of analogies, into a sequence of more specific conjectures about the critical problem and the allied problem of representing geometries over a field or as specific classes of geometries.

(1) Let $\underline{F}$ be a class of finite fields, and let $G(\underline{F})$ be the class of geometries

representable over all fields in the class $\underline{F}$. There is a finite number of obstructions, that is, forbidden minors, for a geometry to belong to the class $G(\underline{F})$.

(2) For a given positive integer c, and a given field F, there is a finite number of obstructions for a geometry representable over F to have critical number c or less.

(3) Let us define a *hereditary class* of geometries as a family of geometries closed under products and minors; for example, unimodular geometries, contractions or cocontractions of graphs, or geometries in the class $G(\underline{F})$. The critical problem for such a hereditary class should be interpretable intrinsically, that is, independently of any embedding of the geometry in a projective space. For each hereditary class a conjecture analogous to (2) can be made.

Notice the analogy between our notion of a hereditary class and the notion of variety in universal algebra. Can something be made of it?

(4) There is a wide class of (hereditary) classes for which the critical problem can be interpreted as known combinatorial problems. We have listed some of them; other candidates are transversal geometries, and the little-investigated simplicial geometries, which offer a natural setting for higher-dimensional coloring problems.

(5) It seems reasonably certain that strong maps are the natural class of morphisms for geometries. Are there any interesting functors of the resulting category—which, to be sure, has not been subjected to serious investigation—into Abelian categories? We have in mind the solution of the homology problem outlined above, and even some applications of "spectral sequences".

(6) It may be worthwhile to reverse the critical problem, and construct interesting hereditary classes of geometries for which the critical problem can be completely solved. This might serve to reveal its mechanism. For series-parallel networks, this has been done by Brylawski.

(7) The "reductions" of the coloring problem à la Birkhoff-Lewis are much too special to yield insight into other critical problems, or even into the coloring problem itself. They should be translated into the language of geometries and thereby made to reveal their significance. Stanley's modular-elements theorem is a successful instance in point, which does away in one fell swoop with a number of "Holzwege".

In closing, we must warn the reader that some of the conjectures just made, general as they are, may need considerable retouching before the attack. But on the whole, we would not have stated them, had we not felt that they are in the right track.

Bibliography

H. Crapo and G.-C. Rota, 1970, *Combinatorial Geometries*, preliminary edition (MIT Press, Cambridge, Mass.).

J. N. Srivastava et al., eds., *A Survey of Combinatorial Theory*
© North-Holland Publishing Company, 1973

CHAPTER 25

Recent Developments in India in the Construction of Confounded Asymmetrical Factorial Designs

K. KISHEN

Statistical Section, Directorate of Agriculture, U.P., Krishi Bhavan, Lucknow, India

and

B. N. TYAGI

Institute of Agricultural Research Statistics, New Delhi 12, India

1. Introduction

It is a pleasure to contribute an article to this volume which is being published in honour of Professor R. C. Bose with whom the senior author had the privilege of working from 1938 to 1940 when both were on the staff of the Indian Statistical Institute, Calcutta. As a result of their collaboration during that period, the combinatorial problem of constructing confounded plans for symmetrical factorial experiments of the type s^m, where s is a prime power and m any positive integer, was tackled and solved (Bose and Kishen [1940]) by use of finite geometries. However, there arise a number of situations when an experimenter has to use factors at different levels. The problem of obtaining confounded plans for such cases has recently received a good deal of attention in India. However, unlike the symmetrical factorial experiments, the asymmetrical factorial designs have so far defied all attempts at finding a general solution. The purpose of this paper is to give a brief account of the various methods developed in recent years in India for obtaining confounded plans of the asymmetrical factorial experiments.

Yates [1937], by trial and hit methods, obtained confounded plans for experiments of the type $3^m \times 2^n$, where m and n are any positive integers. Li [1944] also employed similar methods and obtained plans for confounding in the asymmetrical factorial experiments $4 \times 2^2, 5 \times 2^2, 4 \times 3 \times 2, 4^2 \times 2, 4 \times 3^2, 4^2 \times 3$ and $4^2 \times 2$. It was, however, the work of Nair and Rao [1941, 1942, 1948] which yielded a number of useful plans. They developed a set of sufficient combinatorial conditions for this purpose. The methods of constructing confounded plans for asymmetrical factorial designs developed subsequently can be classified into three broad categories, viz. (i) the methods of finite fields and finite geometries; (ii) the methods of incomplete block designs, and (iii) the methods of fractional replicates of symmetrical factorial designs and others. In this paper, we shall discuss these three methods separately.

2. Definitions

The balanced and partially balanced asymmetrical factorial designs are defined as follows:

2.1. Balanced asymmetrical factorial designs

A factorial design in incomplete blocks confounding certain d.f. belonging to main effects and/or interactions is said to be balanced if the loss of information on each single d.f. belonging to a particular effect is the same (Kishen and Tyagi [1964a]). Suppose, in a factorial experiment, that the main effect of A with m d.f.'s and interaction ABC with n d.f.'s are partially confounded. The design would be called balanced if the loss of information on each d.f. belonging to A is equal, say $L(A)$, and that on each d.f. belonging to ABC is also equal, say $L(ABC)$, where $L(A)$ and $L(ABC)$ may not be equal.

2.2. Partially balanced asymmetrical factorial designs

Kishen and Tyagi [1964b] have defined a confounded asymmetrical design to be partially balanced of order m if the losses of information on single degrees of freedom belonging to a particular confounded effect can be divided into m distinct classes such that the losses of information in the ith class of all effects are all equal to, say, L_i ($i = 1, 2, \ldots, m$). Although the case of three-factor experiments was discussed by Kishen and Tyagi [1964b] and the three-factor interaction was taken into account for defining a partially balanced asymmetrical factorial design, the concept can be easily extended to more complex cases.

3. The methods of finite fields and finite geometries

The method of finite geometries first developed by Bose and Kishen [1940] for solving the problem of confounding in the general symmetrical factorial design s^m, where s is a prime positive integer or a power of a prime and m any positive integer, was extended by Kishen and Srivastava [1959a,b] to the construction of confounded balanced asymmetrical factorial (BAF) designs of the type $s_1 \times s_2 \times \cdots \times s_m$, where s_1 is a prime positive integer or a power of a prime, m any positive integer, the s_i's ($i = 1, 2, \ldots, m$) are not all equal ($s_i \leqslant s_1$, for $i = 2, 3, \ldots, m$). This method of constructing BAF designs is briefly as follows:

Let $\alpha_0, \alpha_1, \ldots, \alpha_{s-1}$ be the s elements of the Galois field GF(s). The equation

$$f_1(x_1) + \alpha_{\mu_2} f_2(x_2) + \alpha_{\mu_3} f_3(x_3) + \cdots \alpha_{\mu_m} f_m(x_m) = \alpha_r \tag{3.1}$$

represents a hypersurface in the m-dimensional Euclidean geometry EG(m, s), $\alpha_{\mu_2}, \alpha_{\mu_3}, \ldots, \alpha_{\mu_m}, \alpha_r$ being any elements of GF(s) and

$$f_i(x_i) = \alpha_{i_1} x_i + \alpha_{i_2} x_i^2 + \cdots + \alpha_{i,s-1} x_i^{s-1}, \tag{3.2}$$

where α_{ij} ($i = 1, 2, \ldots, m; j = 1, 2, \ldots, s-1$) is also an element of GF(s). Kishen and Srivastava [1959b] have shown that it is always possible to obtain an appropriate polynomial $f_i(x_i)$ such that for $x_i = \alpha_0, \alpha_1, \ldots, \alpha_{s-1}$, $f_i(x_i)$ assumes k ($k \leqslant s$) distinct values. Consider now m factors, $A_1, A_2, \ldots, A_m$ at levels $s_1, s_2, \ldots, s_m$, respectively. If $f_i(x_i)$ is a polynomial which takes s_i distinct values, the m-factor interaction would be confounded if we take the pencil of hypersurfaces represented by

$$x_1 + \alpha_{\mu_2} f_2(x_2) + \alpha_{\mu_3} f_3(x_3) + \cdots + \alpha_{\mu_m} f_m(x_m) = \alpha_r \qquad (3.3)$$

in EG(m, s_1), where $\mu_2, \mu_3, \ldots, \mu_m$ are fixed for $r = 0, 1, 2, \ldots, s_1 - 1$. Each of the s_1 hypersurfaces in (3.3) would contain $s_2 \times s_3 \times \cdots \times s_m$ points to which would correspond the $s_2 \times s_3 \times \cdots \times s_m$ treatment combinations. There are two types of interactions confounded in the replication provided by the pencil of hypersurfaces (3.3), viz. (a) the m-factor interaction corresponding to (3.3), and (b) other interactions which would be confounded, in general, automatically owing to the fact that the number of combinations of levels of factors to which they relate is not equal to, or a factor of, the block size. Thus, in the replication (3.3), the main effects of A_1 and all the interactions in which it enters will be partially confounded if

$$s_1 > s_i \qquad (i = 2, \ldots, m).$$

The complete design would be obtained by varying $\mu_2, \mu_3, \ldots, \mu_m$ over $1, 2, \ldots, s_1 - 1$ in (3.3), thus giving $(s_1 - 1)^{m-1}$ replications. Thus, a 5×2^2 BAF design may be constructed in 20 blocks of 4 plots each, as shown in the Illustrative Example 1.

A large number of useful BAF designs can thus be obtained by employing the method of finite geometries. One disadvantage with all such designs is, however, that if none of the s_i's ($i = 2, 3, \ldots, m$) is equal to s_1, the main effects are confounded, which is generally not desired. The second disadvantage is that the confounded plans cannot be obtained if s_1 is not a prime power. The number of replications required for balancing is sometimes very large, which is another disadvantage.

Illustrative Example 1

The $5^0 \times 2^2$ BAF design given by the pencil
$$x_1 + \alpha(x_2^4 + x_3^4) = c \ (\alpha = 1, 2, 3, 4; c = 0, 1, \ldots, 4)$$
is given below:

Levels of BC	Levels of A																			
	Rep. I					Rep. II					Rep. III					Rep. IV				
0 0	0	1	2	3	4	0	1	2	3	4	0	1	2	3	4	0	1	2	3	4
0 1	4	0	1	2	3	3	4	0	1	2	2	3	4	0	1	1	2	3	4	0
1 0	4	0	1	2	3	3	4	0	1	2	2	3	4	0	1	1	2	3	4	0
1 1	3	4	0	1	2	1	2	3	4	0	4	0	1	2	3	2	3	4	0	1

In this design, the main effect A and the interactions AB, AC and ABC are partially confounded.

4. The use of incomplete block designs

4.1. BAF designs associated to PB designs

To overcome the drawbacks of using the finite geometries, Kishen and Srivastava [1959b] proposed the utilization of balanced incomplete block (BIB) designs. Subsequently, Kishen [1960], Kishen and Tyagi [1961, 1964a], Tyagi [1971, 1972] developed the method of constructing confounded plans for BAF designs with the help of incomplete block designs. They not only made extensive use of BIB designs but fully used the balancing properties of symmetrical unequal block (SUB) designs developed by Kishen [1940] and pairwise balanced (PB) designs. Tyagi [1972] has also used the GD designs for constructing BAF designs. We shall here summarize briefly the technique used for the purpose.

We shall first consider the case of three-factor experiments with factors A, B, C, these being at levels q, 2, 2, respectively. Let there be a PB design symbolized by $(q, k_1, k_2, \ldots, k_b, r, \lambda)$. Let the levels of A be denoted by $0, 1, \ldots, q-1$ and the levels of B and C each by 0, 1. According to Kishen and Srivastava [1959b], a $q \times 2^2$ BAF design in $2b$ blocks of $2q$ plots each is obtained by putting the treatment combinations in the following order:

Consider the jth ($j = 1, 2, \ldots, b$) block of the BIB design. If the ith ($i = 0, 1, \ldots, q-1$) treatment falls in the jth block, the corresponding block of the $q \times 2^2$ BAF design would contain the treatment combinations $i00, i11$; otherwise the treatment combination $i01, i10$ would fall in the block. Thus the replication being of two blocks and each block of $2q$ plots. They also obtained $q \times 3^2$ BAF designs in blocks of $3q$ plots each. Although this method yields a wide class of designs, the number of replications required for balancing is quite large. Kishen [1960], Kishen and Tyagi [1964a] and Tyagi [1971] modified this method so as to obtain confounded plans in a smaller number of replications. For this purpose, the SUB and PB designs have been extensively used. Thus, we have now a 7×2^2 BAF design in 8 blocks of 14 plots each associated to a PB design (Illustrative Example 2) in which the loss of information on each of the degrees of freedom of BC and ABC is $1/7$. Similarly, it has been possible to reduce the number of replications of an 11×2^2 design from 11 to 6 by using a special PB design, though the loss of information on each d.f. belonging to BC and ABC is $1/11$ and the design is not, therefore, optimum in the sense of Kishen and Tyagi [1964a]. But optimality is not always so essential if it is obtained at a very high cost. A large number of cases of three- and four-factor asymmetrical factorial experiments can be reduced to the $q \times 2^2$ or $q \times 3^2$ designs. For example, BAF designs $q \times 4 \times 2$,

$q \times 8 \times 2$, $q \times 9 \times 3$ can be easily derived from the corresponding BAF design $q \times 2^2$ or $q \times 3^2$.

4.2. PBAF designs associated to PBIB designs

The use of SUB and PB designs in getting confounded plans for BAF designs has led to a considerable economy in the experimental resources. There are, however, situations where the experimental resources are very scarce and only two or three replications are desired. To meet such contingencies, Kishen and Tyagi [1961] suggested the use of partially balanced incomplete block (PBIB) designs as developed by Bose and Nair [1939]. Kishen and Tyagi [1964b] discussed in detail the construction of partially balanced asymmetrical factorial (PBAF) designs of the type $q \times 2^2$ and $q \times 3^2$ associated to PBIB designs and those derivable by use of pseudo-factors. Thus, the use of PBIB designs has enabled them to obtain the 12×2^2 PBAF design in 3 replications only, and a large number of $q \times 3^2$ ($q = 4, 6, 8, 9, 10, 12$) PBAF designs in 2 replications only.

4.3. BAF designs associated to GD designs

The reduction in the number of replications was achieved by the use of PB or PBIB designs in the case of asymmetrical designs of the class $q \times 2^2$ and $q \times 3^2$. However, the problem of having blocks of smaller size covering a wide class of BAF designs with varying block sizes was still there. This has been tackled by Tyagi [1971] by use of group divisible (GD) designs with parameters v, b, r, k, λ_1, λ_2, m, n for constructing BAF designs of the type $m \times n$ in blocks of size k. As a number of the GD designs are available in two or three replications, the corresponding BAF designs are also obtained with three replications. An example of a 7×2^2 BAF design in 7 blocks of 12 plots each is given in Illustrative Example 3. Incidentally, it may be mentioned here that most of the BAF designs obtained by other methods happen to be a special case of the designs obtained through the GD designs. Some more confounded plans associated to hierarchical Group Divisible designs have also been obtained.

Illustrative Example 2

Consider a BIB design with parameters $v = b = 7$, $r = k = 3$, $\lambda = 1$. Corresponding to this BIB design we get a 7×2^2 design in 8 blocks of 14 plots each as below:

Levels of A	Levels of B and C															
0	0	0	0	1	0	1	0	1	0	0	0	1	0	0	0	0
0	1	1	1	0	1	0	1	0	1	1	1	0	1	1	1	1
1	0	0	0	0	0	1	0	1	0	1	0	0	0	1	0	0
1	1	1	1	1	1	0	1	0	1	0	1	1	1	0	1	1
2	0	1	0	0	0	0	0	1	0	1	0	1	0	0	0	0
2	1	0	1	1	1	1	1	0	1	0	1	0	1	1	1	1
3	0	0	0	1	0	0	0	0	0	1	0	1	0	1	0	0
3	1	1	1	0	1	1	1	1	1	0	1	0	1	0	1	1
4	0	1	0	0	0	1	0	0	0	0	0	1	0	1	0	0
4	1	0	1	1	1	0	1	1	1	1	1	0	1	0	1	1
5	0	1	0	1	0	0	0	1	0	0	0	0	0	1	0	0
5	1	0	1	0	1	1	1	0	1	1	1	1	1	0	1	1
6	0	1	0	1	0	1	0	0	0	1	0	0	0	0	0	0
6	1	0	1	0	1	0	1	1	1	0	1	1	1	1	1	1

Illustrative Example 3

Consider a GD design with parameters $v = 14$, $r = 3$, $k = 6$, $b = 7$, $m = 7$, $n = 2$, $\lambda_1 = 3$, $\lambda_2 = 1$. Corresponding to this design, we get the 7×2^2 design in 7 blocks of 12 plots each as below:

Blocks	Treatment combinations of A, B and C
1	(000, 011, 010, 001, 100, 111, 101, 110, 300, 311, 301, 310).
2	(100, 111, 101, 110, 200, 211, 201, 210, 400, 411, 401, 410).
3	(200, 211, 201, 210, 300, 311, 301, 310, 500, 511, 501, 510).
4	(300, 311, 301, 310, 400, 411, 401, 410, 600, 611, 601, 610).
5	(400, 411, 401, 410, 500, 511, 501, 510, 000, 011, 001, 010).
6	(500, 511, 501, 510, 600, 611, 601, 610, 100, 111, 110, 101).
7	(600, 611, 601, 610, 000, 011, 001, 010, 200, 211, 210, 201).

5. The method of fractional replication of symmetrical factorial designs

This section is based on the results obtained by Das and his coworkers. The method given by them in obtaining BAF designs by taking fractional replication of symmetrical factorial designs is briefly described here.

We shall first consider the case of asymmetrical factorial designs of the type $s \times s_1^m$, where s_1 is a prime number or a power of a prime number and $s_1^{p-1} < s < s_1^p$ where p is any positive integer. Let the first factor with levels s be replaced by p pseudo-factors, each at s_1 levels. Then we can have a confounded plan for s_1^{p+m} experiments in blocks of size s_1^{p+k}. From this s_1^{p+m} design in blocks of s_1^{p+k} plots, we omit those treatment combinations which contain a given number $y = s^1 - s$ of the treatment combinations of pseudo-factors. We are thus left with a confounded design $s \times s_1^m$ in blocks of size $(s_1^p - y)s_1^k$. With the help of different confounding sets for s_1^{p+m} design, we get different replications of $s \times s_1^m$. Let us call the factor with level s as X. If the confounded interaction in s_1^{p+m} design contains a set of $p' > 0$ of p pseudo-factors together with the other m factors, the corresponding confounded design $s \times s_1^m$ would confound an interaction containing X. More than one interaction of s_1^{m+p} design would correspond to the same interaction of the asymmetrical design unless it contains only the other m factors.

The technique can easily be extended to the case of $s \times s' \times s_1^m$ design in blocks of $s \times s' \times s_1^k$ plots by suitably taking two sets of pseudo-factors X_s and Y_s such that for X sets we have $s_1^{p_1-1} < s' < s_1^{p_2}$. Then, by omitting $s_1^{p_1} - s'$ treatment combinations of X pseudo-factors and $s_1^{p_2} - s'$ treatment combinations from the Y sets of pseudo-factors, we get a confounded plan for $s \times s' \times s_1^m$ from the corresponding symmetrical factorial design $s_1^{m+p_1+p_2}$.

Das [1960] has given the criteria of choosing suitable interactions confounded in each replication so that the complete design becomes balanced and is easily amenable to statistical analysis. These designs use components of interactions such as AB^2, etc., which may not be useful in situations when other contrasts of interactions, such as linear, quadratic, etc., are to be tested. Das and Rao [1967] gave an alternative series of confounded asymmetrical factorial designs with factors at two and three levels which confound interactions involving linear and quadratic components of the factors. Banerjee and Das [1969] have also developed further methods for constructing BAF designs derivable from the 2^n series.

In order to obtain confounded asymmetrical factorial designs with the smallest number of replications which can provide mutually orthogonal estimates of all effects, Banerjee [1968] advanced a method of construction of such designs by linking the main effect contrasts of each factor in the asymmetrical design with some or all of the main effect and interaction contrasts of a suitable group of factors, each at 2 levels, forming a symmetrical design. This method provides BAF designs even in one replication. However, sometimes even one replication of the symmetrical design may be too large as compared to the number of distinct treatment combinations of the asymmetrical design. For example, a 9×2^2 design is obtained with 64 treatment combinations with Banerjee's method, as against 36 distinct treatment combinations. The problem of using as small a number of treatment combinations as possible was solved by Bohra [1970] who obtained a $q \times 2^2$ design in 4 blocks of $(q+1)$ plots each in a single replication. The 9×2^2 design has been obtained in 4 blocks of 10 plots each, i.e., in only 40 treatment combinations. An example of 5×2^2 design in 4 blocks of 6 plots each is given in Illustrative Example 4.

Illustrative Example 4

We give below 5×2^2 design obtained by Bohra [1970] is one replication only.

	I			II			III			IV	
X	A	B	X	A	B	X	A	B	X	A	B
0	0	0	0	0	1	0	1	0	0	1	1
4	1	1	4	1	0	4	0	1	4	0	0
2	1	0	2	1	1	2	0	0	2	0	1
1	1	1	1	1	0	1	0	1	1	0	0
4	0	0	4	0	1	4	1	0	4	1	1
3	0	1	3	0	0	3	1	1	3	1	0

The *AB*, *XA*, *XB* and *XAB* effects are partially confounded.

6. Some further problems

As would appear from the foregoing section, considerable attention has been devoted by workers in India in recent years in developing methods of construction of asymmetrical factorial designs, and we have broadly reviewed three of the most important methods developed for the purpose. However, the problem of construction of confounded asymmetrical factorial designs is a complex one and a great deal of work still remains to be done in constructing fractional replicates of asymmetrical factorial designs which are of considerable practical importance in view of the economy in the experimental resources which these bring about. This problem is receiving increasing attention in this country, and main effect plans for asymmetrical factorial experiments are being developed. The results of this investigation will be reported in a separate communication.

References

A. K. Banerjee, 1968, On a method of construction and analysis of any confounded asymmetrical factorial design, Thesis submitted for the Diploma in Agricultural Statistics, Inst. of Agr. Res. Statist., New Delhi.

A. K. Banerjee and M. N. Das, 1969, On a method of construction of confounded asymmetrical factorial designs, *Calcutta Statist. Assoc. Bull.* **18** (71, 72), 163–177.

R. K. Bohra, 1970, On a method of construction and analysis of confounded asymmetrical factorial designs in single replicate, Thesis submitted for the Diploma in Agricultural Statistics, Inst. of Agr. Res. Statist., New Delhi.

R. C. Bose and K. Kishen, 1940, On the problem of confounding in the general symmetrical factorial design, *Sankhya* **5**, 21–36.

R. C. Bose and K. R. Nair, 1939, Partially balanced incomplete block designs, *Sankhya* **4**, 337–373.

M. N. Das, 1960, Fractional replicate as asymmetrical factorial designs, *J. Indian Soc. Agr. Statist.* **12,** 159.

M. N. Das and P. S. Rao, 1967, Construction and analysis of some new series of confounded asymmetrical factorial designs, *Biometrics* **23** (4), 813–822.

K. Kishen, 1960, On a class of asymmetrical factorial designs, *Current Sci. (India)* **29,** 465–466.

K. Kishen and J. N. Srivastava, 1959a, Confounding in asymmetrical factorial designs in relation to finite geometries, *Current Sci. (India)* **28,** 98–100.

K. Kishen and J. N. Srivastava, 1959b, Mathematical theory of confounding in asymmetrical and symmetrical factorial designs, *J. Indian Soc. Agr. Statist.* **11,** 73–110.

K. Kishen and B. N. Tyagi, 1961, On some method of construction of asymmetrical factorial designs, *Current Sci. (India)* **30,** 407–409.

K. Kishen and B. N. Tyagi, 1964a, On the construction and analysis of some balanced asymmetrical factorial designs, *Calcutta Statist. Assoc. Bull.* **14,** 123–149.

K. Kishen and B. N. Tyagi, 1964b, Partially balanced asymmetrical factorial designs, *Contributions to Statistics*, presented to Professor P. C. Mahalanobis on his 70th birthday.

Jerome C. R. Li, 1944, Design and statistical analysis of some confounded factorial experiments, Res. Bull. No. 333, Iowa State College of Agriculture.

K. R. Nair and C. R. Rao, 1941, Confounded designs for asymmetrical factorial experiments, *Sci. Cult. (Calcutta)* **7,** 361–362.

K. R. Nair and C. R. Rao, 1942, A general class of quasi-factorial designs leading to confounded designs for factorial experiments, *Sci. Cult. (Calcutta)* **7,** 457–458.

K. R. Nair and C. R. Rao, 1948, Confounding in asymmetrical factorial experiments, *J. Roy. Statist. Soc.* **10B,** 109–131.

B. N. Tyagi, 1971, Confounded asymmetrical factorial designs, *Biometrics* **27,** 229–232.

B. N. Tyagi, 1972, Balanced confounded asymmetrical factorial designs, Presented at the 25th Ann. Conf. of the Indian Soc. of Agr. Statist. (unpublished).

F. Yates, 1937, The design and analysis of factorial experiments, Imp. Bur. of Soil Sci., Tech. Commun. No. 35.

J. N. Srivastava et al., eds., *A Survey of Combinatorial Theory*
© North-Holland Publishing Company, 1973

CHAPTER 26

Combinatorial Problems in Communication Networks †

BENNET P. LIENTZ‡

University of Southern California, Los Angeles, Calif. 80007, U.S.A.

Abstract

This paper presents some problems in communication networks which can be addressed by combinatorial methods. Of particular interest are discrimination problems in radar networks, partitioning of core storage, and data inconsistency.

1. Introduction

In recent years, communication networks have become more complex and far-reaching. With satellites and advanced ground relay stations the amount of information transmitted and processed has greatly increased. The impact of this has been large delays in message transmission and processing, especially for low priority and routine traffic. However, the problem has not been restricted to military systems. It is occurring in time-sharing operations, hospital information systems, and criminal justice.

The problems associated with communication networks are magnified if computing and processing facilities are inserted at various network locations. Then there is the problem of compatibility of hardware and operating systems. Each center or node is viewed as a processor for routine local jobs. More complex tasks which require special facilities are routed over the network, processed, and then returned.

In this paper we seek to provide an overview of some of the situations which relate to combinatorial analysis. We will first describe a basic network with its processing characteristics (Section 2). In Section 3 we consider combinatorial problems in data processing in a defensive radar-type system. Section 4 presents problems in core storage partitioning and preprocessing. In these sections the aim is to describe the physical situation and then attempt to relate it to specific areas of graph theory and combinatorics. A summary is given in Section 5 along with a brief discussion of related topics.

† This paper was presented at the International Symposium on Combinatorial Mathematics, held at Colorado State University, Fort Collins, Colorado.

‡ This work was partially supported by the Air Force Office of Scientific Research (AFSC) under Grant No. 71–2025, Project No. 9749.

2. Description of networks

By now, network terminology has become somewhat standardized. Here we review the structure of communication networks and examine some of the problems still present.

Suppose that each node or vertex in a network or graph represents a computing facility which can transmit and receive messages to and from other centers. We allow for either a batch or an on-line processing system. Each node in this setting is complex since it is composed of many hardware and software elements. These produce a value of the power, capacity, and other characteristics associated with a center. Some of the ingredients of a node could include disc drives, storage, terminal type and speed, throughput of the mainframe computer, operating systems, and compiler methods. It is then possible to view each center as a super-node composed of nodes, each of which represents a component of the center.

The link or edge structure of the network is given by a distance file or matrix, a line size description, and other properties (such as those pertaining to satellite and microwave relay and routing). Line size here is a measure of capacity on a given link in the network.

A simple example is shown in Fig. 1. In this drawing the equipment at each node and line size are shown. If this network were an on-line system, the complexity would grow with the intertie to the network from each machine. The amount of data that must be stored to describe the network before examining message traffic can be sizeable.

Remaining with this exclusively topological description of the network, it would be useful to be able to derive an array of topological and combinatorial network descriptors which give an identity or signature to a given network. The value of such descriptors is evident in the testing and validation of new methods and procedures on varieties of network configurations. Descriptors could aid in the selection of candidate networks.

Let us consider some of the descriptors that have developed in the past. These include the following:

(i) radius: distance (in links traversed) from a central node to a distant node on the boundary;

(ii) diameter: distance (in links traversed) between the two most distant nodes;

(iii) variance of link to node ratio: $N^{-1}(\sum_{i=1}^{n}(n_i - m)^2)^{\frac{1}{2}}$.

For the variance, n_i is the number of links emanating from node i, $N/2$ is the number of links in the network, m is N/n, n is the number of nodes in the network.

These characteristics have definite meanings in communication networks. For example, variance is a measure of network clumpiness. Also, in modern communications queueing occurs at the nodes or centers so that message and job delays across a network are related to the number and type of nodes each message traverses. This gives significance to radius and diameter which are independent of physical distance.

There are a number of problems that have not really been considered. Some of these are general clumpiness of a network, decomposition, and interconnection. Network properties have been examined from an applications viewpoint by Frank and Frisch [1971] and Berge [1962].

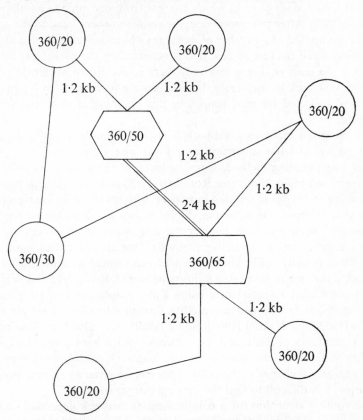

Fig. 1. Example of computation and communication network

Having considered some of the topological properties of a network, we can now add a traffic or commodity flow across the network. The traffic is usually entered in terms of mean values of parameters. Some of these parameters are:

(i) processing power of the mainframe computer (modified bits/sec);
(ii) specifications of peripheral equipment;
(iii) message arrival rate matrix;

 (iv) message size matrix;
 (v) job size matrix;
and other constraints. The (i, j)th entry of the message arrival rate matrix
is the number of messages which go from node i to node j in a given period
of time. The (i, j)th entry of the message size matrix is the mean size of
messages from i to j. The (i, j)th entry of the job size matrix is the mean size
of jobs from i to j. Some explanation of these quantities is needed here.
A job is a unified set of tasks which are sent from one node to another node
for processing. After this processing, the results are sent back to the source.
A job is composed of a number of messages which then may be decomposed
into many small packets of bits. If this occurs, then the packet size must be
specified. As each packet is received, there is usually an acknowledgement
message sent back to the source. If messages are received at an intermediate
node and collected for retransmission, this is termed a store and forward
situation.

For the given input to a network it is then possible to route the traffic
using an out-of-kilter algorithm (see Ford and Fulkerson [1962]) or an
integer programming method. Combinatorial methods have been the basis
for several routing methods (see Rothfarb [1968] and Tang [1963]). Note that
the routing methods are selected in part on whether the links are directional.
In most communication networks, duplex or two way lines are encountered.
In complex situations with high traffic and many small processing jobs,
there are difficulties with existing methods. One area is in multicommodity
flow. Since priority traffic routing can be represented as a problem in this
context, it is of major interest in military networks. Rothschild and Winston
[1966] considered a special case using Euler graphs for two commodities.
Other methods for the two commodity flow problem have been developed
by Hu [1963] and Hakimi [1962]. Some results have also been obtained for
three commodity problem and for networks which have constraints on the
number of disjoint paths between any two nodes. Of course, the flow problem
for a general number of commodity types can be given as a linear program.
However, it is difficult to find the optimal integer solution.

The routing algorithm for a communication network depends in part on
the queueing conditions at the nodes. Usually it is assumed that interarrival
times for messages are independent of message length and that both can be
described by Poisson statistics. In cases where these conditions are not met,
the problem appears to be intractable unless simulation methods are used.
Another common assumption is easily satisfied. This involves messages
arriving at nodes. When this occurs, messages are stored in buffers to be
processed later. Since operators can mount as many discs as available, it can
be argued that buffer size is infinite so that the queueing length can be
assumed to be infinite.

3. Combinatorial problems in real time communications

A number of military and commercial networks operate in real or near-real time environments. Examples are radar tracking networks, command and control, logistics systems, and time sharing operations. These have been projected to grow sizeably and in depth over the next decade. To consider some combinatorial problems in this context a radar-type network can be examined.

In radar or anti-submarine tracking, the field radar locations collect observations and send them to a central processing center (CPC). The network configuration then is that of a star. That is, there is one central location and peripheral locations which input information into the central location. Messages are received at the CPC and put through preprocessing. Here messages are formatted, initially correlated, and filtered. The next step is for these preprocessed messages to be sent to the central computer. Since we wish to concentrate on combinatorial problems, we will neglect the various configurations based on parallel processing and streaming and center on an example situation such as given in Fig. 2.

Field Radar Sites

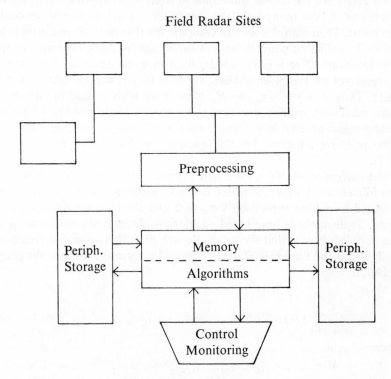

Fig. 2. Sample radar network.

In Fig. 2 the CPC is governed by a control and monitoring unit which allows for manual override and interdiction. Memory in the CPC is divided between peripheral and central core storage. The purpose of the CPC is to develop estimate tracks for various objects as well as to perhaps estimate other properties of the objects being targeted.

Unfortunately, there are several restrictions on the system which can create difficulties. With many radar sites, a given object may be tracked or observed by several sites. Usually each object is allotted certain files in active or central core storage. With noise, instrument error, and data contamination it is likely that information on some objects will be conflicting. If many objects are being observed, then file space is limited. Thus, if there is message conflict and only one file on a given object, then data overlay can occur repeatedly over a given time interval. Important parameters are the determination of when data conflict is significant, the thresholds of importance, and an estimate of the object's path and properties. We will consider combinatorics in this area. The question of how to detect overlay is ignored here since it is a sequential sampling procedure.

To determine the critical thresholds of overlay, an approach is to see how much and in how many ways the maximum amount of overlay or conflict can occur. Then with this we can compute the threshold. Suppose that there are $s+1$ conflicting paths which are supported by input observations from field locations. Over a given standardized time interval, say $[0, t]$, there are N_t messages on the given subject. Suppose that r_i support path $i (1 \leqslant i \leqslant s+1)$. Thus, $r_1 + \cdots + r_{s+1} = N_t$. Now if we wish to find the number of ways maximum overlay can occur, we compute the number of message arrangements arrayed in a string in such a way that no two messages of the same path are adjacent. Let this quantity be denoted by $M(r_1, \ldots, r_{s+1}; s+1)$.

The derivation of $M(\cdots)$ was first given attention by N. V. Smirnov in a multidimensional order statistics context. Sarmanov and Zaharov [1967] obtained a limiting expression for $s = 2$ and for the case where all r_i are equal. Their methods were based on Markov chains and a stationary point. The problem here is that the results do not generalize to higher dimensions easily. Alter and Lientz [1969, 1970] obtained a recursive formula for general s given by

$$M(r_1, \ldots, r_{s+1}; s+1) =$$

$$= \sum_{i=1}^{s+1} \sum_{j=1}^{r_t} (-1)^{j+1} M(r_1, \ldots, r_{i-1}, r_i - j, r_{i+1}, \ldots, r_{s+1}; s+1), \quad (3.1)$$

where

$$M(r_1, \ldots, r_{i-1}, 0, r_{i+1}, \ldots, r_{s+1}; s+1) =$$
$$= M(r_1, \ldots, r_{i-1}, r_{i+1}, \ldots, r_{s+1}; s+1).$$

A table of values for $s = 2$ was computed and found to closely agree with the limiting formula for $N_t \geqslant 25$.

This problem can be given a graph-theoretic interpretation (Alter and Lientz [1970]). Each message is interpreted as a vertex in a tree or a connected, acyclic graph. If a vertex is selected for the beginning of the tree, then this vertex is the root of the tree. The different classes of messages can be interpreted as color classes. Thus, the constraints on adjacency lead to the following definition: A chromatic tree for colored graphs is a tree in which no two adjacent vertices have the same color. A path is then one array of messages. Hence, since all messages must be considered in any array, the tree is a uniform n-tree in which the shortest path from the root to any terminal vertex is n.

Another formulation is to view the problem in a k-colored graph context. There, the number of Hamiltonian paths are sought on a given k-colored graph. This problem can then be approached via generating functions.

An additional area associated with tracking is the resolution of conflict by the selection of one path as a best estimate. This is critical in situations where response must be taken before a critical time barrier. It would be preferable to perform a sequential analysis procedure based on past data, errors, and other factors. This is not possible in situations where the information on many objects conflicts. To resolve this problem, a sequential procedure based on fluctuations can be developed. An interpretation of this problem will now be given.

Suppose at time u there have been N_u messages received with a_{iu} favoring path class i ($1 \leqslant i \leqslant s+1$). The time of last response is T so that $N_T = a_{T1} + \cdots + a_{T,s+1}$. The known results for $s > 1$ are limited. For the case $s = 1$ we can draw upon the results of Takacs [1967, 1965]. The assumption is made that all $\binom{N_T}{a_{T1}}$ combinations are equally probable. Takacs has obtained an expression for $P_j(a_{T1}, a_{T2})$ where

$$P_j(a_{T1}, a_{T2}) = P(a_{u1} > a_{T1}a_{u2}/a_{T2} \text{ for } j \text{ subscripts } 1 \leqslant u \leqslant N_T). \quad (3.2)$$

In this particular setting, a_{T1} and a_{T2} are not known explicitly. Messages are input usually deterministically in a store and forward mode. However, if the distribution of a_{T1} could be estimated, then we could obtain conditional probabilities similar to (3.2). The question that arises is to the criteria for fixing on a given track. Adopting a sequential method, we could terminate the sampling if either

$$a_{u1} > \gamma_1 a_{T1}a_{u2}/a_{T2} \text{ exactly } j_1 \text{ times}$$

or

$$a_{u2} > \gamma_2 a_{T2}a_{u1}/a_{T1} \text{ exactly } j_2 \text{ times}.$$

Formulas obtained in the past relating to other fluctuation events could be modified to determine j_1, j_2, γ_1, and γ_2.

4. Core storage and preprocessing

One part of the radar network example considered in the previous section is that of partitioning core storage, sorting, and preprocessing. The assignment of core storage to various objects can be given as an integer programming problem (Lientz [1972]). Since speed is necessary for at least near real-time environment, exact optimality must be sacrificed for efficiency. The goal is to obtain very good solutions quickly with a small amount of storage.

In a programming context, the objective function is related to the cost if multiple tracks are not maintained on individual subjects and can be formulated linearly. The constraints are on the number of files available. Since each track is assumed to occupy the same file length, the unit of the problem can be taken to be an individual file. Combinatorial methods in integer programming have been given some attention in the past. A survey of general integer programming appears in Balinski [1968]. Applications to file access and establishment are explored by Day [1965]. Combinatorial programming methods are given for the set-covering and other classes of problems in Pierce [1968] and Walker [1960]. Combinatorics and graph theory are of use here for the enumeration of potential solutions in a tree structure. Most of the past methods are of a deterministic combinatorial nature. Lientz [1971] considers a stochastic based method. With the availability of minicomputers, intelligent terminals, and CRT devices it would appear that more combinatorial based methods could be developed and used to partition storage. This would aid in the implementation of implicit enumeration techniques. For some situations it would appear that some combinatorial methods could be realized in analog or hybrid devices to enhance speed.

Core storage partitioning has an impact on preprocessing and sorting of data before entering the central storage area. Those objects which are not permitted multiple tracks will have all data filtered in preprocessing and one track input to central storage. The CPC will then recognize this one track exclusively.

The partitioning method is related to the overall data management, storage, and retrieval system. The method of file access will in part determine which files and blocks are most significant. This would be especially true in cases where the users access and manipulate files through an operational period.

5. Related problems and technological advances

The preceding sections have focused attention on communication networks in general and radar-type networks in particular. There are other areas where combinatorial analysis has a role. Some of these are tracking methods themselves, public transportation systems, and ballistic missile defense. The latter will now be considered. Matlin [1968] reviewed the allocation area and related missile defense topics. The assignment problem is to allocate weapons to targets for a given weapon force and a set of targets. Many methods assume known probabilities of destruction. With this and other assumptions, there has been very little work in sizing the problem upward. A combinatorial allocation method could be beneficial here since computation would increase too rapidly with size. The combinatorial methods used would probably lie in networks with stochastic parameters. The level in the network or tree could be associated with point to subarea to sector levels. Communication network results could be applied in cases where certain classes of targets possess priorities. Additional factors that could add levels to a network model are the type of defense, intelligence, and tolerances. We ignore here damage assessment and sophisticated counter-measure devices. The methods that have been used in the past include Lagrange multipliers, nonlinear programming, and dynamic programming.

A more mathematical problem is that of analyzing parallel processing. Since many military networks are considering or using this, it merits attention. Parallel processing is intended to enhance processing speed of a system as compared to the usually digital computer operation. Logical parallel operations include solving independent subproblems and circuit design. Parallel processing can easily be given a graph-theoretic interpretation. Each step in a method is a node in a directed tree. Links are then transmission branches between steps. An additional factor is a dependence relationship between inputs and outputs at a given node. A special case based on cardinality at a given node has been given by Karp and Miller [1966]. Under general conditions, only initial steps have been taken. With the popularity of parallelism this would appear to be an interesting avenue of approach.

Other technological advances are predicted in hardware with large memories and more efficient input/output devices. Combinatorial methods could be used in conjunction with computer simulation programs to test various configurations.

In conclusion, this paper has presented some areas where combinatorial methods would be valuable. Attention has been restricted to several military classes of networks. The parts of combinatorics that appear most useful are graph theory and the theory of trees.

References

R. Alter and B. P. Lientz, 1969, Applications of a generalized combinatorial problem of Smirnov, *Naval Res. Logist. Quart.* **16**, 543–547.

R. Alter and B. P. Lientz, 1970, A note on a problem of Smirnov: a graph-theoretic interpretation, *Naval Res. Logist. Quart.* **17**, 407–408.

R. Alter and B. P. Lientz, 1972, Remarks on a combinatorial problem in orderings, *Sankhya* Ser. A (to appear).

M. L. Balinski, 1968, Integer programming: methods, uses, and computation, *Management. Sci.* **15**, 191–209.

C. Berge, 1962, *The Theory of Graphs* (Methuen, London).

R. H. Day, 1965, On optimal extracting from a multiple file data storage system: an application of integer programming, *Operations Res.* **13**, 482–494.

L. R. Ford, Jr. and D. R. Fulkerson, 1962, *Flows in Networks* (Princeton University Press, Princeton, N. J.).

H. Frank and I. T. Frisch, 1971, *Communication, Transmission, and Transportation Networks* (Addison-Wesley, Reading, Mass.).

S. L. Hakimi, 1962, Simultaneous flows through communication networks, *IRE Trans. Circuit Theory* **CT-9**, 169–175.

T. C. Hu, 1963, Multicommodity network flows, *Operations Res.* **11**, 344–360.

R. M. Karp and R. E. Miller, 1966, Properties of a model for parallel computations: determinacy, termination, and queueing, *SIAM J. Appl. Math.* **14**, 1390–1411.

B. P. Lientz, 1971, Allocation of components to maximize reliability I, submitted for publication.

B. P. Lientz, 1972, Stochastic allocation of spare components, *Proc. Symp. on Optimizing Methods* (J. Rustagi, ed.; Academic Press, New York, to appear).

S. Matlin, 1968, A review of the literature on the missile allocation problem, General Electric Tech. Inform. Series No. 68 SD 236.

J. F. Pierce, 1968, Application of combinatorial programming to a class of all-zero-one integer programming problems, *Management Sci.* **15**, 191–209.

B. Rothfarb, 1968, Combinatoric methods for multicommodity flows, Ph.D. Thesis, University of California (unpublished).

B. Rothschild and A. Whinston, 1966, On two commodity flow networks, *Operations Res.* **14**, 377–387.

O. V. Sarmanov and V. K. Zaharov, 1967, A combinatorial problem of Smirnov, *Dokl. Akad. Nauk SSSR* **176**, 1147–1150.

L. Takacs, 1965, Application of ballot theorems in the theory of queues, *Proc. Symp. on Congestion Theory* (Univ. of North Carolina Press, Chapel Hill, N.C.).

L. Takacs, 1967, *Combinatorial Methods in the Theory of Stochastic Processes* (Wiley, New York).

D. T. Tang, 1963, Optimal trees for simultaneous flow requirements, *Proc. Natl. Electron. Conf.* **19**, 28–32.

R. L. Walker, 1960, An enumerative technique for a class of combinatorial problems, *Am. Math. Soc. Proc. Symp. Appl. Math.* **10**, 91–94.

J. N. Srivastava et al., eds., *A Survey of Combinatorial Theory*
© North-Holland Publishing Company, 1973

CHAPTER 27

On Hadamard Difference Sets

HENRY B. MANN and ROBERT L. McFARLAND

Department of Mathematics, University of Arizona, Tucson, Ariz. 85721, U.S.A.
Route 8, Lexington, Ohio 44904, U.S.A.

Let D be a difference set in a multiplicative Abelian group G with the parameters

$$(v, k, \lambda) = (4p^2, 2p^2 - p, p^2 - p), \tag{1}$$

where p is an odd prime. We prove:

Theorem. *If $\chi(D) \equiv 0 \pmod{p}$ for all nonprincipal characters χ of G, then* $p = 3$.

Corollary. *If the Sylow 2-subgroup of G is of type $(2, 2)$, or if $p \equiv 3 \pmod{4}$, then* $p = 3$.

This Theorem and part of the Corollary were first established, via a somewhat computational proof, by Turyn [1965, Theorem 12, p. 339, and its Corollary, p. 345].

Proof of Corollary. The hypotheses of the corollary imply that the numbers $\chi(g)$, χ a character of G and g in G, all lie in a field in which p becomes the power of a prime ideal. Hence $\chi(D) \equiv 0 \pmod{p}$ for all nonprincipal characters χ. Whence the corollary.

Proof of Theorem. The hypothesis of the theorem implies $|\chi(D)/p| = 1$ for all nonprincipal characters χ. Hence by a theorem of Kronecker,

$$\chi(D) = p\zeta$$

where ζ is a root of unity which depends on χ. For the principal character χ_0 we have

$$\chi_0(D) = 2p^2 - p.$$

Let $G_4 = \{I, A, B, AB\}$ and G_{p^2} be the Sylow subgroups of G of respective orders 4 and p^2. Express D in the form

$$D = \Sigma_I + A\Sigma_A + B\Sigma_B + AB\Sigma_{AB},$$

where each Σ_X is an element of the group ring of G_{p^2} over the rational integers. The character group of G is the direct product of the character group of G_4 and that of G_{p^2}. Hence, if χ is a character of G_{p^2}, we have either

$$\chi(\Sigma_I) + \varepsilon_A \chi(\Sigma_A) + \varepsilon_B \chi(\Sigma_B) + \varepsilon_A \varepsilon_B \chi(\Sigma_{AB}) = p\zeta_{\varepsilon_A, \varepsilon_B} \tag{2}$$

for $(\varepsilon_A, \varepsilon_B) = (1, 1), (1, -1), (-1, 1), (-1, -1)$ and $\zeta_{\varepsilon_A, \varepsilon_B}$ a root of unity, or

$$\chi(\Sigma_I) + \varepsilon\chi(\Sigma_A) + \varepsilon^2\chi(\Sigma_{A^2}) + \varepsilon^3\chi(\Sigma_{A^3}) = p\zeta_\varepsilon \qquad (2a)$$

for $\varepsilon = 1, -1, i, -i$, and ζ_ε a root of unity.

In both cases we get by summing over the four equations of (2) or (2a)

$$p^{-1} \cdot 4\chi(\Sigma_I) = \zeta_1 + \zeta_2 + \zeta_3 + \zeta_4 = 4\rho, \qquad (3)$$

where the ζ_i's are roots of unity and ρ is an algebraic integer. But (3) shows that no conjugate of ρ has absolute value exceeding one. Hence either $\chi(\Sigma_I) = 0$, or $\chi(\Sigma_I) = p\zeta$ and $\zeta_1 = \zeta_2 = \zeta_3 = \zeta_4 = \zeta$. The same argument can be applied to $\Sigma_A, \Sigma_B, \Sigma_{AB}$, and at least one of the $\chi(\Sigma_X)$ must be nonzero. But then (2) or (2a) shows that $\chi(\Sigma_Y) = 0$ for $Y \neq X$.

An analogous argument gives $\chi_0(\Sigma_X) = \frac{1}{2}(p^2 + p)$ for one of the four values $X = I, A, B, AB$, and $\frac{1}{2}(p^2 - p)$ for the other three.

Now let χ be a character of G_{p^2} whose kernel has order p. Without loss of generality, we may assume $\chi(\Sigma_I) = \pm p$. Now Σ_I has either $\frac{1}{2}(p^2 + p)$ or $\frac{1}{2}(p^2 - p)$ nonzero terms whose characters are either 1 or a primitive pth root of unity. Hence

$$\chi(\Sigma_I) = a + \sum_{j=1}^{p-1} a_j \zeta^j,$$

where a and the a_j's are integers satisfying

$$0 \leqslant a \leqslant p, \quad 0 \leqslant a_j \leqslant p.$$

Since $\chi(\Sigma_I) = \pm p$, we must have

$$a_1 = a_2 = \cdots = a_{p-1} = b.$$

Hence

$$\chi(\Sigma_I) = a - b = \pm p, \quad \chi_0(\Sigma_I) = a + (p-1)b = \frac{1}{2}(p^2 \pm p).$$

If $a - b = p$, then $a = p, b = 0$. Thus

$$\chi_0(\Sigma_I) = p = \frac{1}{2}(p^2 \pm p),$$

so $p = 3$. While if $a - b = -p$, then $a = 0, b = p$. Hence

$$\chi_0(\Sigma_I) = (p-1)p = \frac{1}{2}(p^2 \pm p),$$

so again $p = 3$. This completes the proof.

There are difference sets with the parameters (1) for $p = 2$ and 3; see Menon [1962] and Turyn [1965]. It is not yet known whether there can exist a difference set with the parameters (1) for a prime $p \neq 2, 3$.

References

P. K. Menon, 1962, On difference sets whose parameters satisfy a certain relation, *Proc. Am. Math. Soc.* **13**, 739–745.

R. J. Turyn, 1965, Character sums and difference sets, *Pacif. J. Math.* **15**, 319–346.

J. N. Srivastava et al., eds., *A Survey of Combinatorial Theory*
© North-Holland Publishing Company, 1973

CHAPTER 28

On the Randomization of Block Designs

JUNJIRO OGAWA and SADAO IKEDA†

Department of Mathematics, Statistics and Computing Science
University of Calgary, Calgary, Alberta, Canada

In designing experimentation, experimental units or plots are grouped into blocks in such a way that unit-effects within blocks are as homogeneous as possible; for instance, experimental animals from one same litter are grouped into a block because their genetic constitutions are supposed to be similar. However, it has long been recognized that there still remains a substantial heterogeneity among the units in a block. In order to control this heterogeneity, the so-called "randomization procedure" was initiated by R. A. Fisher [1926]. Mathematical treatments of the Fisher randomization were done by B. L. Welch [1937], E. J. G Pitman [1937] and M. B. Wilk [1955]. The underlying models in those works may be called "the Fisher model", i.e., the linear model taking only the unit-errors on plot-effects into account and totally lacking the technical errors. J. Neyman *et al.* [1935] pointed out that there are instances in which a linear model taking both the technical errors and the unit-errors into account would be more adequate by the very nature of the problem under consideration. A model of this sort is called "the Neyman model". M. D. McCarthy [1939] investigated the null-distribution of the analysis of variance test statistic in a randomized block design under the Neyman model and justified the use of the central F-distribution in the asymptotic sense. However, his formulation of the randomization procedure was not completely satisfactory. Almost all books of mathematical statistics, to the best of our knowledge, dealing with the analysis of block designs devote a couple of pages to emphasizing the importance of the randomization, then they give the analysis of variance based on the normal regression model, i.e. a linear model containing only the technical errors, *as if the unit errors have disappeared by the act of randomization* (Mann [1949]). *From the theoretical point of view, this should be justified.* One of the authors, Ogawa, has initiated a series of investigations on this problem in 1960 (Ogawa [1961, 1962, 1963]) and now with the cooperation of S. Ikeda (Ogawa *et al.* [1964, 1967], Ogawa and Ikeda [1970]), he is in a position to present a fairly satisfactory answer to this question.

† Dr. Ikeda was a former colleague at Nihon University and is a professor of statistics at Soka University since March 1971.

In order to emphasize our main ideas and to let the reader be bothered by mathematical complications as less as possible, we are concerned with the mathematically simplest case, i.e., a randomized complete block design. We are given b blocks of size k each, and hence there are $n = bk$ experimental units on the whole. We adopt a special labeling system of the whole experimental units in such a way that the ith unit in the pth block bears the number $f = (p-1)k+i$. This labeling system will turn out to be very convenient for the description of the randomization later on.

We define the incidence vectors of the $v(=k)$ treatment to be tested by experimentation as follows:

$$\zeta_\alpha = \begin{pmatrix} \zeta_{\alpha 1} \\ \zeta_{\alpha 2} \\ \cdot \\ \cdot \\ \cdot \\ \zeta_{\alpha n} \end{pmatrix}, \quad \text{where} \quad \zeta_{\alpha f} = \begin{cases} 1, & \text{if the } \alpha \text{th treatment occurs in the } f \text{th unit,} \\ 0, & \text{otherwise.} \end{cases}$$

The $n \times v$ matrix $\Phi = \|\zeta_1 \, \zeta_2 \cdots \zeta_v\|$, where $v = k$ in our case, is called the incidence matrix of the treatment. It is almost evident that

$$\Phi'\Phi = \begin{Vmatrix} r_1 & & & & 0 \\ & r_2 & & & \\ & & \cdot & & \\ & & & \cdot & \\ 0 & & & & r_v \end{Vmatrix}$$

where r_α is the number of the replications of the αth treatment. In our case of a complete block design, $r_1 = r_2 = \cdots = r_v = b$.

Let

$$T = \|t_{fg}\| = \Phi\Phi', \quad f, g = 1, \ldots, n,$$

then one can see that

$$t_{fg} = \begin{cases} 1, & \text{if the } f \text{th and } g \text{th units receive the same treatment,} \\ 0, & \text{otherwise} \end{cases}$$

and

$$T^2 = bT \quad \text{or} \quad \left(\frac{1}{b}T\right)^2 = \frac{1}{b}T.$$

Likewise, one defines the incidence vectors and matrix of blocks as follows:

$$\eta_p = \begin{pmatrix} \eta_{p1} \\ \eta_{p2} \\ \cdot \\ \cdot \\ \cdot \\ \eta_{pn} \end{pmatrix}, \quad \text{where} \quad \eta_{pf} = \begin{cases} 1, & \text{if the } f \text{th unit belongs to the } p \text{th block,} \\ 0, & \text{otherwise.} \end{cases}$$

The $n \times b$ matrix $\Psi = \|\mathbf{\eta}_1 \, \mathbf{\eta}_2 \cdots \mathbf{\eta}_b\|$ is called the incidence matrix of blocks. It is easy to see that

$$\Psi'\Psi = \begin{Vmatrix} k & & & 0 \\ & k & & \\ & & \cdot & \\ & & & \cdot \\ & & & \cdot \\ 0 & & & k \end{Vmatrix}, \quad \text{and} \quad B = \Psi\Psi' = \begin{Vmatrix} G_k & & & 0 \\ & G_k & & \\ & & \cdot & \\ & & & \cdot \\ & & & \cdot \\ 0 & & & G_k \end{Vmatrix},$$

where G_k is the $k \times k$ matrix whose elements are all unity. Hence

$$B^2 = kB \quad \text{or} \quad \left(\frac{1}{k}B\right)^2 = \frac{1}{k}B.$$

The $v \times b$ matrix

$$N = \|n_{\alpha p}\| = \Phi'\Psi, \quad \alpha = 1, \ldots, v, \quad p = 1, \ldots, b$$

is the incidence matrix of the design under consideration and

$$n_{\alpha p} = \begin{cases} 1, & \text{if the } \alpha\text{th treatment occurs in the } p\text{th block,} \\ 0, & \text{otherwise.} \end{cases}$$

In our case of a complete block, it is clear that

$$N = \begin{Vmatrix} 1 & 1 & \cdots & 1 \\ 1 & 1 & \cdots & 1 \\ \cdot & \cdot & \cdots & \cdot \\ 1 & 1 & \cdots & 1 \end{Vmatrix} \quad \text{and hence } NN' = bG_k.$$

Since it follows that

$$TB = \Phi\Phi'\Psi\Psi' = \Phi N\Psi' = G_n,$$
$$BT = \Psi\Psi'\Phi\Phi' = \Psi N'\Phi' = G_n,$$

the four matrices $I = I_n$, $G = G_n$, B, T make up a commutative linear algebra, called the relationship algebra (Ogawa and Ishii [1944]), of the complete block design.

$$\frac{1}{n}G = \frac{1}{n}\mathbf{11}', \; \frac{1}{k}B - \frac{1}{n}G, \; \frac{1}{b}T - \frac{1}{n}G, \quad \text{and} \quad I - \frac{1}{k}B - \frac{1}{b}T + \frac{1}{n}G$$

are mutually orthogonal idempotent matrices with respective ranks

$$\alpha_0 = 1, \quad \alpha_1 = b-1, \quad \alpha_2 = k-1 \quad \text{and} \quad \alpha_3 = (b-1)(k-1).$$

The Neyman model is given by

$$\mathbf{x} = \gamma\mathbf{1} + \Phi\mathbf{\tau} + \Psi\mathbf{\beta} + \mathbf{\pi} + \mathbf{e}, \tag{1}$$

or

$$x_f = \gamma + \sum_{\alpha=1}^{v} \zeta_{\alpha f}\tau_\alpha + \sum_{p=1}^{b} \eta_{pf}\beta_p + \pi_f + e_f, \; f = 1, 2, \ldots, n,$$

where $\mathbf{x}' = (x_1, x_2, \ldots, x_n)$ is the observation vector, $\mathbf{1}' = (1, 1, \ldots, 1)$, $\mathbf{\tau}' = (\tau_1, \ldots, \tau_v)$ and $\mathbf{\beta}' = (\beta_1, \ldots, \beta_b)$ are the treatment-effects and block-effects being subject to the restrictions $\tau_1 + \cdots + \tau_v = 0$ and $\beta_1 + \cdots + \beta_b = 0$,

12

respectively, $\pi' = (\pi_1\pi_2 \cdots \pi_n)$ is its unit-errors being subject to the restriction $\Psi'\pi = 0$ and finally $e' = (e_1e_2 \cdots e_n)$ stands for the technical-errors distributed as $\mathcal{N}(0, \sigma^2 I)$, where σ^2 is the unknown variance.

We are to consider the sampling distribution of the F-statistic occurring in the analysis of variance of the complete block design under the Neyman model when the randomization procedure is applied.

Let us put

$$\mathbf{t} = (T_1, \ldots, T_k)',$$

where $T_\alpha = \zeta_\alpha'\mathbf{x}$ is the treatment sum, $\alpha = 1, \ldots, k$, and further let

$$\mathbf{b} = (B_1, \ldots, B_b)',$$

where $B_p = \eta_p'\mathbf{x}$ is the block sum, $p = 1, \ldots, b$. Then one can see that

$$\mathbf{x}'\left(\frac{1}{b}T - \frac{1}{n}G\right)\mathbf{x} = \frac{1}{b}(T_1^2 + \cdots + T_k^2) - n\bar{x}^2, \text{ sum of squares due to treatments,}$$

$$\mathbf{x}'\left(\frac{1}{k}B - \frac{1}{n}G\right)\mathbf{x} = \frac{1}{k}(B_1^2 + \cdots + B_b^2) - n\bar{x}^2, \text{ sum of squares due to blocks}$$

and

$$\mathbf{x}'\left(I - \frac{1}{k}B - \frac{1}{b}T + \frac{1}{n}G\right)\mathbf{x} = \mathbf{x}'\left(I - \frac{1}{n}G\right)\mathbf{x} - \mathbf{x}'\left(\frac{1}{k}B - \frac{1}{n}G\right)\mathbf{x} - \mathbf{x}'\left(\frac{1}{b}T - \frac{1}{n}G\right)\mathbf{x}$$

$$= \sum_{f=1}^{n} x_f^2 - \frac{1}{k}\sum_{p=1}^{b} B_p^2 - \frac{1}{b}\sum_{\alpha=1}^{k} T_\alpha^2 + n\bar{x}^2; \text{ sum of squares due to error.}$$

In the first place, we consider the probability distribution of the F-statistic given by

$$F = (b-1)\frac{\mathbf{x}'\left(\dfrac{1}{b}T - \dfrac{1}{n}G\right)\mathbf{x}}{\mathbf{x}'\left(I - \dfrac{1}{k}B - \dfrac{1}{b}T + \dfrac{1}{n}G\right)\mathbf{x}}$$

before the randomization under the Neyman model.

Since

$$\left(\frac{1}{b}T - \frac{1}{n}G\right)\mathbf{x} = \Phi\left(\tau + \frac{1}{b}\Pi\right) + \left(\frac{1}{b}T - \frac{1}{n}G\right)e,$$

where $\Pi = (\Pi_1, \ldots, \Pi_k)'$ with $\Pi_\alpha = \zeta_\alpha'\pi$, the variate

$$\chi_1^2 = \mathbf{x}'\left(\frac{1}{b}T - \frac{1}{n}G\right)\mathbf{x}/\sigma^2$$

is distributed as the non-central chi-square distribution of degrees of freedom $(k-1)$ with the non-centrality parameter $\delta_1/2\sigma^2$, where

$$\delta_1 = b\left(\tau + \frac{1}{b}\Pi\right)'\left(\tau + \frac{1}{b}\Pi\right).$$

Hence its probability element is given by

$$\exp\left(-\delta_1/2\sigma^2\right) \sum_{\mu=0}^{\infty} \frac{(\delta_1/2\sigma^2)^{\mu}}{\mu!} \frac{(\chi_1^2/2)^{(k-1)/2+\mu-1}}{\Gamma[(k-1)/2+\mu]} e^{-\frac{1}{2}\chi_1^2}\, d(\chi_1^2/2). \qquad (3)$$

In a similar manner one can see that the variate

$$\chi_2^2 = \mathbf{x}'\left(I - \frac{1}{k}B - \frac{1}{b}T + \frac{1}{n}G\right)\mathbf{x}/\sigma^2$$

is distributed as the non-central chi-square distribution of degrees of freedom $(b-1)(k-1)$ with the non-centrality parameter $\delta_2/2\sigma^2$, where

$$\delta_2 = \boldsymbol{\pi}'\boldsymbol{\pi} - \frac{1}{b}\boldsymbol{\Pi}'\boldsymbol{\Pi} = \Delta - \frac{1}{b}\boldsymbol{\Pi}'\boldsymbol{\Pi},$$

hence its probability element is given by

$$\exp\left(-\delta_2/2\sigma^2\right) \sum_{v=0}^{\infty} \frac{(\delta_2/2\sigma^2)^{v}}{v!} \frac{(\chi_2^2/2)^{(b-1)(k-1)/2+v-1}}{\Gamma[(b-1)(k-1)/2+v]} e^{-\frac{1}{2}\chi_2^2}\, d(\chi_2^2/2)$$

Since the two variates χ_1^2 and χ_2^2 are stochastically independent, the probability element of the F before the randomization under the Neyman model is given by

$$\exp\left(-\xi/2\sigma^2\right) \sum_{l=0}^{\infty} \frac{(\xi/2\sigma^2)^{l}}{l!} \sum_{\mu+v=l} \frac{l!}{\mu!v!}\eta^{\mu}(1-\eta)^{v}$$

$$\times \frac{\Gamma[b(k-1)/2+\mu+v]}{\Gamma[(k-1)/2+\mu]\,\Gamma[(b-1)(k-1)/2+v]}$$

$$\times \left(\frac{F}{b-1}\right)^{(k-1)/2+\mu-1} \left(1+\frac{F}{b-1}\right)^{-b(k-1)/2-\mu-v} d\left(\frac{F}{b-1}\right), \qquad (5)$$

where we have put

$$\xi = \Delta + b\boldsymbol{\tau}'\boldsymbol{\tau} + 2\mathcal{T}, \qquad \eta = 1 - \frac{\Delta - H}{\Delta + b\boldsymbol{\tau}'\boldsymbol{\tau} + 2\mathcal{T}} = \frac{b\boldsymbol{\tau}'\boldsymbol{\tau} + 2\mathcal{T} + H}{\Delta + \boldsymbol{\tau}'\boldsymbol{\tau} + 2\mathcal{T}} \qquad (6)$$

$$\mathcal{T} = \boldsymbol{\tau}'\boldsymbol{\Pi}, \qquad H = \frac{1}{b}\boldsymbol{\Pi}'\boldsymbol{\Pi}.$$

Now under the normal regression model

$$\mathbf{x} = \gamma\mathbf{1} + \boldsymbol{\Phi}\boldsymbol{\tau} + \boldsymbol{\Psi}\boldsymbol{\beta} + \mathbf{e},$$

the probability element of the F is given by

$$\frac{\Gamma[b(k-1)/2]}{\Gamma[(k-1)/2]\Gamma[(b-1)(k-1)/2]}\left(\frac{F}{b-1}\right)^{(k-1)/2-1}\left(1+\frac{F}{b-1}\right)^{-b(k-1)/2} d\left(\frac{F}{b-1}\right). \qquad (7)$$

One wants to show that after the randomization (5) is asymptotically equivalent to (7) in the sense of the type $(\mathcal{M})_d$ as $b \to \infty$, provided certain uniformity conditions on the unit-errors are satisfied. The probability element of F should be obtained by taking the mathematical expectation of (5) with respect to the permutation distribution of (ξ, η) due to randomization. For the sake of mathematical simplicity, we shall be concerned, henceforth, with the null-distribution under the null-hypothesis $H_0 : \tau = 0$ to be tested. The probability element of the null-distribution of the F before the randomization is given by

$$\exp\left(-\Delta/2\sigma^2\right) \sum_{l=0}^{\infty} \frac{(\Delta/2\sigma^2)^l}{l!} \sum_{\mu+\nu=l} \frac{l!}{\mu!\nu!} \theta^\mu(1-\theta)^\nu \, h_{\mu\nu}(F) \, dF. \tag{8}$$

where

$$h_{\mu\nu}(F) = \frac{\Gamma[b(k-1)/2+\mu+\nu]}{\Gamma[(k-1)/2+\mu] \cdot \Gamma[(b-1)(k-1)/2+\nu]}$$

$$\times \left(\frac{F}{b-1}\right)^{(k-1)/2+\mu-1} \left(1+\frac{F}{b-1}\right)^{-b(k-1)/2-\mu-\nu} \frac{1}{b-1} \tag{9}$$

and

$$\theta = \frac{1}{b}\mathbf{\Pi}'\mathbf{\Pi}/\Delta. \tag{10}$$

The randomization procedure is described in mathematical terms. Let

$$\sigma_p = \begin{pmatrix} 1 & \cdots & 2 & \cdots & k \\ \sigma_p(1) & \sigma_p(2) & \cdots & \sigma_p(k) \end{pmatrix}, \qquad p = 1, \ldots, b$$

be the permutations in p blocks, and let S_{σ_p} be the permutation matrix corresponding to σ_p, i.e.,

$$S_{\sigma_p} \begin{bmatrix} 1 \\ 2 \\ \cdot \\ \cdot \\ \cdot \\ k \end{bmatrix} = \begin{bmatrix} \sigma_p(1) \\ \sigma_p(2) \\ \cdot \\ \cdot \\ \cdot \\ \sigma_p(k) \end{bmatrix}$$

and let

$$\mathcal{U}_\sigma = \left\| \begin{matrix} S_{\sigma_1} & & & 0 \\ & S_{\sigma_2} & & \\ & & \cdot & \\ & & & \cdot \\ 0 & & & S_{\sigma_d} \end{matrix} \right\|$$

be the permutation matrix corresponding to the randomization in the whole b blocks.

Starting with any fixed incidence matrix Φ_0 of treatments, the randomization makes the incidence matrix Φ a random matrix such that

$$P\{\Phi = \mathcal{U}_\sigma\Phi_0\} = 1/(k!)^b \tag{11}$$

for all $\mathcal{U}_\sigma$.

Let
$$\Pi_p^\sigma = (\Pi_{1p}^\sigma \cdots \Pi_{kp}^\sigma)'.$$
where
$$\Pi_{\alpha p}^\sigma = \sum_{i=1}^{k} \zeta_{\alpha,(p-1)k+i} \, \pi_{\sigma_p(i)}^{(p)}, \qquad \alpha = 1, \ldots, k, \, p = 1, \ldots, b,$$

$$\Pi^\sigma = \mathcal{U}_\sigma \Pi = \sum_{p=1}^{b} \Pi_p^\sigma.$$

It is not difficult to see that

$$E(\Pi_p^\sigma) = 0 \quad \text{and} \quad E(\Pi_p^\sigma \Pi_p^{\sigma\prime}) = \frac{\Delta_p}{k(k-1)} \Lambda_p, \tag{12}$$

where E stands for the expectation with respect to the permutation distribution due to the randomization and

$$\Lambda_p = \begin{Vmatrix} (k-1)n_{1p} & -n_{1p}n_{2p} & \cdots & -n_{1p}n_{kp} \\ -n_{2p}n_{1p} & (k-1)n_{2p} & \cdots & -n_{2p}n_{kp} \\ \vdots & \vdots & & \vdots \\ -n_{kp}n_{1p} & -n_{kp}n_{2p} & \cdots & (k-1)n_{kp} \end{Vmatrix} = \begin{Vmatrix} k-1 & -1 & \cdots & -1 \\ -1 & k-1 & \cdots & -1 \\ \vdots & \vdots & & \vdots \\ -1 & -1 & \cdots & k-1 \end{Vmatrix}$$

It should be noticed that

$$\sum_{p=1}^{b} \Lambda_p = bkI_k - NN' = b(kI_k - G_k). \tag{13}$$

We now state the uniformity conditions to be imposed on the unit-errors:

$$\left.\begin{aligned} \bar{\Delta} &= \frac{1}{b} \sum_{p=1}^{b} \Delta_p \to \Delta_0 \\ \frac{1}{\bar{\Delta}^2} \frac{1}{b} \sum_{p=1}^{b} (\Delta_p - \bar{\Delta})^2 \text{ bounded} \end{aligned}\right\} \text{as } b \to \infty, \tag{14}$$

where
$$\Delta_p = \sum_{i=1}^{k} \pi_i^{(p)2}; \; \pi_i^{(p)} = \pi_{(p-1)k+i}, \, p = 1, \ldots, b.$$

Now we refer to the central limit theorem due to W. Feller (Rao [1965]).

Theorem. Let $\mathcal{X}_1^{(n)}, \ldots, \mathcal{X}_n^{(n)}$ be a sequence of independent random variables of k dimensions such that
$$E(\mathcal{X}_i^{(n)}) = 0 \quad \text{and} \quad E(\mathcal{X}_i^{(n)} \mathcal{X}_i^{(n)\prime}) = \Lambda_i^{(n)}, \, i = 1, \ldots, n. \tag{15}$$

Suppose that the conditions
$$\frac{1}{n} \sum_{i=1}^{n} \Lambda_i^{(n)} \to \Lambda(\neq 0) \quad \text{and} \quad \frac{1}{n} \sum_{i=1}^{n} \int_{\|\mathbf{x}\| > \varepsilon\sqrt{n}} \|\mathbf{x}\|^2 \, dF_i^{(n)} \to 0 \quad \text{as } n \to \infty \tag{16}$$

are satisfied, where $F_i^{(n)}$ denotes the distribution function of $\mathcal{X}_i^{(n)}$ and $\|\mathbf{x}\|$ is the Euclidean norm of the vector $\mathbf{x}$. Then

$$\mathcal{L}\left(\frac{1}{\sqrt{n}} \sum_{i=1}^{n} \mathcal{X}_i^{(n)}\right) \to \mathcal{N}(0, \Lambda) \quad \text{as } n \to \infty.$$

i.e., $(1/\sqrt{n}) \sum_{i=1}^{n} \mathcal{X}_i^{(n)}$ converges in law to $N(0, \Lambda)$.

Now take $n = b$ and $\mathscr{X}_i^{(n)} = \mathbf{\Pi}_i^\sigma$, $i = 1, \ldots, b$, then

$$E(\mathscr{X}_i^{(n)}) = E(\mathbf{\Pi}_i^\sigma) = 0, \quad E(\mathscr{X}_i^{(n)}\mathscr{X}_i^{(n)'}) = E(\mathbf{\Pi}_i^\sigma\mathbf{\Pi}_i^{\sigma'}) = \frac{\Delta_i}{k(k-1)}\Lambda_i.$$

and hence

$$\frac{1}{b} \sum_{p=1}^{b} E(\mathscr{X}_p^{(n)}\mathscr{X}_p^{(n)'}) = \frac{1}{bk(k-1)} \sum_{p=1}^{b} \Delta_p\Lambda_p$$

$$= \frac{1}{k(k-1)} \frac{1}{b}\left[\bar{\Delta} \sum_{p=1}^{b} \Lambda_p + \sum_{p=1}^{b} (\Delta_p - \bar{\Delta})\Lambda_p \right]$$

$$\to \frac{\Delta_0}{k(k-1)}(kI_k - G_k) \equiv \Lambda \text{ as } b \to \infty$$

In fact, the (ij)-element of the matrix $(1/\bar{\Delta}b) \sum_{p=1}^{b} (\Delta_p - \bar{\Delta}_p)\Lambda_p$ is

$$\left(\frac{1}{\bar{\Delta}b} \sum_{p=1}^{b} (\Delta_p - \bar{\Delta})\Lambda_p \right)_{ij} = \frac{1}{\bar{\Delta}b} \sum_{p=1}^{b} (\Delta_p - \bar{\Delta})(\Lambda_p)_{ij} = 0.$$

Furthermore, since $|\mathbf{x}|^2 = \mathbf{\Pi}_p^{\sigma'}\mathbf{\Pi}_q^\sigma = \Delta_p$ for all σ, in our case

$$\frac{1}{b} \sum_{p=1}^{b} \int_{\mathbf{x} > \varepsilon\sqrt{b}} |\mathbf{x}|^2 \, dF_p^{(b)} = \frac{1}{b} \sum_{p=1}^{b} \Delta_p P\left\{ \frac{1}{b}\mathbf{\Pi}_p^{\sigma'}\mathbf{\Pi}_p^\sigma > \varepsilon^2 \right\}$$

$$\leqq \frac{1}{\varepsilon^2 b^2} \sum_{p=1}^{b} \Delta_p E(\mathbf{\Pi}_p^{\sigma'}\mathbf{\Pi}_p^\sigma) = \frac{1}{\varepsilon^2 b^2} \sum_{p=1}^{b} \Delta_p^2$$

$$= \frac{1}{\varepsilon^2 b^2}\left[\sum_{p=1}^{b} (\Delta_p - \bar{\Delta})^2 + b\bar{\Delta}^2 \right]$$

$$= \frac{1}{\varepsilon^2 b} \bar{\Delta}^2\left[1 + \frac{1}{b\bar{\Delta}^2} \sum_{p=1}^{b} (\Delta_p - \bar{\Delta})^2 \right] \to 0 \text{ as } b \to \infty$$

Thus all conditions (15), (16) are satisfied and therefore

$$\mathscr{L}\left(\frac{1}{\sqrt{b}}\mathbf{\Pi}^\sigma \right) \to \mathscr{N}\left(0, \frac{\Delta_0}{k(k-1)}(kI_k - G_k) \right) \text{ as } b \to \infty. \tag{17}$$

Whence one obtains

$$\mathscr{L}\left(\frac{k-1}{\Delta_0} \frac{1}{b}\mathbf{\Pi}^{\sigma'}\mathbf{\Pi}^\sigma \right) \to \chi_{k-1}^2 \text{ as } b \to \infty.$$

and hence by Theorem I.1, which will be mentioned later,

$$\mathscr{L}(b(k-1)\theta) \to \chi_{k-1}^2 \text{ as } b \to \infty. \tag{18}$$

We have to refer to the idea of the asymptotic equivalence of random variables expounded by one of the authors (Ikeda [1963, 1968]).

Two sequences of n_s-dimensional random variables $\{\mathscr{X}_s^{(n_s)}\}$ and $\{\mathscr{Y}_s^{(n_s)}\}$, where the dimensions of the random variables may depend on s, are said to be *asymptotically equivalent in the sense of* $(\mathscr{C})_d$ *as* $s \to \infty$, if

$$d(\mathscr{X}_s^{(n_s)}, \mathscr{Y}_s^{(n_s)}; \mathscr{C}) = \sup_{E \in \mathscr{C}} | P^{\mathscr{X}_s^{(n_s)}}(E) - P^{\mathscr{Y}_s^{(n_s)}}(E) | \to 0 \text{ as } s \to \infty, \quad (19)$$

where $\mathscr{C}$ is any given non-empty subclass of the usual Borel field of subsets of the n_s-dimensional Euclidean space R_{n_s}. We denote this asymptotic equivalence by

$$\mathscr{X}_s^{(n_s)} \sim \mathscr{Y}_s^{(n_s)}(\mathscr{C})_d \, (s \to \infty).$$

Let $\mathscr{M}$ be the class of all subsets of the form

$$M = \{\mathbf{x} = (x_1, \ldots, x_{n_s}); \, -\infty \leqslant x_i < a_i, \, i = 1, \ldots, n_s\},$$

where the a_i's are extended real values. $M = \emptyset$ (empty set) if $a_i = -\infty$ for some i, and $M = R_{n_s}$ if $a_i = \infty$ for all i. Let $\mathscr{S}$ be the class of all subsets of the form

$$S = \{\mathbf{x} = (x_1, \ldots, x_{n_s}): \, b_i \leq x_i \leq a_i, \, i = 1, \ldots, n_s\},$$

where the a_i's and b_i's are extended real values. $S = \emptyset$ if $a_i = b_i$ for some i and $S = R_{n_s}$ if $b_i = -\infty$ and $a_i = \infty$ for all i. Finally, let $\mathscr{B}$ be the Borel field containing the class $\mathscr{M}$. Evidently

$$\mathscr{M} \subseteq \mathscr{S} \subseteq \mathscr{B}.$$

A sequence of random variables $\{\mathscr{X}_s^{(n_s)}\}$ is said to have the property $B(\mathscr{S})$ if for any given $\varepsilon > 0$, there exist a bounded subset $S \in \mathscr{S}$ and a positive integer s_0 such that

$$P^{\mathscr{X}_s^{(n_s)}}(S) > 1 - \varepsilon \qquad \text{for all } s \geq s_0$$

Also it is said to have the property $C(\mathscr{S})$, if for any given $\varepsilon > 0$, there can be found a positive number $\delta = \delta(\varepsilon)$ and a positive integer $s_0 = s_0(\varepsilon)$ such that $\mu(E) < \delta$, $E \in \mathscr{S}$ implies

$$P^{\mathscr{X}_s^{(n_s)}}(E) < \varepsilon \qquad \text{for all } s \geq s_0,$$

Now we quote three theorems established by Ikeda that are necessary to show that the null-distribution of the F-statistic is asymptotically equivalent to the familiar central F-distribution in the sense of the type $(\mathscr{M})_d$, as $b \to \infty$, if the design is randomized.

Theorem I.1. *If* $\{\mathscr{X}_s^{(n_s)}\}$ *has the properties* $B(\mathscr{S})$ *and* $C(\mathscr{S})$ *simultaneously, then for any sequence of real numbers* $\{c_1^s, \ldots, c_{n_s}^s\}$ *such that* $c_i^s \to 1$ *as* $s \to \infty$ *for all i, the following holds:*

$$\mathscr{X}_s^{(n_s)} = (X_1^s, \ldots, X_{n_s}^s)' \sim \mathscr{Y}_s^{(n_s)} = (c_1^s Y_1^s, \ldots, c_{n_s}^s Y_{n_s}^s)'(\mathscr{M})_d, \, s \to \infty. \quad (20)$$

Theorem I.2. *Suppose* $\mathscr{X}_s^{(n_s)}$ *and* $\mathscr{Y}_s^{(n_s)}$ *have the probability density functions* $f_s(x)$ *and* $g_s(y)$, *respectively. If the Kullback-Leibler mean information converges to zero, i.e.,*

$$I(f_s; g_s) = E_{f_s}\left(\log \frac{f_s}{g_s}\right) \to 0 \quad as \quad s \to \infty, \quad (21)$$

then

$$\mathscr{X}_s^{(n_s)} \sim \mathscr{Y}_s^{(n_s)}(\mathscr{M})_d, \ s \to \infty.$$

Theorem I.3. *Let* $\{(X_s, Y_s)\}$ *be a sequence of 2-dimensional random variables, and let* $F_s(x)$ *and* $p_s(y|x)$ *be the cumulative distribution function of* X_s *and the conditional probability density function of* Y_s *given* $X_s = x$, *respectively. Further let* $\{X_s^*\}$ *be a sequence of random variables with c.d.f.* $F_s^*(x)$. *It is seen that* Y_s *has the p.d.f.*

$$h_s(y) = \int_{R_1} p_s(y \mid x) \, dF_s(x)$$

Let a random variable with p.d.f.

$$h_s^*(y) = \int_{R_1} p_s(y \mid x) \, dF_s^*(x)$$

be Y_s^*. *Then* $X_s \sim X_s^*(\mathscr{M})_d, \ s \to \infty$, *implies* $Y_s \sim Y_s^*(\mathscr{M})_d, \ s \to \infty$, *if the following conditions are satisfied*:

(i) *there exists a sequence* $\{(c_s, d_s)\}$ *such that the sequence* $\{(X_s - d_s)/c_s\}$ *has the property* $B(\mathscr{S})$,

(ii) $q_s(y|x) = p_s(y|c_s x + d_s)$ *is differentiable with respect to* x, *and*

(iii) *for any bounded subset* $E \in \mathscr{S}$,

$$\sup_{x \in E} \int_{R_1} \left| \frac{\partial p_s(y \mid c_s x + d_s)}{\partial x} \right| dy$$

is bounded for all values of s *which are sufficiently large.*

Theorem I.1 assures us that

$$b(k-1)\theta = (k-1)\mathbf{\Pi}^{\sigma'}\mathbf{\Pi}^{\sigma}/(b\bar{\Delta}) \sim (k-1)\mathbf{\Pi}^{\sigma'}\mathbf{\Pi}^{\sigma}/(b\Delta_0)(\mathscr{M})_d, \ b \to \infty, \quad (22)$$

because

$$(k-1)\mathbf{\Pi}^{\sigma'}\mathbf{\Pi}^{\sigma}/(b\Delta_0) = \frac{\Delta_0}{\bar{\Delta}}(k-1)\mathbf{\Pi}^{\sigma'}\mathbf{\Pi}^{\sigma}/(b\bar{\Delta}) \quad \text{and} \quad \Delta_0/\bar{\Delta} \to 1 \text{ as } b \to \infty.$$

Lemma. *The sequence of random variables* $\{X_b\}$ *having the probability density function*

$$f_b(x) = \frac{\Gamma(\tfrac{1}{2}b(k-1))}{\Gamma(\tfrac{1}{2}(k-1))\Gamma(\tfrac{1}{2}(b-1)(k-1))} x^{\frac{1}{2}(k-1)-1} (1-x)^{\frac{1}{2}(b-1)(k-1)-1}$$

$$\text{for} \quad 0 < x < 1$$

and the sequence of random variables $\{\theta\}$ *having the probability density function*

$$g_b(x) = \frac{(\tfrac{1}{2}b(k-1))^{\frac{1}{2}(k-1)}}{\Gamma(\tfrac{1}{2}(k-1))} x^{\frac{1}{2}(k-1)-1} e^{-\frac{1}{2}b(k-1)x} \quad \text{for} \quad 0 < x < \infty$$

are asymptotically equivalent in the sense of the type $(\mathscr{M})_d$ *as* $b \to \infty$.

Proof. According to Theorem I.2, one has to show that
$$I(f:g) \to 0 \quad \text{as} \quad b \to \infty.$$

Now
$$
\begin{aligned}
I(f:g) &= \log \Gamma(\tfrac{1}{2}b(k-1)) - \log \Gamma(\tfrac{1}{2}(b-1)(k-1)) - \tfrac{1}{2}(k-1)\log(\tfrac{1}{2}b(k-1)) \\
&\quad + [\tfrac{1}{2}(b-1)(k-1)-1]E_f[\log(1-X_b)] + \tfrac{1}{2}b(k-1) - E_f(X_b) \\
&\sim [\tfrac{1}{2}b(k-1)+\tfrac{1}{2}]\log(\tfrac{1}{2}b(k-1)) - \tfrac{1}{2}b(k-1) - [\tfrac{1}{2}(b-1)(k-1)+\tfrac{1}{2}] \\
&\quad \times \log(\tfrac{1}{2}(b-1)(k-1)) + \tfrac{1}{2}(b-1)(k-1) - \tfrac{1}{2}(k-1)\log(\tfrac{1}{2}b(k-1)) \\
&\quad + [\tfrac{1}{2}(b-1)(k-1)-1]\left[\frac{\Gamma'(\tfrac{1}{2}(b-1)(k-1))}{\Gamma(\tfrac{1}{2}fb-1)(k-1))} - \frac{\Gamma'(\tfrac{1}{2}b(k-1))}{\Gamma(\tfrac{1}{2}b(k-1))}\right] + \tfrac{1}{2}(k-1).
\end{aligned}
$$

Since
$$\frac{\Gamma'(p)}{\Gamma(p)} = \log p - \frac{1}{2p} + O\left(\frac{1}{p^2}\right) \text{ as } p \to \infty,$$

one can see immediately that
$$I(f:g) \to 0 \quad \text{as} \quad b \to \infty.$$

And hence
$$\theta \sim X_b(\mathcal{M})_d, \qquad b \to \infty. \tag{23}$$

Thus one can integrate out θ from (8) with respect to the beta-distribution with p.d.f.

$$f_b(x) = \frac{\Gamma(\tfrac{1}{2}b(k-1))}{\Gamma(\tfrac{1}{2}(k-1))\Gamma(\tfrac{1}{2}(b-1)(k-1))} x^{\frac{1}{2}(k-1)-1}(1-x)^{\frac{1}{2}(b-1)(k-1)-1}$$

$$\text{for } 0 < x < 1.$$

to obtain the central F-distribution (7). In order to show that this distribution is asymptotically equivalent to the true distribution of the F-statistic in the sense of the type $(\mathcal{M})_d$, $b \to \infty$, one has to show that the three conditions of Theorem I.3 are satisfied in our case. As a matter of fact, it would be enough to check the condition (iii), since the conditions (i) and (ii) are apparently satisfied in our case:

$$p_b(F \mid \theta) = \exp(-\Delta/2\sigma^2) \sum_{l=0}^{\infty} \frac{(\Delta/2\sigma^2)^l}{l!} \sum_{\mu+\nu=l} \frac{l!}{\mu!\nu!} \theta^\mu (1-\theta)^\nu h_{\mu\nu}(F),$$

$$\frac{\partial q_b(F \mid \theta)}{\partial \theta} = \left[\frac{1}{b(k-1)} \frac{\partial}{\partial \theta} p_b(F \mid \theta)\right]_{\theta=\theta/b(k-1)}$$

$$= \frac{1}{b(k-1)} \exp(-\Delta/2\sigma^2) \sum_{l=1}^{\infty} \frac{(\Delta/2\sigma^2)^l}{l!} \sum_{\mu+\nu=l} \frac{l!}{\mu!\nu!}$$

$$\times \left[\mu\left(\frac{\theta}{b(k-1)}\right)^{\mu-1}\left(1-\frac{\theta}{b(k-1)}\right)^\nu - \nu\left(\frac{\theta}{b(k-1)}\right)^\mu\left(1-\frac{\theta}{b(k-1)}\right)^{\nu-1}\right] h_{\mu\nu}(F)$$

hence

$$\int_{R_1} \left| \frac{\partial q_b(F \mid \theta)}{\partial \theta} \right| dF \leqq \frac{1}{b(k-1)} \exp\left(-\Delta/2\sigma^2\right) \sum_{l=1}^{\infty} \frac{(\Delta/2\sigma^2)^l}{l!} \sum_{\mu+\nu=l} \frac{l!}{\mu!\nu!}$$

$$\times \left[\mu\left(\frac{\theta}{b(k-1)}\right)^{\mu-1}\left(1-\frac{\theta}{b(k-1)}\right)^{\nu} + \nu\left(\frac{\theta}{b(k-1)}\right)^{\mu}\left(1-\frac{\theta}{b(k-1)}\right)^{\nu-1} \right]$$

$$= \frac{\Delta}{b(k-1)\sigma^2} \rightarrow \frac{\Delta_0}{(k-1)\sigma^2} \text{ as } b \rightarrow \infty,$$

consequently

$$\sup_{\theta} \int_{R_1} \left| \frac{\partial q_b(F/\theta)}{\partial \theta} \right| dF$$

is bounded for sufficiently large b.

Thus we reached our final conclusion that the null-distribution, under the Neyman model, of the F-statistic occurring in the analysis of variance of a randomized block design is asymptotically equivalent to the familiar central F-distribution in the sense of the type $(\mathcal{M})_d$ as $b \rightarrow \infty$, provided that the uniformity conditions on the unit errors

$$\bar{\Delta} = \frac{1}{b} \sum_{p=1}^{b} \Delta_p \rightarrow \Delta_0 > 0, \quad \text{and} \quad \frac{1}{\bar{\Delta}^2} \frac{1}{b} \sum_{p=1}^{b} (\Delta_p - \bar{\Delta})^2 = o(b) \text{ as } b \rightarrow \infty$$

are satisfied.

References

R. A. Fisher, 1926, The arrangement of field experiment, *J. Min. Agr.* **23**, 503–513.

S. Ikeda, 1963, Asymptotic equivalence of probability distributions with applications to some problems of asymptotic independence, *Ann. Inst. Statist. Math. Tokyo* **15**, 87–116.

S. Ikeda, 1968, Asymptotic equivalence of real probability distributions, *Ann. Inst. Statist. Math. Tokyo* **20**, 339–362.

H. B. Mann, 1949, *Analysis and Design of Experiments* (Dover Publ., New York).

M. D. McCarthy, 1939, On the application of the z-test to randomized blocks, *Ann. Math. Statist.* **10**, 337–359.

J. Neyman, with the cooperation of K. Iwaskiewicz and S. Kolodziejczyk, 1935, Statistical problems in agricultural experimentation, *J. Roy. Statist. Soc. Suppl.* **2**, 107–154. Discussions, *Ibid.* **2**, pp. 154–180.

J. Ogawa, 1961, The effect of randomizations on the analysis of randomized block design, *Ann. Inst. Statist. Math. Tokyo* **13**, 105–117.

J. Ogawa, 1962, On the randomization in Latin-square design under the Neyman model, *Proc. Inst. Statist. Math. Tokyo* **10**, 1–16.

J. Ogawa, 1963, On the null-distribution of the F-statistic in a randomized BIB design under the Neyman model, *Ann. Math. Statist.* **34**, 1558–1568.

J. Ogawa and S. Ikeda, 1970, The asymptotic non-null distribution of the F-statistic for testing a partial null-hypothesis in a randomized PBIB design with m associate classes under the Neyman model (to be published).

J. Ogawa and G. Ishii, 1964, The relationship algebra and the analysis of a PBIB design, *Ann. Math. Statist*. **36,** 1815–1828.

J. Ogawa, S. Ikeda and M. Ogasawara, 1964, On the null-distribution of the F-statistic in a randomized PBIB design with two associate classes under the Neyman model, Essays in Probability and Statistics, The University of North Carolina Monograph Series in Probability and Statistics No. 3, pp. 517–548.

J. Ogawa, S. Ikeda and M. Ogasawara, 1967, On the null-distribution of the F-statistic for testing a partial null-hypothesis in a randomized PBIB design with m associate classes under the Neyman model, *Ann. Inst. Statist. Math. Tokyo* **19,** 313–330.

E. J. G. Pitman, 1937, Significance tests which can be applied to samples from any populations, IV. The analysis of variance test, *Biometrika* **29,** 70–90.

C. R. Rao, 1965, *Linear Statistical Inference and its Applications* (Wiley, New York), p. 118, 4, 7, Multivariate central limit theorem.

B. L. Welch, 1937, On the z-test in randomized blocks and Latin squares, *Biometrika* **29,** 21–52.

M. B. Wilk, 1955, The randomized analysis of generalized randomized block designs, *Biometrika* **42,** 70–79.

J. N. Srivastava et al., eds., *A Survey of Combinatorial Theory*
© North-Holland Publishing Company, 1973

CHAPTER 29

Some Combinatorial Problems of Arrays and Applications to Design of Experiments†

C. RADHAKRISHNA RAO

Indian Statistical Institute, Calcutta, India

1. Introduction

The combinatorial arrangements known as hypercubes, and more generally orthogonal arrays of strength d were introduced exactly twenty-five years ago (Rao [1946a]). They were applied in the construction of confounded symmetrical and asymmetrical factorial designs, multifactorial designs (fractional replications) and so on (Rao [1946b, 1947, 1949] and Nair and Rao [1948]). Later, orthogonal arrays of strength 2 were found useful in the construction of other combinatorial arrangements. Bose, Parker and Shrikhande [1960] used it in the disproof of Euler's conjecture. Recently, Ray-Chaudhuri and Wilson [1971a,b] used orthogonal arrays of strength 2 to generate resolvable balanced incomplete block designs as an intermediate step in the solution of Kirkman's school girl problem.

The object of the present paper is to consider arrays with a number of other combinatorial structures slightly weaker than that of orthogonal arrays and indicate their use in the design of experiments. The use of orthogonal arrays with a variable number of symbols in the rows is discussed in some detail.

We define a rectangular array, denoted by $(N, r, s_1 \times \cdots \times s_r)$ as an $r \times N$ matrix with entries in the ith row from a set S_i of s_i elements. The product $s_1 \times \cdots \times s_r$ indicates the totality of column vectors with the ith component $\in S_i$, $i = 1, \ldots, r$ (or the number of treatment combinations of r factors at levels $s_1, \ldots, s_r$ in the terminology of factorial experiments) from which N columns (treatment combinations) of the array are chosen. When S_i are all the same with cardinality s, then the array is denoted by (N, r, s^r) or simply (N, r, s). Similarly, $(N, r, s_1^{r_1} \times s_2^{r_2})$, $r_1 + r_2 = r$ will be an array with two different sets of symbols and so on.

For a given selection of d rows $\alpha_1, \ldots, \alpha_d$ we denote by $n(i_1, \ldots, i_d)$ the number of times the column vector $(i_1, \ldots, i_d)'$, $i_1 \in S_{\alpha_1}, \ldots, i_d \in S_{\alpha_d}$, occurs in the $d \times N$ submatrix specified by the selected rows.

---

† Paper read at the International Symposium on Combinatorial Mathematics and its Applications, Fort Collins, Colorado, September 1971.

2. Orthogonal array of strength d

Definition 1. A (N, r, s) array is said to be an *orthogonal array of strength d* if

$$n(i_1, \ldots, i_d) = \lambda \text{ (constant)} \tag{2.1}$$

for all possible combinations $i_1, \ldots, i_d$ $(i_j \in S)$, and for any selection of d rows. Such an array is denoted by (N, r, s, d). The constant λ is called the index of the orthogonal array.

By definition, an orthogonal array of strength d is also an orthogonal array of strength $i < d$. The following global inequalities for the parameters of an orthogonal array have been obtained in Rao [1947, 1949]:

$$N-1 \geqslant \binom{r}{1}(s-1)+ \cdots +\binom{r}{t}(s-1)^t \qquad\qquad\text{if } d = 2t \tag{2.2}$$

$$N-1 \geqslant \binom{r}{1}(s-1)+ \cdots +\binom{r}{t}(s-1)^t+\binom{r-1}{t}(s-1)^{t+1} \quad\text{if } d = 2t+1 \tag{2.3}$$

The inequalities (2.2) and (2.3) have been used to obtain the volume bound or the sphere packing bound in coding theory by Hamming [1951]. (See Berlekamp [1968], Chapter 13, section on Hamming-Rao high rate volume bound.)

The inequalities do not involve the index of the array and they could be improved in particular cases by considering the index number (Bose and Bush [1952], Bush [1952]).

Methods of constructing orthogonal arrays due to Rao [1946b], Bose and Bush [1952], Bush [1952], Seiden [1954], Addleman and Kempthorne [1961], Shrikhande [1964], Seiden and Zemach [1966], Shrikhande and Bhagwan Das [1970] and others are described in Raghavarao [1971].

The use of orthogonal arrays of strength d in the construction and analysis of multifactorial designs (fractional replicated designs) when interactions above a certain order are assumed away is discussed by Rao [1947], Box and Wilson [1951] and others.

3. Semi orthogonal array

Definition 2. A (N, r, s) array is said to be a *semi orthogonal array of strength d* if

$$n(i_1, \ldots, i_d) \leqslant 1 \tag{3.1}$$

for any combination of $i_1, \ldots, i_d$ $(i_j \in S)$ and any selection of d rows.

Lemma 2.1. *Let $n(i_1) = k$, constant for all $i_1 \in S$ and all rows. Then the following relationships hold among the parameters of a semi orthogonal array of strength d:*

$$N = ks, \tag{3.2}$$

$$r \leqslant \frac{(d-1)(N-1)}{(k-1)}. \tag{3.3}$$

The result (3.2) follows from the condition $n(i_1) = k$. To prove (3.3), let $a_1, \ldots, a_r$ be the symbols in the first column of (N, r, s). The symbol a_i in the ith row will appear in $(k-1)$ other columns. In view of (3.1), the number of rows in which any given column has the same symbol with the first column is less than $(d-1)$. Then we must have $(d-1)(N-1) \geqslant r(k-1)$, giving the inequality (3.3).

Lemma 2.2. *A semi orthogonal array of strength d with $n(i_1) = k$ and maximum number of constraints (i.e., when the equality* (3.3) *holds) implies the existence of a resolvable balanced incomplete block design* (RBIBD) *with the parameters,* $v = N, r, k, \lambda = d-1$.

We identify the columns with varieties. Then each row provides s blocks by considering the varieties (columns) with the same symbol as belonging to a block. It is easy to see that if the array is semi orthogonal of strength d, then the resulting arrangement in blocks formed from all the rows is a RBIBD. Similarly a RBIBD gives rise to a semi orthogonal array.

4. Balanced array

Chakravarti, in a thesis written at the Indian Statistical Institute, imposed a weaker combinatorial structure on an array by withdrawing the condition of constancy in the equation (2.1) but retaining some symmetry for the different values which $n(i_1, \ldots, i_d)$ can take (see Chakravarti [1956]).

Definition 3. A (N, r, s) array is said to be a *balanced array of strength d* if $n(i_1, \ldots, i_d)$ is constant for all permutations of $i_1, \ldots, i_d$, and for any selection of d rows.

Chakravarti called an array satisfying definition 3 as partially balanced while Srivastava renamed it as a balanced array. It may be seen that a balanced array is characterized by an index set $\lambda_1, \lambda_2, \ldots$, each λ_i depending on a particular combination of d elements with repetitions allowed. Balanced arrays satisfy the same properties as orthogonal arrays when used as fractional replicated factorial designs in terms of estimability of main effects and interactions, but the estimates of main effects and interactions may have different precisions besides being correlated. The construction and use of such designs have been indicated in Chakravarti [1956, 1961, 1963] and extensively investigated by Srivastava [1971], Srivastava and Anderson [1970], and Srivastava and Chopra [1971a,b,c] in the special case $s = 2$, i.e. S has two symbols 0 and 1. It may be noted that when $s = 2$, the function $n(i_1, \ldots, i_d)$ can take only $(d+1)$ distinct values corresponding to the $(d+1)$ distinct weights which the vector $(i_1, \ldots, i_d)$ can have, the weight of a vector being the number of unities in it.

5. Semi balanced array

Definition 4. A (N, r, s) array is said to be a *semi balanced array of strength* d if for any selection of d rows:

(i) $n(i_1, \ldots, i_d) = 0$ if any two i_j are equal, and

(ii) $\sum_P n(i_1, \ldots, i_d) = \lambda$ (constant),

where P represents summation over all permutations of distinct elements $i_1, \ldots, i_d$.

A semi balanced array of strength 2 is useful in the construction of some combinatorial arrangements. The minimum value of N in such a case is $s(s-1)/2$, with $\lambda = 1$. Lemma 5.1 gives the construction of $(s(s-1)/2, s, s, 2)$ which is semi balanced and has the maximum number of rows (Rao [1961a]).

Lemma 5.1. *Let s be an odd prime or an odd prime power in which case there exists a Galois field with s elements, GF(s). Then $(s(s-1)/2, s, s, 2)$ semi balanced array exists.*

Let $\alpha_0 = 0$, $\alpha_1, \ldots, \alpha_{s-1}$ be the elements of GF(s). Since s is odd, the elements can also be written as $\alpha_0 = 0$, $\beta_1, -\beta_1, \ldots, \beta_{(s-1)/2}, -\beta_{(s-1)/2}$. Consider the vectors

$$(\beta_i\alpha_0, \beta_i\alpha_1, \ldots, \beta_i\alpha_{s-1}), \qquad i = 1, \ldots, (s-1)/2. \tag{5.1}$$

The symmetric differences of elements occurring in the $(r+1)$th and the $(u+1)$th positions in (5.1) are

$$\beta_i(\alpha_r - \alpha_u), \quad -\beta_i(\alpha_r - \alpha_u), \qquad i = 1, \ldots, (s-1)/2.$$

Since $(\alpha_r - \alpha_u) \neq 0$, the differences (5.2) include all non-zero elements of GF(s) exactly once each. Hence by an application of the theorem of differences due to Bose [1939] it follows that the totality of sets

$$(\beta_i\alpha_0 + \alpha_j, \beta_i\alpha_1 + \alpha_j, \ldots, \beta_i\alpha_{s-1} + \alpha_j),$$
$$j = 0, 1, \ldots, s-1, \quad i = 1, 2, \ldots, (s-1)/2, \tag{5.3}$$

are such that in any two positions, all the combinations of s elements taken two at a time occur exactly once each. The vectors (5.3) written as columns provide the semi balanced array $(s(s-1)/2, s, s, 2)$.

For example, if $s = 5$, the residue classes $0, 1, \ldots, 4 \pmod 5$ form a field. We write the 5 elements of GF(5) as $0, \pm, 1\pm2$ and hence the key sets are, using the formula (5.1),

$$(0, 1, 2, 3, 4), \qquad (0, 2, 4, 1, 3), \tag{5.4}$$

where the second vector is obtained from the first on multiplying by 2. Writing (5.4) vertically (shown in blocks) and generating the other columns by the addition of elements of GF(5) as indicated in (5.3) we obtain the 10 columns

0	1	2	3	4
1	2	3	4	0
2	3	4	0	1
3	4	0	1	2
4	5	1	2	3

0	1	2	3	4
2	3	4	0	1
4	0	1	2	3
1	2	3	4	0
3	4	0	1	2

which is a semi balanced array $(10, 5, 5, 2)$. The same arrangement happens to be a semi balanced array $(10, 5, 5, 3)$ of strength 3. Considering only the first 3 rows, we obtain a semi balanced array $(10, 3, 5, 2)$, which has been used in deriving a cyclic solution to the resolvable balanced incomplete block design with the parameters

$$v = 36, \qquad b = 84, \qquad r = 14, \qquad k = 6, \qquad \lambda = 2$$

The actual design is given in Rao [1961b].

6. Partially balanced array

Definition 5. Let $T_1, \ldots, T_k$ be disjoint sets of selections of d rows from r rows of (N, r, s). Consider a selection of d rows belonging to a set T_j (say) and the corresponding $d \times N$ matrix. These d rows are said to be *balanced* if $n(i_1, \ldots, i_d)$ is invariant for permutations of $i_1, \ldots, i_d$, in which case we may denote its index set by $(\lambda_1^{(j)}, \lambda_2^{(j)}, \ldots)$. For a balanced array as in Definition 3, the index set is independent of j. Otherwise the array is said to be *partially balanced of strength d*.

Definition 5 as it stands is too general. We consider some special cases by suitable choices of $T_1, \ldots, T_k$ when $d = 2$.

We may impose a partially balanced incomplete block (PBIB) design structure on the rows of (N, r, s). Given any row, we select n_1 rows as its first associates, n_2 as second associates, and so on. The association scheme is subject to the restriction: for any two rows which are ith associates, there is a fixed number of rows common to jth associates of one and kth associates of the other. The set T_i in such a case consists of pairs of rows which are ith associates.

An example of such a partially balanced array of strength 2 and two symbols is the incidence matrix of any PBIB design.

Another type of partially balanced array is obtained by classifying the rows into k groups and considering the k^2 sets T_{ij}, $i, j = 1, \ldots, k$, with T_{ij} containing pairs of rows one from group i and the other from group j.

An example of such a partially balanced array of strength 2 with two symbols is the incidence matrix of an intra and inter group balanced (IIGB) design introduced by Nair and Rao [1942].

Partially balanced arrays can be used as fractional replicated designs but the analysis will be more complicated than in the case of balanced arrays.

7. Orthogonal arrays with a variable number of symbols in rows

Definition 6. A $(N, r, s_1 \times \cdots \times s_r)$ array is said to be an *orthogonal array of strength d* if the following condition is true. For any selection of d rows, say the α_1-th, . . ., α_d-th, which defines a $d \times N$ submatrix, $n(i_1, \ldots, i_d)$ is constant for all combinations $i_1 \in S_{\alpha 1}, \ldots, i_d \in S_{\alpha d}$; the constant may, however, depend on the set of selected rows.

Orthogonal arrays with a variable number of symbols in the rows seem to play an important role in the construction of asymmetrical multifactorial experiments and also in factorial experiments involving qualitative and quantitative factors. We discuss some methods of constructing such arrays.

7.1. Some inequalities

Consider an orthogonal array $(N, r, s_1 \times \cdots \times s_r, d)$ of strength d. Using arguments similar to those in Rao [1947, 1949], we obtain the inequalities

$$N-1 \geqslant P_1 + \cdots + P_t, \qquad \text{when } d = 2t, \tag{7.1}$$

and

$$N-1 \geqslant P_1 + \cdots + P_t + Q, \quad \text{when } d = 2t+1. \tag{7.2}$$

In (7.1) and (7.2),

$$P_k = \sum(s_{i_1} - 1) \cdots (s_{i_k} - 1),$$

where the summation is taken over all combinations of k elements out of $s_1, \ldots, s_r$ and $Q = \max Q_i$, where

$$Q_i = (s_i - 1)\sum(s_{i_1} - 1) \cdots (s_{i_t} - 1) \tag{7.3}$$

and the summation is taken over all combination of t elements out of $s_1, \ldots, s_r$ excluding s_i.

7.2. Construction of arrays of strength 2

An array $(N, r, p \times s^{r-1}, 2)$ of strength 2, where $p = s^k$, is of some interest in design of experiments. In such a case, the minimum value of N is s^{k+1}. Applying the inequality (7.1) we have

$$s^{k+1} - 1 \geqslant (s^k - 1) + (r-1)(s-1) \quad \text{or} \quad s^k + 1 \geqslant r, \tag{7.4}$$

so that the maximum number of rows is $s^k + 1$.

To construct such an array with the maximum value of r equal to $s^k + 1$, we proceed as follows: First construct an orthogonal array

$$\left(s^{k+1}, \frac{s^{k+1} - 1}{s-1}, s, 2\right) \tag{7.5}$$

of strength 2 with the same number of s symbols in each row. Choose k rows of this array such that the $k \times s^{k+1}$ submatrix is of strength k. Retain these k initial rows and only these, from the rest each of which together with the k initial rows provides a $(k+1) \times s^{k+1}$ submatrix of strength $(k+1)$. There are exactly s^k rows among the rest which satisfy this condition. The next step is to replace the k initial rows by a single row by attaching to each combination of k elements (k-vector) a particular symbol. There are s^k different vectors each occurring s times in the initial k rows which provide a row with entries belonging to a set of s^k symbols. This row together with the s^k rows already chosen (from the rows excluding the initial k) give rise to the array

$$(s^{k+1}, s^k + 1, p \times s^k, 2), \, p = s^k. \tag{7.6}$$

As an example let us consider the orthogonal array $(2^3, 7, 2, 2)$ of strength 2 given in Table 1. If we omit row 4, then each of the rows 3, 5, 6, 7 forms with rows 1 and 2 an array of strength 3. Writing a for $\binom{1}{1}$, b for $\binom{1}{0}$, c for $\binom{0}{1}$ and d for $\binom{0}{0}$ in the first two rows and retaining rows 3, 5, 6, 7, we obtain an

TABLE 1

Row no.	(8, 7, 2, 2) orthogonal array							
1	1	1	1	1	0	0	0	0
2	1	1	0	0	1	1	0	0
3	1	0	1	0	1	0	1	0
4	1	1	0	0	0	0	1	1
5	1	0	0	1	1	0	0	1
6	1	0	1	0	0	1	0	1
7	1	0	0	1	0	1	1	0

TABLE 2

Row no.	(8, 5, 4×2^4, 2) orthogonal array							
1	a	a	b	b	c	c	d	d
2	1	0	1	0	1	0	1	0
3	1	0	0	1	1	0	0	1
4	1	0	1	0	0	1	0	1
5	1	0	0	1	0	1	1	0

orthogonal array of strength 2 with 5 rows, the first row containing 4 symbols and others only 2 symbols. We denote such an array given in Table 2 by $(8, 5, 4 \times 2^4, 2)$, where 4×2^4 indicates that there is one row with 4 symbols and 4 rows with 2 symbols each.

From Table 2, by omitting the first two columns we obtain the array $(6, 5, 3 \times 2^4, 2)$ given in Table 3 which is of strength 2 and is only partially balanced. For instance, the pair of rows consisting of the first and any other is orthogonal while a pair of two rows which do not include the first row is only balanced.

TABLE 3

Row no	$(6, 5, 3 \times 2^4, 2)$ partially balanced array					
1	b	b	c	c	d	d
2	1	0	1	0	1	0
3	0	1	1	0	0	1
4	1	0	0	1	0	1
5	0	1	0	1	1	0

Thus a variety of arrays can be generated from the orthogonal array of Table 1, all of which can be used as designs for factorial experiments.

Similarly, from the orthogonal array $(16, 15, 2, 2)$ one can construct the orthogonal array $(16, 9, 8 \times 2^8, 2)$ and partially balanced arrays of various types.

7.3. Practical applications

W. J. Youden suggested an experiment which uses an array of a type different from those discussed in Section 7.2 as a design. There were three qualitative factors (two operators K, Y; two burette fillings F_1, F_2; two halves (top and bottom) of the burette, T, B) and a quantitative factor (involving the measurement of volume). The design and the response measurements were as in Table 4.

Youden remarked: "The eight samples are suitable not only for fitting a straight line so that the slope may be used as a check on the method but furnish orthogonal comparisons of operators, burette fillings, and the two halves of the burette".

Youden's design is extremely interesting and does provide orthogonal comparisons under the assumption that the regression is linear. However, a design based on an array such as the one given in Table 2 is more flexible

TABLE 4

Sample no.	Combination of qualitative factors			Volume of liquid	Response wt. of BaSO$_4$
1	F_1	K	T	9	41.8
2	F_1	K	T	16	77.4
3	F_1	Y	B	10	46.7
4	F_1	Y	B	15	70.2
5	F_2	Y	T	11	50.6
6	F_2	Y	T	14	66.7
7	F_2	K	B	12	56.9
8	F_2	K	B	13	63.1

and furnishes orthogonal comparisons for a maximum of 4 qualitative factors (A, B, C, D) and also linear, quadratic and cubic components of the regression of the response variable on an independent variable which can be controlled at four different values a, b, c and d. The analysis of variance for the design based on $(8, 5, 4 \times 2^4, 2)$ is given in Table 5. However, there are no degrees of freedom for the estimation of error unless the quadratic and cubic components are negligible or there is an independent estimate of error as in certain chemical and physical determinations.

TABLE 5

	Main effects				Regression on independent variable			Total
	A	B	C	D	linear	quadratic	cubic	
Due to D.F.	1	1	1	1	1	1	1	7

The partially balanced array $(6, 4, 3 \times 2^3, 2)$ constructed in Table 3 can be used after omitting the last row as a multifactorial design with 3 qualitative factors each at two levels and a quantitative factor which can be controlled at 3 values. The design allows the estimation of main effects of the qualitative factors, and the linear and quadratic coefficients in the regression of the response variable on an independent variable (quantitative factor). The estimates of main effects are no longer orthogonal to each other but are orthogonal to estimates of regression coefficients. It may be noted that if all the five rows of Table 3 are used in a design with 4 qualitative factors and one quantitative factor, all the main effects would not be separately estimable.

The array $(8, 5, 4 \times 2^4, 2)$ of Table 2 can also be used as a fractional replicated 4×2^4 asymmetrical factorial design (with five factors) using only 8 assemblies. From such a design we have orthogonal estimates of main effects of all the factors under the assumption that the interactions are absent.

Similarly one can use $(16, 9, 8 \times 2^8, 2)$ in a variety of ways. As a fractional replicated 8×2^8 asymmetrical factorial design, involving 9 factors, one at 8 levels and the rest at two levels each, it provides orthogonal estimates of main effects. It can also be used as a multifactorial design with 8 factors at 2 levels each and a quantitative factor which may have a multivariate structure to enable multiple regression analysis. Thus, if we wish to study the regression of the response variable on two independent variables, independently of the effects of as many as 8 qualitative factors, we need only associate with each of the 8 symbols in the first row a pair of values (x, y) which the independent variables can assume. The 15 degrees of freedom in such a case can be decomposed into 8 degrees of freedom one for each main effect of the qualitative factors and 7 degrees of freedom for regression coefficients and error.

References

S. Addleman and O. Kempthorne, 1961, Some main effects plan and orthogonal arrays of strength two, *Ann. Math. Statist.* **32**, 1167–1176.

R. Berlekamp, 1968, *Algebraic Coding Theory* (McGraw-Hill, New York).

R. C. Bose, 1939, On the construction of balanced incomplete block designs, *Ann. Eugenics (London)* **9**, 353–399.

R. C. Bose and K. A. Bush, 1952, Orthogonal arrays of strength two and three, *Ann. Math. Statist.* **23**, 508–524.

R. C. Bose, S. S. Shrikhande and E. T. Parker, 1960, Further results on the construction of mutually orthogonal Latin squares and the falsity of Euler's conjecture, *Canad. J. Math.* **12**, 189–203.

G. E. P. Box and K. B. Wilson, 1951, On the experimental attainment of optimum conditions, *J. Roy. Statist. Soc.* B **13**, 1–45.

K. A. Bush, 1952, Orthogonal arrays of index unity, *Ann. Math. Statist.* **23**, 426–434.

I. M. Chakravarti, 1956, Fractional replication in asymmetrical factorial designs and partially balanced arrays, *Sankhya* **17**, 143–164.

I. M. Chakravarti, 1961, On some methods of construction of partially balanced arrays, *Ann. Math. Statist.* **32**, 1181–1185.

I. M. Chakravarti, 1963, Orthogonal arrays and partially balanced arrays and their applications in design of experiments, *Metrika* **7**, 231–243.

D. V. Chopra and J. N. Srivastava, 1971, Optimal balanced resolution V factorial designs of the 2^7 series, $29 \leqslant N \leqslant 42$ (submitted).

K. R. Nair and C. R. Rao, 1942, Incomplete block designs for experiments involving several groups of varieties, *Sci. Cult. (Calcutta)* **7**, 625.

K. R. Nair and C. R. Rao, 1948, Confounding in asymmetrical factorial experiments, *J. Roy. Statist. Soc.* B **10**, 109–131.

D. Raghavarao, 1971, *Construction and Combinatorial Problems in Design of Experiments* (Wiley, New York).

C. Radhakrishna Rao, 1946a, Difference sets and combinatorial arrangements derivable from finite geometries, *Proc. Natl. Inst. Sci.* **12**, 123–135.

C. Radhakrishna Rao, 1946b, Hypercubes of strength d leading to confounded designs in factorial experiments, *Bull. Calcutta Math. Soc.* **38**, 67–78.

C. Radhakrishna Rao, 1947, Factorial experiments derivable from combinatorial arrangements of arrays, *J. Roy. Statist. Soc., Suppl.* **9**, 128–139.

C. Radhakrishna Rao, 1949, On a class of arrangements, *Edinburgh Math. Soc.* **8**, 119–125.

C. Radhakrishna Rao, 1950, The theory of fractional replication in factorial experiments, *Sankhya* **10**, 229–256.

C. Radhakrishna Rao, 1961a, Combinatorial arrangements analogous to orthogonal arrays, *Sankhya* **23**, 283–286.

C. Radhakrishna Rao, 1961b, A combinatorial assignment problem, *Nature*, 100.

D. K. Ray-Chaudhuri and R. M. Wilson, 1971a, Solution of Kirkman's school girl problem, *Proc. Symp. in Pure Mathematics, Combinatorics, Am. Math. Soc.* **19** 187–204.

D. K. Ray-Chaudhuri and R. M. Wilson, 1971b, Existence of resolvable block designs, *Proc. Int. Symp. on Combinatorial Mathematics and its Applications*.

E. Seiden, 1954, On the problems of construction of orthogonal arrays, *Ann. Math. Statist.* **25**, 151–156.

E. Seiden and R. Zemach, 1966, On orthogonal arrays, *Ann. Math. Statist.* **37**, 1355–1370.

S. S. Shrikhande, 1964, Generalized Hadamard matrices and orthogonal arrays of strength 2, *Canad. J. Math.* **16**, 736–740.

S. S. Shrikhande and Das Bhagwan, 1969, A note on embedding of orthogonal arrays of strength 2, *Combinatorial Mathematics and its Applications* (University of North Carolina Press, Chapel Hill, N. Car.), pp. 256–273.

S. S. Shrikhande and Das Bhagwan, 1970, A note on embedding of Hadamard matrices, *Essays in Probability and Statistics* (University of North Carolina Press, Chapel Hill, N. Car.), pp. 673–688.

J. N. Srivastava, 1972, Some general existence conditions for balanced arrays of strength t and 2 symbols, *J. Combin. Theory*.

J. N. Srivastava and D. A. Anderson, 1970, Optimal fractional factorial plans for main effects orthogonal to two-factor interactions; 2^m series, *J. Am. Statist. Assoc.* **65**, 828–843.

J. N. Srivastava and D. V. Chopra, 1971a, On the characteristic roots of the information matrix for balanced fractional 2^m factorial designs of resolution V, with applications, *Ann. Math. Statist.* **42**, 722–736.

J. N. Srivastava and D. V. Chopra, 1971b, Some new results in the combinatorial theory of balanced arrays of strength four with $2 \leqslant \mu_2 \leqslant 6$ (A.R.L. Technical Report 71–0072).

J. N. Srivastava and D. V. Chopra, 1971c, Optimal balanced 2^m factorial designs of resolution V, $m \leqslant 6$, *Technometrics* **13**, 257–269.

J. N. Srivastava et al., eds., *A Survey of Combinatorial Theory*
© North-Holland Publishing Company, 1973

CHAPTER 30

The Existence of Resolvable Block Designs

D. K. RAY-CHAUDHURI† and RICHARD M. WILSON†

Department of Mathematics, Ohio State University, Columbus, Ohio 43210, U.S.A.

1. Introduction

Let v and λ be positive integers and K a set (finite or infinite) of positive integers. A (v, K, λ)-*pairwise balanced design* (PBD) is a pair $(X, \mathscr{B})$ where X is a set of *points*, $\mathscr{B} = \{B_i : i \in I\}$ is a family of subsets of X, called *blocks*, such that

(i) $|X| = v$,

(ii) $2 \leqslant |B_i| \in K$, for every $i \in I$,

(iii) the number of indices $i \in I$ for which $\{x, x'\} \subseteq B_i$ is exactly λ, for each pair $x, x' \in X$, $x \neq x'$.

Here, $|S|$ denotes the cardinality of a finite set S. When $K = \{k\}$ for a positive integer k, a (v, K, λ)-PBD is traditionally called a (v, k, λ)-*balanced incomplete block design* (bibd).

A subfamily $\mathscr{B}' = \{B_i : i \in I'\}$ of the blocks $\mathscr{B} = \{B_i : i \in I\}$ of a (v, K, λ)-PBD $(X, \mathscr{B})$ is said to be a *parallel class* of blocks iff for each point $x \in X$, there is precisely one index $i \in I'$ with $x \in B_i$, i.e., $\mathscr{B}'$ is a partition of X. A (v, k, λ)-bibd $(X, \mathscr{B})$ with $v > 1$ is said to be *resolvable* iff there exists a partition

$$I = I_1 \cup I_2 \cup \cdots \cup I_r$$

such that each subfamily $\mathscr{B}_j = \{B_i : i \in I_j\}$ is a parallel class, $j = 1, 2, \ldots, r$.

The following is the simplest non-trivial bibd (here $v = 7$, $k = 3$, $\lambda = 1$):

$$X = \{1, 2, 3, 4, 5, 6, 7\},$$

$B_1 = \{1, 2, 4\}, \quad B_2 = \{2, 3, 5\}, \quad B_3 = \{3, 4, 6\}, \quad B_4 = \{4, 5, 7\},$
$B_5 = \{5, 6, 1\}, \quad B_6 = \{6, 7, 2\}, \quad B_7 = \{7, 1, 3\}.$

The following is the simplest non-trivial resolvable bibd (here $v = 9$, $k = 3$, $\lambda = 1$, and there are $r = 4$ parallel classes):

$$X = \{1, 2, 3, 4, 5, 6, 7, 8, 9\}$$

$\mathscr{B}_1$	$\mathscr{B}_2$	$\mathscr{B}_3$	$\mathscr{B}_4$
$B_1 = \{1, 2, 3\}$	$B_4 = \{1, 4, 7\}$	$B_7 = \{1, 5, 9\}$	$B_{10} = \{1, 6, 8\}$
$B_2 = \{4, 5, 6\}$	$B_5 = \{2, 5, 8\}$	$B_8 = \{2, 6, 7\}$	$B_{11} = \{2, 4, 9\}$
$B_3 = \{7, 8, 9\}$	$B_6 = \{3, 6, 9\}$	$B_9 = \{3, 4, 8\}$	$B_{12} = \{3, 5, 7\}$

† This research was supported in part by the NSF Research Grant No. GP–9375.

Note that when $\lambda = 1$, it is impossible that $B_i = B_{i'}$, for distinct indices $i \neq i'$ in a $(v, K, 1)$-pairwise balanced design $(X, \mathscr{B})$; and thus $\mathscr{B}$ is a set of blocks, rather than a family of blocks.

We shall be concerned with the question of the characterization of those parameter triples $(v, k, 1)$ for which resolvable $(v, k, 1)$-bibd's exist. We offer the following partial answer, to be made precise in the next section: Let $k \geqslant 2$ be given. Then, with finitely many exceptions, resolvable $(v, k, 1)$-bibd's exist whenever it is not "unreasonable".

Before we proceed to the general discussion, we consider the case $k = 2$. Given a set X of v points, there is a unique $(v, 2, 1)$-bibd $(X, \mathscr{B})$; namely, where $\mathscr{B}$ is the class of all subsets of X with cardinality 2. It is unreasonable to expect that this design is always resolvable, for the existence of a single parallel class of blocks of size 2 implies that $v \equiv 0 \pmod 2$. But this necessary condition, that v be even, is in fact sufficient as can be seen from the following construction.

Suppose $v = 2t+2$, $t \geqslant 0$. Take $X = Z_{2t+1} \cup \{\infty\}$ where Z_{2t+1} is the cyclic (additive) group of residues modulo $2t+1$. For each $j \in Z_{2t+1}$, let

$$\mathscr{B}_j = \{\{\infty, j\}, \{j+1, j-1\}, \{j+2, j-2\}, \ldots, \{j+t, j-t\}\}.$$

Then each $\mathscr{B}_j$ is a parallel class and $(X, \mathscr{B}_0 \cup \mathscr{B}_1 \cup \cdots \cup \mathscr{B}_{2t})$ is the $(v, 2, 1)$-bibd.

The $(v, 2, 1)$-bibd is thus resolvable if and only if v is even. The above construction is given by Kraitchik [1953] as the solution of a puzzle similar to the Kirkman school girl problem $(k = 3)$.

We state some necessary conditions, historical remarks, and our main result in Section 2. The proof of the main theorem comprises Sections 3 and 4. We give an application of this result to the construction of bibd's with "maximum" subdesigns in Section 5 and close with the statement of some unsolved problems in Section 6.

2. Necessary conditions and some historical remarks

Let v, k, λ be positive integers, $k \geqslant 2$, and let $(X, \mathscr{B})$ be a (v, k, λ)-bibd. It is well known that in such a bibd every point is contained in $r = \lambda(v-1)/(k-1)$ blocks and the total number of blocks is $b = \lambda v(v-1)/k(k-1)$. Since r and b must be integers, we have the necessary conditions

$$\lambda(v-1) \equiv 0 \pmod{k-1}$$
$$\lambda v(v-1) \equiv 0 \pmod{k(k-1)}$$

(1)

for the existence of a (v, k, λ)-bibd.

The necessary conditions on v for the existence of a $(v, 3, 1)$-bibd reduce to $v \equiv 1$ or $3 \pmod 6$. The question of whether this is also sufficient was raised by Steiner [1853]. Therefore $(v, 3, 1)$-bibd's are called Steiner triple systems. However, the problem had been considered by Kirkman [1847] and also solved by him in the same paper. Independent solutions were given by Reiss [1859] and Moore [1893]. In the twentieth century, the first significant contributions to the problem of existence of bibd were made by Bose [1939] and Hanani. Bose introduced the difference method which enabled one to construct several infinite families of (v, k, λ)-bibd's. Bose and Shrikhande [1960] and Hanani [1961] introduced the composition method which composes several designs of small order into a design of large order. Hanani proved that the necessary conditions (1) are also sufficient for $k = 3$ and 4. For $k = 5$, Hanani showed that the condition is sufficient except for $(v, k, \lambda) = (15, 5, 2)$ in which case the design does not exist (Hanani [1961, 1965]). R. M. Wilson in his Ph.D. dissertation proved that if k is a prime power, then the necessary conditions (1) are "asymptotically sufficient". More precisely, and considering only $\lambda = 1$ for simplicity, for each k there exists a constant $C(k)$ such that if $v \geqslant C(k)$ and $v \equiv 1$ or $k \pmod{k(k-1)}$, then a $(v, k, 1)$-bibd exists. If k is a prime power, the conditions (1) with $\lambda = 1$ reduce to $v \equiv 1$ or $k \pmod{k(k-1)}$.

If a (v, k, λ)-resolvable bibd exists, clearly we must also have

$$v \equiv 0 \pmod{k}. \tag{2}$$

For $\lambda = 1$, the conditions (1) and (2) reduce to $v \equiv k \pmod{k(k-1)}$. In this paper we prove that this necessary condition is "asymptotically sufficient" for the existence of $(v, k, 1)$-resolvable bibd's:

Main Theorem. *Given $k \geqslant 2$, there exists a constant $C(k)$ such that if $v \geqslant C(k)$ and $v \equiv k \pmod{k(k-1)}$, then a $(v, k, 1)$-resolvable bibd exists.*

The proof is given in Section 4.

The first introduction of resolvable bibd was as Query 6 on page 48 of the *Lady's and Gentleman's Diary of 1850* by Rev. Thomas P. Kirkman.

Query 6. *Fifteen young ladies in a school walk out three abreast for seven days in succession: it is required to arrange them daily, so that no two will walk twice abreast.*

Clearly, such an arrangement can be derived from a $(15, 3, 1)$-resolvable bibd if we identify the fifteen young ladies with the fifteen points of the design. The seven parallel classes correspond to the seven days of the week. Cayley [1850] had a solution for this problem.

Kirkman had solutions for $(9, 3, 1)$ and $(15, 3, 1)$-resolvable bibds. Kirkman in his paper [1850] mentions that Rev. James Mease, A.M. of Freshford, Kilkenny, sent him a solution for a $(27, 3, 1)$-resolvable bibd and that he generalized the method to construct $(3^m, 3, 1)$-resolvable bibd's. Kirkman in his paper [1850] without the benefit of the knowledge of projective and affine geometry constructed $(q^m, q, 1)$-resolvable bibd's for q a prime. The necessary condition for the existence of a $(v, 3, 1)$-resolvable bibd is $v \equiv 3 \pmod{6}$.

A large number of papers had been written on the subject of $(v, 3, 1)$-resolvable bibd. But the existence problem was not settled until 1968. The present authors (Ray-Chaudhuri and Wilson [1971]) proved that a $(v, 3, 1)$-resolvable bibd exists iff $v \equiv 3 \pmod 6$. A bibliography of the papers on $(v, 3, 1)$-resolvable bibd's (up to 1912) can be found in Eckenstein [1912].

Recently, the present authors and H. Hanani proved that the necessary conditions (1) and (2) are also sufficient in the case $k = 4$ (Hanani et al. [to appear]): $(v, 4, 1)$-resolvable bibd's exist iff $v \equiv 4 \pmod{12}$.

But it is known that the necessary condition $v \equiv k \pmod{k(k-1)}$ is not always sufficient. While $v = k^2$ satisfies this congruence, the Bruck-Ryser Theorem (see Hall [1967]) shows that there are infinitely many integers k for which $(k^2, k, 1)$-resolvable bibd's (finite affine planes of order k) do not exist, the smallest of which is $k = 6$. We conjecture, however, that $v \equiv 5 \pmod{20}$ is sufficient for the existence of a $(v, 5, 1)$-resolvable bibd.

3. Completed designs and some preliminary results

Given an integer $k \geqslant 2$, define two sets R_k and R_k^* of positive integers by

$$R_k = \{r \colon ((k-1)r+1, k, 1)\text{-bibd exists}\},$$
$$R_k^* = \{r \colon ((k-1)r+1, k, 1)\text{-resolvable bibd exists}\}.$$

Note that the necessary condition (2) implies that $r \equiv 1 \pmod k$ for every $r \in R_k^*$ and that our main theorem is equivalent to the assertion of the existence of a constant $C^*(k)$ such that

$$\{r \colon r \geqslant C^*(k), r \equiv 1 \pmod k\} \subseteq R_k^*.$$

Let k and t be positive integers, $k \geqslant 2$. By a (k, t)-*completed resolvable design* we mean a $(kt+1, \{k+1, t\}, 1)$-pairwise balanced design $(X, \mathscr{B})$ in which there is precisely one block of size t and the other blocks have size $k+1$ (unless $t = 1$ or $t = k+1$, where all blocks are to have size $k+1$). The block of size t will be called the *block at infinity*. (If $t = k+1$, any block may be chosen as the block at infinity; if $t = 1$, the block at infinity is to be some singleton subset of X.) The relation of resolvable designs to completed resolvable designs is the same as that of finite affine planes to finite projective planes.

Lemma 1. $t \in R_k^*$ *iff a* (k, t)-*completed resolvable design exists.*

Proof. Assume that $t \in R_k^*$ and let $(X, \mathscr{B})$ be a $((k-1)t+1, k, 1)$-resolvable bibd. To construct a (k, t)-completed resolvable design, for the point set we take $X^* = X \cup \{\theta_1, \theta_2, \ldots, \theta_t\}$. Let $\mathscr{B}_i$, $i = 1, 2, \ldots, t$, be the t parallel classes of the resolvable bibd. For each $B \in \mathscr{B}_i$, define $B^* = B \cup \{\theta_i\}$. Let $B_\infty = \{\theta_1, \theta_2, \ldots, \theta_t\}$ and define $\mathscr{B}^* = \{B^* \colon B \in \mathscr{B}\} \cup \{B_\infty\}$. Then it is easily checked that $(X^*, \mathscr{B}^*)$ is a (k, t)-completed resolvable design.

Conversely, assume that $(X^*, \mathscr{B}^*)$ is a (k, t)-completed resolvable design. From the definition of a completed resolvable design, $\mathscr{B}^*$ contains a block $B_\infty = \{\theta_1, \theta_2, \ldots, \theta_t\}$ of size t. Let x be a point other than θ_i, $i = 1, 2, \ldots, t$. Then there is a unique block $B_i \in \mathscr{B}^*$ of size $k+1$ containing both x and θ_i, $i = 1, 2, \ldots, t$. For different choices of i, the sets $B_i - \{x\}$ are disjoint and each consists of k elements. Thus $|\bigcup_{i=1}^{t} B_i| = kt+1 = |X^*|$, so there can be no other blocks containing x. This argument shows that every block B^* of $\mathscr{B}^*$, other than B_∞, intersects B_∞ in precisely one point. To construct a $((k-1)t+1, k, 1)$-resolvable bibd, for the point set take $X = X^* - B_\infty$. For each block $B^* \in \mathscr{B}^*$, $B^* \neq B_\infty$, define $B = B(B^*) = B^* - (B^* \cap B_\infty)$ and put $\mathscr{B} = \{B : B^* \in \mathscr{B}^*, B^* \neq B_\infty\}$. Clearly, $(X, \mathscr{B})$ is a $((k-1)t+1, k, 1)$-bibd. Let $\mathscr{B}_i^* = \{B^* : \theta_i \in B^*, B^* \neq B_\infty\}$ and let $\mathscr{B}_i = \{B : B^* \in \mathscr{B}_i^*\}$ for each $i = 1, 2, \ldots, t$. Then we see that $\mathscr{B}_1, \mathscr{B}_2, \ldots, \mathscr{B}_t$ are t parallel classes, and this completes the proof of the lemma.

Given a set K (finite or infinite) of positive integers, let

$$\mathbf{B}(K) = \{v > 0 : (v, K, 1)\text{-PBD exists}\}.$$

The mapping $K \to \mathbf{B}(K)$ is a closure operation on the subsets of the positive integers, i.e.,

   (i) $K \subseteq \mathbf{B}(K)$,
   (ii) $K_1 \subseteq K_2$ implies $\mathbf{B}(K_1) \subseteq \mathbf{B}(K_2)$,
   (iii) $\mathbf{B}(\mathbf{B}(K)) = \mathbf{B}(K)$.

The first property is easily seen since given $k \geqslant 2$, $k \in K$, the pair $(X, \{X\})$, where $|X| = k$, is a $(k, K, 1)$-PBD and hence $k \in \mathbf{B}(K)$. This shows that all of K, with the possible exception of 1, is contained in $\mathbf{B}(K)$. (Whether or not $1 \in K$ is irrelevant as far as the set $\mathbf{B}(K)$ is concerned.) But we always have $1 \in \mathbf{B}(K)$ because of the trivial $(1, K, 1)$-PBD $(\{x\}, \emptyset)$.

Property (ii) is immediate.

To prove property (iii), it suffices to show that $\mathbf{B}(\mathbf{B}(K)) \subseteq \mathbf{B}(K)$, for property (i) gives the opposite inclusion. Given $v \in \mathbf{B}(\mathbf{B}(K))$, there exists a $(v, \mathbf{B}(K), 1)$-PBD $(X, \mathscr{A})$. Then for each $A \in \mathscr{A}$, $|A| \in \mathbf{B}(K)$, and hence there exists an $(|A|, K, 1)$-PBD $(A, \mathscr{B}_A)$. With $\mathscr{B} = \bigcup_{A \in \mathscr{A}} \mathscr{B}_A$, it is easily checked that $(X, \mathscr{B})$ is a $(v, k, 1)$-PBD and hence $v \in \mathbf{B}(K)$.

A set K of positive integers is said to be a *closed set* iff $\mathbf{B}(K) = K$. The set of all positive integers, and the set $\{1\}$, are trivial examples of closed sets. For any set J, $K = \mathbf{B}(J)$ is a closed set (the *closure* of J).

Hanani [1961] has shown that the set R_k is always a closed set (also see Wilson [1972a]). We have a similar result for R_k^*.

Theorem 1. $\mathbf{B}(R_k^*) = R_k^*$, i.e. R_k^* is a closed set.

Proof. As remarked above, we always have $K \subseteq \mathbf{B}(K)$ and in particular,

$R_k^* \subseteq \mathbf{B}(R_k^*)$. Now let $v \in \mathbf{B}(R_k^*)$ be given. That is, there exists a PBD $(X, \mathscr{S})$ on v points such that $|S| \in R_k^*$, for all $S \in \mathscr{S}$. To prove the theorem, because of Lemma 1 it will be sufficient to construct a (k, v)-completed resolvable design. For the point set we take $X^* = X \times I_k \cup \{\theta\}$. Since for each $S \in \mathscr{S}$ we have $|S| \in R_k^*$, there exists by Lemma 1 a $(k, |S|)$-completed resolvable design. Construct such a design on the point set $S \times I_k \cup \{\theta\}$. Write $S = \{a_1, a_2, \ldots, a_t\}$. Without loss of generality, we can take the block at infinity to be $\{(a_1, 1), (a_2, 1), \ldots, (a_t, 1)\}$, and the blocks through θ to be $\{\theta, (a_i, 1), (a_i, 2), \ldots, (a_i, k)\}$, $i = 1, 2, \ldots, t$. In other words, we assume the following configuration:

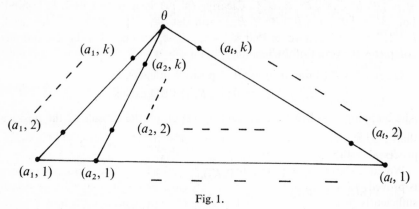

Fig. 1.

By $\mathscr{B}_s$ we denote the remaining blocks of the $(k, |S|)$-completed resolvable design on $S \times I_k \cup \{\theta\}$. This is done for each block $S \in \mathscr{S}$.

Let $B_\infty = \{(a, 1) : a \in X\}$ and let $\mathscr{B}_0$ be the class of blocks

$$\{\theta, (a, 1), (a, 2), \ldots, (a, k)\}, \qquad a \in X.$$

Put $\mathscr{B}^* = (\bigcup_{S \in \mathscr{S}} \mathscr{B}_S) \cup \{B_\infty\} \cup \mathscr{B}_0$. We now check that $(X^*, \mathscr{B}^*)$ is the desired (k, v)-completed resolvable design. The pairs of points of the type $\{(a, i), (a', j)\}$, $a \neq a'$, $i \neq j$, occur in one block of $\mathscr{B}_S$ where S is the unique block containing a and a'. The pairs of the type $\{(a, i), (a, j)\}$ occur in a block of $\mathscr{B}_0$. Pairs of the type $\{(a, i), (a', 1)\}$ occur only in B_∞, and pairs of the type $\{(a, i), (a', i)\}$, $i \neq 1$, occur in one block of $\mathscr{B}_S$, where S is the block of $\mathscr{S}$ containing a and a'. Finally, all pairs of the type $\{\theta, (a, i)\}$ occur only in one block of $\mathscr{B}_0$. This completes the proof of the theorem.

4. Special constructions; proof of the main theorem

The assertion of Theorem 1, that R_k^* is a closed set, already yields much information about the structure of R_k^* in view of the following Theorem A, proved in Wilson [1972a].

Let K be a subset of the positive integers, finite or infinite. Let $\beta(K)$ denote the greatest common divisor of the integers $k(k-1), k \in K$ (the non-negative generator of the ideal in the ring of integers which is generated by $\{k(k-1) : k \in K\}$). By a *fiber* of K we mean a residue class f modulo $\beta(K)$ for which there exists at least one $k \in K$ with $k \equiv f \pmod{\beta(K)}$. A fiber f of K is said to be *eventually complete* iff there exists a constant $C = C(f)$ such that

$$\{v : v \geqslant C, v \equiv f \pmod{\beta(K)}\} \subseteq K.$$

Theorem A. *Let K be a closed set. Then every fiber of K is eventually complete.*

In Wilson [1972a], this property of closed sets K is expressed by asserting that K is "eventually periodic with period $\beta(K)$".

We shall prove the following theorem in this section.

Theorem 2. *Let $k \geqslant 2$ be given.*

(i) *If k is even, then $\beta(R_k^*) = k$ and the residue class 1 modulo k is a fiber of R_k^*.*

(ii) *If k is odd, then $\beta(R_k^*) = 2k$ and the residue classes 1 and $k+1$ modulo $2k$ are fibers of R_k^*.*

Of course, 1 is surely a fiber of R_k^* for the simple reason that $1 \in R_k^*$. But we shall exhibit infinitely many elements $v \in R_k^*$ with $v \equiv 1 \pmod{\beta(R_k^*)}$ in the process of establishing Theorem 2.

We have shown in Section 3 that R_k^* is a closed set. Thus once Theorem 2 is proved, an application of Theorem A yields the result that R_k^* contains all sufficiently large integers $r \equiv 1 \pmod{k}$ (in both cases k even and k odd). This proves our main theorem:

Theorem 3. *Let $k \geqslant 2$ be given. $\mathbf{B}^*(k)$ contains all sufficiently large integers $v \equiv k \pmod{k(k-1)}$.*

Here, and in what follows, let

$$\mathbf{B}^*(k) = \{v > 1 : (v, k, 1)\text{-resolvable bibd exists}\}.$$

The remainder of this section is devoted to the proof of Theorem 2. Our first step is to construct infinitely many resolvable bibd's with the aid of finite fields. We prove

Lemma 2. *Let $k \geqslant 2$ be given and let t be an integer such that*

$$q = k(k-1)t+1$$

is a power of a prime number and

$$q > \{\tfrac{1}{2}k(k-1)\}^{k(k+1)}.$$

Then $kq \in \mathbf{B}^*(k)$, or equivalently, $k^2 t + 1 \in R_k^*$.

For the proof of Lemma 2 we require a result (Theorem B below, proved in Wilson [1972b]) concerning the existence of certain "configurations" in finite fields.

Let q be a prime power and e a divisor of $q-1$. The cyclic multiplicative group H of the Galois field $GF(q)$ with q elements has a unique subgroup H_0 of index e, namely the group of eth powers of non-zero field elements. If we write $q = ef+1$ and let ω be a primitive element of $GF(q)$, then

$$H_0 = \{\omega^{ei} : i = 0, 1, 2, \ldots, f-1\}.$$

We define the multiplicative cosets $H_0, H_1, H_2, \ldots, H_{e-1}$ of H_0 in H by

$$H_j = \omega^j H_0 = \{\omega^{ei+j} : i = 0, 1, \ldots, f-1\}.$$

Theorem B. *Let $q = ef+1$ be a prime power and r a positive integer. Suppose that $q > e^{r(r-1)}$. Then for any choice of a mapping C assigning to each ordered pair (i, j) of integers with $1 \leqslant i < j \leqslant r$ a multiplicative coset*

$$C(i, j) \in \{H_0, H_1, \ldots, H_{e-1}\}$$

modulo the group H_0 of e-th powers in $GF(q)$, there exists an r-tuple $(a_1, a_2, \ldots, a_r)$ of field elements such that $a_j - a_i \in C(i, j)$ for every pair (i, j), $1 \leqslant i < j \leqslant r$.

Proof of Lemma 2. Let $q = k(k-1)t+1$ be a prime power with

$$q > \{\tfrac{1}{2}k(k-1)\}^{k(k+1)}.$$

We shall apply Theorem B with $e = \tfrac{1}{2}k(k-1)$ and $r = k+1$. Define a choice $C : \{(i, j) : 0 \leqslant i < j \leqslant k\} \to \{H_0, H_1, \ldots, H_{e-1}\}$ in any way such that the $e = \tfrac{1}{2}k(k-1)$ cosets $C(i, j)$, $1 \leqslant i < j \leqslant k$, are, in some order, precisely the cosets $H_0, H_1, \ldots, H_{e-1}$, and such that the cosets $C(0, j)$, $1 \leqslant j \leqslant k$, are distinct. (This cannot be done if $k = 2$, but the assertion of the Lemma is trivial in this case.) By Theorem B, there exists a $(k+1)$-tuple $(a_0, a_1, \ldots, a_k)$ of elements of $GF(q)$ such that $a_j - a_i \in C(i, j)$, $0 \leqslant i < j \leqslant k$.

Put $b_1 = a_1 - a_0, b_2 = a_2 - a_0, \ldots, b_k = a_k - a_0$. Then

$$\{b_j - b_i : 1 \leqslant i < j \leqslant k\}$$

is a system of representatives for the $e = \tfrac{1}{2}k(k-1)$ cosets $H_0, H_1, \ldots, H_{e-1}$, and $b_1, b_2, \ldots, b_k$ lie in distinct cosets modulo H_0. The latter assertion shows that the sets in the union

$$S = \bigcup_{j=1}^{k} \{\omega^{em}b_j : m = 0, 1, \ldots, t-1\}$$

are disjoint, i.e., S consists of kt distinct field elements. Then $U = GF(q) - S$ contains $q - kt$ elements.

We construct a $(qk, k, 1)$-resolvable bibd on the set of points $X = GF(q) \times \{1, 2, \ldots, k\}$. The construction is an application of the "method of mixed differences" introduced by R. C. Bose [1939] (also see Hall [1967]).

For $u, c \in GF(q)$ and $A \subseteq X$, say $A = \{(c_1, l_1), (c_2, l_2), \ldots, (c_n, l_n)\}$, $c_i \in GF(q), l_i \in \{1, 2, \ldots, k\}$, we denote by $uA + c$ the subset

$$uA + c = \{(uc_1 + c, l_1), (uc_2 + c, l_2), \ldots, (uc_n + c, l_n)\}$$

of X. A special case of the method of mixed differences consists of the observation that, given "base blocks" $A_1, A_2, \ldots, A_m \subseteq X$, the pair

$$\left(X, \bigcup_{i=1}^{m} \{A_i + c : c \in GF(q)\}\right)$$

is a pairwise balanced design iff

(i) for every $l \in \{1, 2, \ldots, k\}$, the list of differences $a-b$, where (a, l) and (b, l) occur together in some base block A_i, contains every non-zero element of $GF(q)$ precisely once, and

(ii) for each $l_1, l_2 \in \{1, 2, \ldots, k\}$, $l_1 \neq l_2$, the list of differences $a-b$, where (a, l_1) and (b, l_2) occur together in some base block A_i, contains every element of $GF(q)$ precisely once.

This is easily verified.

To continue with our construction, define

$$B^* = \{(b_1, 1), (b_2, 2), \ldots, (b_k, k)\},$$
$$B_1 = \{(b_1, 1), (b_2, 1), \ldots, (b_k, 1)\},$$
$$B_2 = \{(b_1, 2), (b_2, 2), \ldots, (b_k, 2)\},$$
$$\vdots$$
$$B_k = \{(b_1, k), (b_2, k), \ldots, (b_k, k)\},$$

where $b_1, b_2, \ldots, b_k$ are as above. We take as base blocks the collection

$$uB^* \qquad (u \in GF(q)),$$
$$(\omega^{em}b_1)B_1 \qquad (m = 0, 1, \ldots, t-1),$$
$$(\omega^{em}b_2)B_2 \qquad (m = 0, 1, \ldots, t-1),$$
$$\vdots$$
$$(\omega^{em}b_k)B_k \qquad (m = 0, 1, \ldots, t-1).$$

For $l_1, l_2 \in \{1, 2, \ldots, k\}$, $l_1 \neq l_2$, the list of differences $a-b$, where $\{(a, l_1), (b, l_2)\}$ is contained in a base block, is $u(b_{l_1} - b_{l_2})$, $u \in GF(q)$, and clearly each element of $GF(q)$ is obtained precisely once. For $l \in \{1, 2, \ldots, k\}$, the list of differences $a-b$, where $\{(a, l), (b, l)\}$ is contained in a base block, is

$$\omega^{em}b_l d_{ij}, \qquad m = 0, 1, \ldots, t-1, \quad i,j = 1, 2, \ldots, k,$$

where $d_{ij} = b_j - b_i$. Since $d_{ij} = -d_{ji}$ and $\omega^{et} = -1$, the non-zero elements of this list are

$$\omega^{em}b_l d_{ij}, \qquad m = 0, 1, \ldots, 2t-1, \quad 1 \leqslant i \leqslant j \leqslant k,$$

or

$$\bigcup_{1 \leqslant i < j \leqslant k} b_l d_{ij} H_0.$$

Since $\{d_{ij} : 1 \leqslant i < j \leqslant k\}$ is a system of representatives for the cosets of H_0, so is $\{b_l d_{ij} : 1 \leqslant i < j \leqslant k\}$ and it is now clear that every nonzero element of $GF(q)$ is obtained precisely once as such a difference.

This shows that the system $(X, \mathscr{A})$, where

$$\mathscr{A} = \bigcup_{j=1}^{k} \{(\omega^{em}b_j)B_j + c : c \in GF(q), m = 0, 1, \ldots, t-1\}$$
$$\cup \{uB^* + c : u, c \in GF(q)\},$$

is a $(qk, k, 1)$-bibd. To complete the proof of Lemma 2, it remains to show

13

that this design is resolvable. This can be done as follows. For each $c \in \mathrm{GF}(q)$, let

$$\mathscr{A}_c = \bigcup_{j=1}^{k} \{(\omega^{em}b_j)B_j + c : m = 0, 1, \ldots, t-1\} \cup \{uB^* + c : u \in U\},$$

and for each $u \in S$, let

$$\mathscr{B}_u = \{uB^* + c : c \in \mathrm{GF}(q)\}.$$

Then

$$\mathscr{A} = \left(\bigcup_{c \in GF(q)} \mathscr{A}_c \right) \cup \left(\bigcup_{u \in S} \mathscr{B}_u \right)$$

is a partition of $\mathscr{A}$, and it is not hard to verify that each $\mathscr{A}_c$ and $\mathscr{B}_u$ is a parallel class on X.

The simplest example of the construction of Lemma 2 occurs when $k = 3$, $q = 7$ (so $t = e = 1$). Here we may take $b_1 = 2$, $b_2 = 3$, $b_3 = 5$ in GF(7). Our construction produces the base blocks

$$0B^* = \{0_1, 0_2, 0_3\},$$

$$1B^* = \{2_1, 3_2, 5_3\}, \quad 2B^* = \{4_1, 6_2, 3_3\}, \quad 3B^* = \{6_1, 2_2, 1_3\},$$

$$4B^* = \{1_1, 5_2, 6_3\}, \quad 5B^* = \{3_1, 1_2, 4_3\}, \quad 6B^* = \{5_1, 4_2, 2_3\},$$

$$2B_1 = \{4_1, 6_1, 3_1\}, \quad 3B_2 = \{6_2, 2_2, 1_2\}, \quad 5B_3 = \{3_3, 1_3, 4_3\},$$

for a (21, 3, 1)-resolvable bibd on the set of points $\mathrm{GF}(7) \times \{1, 2, 3\}$, where we have written c_i for the ordered pair (c, i). The blocks $\{0B^*, 1B^*, 4B^*, 6B^*, 2B_1, 3B_2, 5B_3\}$ form a parallel class and seven parallel classes are obtained by developing this basic one. Each of the remaining base blocks $2B^*, 3B^*, 5B^*$, when developed modulo 7, produces an additional parallel class for a total of ten classes.

As another example, take $k = 5$, $q = 41$ (so $t = 2$, $e = 10$). The subset $\{b_1, b_2, b_3, b_4, b_5\} = \{1, 10, 16, 18, 37\}$ has the properties that the elements lie in distinct cosets modulo $H_0 = \{\omega^0, \omega^{10}, \omega^{20}, \omega^{30}\}$ and $\{b_j - b_i : 1 \leqslant i < j \leqslant 5\}$ is a system of representatives for the 10 cosets of H_0. This may be verified with the aid of the tables of Oklahoma Math. Tables Project [1962]. The construction of Lemma 2 produces a (205, 5, 1)-resolvable bibd.

We require one further construction for the proof of Theorem 2 (Lemma 3 below). This is a special case of Theorem 3 of Ray-Chaudhuri and Wilson [1971].

A *transversal design* with k groups of size n, in brief a TD(k, n), is a triple $(X, \mathscr{G}, \mathscr{A})$, where X is a set of kn points, $\mathscr{G} = \{G_1, G_2, \ldots, G_k\}$ is a partition of X into k subsets G_i (called *groups*) each consisting of n points, and $\mathscr{A}$ is a class of subsets of X (called *blocks* or *transversals*) each of which meets each group G_i in precisely one point (and so consists of k points) and such that any pair x, y of elements of X which do not lie in the same group are contained together in precisely one of the blocks $A \in \mathscr{A}$.

It is well known that the existence of a TD(k, n) is equivalent to the existence of $k-2$ mutually orthogonal Latin squares of order n (Bose et al. [1960]; Hall [1967]; Wilson [1971]). And thus by a result of Chowla et al. [1960] (also see Hall [1967]), there exists a constant L_k such that TD(k, n)'s exist whenever $n > L_k$ (we may take $L_k = (k+2)^{17}$ for sufficiently large k as shown in Wilson [1971]).

Lemma 3. Let $k \geqslant 2$ be given. If $v \in \mathbf{B}^*(k)$ and $v > L_{k+1}$, then $kv \in \mathbf{B}^*(k)$. Hence, if $r \in R_k^*$ is sufficiently large, then $kr+1 \in R_k^*$.

Proof. Let $v \in \mathbf{B}^*(k)$ be given with $v > L_{k+1}$ and let $(X, \mathcal{G}, \mathcal{A})$ be a TD$(k+1, v)$, where $\mathcal{G} = \{G_0, G_1, G_2, \ldots, G_k\}$. We construct a $(kv, k, 1)$-resolvable bibd on the set of points $X^* = G_1 \cup G_2 \cup \cdots \cup G_k$.

For each i, $1 \leqslant i \leqslant k$, construct a $(v, k, 1)$-resolvable bibd $(G_i, \mathcal{B}^{(i)})$ on the set of points G_i with $\mathcal{B}^{(i)} = \mathcal{B}_1^{(i)} \cup \mathcal{B}_2^{(i)} \cup \cdots \cup \mathcal{B}_r^{(i)}$, say, a partition of $\mathcal{B}^{(i)}$ into parallel classes (here $r = (v-1)/(k-1)$). Let $\mathcal{B}_j = \bigcup_{i=1}^{k} \mathcal{B}_j^{(i)}$ for $1 \leqslant j \leqslant r$ and note that each $\mathcal{B}_j$ is a parallel class on X^*. For each point $z \in G_0$, let $\mathcal{A}_z = \{A \cap X^* : z \in A, A \in \mathcal{A}\}$ and note that each $\mathcal{A}_z$ is a parallel class of blocks of size k on X^*.

Then $(X^*, \mathcal{B})$, where $\mathcal{B} = (\bigcup_{j=1}^{r} \mathcal{B}_j) \cup (\bigcup_{z \in G_0} \mathcal{A}_z)$, is a $(kv, k, 1)$-resolvable bibd and thus $kv = (k-1)(kr+1)+1 \in \mathbf{B}^*(k)$.

Proof of Theorem 2. Let $k \geqslant 2$ be given, and put $\beta = \beta(R_k^*)$. We claim that

$$\beta = \begin{cases} k & \text{if } k \text{ is even,} \\ 2k & \text{if } k \text{ is odd.} \end{cases}$$

Firstly, recall that $r \equiv 1 \pmod{k}$ for all $r \in R_k^*$, so k is a common divisor of all $r(r-1)$, $r \in R_k^*$. Thus k divides $\beta = \text{g.c.d.}\{r(r-1) : r \in R_k^*\}$. Clearly, 2 is a common divisor of $\{r(r-1) : r \in R_k^*\}$, so 2 divides β; thus if k is odd, $2k$ must divide β.

Now define the integer z by $\beta = zk$. We have just seen that z is even if k is odd, and our claim reduces to the assertion that

$$z = 1 \quad \text{if } k \text{ is even,}$$
$$z = 2 \quad \text{if } k \text{ is odd.}$$

By Dirichlet's famous theorem on primes in arithmetic progressions, there exist infinitely many integers m for which $q = k(k-1)zm+1$ is prime. With the exception of finitely many of these integers m, Lemma 2 asserts that $k^2zm+1 \in R_k^*$ and then Lemma 3 shows that

$$k(k^2zm+1)+1 = k^3zm+k+1 \in R_k^*.$$

So for at least one integer m, $k^3zm+k+1 \in R_k^*$ and hence β must be a divisor of $(k^3zm+k+1)(k^3zm+k)$, which is congruent to $k(k+1)$ modulo $\beta = zk$. Thus β divides $k(k+1)$, or equivalently, z is a divisor of $k+1$.

Suppose, for contradiction, that z has an odd prime divisor p. Since z divides $k+1$, p is relatively prime to $k(k-1)$ and hence $p^n \equiv 1 \pmod{k(k-1)}$

for all integers n divisible by $\phi(k(k-1))$ (Euler's phi-function). If such an n is chosen sufficiently large, we write $p^n = k(k-1)t+1$ and apply Lemma 2 to conclude that $k^2t+1 \in R_k^*$. Then $k^2t(k^2t+1)$ is divisible by β and hence by p. But p divides neither k nor t nor $k^2t+1 = p^n+kt$, a contradiction. Thus z has no odd prime divisors.

If k is even, the proof is complete; for we know that z divides $k+1$, which is odd, whence $z = 1$.

Assume now that k is odd. The integers $4k(k-1)$ and $2k(k-1)+1$ are relatively prime, so by Dirichlet's theorem there exist arbitrarily large primes $q = 4k(k-1)t+(2k(k-1)+1)$. By Lemma 2, for some such t, $k^2(4t+2)+1 \in R_k^*$. Then β, and hence z, divides $k^2(4t+2)[k^2(4t+2)+1]$. Then surely z cannot be divisible by 4, whence $z = 2$.

It remains only to show that $k+1$ is a fiber of R_k^* when k is odd. The integers $2k(k-1)$ and $k(k-1)+1$ are relatively prime and again by Dirichlet's theorem there are infinitely many primes $q = 2k(k-1)t+(k(k-1)+1)$. For such q's which are sufficiently large, Lemma 2 asserts $k^2(2t+1)+1 \in R_k^*$. We note that $k^2(2t+1)+1 \equiv k+1 \pmod{2k}$, when k is odd, and the proof is complete.

5. An application: maximum subdesigns

Let $(X, \mathcal{A})$ and $(Y, \mathcal{B})$ be pairwise balanced designs. If $X \subseteq Y$ and $\mathcal{A} \subseteq \mathcal{B}$, we say that $(X, \mathcal{A})$ is a *subdesign* of $(Y, \mathcal{B})$, or that $(X, \mathcal{A})$ is *embedded* in $(Y, \mathcal{B})$.

Now assume that $(X, \mathcal{A})$ is a $(u, k, 1)$-bibd, $(Y, \mathcal{B})$ is a $(v, k, 1)$-bibd, and that $(X, \mathcal{A})$ is a subdesign of $(Y, \mathcal{B})$. Clearly $v \geqslant u$, but if we assume that $v > u$, then a simple observation shows that

$$v \geqslant (k-1)u+1.$$

To derive this, select a point $y \in Y$, $y \notin X$. For each of the u points $x \in X$, there is a block $B_x \in \mathcal{B}$ such that $\{y, x\} \subseteq B_x$, and the blocks B_x, $B_{x'}$ $(x, x' \in X, x \neq x')$ are distinct. Thus the number $r = (v-1)/(k-1)$ of blocks in $\mathcal{B}$ containing the point y cannot be less than u, and the stated inequality follows.

It is known (Doyen and Wilson [to appear]) in the case $k = 3$, that given u, $v \in \mathbf{B}(3)$, the condition $v \geqslant 2u+1$ is in fact sufficient for the existence of a Steiner triple system of order v containing a Steiner triple system of order u as a subdesign. It is also known (Wilson [1972a]) that given any $u \in \mathbf{B}(k)$, there exists a constant $C = C(k, u)$ such that for every $v \in \mathbf{B}(k)$ with $v \geqslant C$, there exists a $(u, k, 1)$-bibd containing a $(u, k, 1)$-bibd as a subdesign.

It is relevant to notice that subdesigns can be "unplugged" and replaced. That is, if $(X, \mathcal{A})$ is a subdesign of $(Y, \mathcal{B})$ and $(X, \mathcal{C})$ is any pairwise balanced design, then $(Y, (\mathcal{B}-\mathcal{A}) \cup \mathcal{C})$ is a pairwise balanced design containing $(X, \mathcal{C})$

as a subdesign. Hence, if there exists a single $(v, k, 1)$-bibd containing a subdesign of order u, then every $(u, k, 1)$-bibd can be embedded into a $(v, k, 1)$-bibd.

Let us consider now the extreme case in the inequality $v \geqslant (k-1)u+1$, when $v = (k-1)u+1$. A somewhat well known construction (Bose and Shrikhande [1960]; Wilson [1972a]) shows that when $k-1$ is a prime power, any $(u, k, 1)$-bibd can be embedded in a $(v, k, 1)$-bibd with $v = (k-1)u+1$. This construction makes use of a projective plane of order $k-1$ and hence cannot be immediately generalized. However, the statement remains asymptotically true:

Theorem 4. Let $k \geqslant 2$ be given. There exists a constant $C = C(k)$ such that whenever $u \geqslant C$, any $(u, k, 1)$-bibd can be embedded in a $(v, k, 1)$-bibd with $v = (k-1)u+1$.

Proof. We take C to be a constant such that R_{k-1}^* contains all integers $r \geqslant C$ with $r \equiv 1 \pmod{k-1}$. The existence of C is established by Theorem 3.

Now let $(X, \mathscr{A})$ be a $(u, k, 1)$-bibd with $u \geqslant C$. Since $u \equiv 1 \pmod{k-1}$, we have $u \in R_{k-1}^*$ and hence, by Lemma 1, there exists a $(k-1, u)$-completed resolvable design $(Y, \mathscr{B})$. $\mathscr{B}$ contains a block B_∞ of size u, all other blocks of $\mathscr{B}$ have size k, and $|Y| = (k-1)u+1$. We may assume that $X \subseteq Y$, and even $X = B_\infty$. With this understanding, $(Y, (\mathscr{B}-\{B_\infty\}) \cup \mathscr{A})$ is a $((k-1)u+1, k, 1)$-bibd containing $(X, \mathscr{A})$ as a subdesign.

6. Generalizations and unsolved problems

Kirkman in his paper, "Note on an unanswered prize question" (Kirkman [1850]) proposed a very interesting generalization of the school girl problem. Let v, k and t be positive integers. Let X be a set of v elements. Let $\mathscr{A}$ be a class of subsets (blocks) of X. The pair $(X, \mathscr{A})$ is said to be a *partial t-design* on v points with blocks of size k iff every block of $\mathscr{A}$ has cardinality k and any t-subset of X is contained in at most one block. Kirkman's general problem can be called the problem of packing parallel classes of blocks in a partial t-design. Let r be the largest integer for which there exist disjoint parallel classes $\mathscr{A}_1, \mathscr{A}_2, \ldots, \mathscr{A}_r$ such that $(X, \mathscr{A}_1 \cup \mathscr{A}_2 \cup \cdots \cup \mathscr{A}_r)$ is a partial t-design on v points with blocks of size k. Clearly, if $r \geqslant 1$, then $v \equiv 0 \pmod{k}$. On the other hand, it is easy to see that the number r of parallel classes can be at most $\binom{v-1}{t-1}/\binom{k-1}{t-1}$. It is quite easy to propose a much more general problem, of packing of t'-designs into a partial t-design $(1 \leqslant t' \leqslant t)$. However, Kirkman's problem of packing parallel classes (i.e., 1-designs) into a partial t-design is already an extremely difficult problem. Essentially nothing is known in the case $t \geqslant 3$. But we wish to call attention to a particular case which admits some hope for a reasonable answer.

Let $t \geqslant 2$ be given and let $P_t(X)$ denote the collection of all t-subsets of a set X (subsets having cardinality t). Suppose that the cardinality of X is divisible by t. Question: Is the trivial t-design $(X, P_t(X))$ always resolvable? That is, does there exist a partition $P_t(X) = \mathscr{A}_1 \cup \mathscr{A}_2 \cup \cdots \cup \mathscr{A}_r$ into parallel classes $\mathscr{A}_i$ (here, necessarily, $r = \binom{v-1}{t-1}$ where $v = |X|$)?

For $t = 2$, the question is answered affirmatively by the construction of Section 1. There remains the problem of determining the number of such resolutions and/or the number of non-isomorphic resolutions. We remark that it is not in general possible to extend a set $\mathscr{A}_1, \mathscr{A}_2, \ldots, \mathscr{A}_m$ ($\mathscr{A}_i \subseteq P_2(X)$, $m < |X| - 1$) of parallel classes (also called perfect matchings) to a full resolution of $P_2(X)$.

We might note that the above question is trivial when $|X| = t$ and also when $|X| = 2t$, where the unique solution is obtained by pairing each t-subset of X with its complement.

The case $t = 3$ is of particular historical interest. On consideration of Kirkman's original Fifteen Schoolgirls Problem, Sylvester noted that it was possible to send the school girls out walking in triples for each day of the quarter (one quarter = 13 weeks = 91 days) in such a way that any 3-subset of the school girls would appear as a triple precisely once. (This is reported by Cayley [1850]. A solution also appears in Kirkman [1850]). That is, for $t = 3$, $|X| = 15$, the answer to our question is yes. It is perhaps not inappropriate to call such resolutions of $P_3(X)$ *Sylvester systems*.

References

R. C. Bose, 1939, On the construction of balanced incomplete block designs, *Ann. Eugen.* **9**, 353–399.

R. C. Bose and S. S. Shrikhande, 1960, On the composition of balanced incomplete block designs, *Canad. J. Math.* **12**, 177–188.

R. C. Bose, S. S. Shrikhande and E. T. Parker, 1960, Further results on the construction of mutually orthogonal Latin squares and the falsity of a conjecture of Euler, *Canad. J. Math.* **12**, 189–203.

A. Cayley, 1850, On the triadic arrangements of seven and fifteen things, *London, Edinburgh and Dublin Phil. Mag. and J. Sci.* **37**, 279–292.

S. Chowla, P. Erdös and E. G. Straus, 1960, On the maximal number of pairwise orthogonal Latin squares of a given order, *Canad. J. Math.* **12**, 204–208.

J. Doyen and R. M. Wilson, Embeddings of Steiner triple systems, *Discrete Math.* **5** (to appear).

O. Eckenstein, 1912, Bibliography of Kirkman's school girl problem, *Messenger of Math.* **41–42**, 33–36.

M. Hall, Jr., 1967, *Combinatorial Theory* (Blaisdell, Waltham, Mass.).

H. Hanani, 1961, The existence and construction of balanced incomplete block designs, *Ann. Math. Statist.* **32**, 361–386.

H. Hanani, 1965, A balanced incomplete block design, *Ann. Math. Statist.* **36**, 711.

H. Hanani, D. K. Ray-Chaudhuri and Richard M. Wilson, 1972, On resolvable designs, *Discrete Math.* **3**, 343–357.

T. P. Kirkman, 1847, On a problem in combinations, *Cambridge and Dublin, Math. J.* **2**, 191–204.

T. P. Kirkman, 1850, Note on an unanswered prize question, *Cambridge and Dublin Math. J.* **5**, 255–262.

M. Kraitchik, 1953, *Mathematical Recreations* (second revised edition; Dover, New York), p. 226.

E. H. Moore, 1893, Concerning triple systems, *Math. Ann.* **43**, 271–285.

D. K. Ray-Chaudhuri and R. M. Wilson, 1971, Solution of Kirkman's school girl problem, *Proc. Symp. in Pure Mathematics, Combinatorics, Am. Math. Soc.* **19**, 187–204.

D. K. Ray-Chaudhuri and R. M. Wilson, On the existence of resolvable balanced incomplete block designs, *Combinatorial Structures and Their Applications* (Gordon and Breach, New York), pp. 331–341.

M. Reiss, 1859, Uber eine Steinersche combinatorische Aufgabe welche . . ., *Crelle's J. Reine Angew Math.* **56**, 326–344.

J. Steiner, 1853, Combinatorische Aufgabe, *Crelle's J. Reine Angew. Math.* **45**, 181–182.

Univ. of Oklahoma Math. Tables Project, 1962, *Table of Indices and Power Residues* (Norton, New York).

R. M. Wilson, 1971, Concerning the number of orthogonal Latin squares, Manuscript, Ohio State University, Department of Mathematics, Columbus, Ohio.

R. M. Wilson, 1972a, An existence theory for pairwise balanced designs, I and II, *J. Combin. Theory* (to appear).

R. M. Wilson, 1972b, Cyclotomy and difference families in elementary abelian groups, *J. Number Theory*, **4**, 17–47.

J. N. Srivastava et al., eds., *A Survey of Combinatorial Theory*
© North-Holland Publishing Company, 1973

Variants of (v, k, λ)-Designs †

H. J. RYSER

California Institute of Technology, Pasadena, Calif., U.S.A.

Summary

The present paper surveys the recent literature on new types of combinatorial designs that are closely related to the classical (v, k, λ)-designs.

1. Introduction

Let $X = \{x_1, x_2, \ldots, x_v\}$ be a set of v elements (a v-set). Subsets $X_1, X_2, \ldots,$ X_v of X are called a (v, k, λ)-*design* (*symmetric block design*) provided:

Each X_i is a k-subset of X. (1.1)

Each $X_i \cap X_j$ for $i \neq j$ is a λ-subset of X. (1.2)

The integers v, k, and λ satisfy $0 < \lambda < k < v-1$. (1.3)

Now let $X = \{x_1, x_2, \ldots, x_m\}$ be an m-set and let $X_1, X_2, \ldots, X_n$ be subsets of X. We set $a_{ij} = 1$ if x_i is a member of X_j and we set $a_{ij} = 0$ if x_i is not a member of X_j. The resulting $(0, 1)$-matrix

$$A = [a_{ij}] \tag{1.4}$$

of size m by n is called the *incidence matrix* for the subsets $X_1, X_2, \ldots, X_n$ of X. It is clear that A characterizes the configuration of subsets. We use the following notation throughout: The matrix A^T denotes the transpose of the matrix A. The matrix I is the identity matrix of a specified order and the matrix J is the matrix of 1's of a specified order. Incidence matrices provide a remarkably efficient scheme for describing combinatorial phenomena. Thus, for example, properties (1.1) and (1.2) of a (v, k, λ)-design are characterized by a $(0, 1)$-matrix A of order $m = n = v$ that satisfies the matrix equation

$$A^\mathrm{T}A = (k-\lambda)I+\lambda J. \tag{1.5}$$

The symmetric block designs and the more general block designs on the parameters b, v, r, k and λ play a basic role in the development of modern combinatorics. We make no attempt to summarize their extensive literature

† This research was supported in part by the Army Research Office, Durham under Grant DA–ARO–D–31–124–G1138 and the Office of Naval Research under Contract N00014–67–A–0094–0010.

here. Such summaries are available, for example, in Dembowski [1968], Hall [1967], and Ryser [1963]. Instead, we discuss "variants" of (v, k, λ)-designs. These new configurations turn out to be of considerable interest in their own right. In this paper we place special emphasis on a number of intriguing questions that remain unanswered. All of the variants discussed are conveniently described in terms of incidence matrices that satisfy modified matrix equations of the general form (1.5).

2. Extensions of a theorem of De Bruijn and Erdös

Let X be an m-set and let $X_1, X_2, \ldots, X_n$ be subsets of X. Suppose that each X_i and X_j with $i \neq j$ intersect in exactly one element of X. We further require that both n and the number of elements in each X_j be greater than one. Then a theorem of De Bruijn and Erdös [1948] asserts that $m \geq n$. Moreover, in the case of equality, $m = n$, the configurations are the finite projective planes (and the degenerate plane or triangle), or else we have $n > 3$ and we may label the elements and the subsets so that

$$X_1 = \{2, 3, \ldots, n\}, \quad X_2 = \{1, 2\}, \quad X_3 = \{1, 3\}, \ldots, X_n = \{1, n\}.$$

Recently, Ryser [1968] and Woodall [1970] attempted to extend the De Bruijn-Erdös theorem so that each X_i and X_j with $i \neq j$ intersect in exactly λ elements of X, where λ is a fixed but arbitrary positive integer. The following is one of their main conclusions. (A *line* of a matrix designates either a row or a column of the matrix.)

Theorem 2.1. *Let A be a $(0, 1)$-matrix of size m by n such that*

$$A^{\mathrm{T}}A = \mathrm{diag}\,[k_1 - \lambda, k_2 - \lambda, \ldots, k_n - \lambda] + \lambda J, \tag{2.1}$$

where $n > 1$ and $k_j > \lambda \geq 1$. Then

$$m \geq n, \tag{2.2}$$

and if equality holds in (2.2) then A satisfies one of the following two requirements:

(X) *Each line sum of A equals a positive integer k.*

(Y) *The matrix A has exactly two distinct row sums r_1 and r_2 and these numbers satisfy*

$$r_1 + r_2 = n + 1. \tag{2.3}$$

We remark that the inequality (2.2) is due to Majumdar [1953] and its proof is based on a modification of the simple proof by Bose [1949] of the Fisher inequality. The case of equality in (2.2) raises a number of important questions. The configurations associated with (X) are precisely the (v, k, λ)-designs with $m = n = v$ (apart from unimportant degeneracies that are entirely matters of definition). The configurations associated with (Y) are called λ-*designs* on n elements.

The following theorem of Ryser [1968] suggests that the λ-designs with $\lambda > 1$ may be much more difficult to tabulate than the λ-designs with $\lambda = 1$.

Theorem 2.2. *A λ-design with $\lambda = 2$ exists only for $n = 7$ and in this case the unique configuration is given by*

$$\{1, 2, 4\}, \{1, 4, 6, 7\}, \{2, 5, 7, 1\}, \{3, 6, 1, 2\},$$
$$\{4, 7, 2, 3\}, \{5, 1, 3, 4\}, \{6, 2, 4, 5\}.$$

Woodall [1970] has shown that the number of λ-designs for each fixed value of $\lambda > 1$ is finite. The corresponding assertion for (v, k, λ)-designs remains a difficult unsolved problem. Let a (v, k, λ')-design have parameters not of the form $v = 4\lambda - 1, k = 2\lambda - 1, \lambda' = \lambda - 1$. In this case an elementary but basic modification of the (v, k, λ')-design allows one to construct a λ-design with $\lambda = k - \lambda'$ and row sums $v - k$ and $k + 1$. Bridges [1970] has called the λ-designs constructed in this prescribed manner (including the λ-designs with $\lambda = 1$) *type 1 λ-designs*. All known λ-designs are of type 1. The main conjecture in the study of λ-designs is that all λ-designs are of type 1. The combined efforts of Bridges [1970], Bridges and Kramer [1970], and Kramer [1969] have verified the validity of this conjecture for $\lambda \leqq 9$. But the possibility of the existence of fascinating new λ-designs not of type 1 remains open.

3. The matrix equation $XY = (k - \lambda)I + \lambda J$

The study of λ-designs was motivated by a modification of the right side of the matrix equation (1.5). Bridges and Ryser [1969] have carried out investigations that involve a modification of the left side of (1.5). Specifically, they prove the following theorem.

Theorem 3.1. *Let X and Y be nonnegative integral matrices of order $n > 1$ such that*

$$XY = (k - \lambda)I + \lambda J, \tag{3.1}$$

where $k \neq \lambda$ and the integers k and λ are relatively prime. Then there exist positive integers r and s such that X has constant line sums r and Y has constant line sums s, where

$$rs = k + (n - 1)\lambda. \tag{3.2}$$

Moreover,

$$YX = XY. \tag{3.3}$$

We remark that the matrix equation (3.1) has a direct set theoretic interpretation. In fact, certain special solutions of (3.1) yield interesting combinatorial configurations that are strongly regular graphs. In this connection, we mention the investigations of Goethals and Seidel [1967] and Seidel [1968].

It is of interest to note that the arithmetical requirement $(k, \lambda) = 1$ in Theorem 3.1 is a necessary part of the hypothesis of the theorem. Thus let

A_1 be the incidence matrix of a (v_1, k_1, λ_1)-design on the Hadamard parameters $v_1 = 4t-1$, $k_1 = 2t-1$, $\lambda_1 = t-1$, $t > 1$ and define

$$A = \begin{bmatrix} t & 0 \cdots 0 \\ \hline t-1 & \\ \cdot & \\ \cdot & A_1 \\ \cdot & \\ t-1 & \end{bmatrix}. \tag{3.4}$$

Then A is a nonnegative integral matrix that satisfies the matrix equation

$$AA^{\mathrm{T}} = tI + t(t-1)J. \tag{3.5}$$

But A does not have constant line sums and $AA^{\mathrm{T}} \neq A^{\mathrm{T}}A$.

The following corollaries imply that Theorem 3.1 gives new insight into the structure of certain (v, k, λ)-designs.

Corollary 3.2. *Let* X *and* Y *be nonnegative integral matrices of order* n^2+n+1 $(n > 1)$ *such that*

$$XY = nI + J, \tag{3.6}$$

and let this factorization be proper in the sense that neither X *nor* Y *is a permutation matrix. Then* $X = Y^T$ *is the incidence matrix of a projective plane of order* n.

Corollary 3.3. *Let the incidence matrix* A *of a* (v, k, λ)-*design be written as a product of* $(0, 1)$-*matrices of order* v

$$A = \prod_{i=1}^{t} A_i \qquad (t > 1), \tag{3.7}$$

where $k > \lambda^2$ *and the integers* k *and* λ *are relatively prime. Then* $t-1$ *of the factors of* A *are permutation matrices and the remaining factor* A_j *is of the form* PAQ, *where* P *and* Q *are permutation matrices.*

The hypothesis in Corollary 3.2 that requires the matrices X and Y to be nonnegative integral is very essential. There is in fact a well known integral matrix A of order 111 that satisfies the matrix equation

$$AA^T = 10I + J. \tag{3.8}$$

But the matrix A is many stages removed as a possible candidate for the incidence matrix of a projective plane of order 10 (see, for example, Hall [1967] and Ryser [1963]). Concerning Corollary 3.3 we remark that no incidence matrix of a (v, k, λ)-design is known that factors "properly" into a product of $(0, 1)$-matrices.

4. Multiplicative designs

Let A be a $(0, 1)$-matrix of order $n \geq 3$ that satisfies the matrix equation

$$A^{\mathrm{T}}A = \mathrm{diag}\,[k_1-\lambda_1, k_2-\lambda_2, \ldots, k_n-\lambda_n] + [\sqrt{\lambda_i}\sqrt{\lambda_j}], \tag{4.1}$$

where $k_i - \lambda_i$ and λ_j are positive. A configuration whose incidence matrix A fulfills these requirements is called a *multiplicative design* on the parameters $k_1, k_2, \ldots, k_n$ and $\lambda_1, \lambda_2, \ldots, \lambda_n$. Our definition of a multiplicative design places heavy restrictions on the parameters $k_1, k_2, \ldots, k_n$ and $\lambda_1, \lambda_2, \ldots, \lambda_n$. But multiplicative designs may be regarded as a natural generalization of (v, k, λ)-designs and λ-designs. The basic generalization involves the replacement of the matrix λJ by a symmetric matrix of rank 1. Ryser [1972] has investigated various properties of multiplicative designs. For example, one may prove that the parameters $k_1, k_2, \ldots, k_n$ and $\lambda_1, \lambda_2, \ldots, \lambda_n$ of a multiplicative design satisfy

$$\left[\frac{k_1^2}{k_1 - \lambda_1} + \cdots + \frac{k_n^2}{k_n - \lambda_n} - n \right] \left[1 + \frac{\lambda_1}{k_1 - \lambda_1} + \cdots + \frac{\lambda_n}{k_n - \lambda_n} \right]$$
$$= \left[\frac{\sqrt{\lambda_1}}{k_1 - \lambda_1} k_1 + \cdots + \frac{\sqrt{\lambda_n}}{k_n - \lambda_n} k_n \right]^2. \quad (4.2)$$

A multiplicative design on the parameters $k_1, k_2, \ldots, k_n$ and $\lambda_1, \lambda_2, \ldots, \lambda_n$ is called a *uniform design* provided

$$k_1 - \lambda_1 = k_2 - \lambda_2 = \cdots = k_n - \lambda_n \equiv c. \quad (4.3)$$

Ryser [1972] established the following duality theorem for uniform designs.

Theorem 4.1. *Let A be the incidence matrix of a uniform design on the parameters $k_1, k_2, \ldots, k_n$ and $\lambda_1, \lambda_2, \ldots, \lambda_n$. Then A^T is also the incidence matrix of a uniform design and satisfies the matrix equation*

$$AA^T = cI + ct[x_i x_j], \quad (4.4)$$

where

$$t = 1 + c^{-1}(\lambda_1 + \cdots + \lambda_n), \quad (4.5)$$

$$tx_i = c^{-1}(\sqrt{\lambda_1} a_{i1} + \cdots + \sqrt{\lambda_n} a_{in}). \quad (4.6)$$

We next give an example of an interesting family of uniform designs. Let A' be the incidence matrix of a (v', k', λ')-design on the Hadamard parameters $v' = 4t - 1$, $k' = 2t$, $\lambda' = t$. Let A^* be the incidence matrix of a (v^*, k^*, λ^*)-design on the projective plane parameters $v^* = t^2 + t + 1$, $k^* = t + 1$, $\lambda^* = 1$. We define

$$A = \begin{bmatrix} J & A' \\ A^* & 0 \end{bmatrix}, \quad (4.7)$$

where J is the matrix of 1's of size v' by v^* and 0 is the zero matrix of size v^* by v'. It is not difficult to verify that A is the incidence matrix of a uniform design. The matrix A is of order $t(t + 5)$ and may be constructed for various values of t, for example, $t = 2^\alpha$. But the general problem of the tabulation of all multiplicative designs has scarcely been touched upon. For example, no multiplicative design is known that possesses more than two distinct values for the parameters $\lambda_1, \lambda_2, \ldots, \lambda_n$.

5. The matrix equation $A^2 = D + \lambda J$

The matrix equation

$$A^2 = J \tag{5.1}$$

has been investigated recently by various authors including Evans [1967], Knuth [1970], Knuth and Bendix [1970], and Mendelsohn [1968]. The results are of interest in the study of universal algebras and directed graphs. We may in fact interpret a solution of (5.1) as the incidence matrix of a directed graph with exactly one path of length two between every pair of points.

In view of the preceding discussion it is natural to investigate the more general matrix equation

$$A^2 = D + \lambda J, \tag{5.2}$$

where D is an arbitrary diagonal matrix and λ is a positive integer. In terms of directed graphs we require exactly λ paths of length two between every pair of *distinct* points.

It is not difficult to verify that the matrix A of (5.1) has constant line sums. It is also evident that the matrix A of (5.2) need no longer have constant line sums. The following matrices with $\lambda = 1$ provide counterexamples:

$$\begin{bmatrix} 1 & 1 & \cdots & 1 \\ 1 & & & \\ \cdot & & & \\ \cdot & & 0 & \\ \cdot & & & \\ 1 & & & \end{bmatrix} (n \geq 2), \tag{5.3}$$

where 0 is the zero matrix of order $n-1$, and

$$\begin{bmatrix} 0 & 1 & \cdots & 1 \\ 1 & & & \\ \cdot & & & \\ \cdot & & Q & \\ \cdot & & & \\ 1 & & & \end{bmatrix} (n \geq 4), \tag{5.4}$$

where Q is a symmetric permutation matrix of order $n-1$.

We call two (0, 1)-matrices A and A' of order n *equivalent* provided that there exists a permutation matrix P of order n such that

$$A' = P^T A P. \tag{5.5}$$

Ryser [1970] established the following theorem that describes the status of the line sums of the matrix A of (5.2).

Theorem 5.1. *Let A be a $(0, 1)$-matrix of order $n > 1$ that satisfies the matrix equation*

$$A^2 = D + \lambda J, \tag{5.6}$$

where D is a diagonal matrix and λ is a positive integer. Then A has constant line sums c except for the $(0, 1)$-matrices A of order n with $\lambda = 1$ equivalent to (5.3) or (5.4) and the $(0, 1)$-matrices A of order 5 with $\lambda = 2$ equivalent to

$$\begin{bmatrix} 0 & 1 & 1 & 1 & 1 \\ 1 & 1 & 1 & 0 & 0 \\ 1 & 0 & 0 & 1 & 1 \\ 1 & 1 & 1 & 0 & 0 \\ 1 & 0 & 0 & 1 & 1 \end{bmatrix}. \tag{5.7}$$

Furthermore, if A has constant line sums c, then

$$A^2 = dI + \lambda J, \tag{5.8}$$

where

$$c^2 = d + \lambda n \tag{5.9}$$

and

$$-\lambda < d \leqq c - \lambda. \tag{5.10}$$

Let the integral parameters n, c, d and λ of Theorem 5.1 satisfy (5.9) and (5.10). Then the determination of the existence or the nonexistence of a solution of (5.8) with constant line sums c is of itself a difficult unsolved problem. A special case of the problem requires the construction of a (v, k, λ)-design with a symmetric incidence matrix. We remark in conclusion that simple constructions are available for $d = 0$ and $d = 1$. In these special cases it is very natural to attempt to classify all of the nonequivalent solutions. But Hoffman [1967] originally posed this classification problem for the solutions of (5.1) and even here it remains unsolved.

References

R. C. Bose, 1949, A note on Fisher's inequality for balanced incomplete block designs, *Ann. Math. Statist.* **20**, 619–620.

W. G. Bridges, 1970, Some results on λ-designs, *J. Combin. Theory* **8**, 350–356.

W. G. Bridges and E. S. Kramer, 1970, The determination of all λ-designs with $\lambda = 3$, *J. Combin. Theory* **8**, 343–349.

W. G. Bridges and H. J. Ryser, 1969, Combinatorial designs and related systems, *J. Algebra* **13**, 432–446.

N. G. De Bruijn and P. Erdös, 1948, On a combinatorial problem, *Indagationes Math.* **10**, 421–423.

P. Dembowski, 1968, *Finite Geometries* (Springer, Berlin).

T. Evans, 1967, Products of points — some simple algebras and their identities, *Am. Math. Monthly* **74**, 362–372.

J. M. Goethals and J. J. Seidel, 1967, Orthogonal matrices with zero diagonal, *Canad. J. Math.* **19**, 1001–1010.

M. Hall, Jr., 1967, *Combinatorial Theory* (Blaisdell, Waltham, Mass.).

A. J. Hoffman, 1967, Research problems, *J. Combin. Theory* **2**, 393.

D. E. Knuth, 1970, Notes on central groupoids, *J. Combin. Theory* **8**, 376–390.

D. E. Knuth and P. B. Bendix, 1970, Simple word problems in universal algebras, *Computational Problems in Abstract Algebra* (Pergamon, Oxford).

E. S. Kramer, 1969, On λ-designs, Dissertation, Univ. of Michigan, Ann Arbor, Mich.

K. N. Majumdar, 1953, On some theorems in combinatorics relating to incomplete block designs, *Ann. Math. Statist.* **24**, 377–389.

N. S. Mendelsohn, 1968, An application of matrix theory to a problem in universal algebra, *Linear Algebra* **1**, 471–478.

H. J. Ryser, 1963, *Combinatorial Mathematics*, Carus Monograph **14** (Wiley, New York).

H. J. Ryser, 1968, An extension of a theorem of De Bruijn and Erdös on combinatorial designs, *J. Algebra* **10**, 246–261.

H. J. Ryser, 1970, A generalization of the matrix equation $A^2 = J$, *Linear Algebra* **3**, 451–460.

H. J. Ryser, 1972, Symmetric designs and related configurations, *J. Combin. Theory* **12**, 98–111.

J. J. Seidel, 1968, Strongly regular graphs with $(-1, 1, 0)$ adjacency matrix having eigenvalue 3, *Linear Algebra* **1**, 281–298.

D. R. Woodall, 1970, Square λ-linked designs, *Proc. London Math. Soc.* (3) **20**, 669–687.

J. N. Srivastava et al., eds., *A Survey of Combinatorial Theory*
© North-Holland Publishing Company, 1973

CHAPTER 32

Coding Need: A System's Concept

FREDERICK D. SCHMANDT

Rome Air Development Center, Griffiss Air Force Base, N.Y. 13440, U.S.A.

1. Introduction

The intent of this paper is to present in an expository fashion some of the major problems encountered within the Air Force in the application of error detection and/or correction coding to increase reliability or throughput.

Communication requirements within the Air Force are many, varied and constantly changing. There is a wide variety of channels being used for communications-satellite, tropospheric scatter, microwave, HF and wire lines among others. Within each of these classes there are numerous different uses. In many cases, communication is not as reliable as desired. With existing systems, a significant fraction of traffic often consists of repeat transmissions. In addition to causing costly delays, message repetition is known to be an inefficient utilization of available capacity. Since some studies involving coding have shown impressive improvements in reliabilities and throughput and present practices often yield unsatisfactory results, coding has a great deal of promise for future systems.

Rather than describe any specific problem, this paper will identify a general problem prevalent in any coding application. The failure to attack this problem can greatly stifle coding's acceptance and application.

I will begin by presenting the concept of using coding in a multiplex environment. This item covers new work and is considered to be meaningful in its own right. Furthermore it helps in describing the general problem.

2. A coder/multiplexer concept

An ever increasing emphasis of digital communications has been evidenced within the Air Force in recent years. Expectations are that in the future, primarily digital traffic will be employed on most communications links. With this increase in volume, a new importance is placed upon the multiplexing and reliable transmission of such traffic.

Data input to a multiplex system consists of many subchannels operating at varying rates. The multiplex requirement is to sample the channels in such a way that time coherence is maintained in each subchannel at the receiving terminal – the multiplex synchronization problem. A similar problem exists when error correction coding is used to improve reliability; to decode, one needs to know where the code block starts.

When these functions are cascaded, simultaneous synchronization of both functions is required for proper data recovery. The encoding operation requires a data input in groups of bits. In multiplexing, several parallel channels furnish bits simultaneously. One is thus led to question: "Can we construct related frame structures for the multiplexing and coding functions so that synchronization of one automatically provides synchronization of the other?"

With this reasoning as a basis, RADC sponsored a contractual exploratory development program (Solomon and Spencer [1971a, b]) by TRW with the objective of developing a technique of combining the operations of Time Division Multiplex (TDM) and Forward Error Correction (FEC) coding into one unit. We specified that the techniques developed should be capable of multiplexing several parallel channels of high speed synchronous digital data into a serial data stream, encoded for error correction. With the parameters shown in Fig. 1 we set out to design a coder/multiplex system. At these rates, the decoding speed can be a very prime consideration. To ensure that the practical aspects of speed and interface capabilities received adequate consideration, we called for the complete logical design of the device. It was specified that interleaved block codes and concatenated block codes should serve as the basic vehicles in the study.

Code Rate ‑ ‑ ‑ ‑ ‑ ‑ ‑ ‑ ‑ ‑ ‑ $\frac{1}{2}$

Subchannel Info Rates ‑ ‑ ‑ ‑ 75×2^n between 1200 bps and 9.6 Kbps

Multiplexed Encoded 75×2^n between 307.2 and 2457.6 Kbps
Bit Rate

Transparency Req. ‑ ‑ ‑ ‑ ‑ ‑ No more than 100 data bits from each multiplexed subchannel per coding frame

Channel ‑ ‑ ‑ ‑ ‑ ‑ ‑ ‑ ‑ ‑ ‑ ‑ ‑ Tropo

Fig. 1. Assumed parameters.

As indicated in Fig. 2, we can define a "natural multiplex frame length", the number of bits necessary for a single complete multiplexing of all subchannels. Setting the sum of the input information rates equal to the output information rate and eliminating the common factor of 1200 yields a natural frame length of 128×2^i where i depends on the modem rate. Thus, if 1200 bits per second is the rate of a particular channel, its contribution to the natural word is one bit. If the rate is $2^j \times 1200$, it will contribute 2^j bits.

How do we relate the code and multiplex frame structures? The most direct way is to make them equal. The alternatives are to make the multiplex frame a submultiple of the coding frame or vice versa. Still another alternative remains: we can somewhat relax the restriction of using the exact natural frame length and accept some minor complications to allow greater flexibility in code choice. The decision on what method to use will be answered by the other restrictions.

$$m_i = \# \text{ Channels with rate } r_i = 1200 \times 2^i \qquad i = 0, 1, 2, 3$$

With the assumed output rates

$$\sum_i m_i r_i = 1200 \times 2^j \qquad j = 7, 8, 9, 10$$

or

$$\sum_i m_i(1200 \times 2^i) = 1200 \times 2^j$$

so

$l = \#$ Bits necessary for a single complete multiplexing
       of all subchannels

$$= \sum_i m_i 2^i = 2^j$$
$$= 128 \times 2^i \qquad i = 0, 1, 2, 3$$

Depending on the modem output rate

Fig. 2. 'Natural Multiplex Frame Length-l'

The imposed transparency limits the maximum information block size to a number on the order of 1600. This restriction upper bounds the size of an interleaved coding block and consequently mitigates against the error randomizing effect. The magic of a super interleaved system which takes care of long fades on tropo channels thus disappears. In spite of this, because of the simplicity and economy of interleaved systems in comparison to concatenated techniques, they were not immediately dismissed. We knew we were losing some of the effectiveness but how much was in question. Was the additional cost of concatenation necessary? We chose some rate $\frac{1}{2}$ codes and upper bounded interleaved performance by assuming random errors. The basic code length here was kept relatively small, however, because the interleaved block length restriction of 1600 prevents the selection of a long code. Under these very favorable conditions, the improvements possible were at best marginal in comparison to what was desired. Realistically expecting much worse performance, interleaving was rejected. It should be pointed out that the rather stringent transparency condition of 100 bits maximum per subchannel was the cause of the inadequacy of interleaving. Under other assumptions, interleaving may well be the best technique.

Having rejected interleaving, henceforth our attention will be restricted to concatenated schemes. As with interleaving, the code length here is restricted by transparency.

We needed a rate $\frac{1}{2}$ concatenated code. Recognizing that cost and complexity were critical, we restricted the inner code to arithmetically decodable Euclidean geometry codes. The outer code complexity was also a matter for concern. The possible codes under these restrictions were listed. From these two basic possibilities were selected for detailed investigation:

(1) A 5 error correcting (63, 36) BCH inner code with a 3 error correcting (48, 42) outer RS code. Overall (3024, 1512) code.

(2) A 4 error correcting (56, 32) shortened BCH inner code with a 2 error correcting (32, 28) outer RS code. Overall (1792,896) code.

The problem in multiplexing is integrating the natural frame length with the coding block length. The first code's length was basically incommensurable with the natural frame length. Techniques for overcoming this difficulty were worked out and a conceptual design of a coder/multiplex device was carried out. The second code's length is more naturally commensurate with the natural frame length. Figure 3 shows the relationships. For the case of the lowest modem rate, we have no multiplex problem. For the highest data rate, we have 8 code word blocks for 7 natural multiplex frames, i.e., every 8 word blocks begin a natural frame; so we need only 3 bits to tell us where in the 8 word blocks we are. How we arrange this and obtain overall sync will soon be described.

Natural Multiplex Frame Lengths $- 128 \times 2^i$
Code Info Bit Length $= 896 = 128 \times 7$
If $i = 0$ (Frame Length $= 128$)
  7 Multiplex Frames $= 1$ Code Word
If $i = 1$ (Frame Length $= 256$)
  7 Multiplex Frames $= 2$ Code Words
If $i = 2$ (Frame Length $= 512$)
  7 Multiplex Frames $= 4$ Code Words
If $i = 3$ (Frame Length $= 1024$)
  7 Multiplex Frames $= 8$ Code Words

Fig. 3. Code 2 Mux Frame vs Code Frame.

With the above discussion you might think the second code wins hands down. This is not true. The mathematics works out nicely but as yet relative performance in terms of error rate improvements has not even been mentioned. For the binary symmetric channel, such a comparison is easy. Using an input probability of error of 10^{-2} gives output probabilities of approximately 10^{-12} for the first code and 5×10^{-10} for the second. Clearly, for almost any conceived application either code would suffice, and the second less powerful code would be selected because of its easier mechanization if (and it is a big if) the channel were a binary symmetric channel. Unfortunately, ours was not. A high speed tropo application was envisioned.

If coding is viewed as the theory of structuring the signal so that the set of error patterns which occur most frequently in the channel can be corrected then the intelligent application of coding requires a knowledge of at least the general characteristics of the error. A satisfactory model suitable for code evaluation for our purpose was nonexistent unless extensive computer time was expended. A study was made and a quite empirical model of the tropo channel burst statistics was obtained. This model can be viewed as an extension of the Bello-Nelin model (Solomon and Spencer [1971a]). It is believed to be fairly accurate in yielding expected performances of concatenated codes. Based upon an analysis of the two codes and the predicted burst statistics, we concluded that the second code should slightly outperform the first. This result was somewhat surprising. Intuition would cause one to believe that the first code, since it is more powerful in the random error case, would be more powerful for our envisioned application. However, analysis of the predicted burst statistics and the codes' correction capabilities shows intuition to be wrong.

At this point we had concluded that while implementation of an interleaved code would be much more economical improvements would be too small. Of the two concatenated codes considered in detail, the second code was more amenable to the natural multiplex frame length and theoretically yields better performance. Thus code 2 was selected as the candidate for implementation.

The next question was, now that we have selected a candidate code: can we truly carry out all the details of design without running into exorbitant costs and complexities?

Figure 4 shows one code frame with a sync code added. This sync is included in a manner very similar to the technique derived by Frey [1965] at IBM known as message framing – the difference being that instead of using a fixed pattern we use a counting pattern. The first 4 bits of the sync code starting from the first row are the binary equivalent of the numbers 0 to 15 and then 15 to 0. This gives us our 32 rows of the code. The remaining information bit position of the sync code gives us 32 bits with which to convey the multiplex frame information. The (24, 5) synchronization code used was a Solomon-Stiffler punctured code. Accepting, for the moment, the fact that the basic details for encoding and decoding the concatenated and sync codes are known, we now consider initial sync acquisition.

At the decoder, 56 consecutive bits are considered at a time. These bits may coincide with a transmitted word of the inner code or, more likely, may be part of two consecutive words. The first 32 bits of this tentative inner codeword are passed through the inner code encoding shift register, and a set of 24 redundant bits are calculated according to the (56, 32) BCH code algorithm. These 24 bits are then added, modulo 2, to the last 24 bits of the tentative word. The result of this operation is a 24 bit word which is decoded according to the algorithm of the (24, 5) punctured code. The information bits, if all is well, should tell their position in the outer word, establishing inner and outer codeword sync. Clearly, choosing a starting position and decoding one sync word tells us nothing. The four information bits that are used to indicate the inner code word position in the outer code are converted to a number which is stored as the first element in a 56 × T column matrix T is an arbitrary but fixed integer.

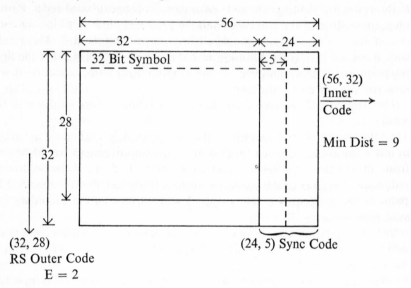

Fig. 4. Code Frame.

The initial entry bit of the serial stream is then dropped and the 57th bit is used with bits 2 through 56 to form a new code word for the (56, 32) decoder. Decoding proceeds as above with the resulting number stored in the second position, first column, of the 56 row by T column matrix. The decoding continues, each time a new bit being entered into the (56, 32) decoder, until the entire $56 \times T$ matrix is filled one column at a time top to bottom. The rows of the matrix are then examined for a sequence of ascending and/or descending numbers i such that $i \leqslant 15$. Once this sequence is found in a row of the matrix, inner word synchronization corresponds to the framed word identified with the matrix elements of the row. Multiplex information is given by the 5th information bit in the sync code.

Correct sync will be obtained if the number of columns, T, in the matrix is large enough to assure that the requisite sequence could not have occurred randomly. If the word is misframed, the 4 information bits from the sync code will describe a random number between 0 and 15. The probability that the next number in the row follows the correct sequence is $\frac{1}{16}$. The probability that T of these numbers is misframed is $(\frac{1}{16})^{(T-1)}$. In practice, rather than requiring T numbers in sequence, a criterium of P out of T numbers is more reasonable.

The sync code itself has powerful error correction capabilities. Correct sync can therefore be obtained even if the correctly framed code word contains errors in the redundant bits that contain the additive sync pattern. If the correctly framed code word has errors in the redundant bit positions only, these errors will translate directly into the sync code word in a one-to-one manner. The (24, 5) code has a five error correction capability. If the code word is correctly framed and there are 5 or less errors in the redundant bit positions, these errors will be corrected and the proper number in the sync pattern sequence will result.

The next case to consider would be the condition where the inner code word is correctly framed but there are errors in the information bit positions. The (56, 32) code has a minimum distance of nine. If a single error occurred in one of the 32 information bit positions then the redundant bits calculated from these information bits must differ from the originally calculated redundant bits in at least eight positions. The correct number in the sync patterns sequence will not be achieved under this condition. The correct sync sequence number can be accidentally obtained only if a particular set of transmission errors causes a new set of redundant bits to be calculated that differ in five or less places from the original set of 24 redundant bits. Thus, if there are any errors in the informration bits, at least 4 errors must occur there to obtain the correct number by accident.

A simulation using our channel model led us to conclude that the addition of the sync code to the redundant bits as described is feasible and effective. The criterion of using a threshold of P out of T numbers in sequence seems to be a reasonable approach to using these codes on a range of transmission channels. These parameters can be made variable in any hardware implementation to suit the existing conditions. This technique should perform well on a fading channel with long term error rates as high as 0.02. For error rates higher than this value, a different synchronization technique would seem to be required. Once original sync is attained, the threshold P can be lowered.

Since the literature contains extensive discussions on the encoding and decoding of the concatenated and sync codes, we shall not go into details here. Suffice it to say that throughout the course of this study many modifications in implementation were considered to reduce complexity and increase speed. Some are summarized below.

(i) The (32, 28) RS code over $GF(2^{32})$ was implemented as four parallel coders each operating over $GF(2^8)$.

The concept of viewing a Reed Solomon code over a smaller subfield is defined in Forney's original [1966] work. Viewing the code in this manner leads to decoding simplifications and can only improve performance. The parallel operation also contributes to the feasibility of high speed operation.

(ii) An easier inversion operation for elements of finite fields was developed.

The chosen scheme uses an outer RS code which can be taken over $GF(2^8)$. The decoding procedures involve multiplying and inverting in this field. To simplify inversion operations, and especially to avoid table look up in large fields, a representation of binary tuples was developed which simplifies division considerably and reduces computation to a smaller field, where the number of elements is the square root of the larger field.

(iii) If one chooses to decode the RS code as a burst code two easily mechanizable procedures were developed.

The procedures for decoding as a burst code were uncovered early in the study but not actually included in the design.

(iv) A simpler decoding of certain BCH codes via majority logic decoding was uncovered.

The majority logic decoding technique developed for decoding the inner code resulted from a study to reduce complexity in this code.

(v) Buffer storage requirements were somewhat reduced.

By combining the coding and multiplex functions, a great deal of duplicate storage is eliminated. The multiplex function is inherently contained within the coding sync. In addition to this, the actual mechanization reduces some further buffer requirements often in evidence in applications.

These modifications among others can contribute considerably to the feasibility of a device.

Late in the study, a strong case was built for using a deletion technique in the inner code and then performing error correction erasure filling in the outer code. The actual implementation design does not contain such decoding although the technology exists and is readily available.

As a result of this effort, we have a design of an integrated multiplexing and coding system which gives fully automatic frame and code synchronization with no bandwidth penalty. Some advantages of such a system are:

1. More powerful codes can be used than could be allowed on an individual channel basis.

2. When a decoding error occurs, the resultant bit errors are amortized among the subchannels with some subchannels receiving correct data.

3. System economics.

The first advantage arises since the delay times experienced by the individual subchannel user are small relative to his much lower data rate. This advantage is particularly evident when the system imposes stringent transparency requirements. With regards to the second advantage without multiplex when a decoding error occurs on a single channel the user is presented data containing a high incidence of errors. With multiplex even if a catastrophic condition arises the subchannels experience a random sprinkling of bit errors over its duration. System economics is evidenced because of the combination of certain functions which reduces a great deal of duplicate equipment.

3. The general problem

Now that I have gone through with the discussion of some aspects of coding/ multiplex, what is the general problem I mentioned at the beginning? It is simply that coding is being studied as a separate entity rather than as a part of a system concept.

Only recently has coding evolved into an economically feasible technology. To utilize this technology, the system design engineer has to be given rules on how to choose a code and shown how coding can be tied to the other functions of the system. It is true that being overly concerned with applications can stifle progress. On the other hand, failure to understand the implications of one's work in terms of how it fits into an overall system concept often leads to misdirected research when it cannot be afforded. The error processing devices being incorporated into our present systems are a result of advances made in basic coding theory in quite recent years. Only now is coding starting to be accepted to any degree. Why has acceptance been so slow? One may argue that it always takes time for a new concept to be accepted, but I believe coding's slow early progress is due to a more fundamental cause. The system designer has to be concerned with the wider aspect of a system. He is interested in applying a solution now. He cannot go through an analysis such as that carried out above. He cannot start from scratch. If he had a problem involving operation in a multiplex environment, he would analyze the situation without coding. If acceptable, he would design the system. If not, he would consider coding as an add-on. Costs and transparency would limit him to a very weak coding technique. The cascading of the functions instead of integrating them would prohibit him from obtaining the listed advantages. When the improvements were predicted, he quite conceivably would discard coding. If familiar with the discussed study, he would have a much higher likelihood of incorporating coding into his system. His parameters would most likely be different, but he has a system model from which he can work.

There are ways to determine the effectiveness of a coding technique once one knows the error pattern characteristics. (Finding these characteristics often involves a great deal of effort in itself.) But such analyses are only a beginning for one interested in the selection of a technique for a given application. System constraints weigh heavily on the choice.

In an actual application, one is not usually interested in obtaining the theoretically minimum probability of error attainable within the communication system constraints. Rather he is interested in obtaining the most economical technique available which gives an acceptable level of performance, that is, a level of performance which meets all the requirements of the user.

If he is looking for very long range solutions, he can be quite cavalier in his possible solutions. He has time to study, check out and modify approaches. Here the full theory of coding can be brought into play. Here coding is usually being adequately considered (at least the existing proved-out theory). But when one is interested in a near time framework, whether it be to upgrade an existing system or to design a new system, he usually considers only well demonstrated and cost effective solutions. Here coding is often not receiving the attention it deserves. Why? Mainly, because the design engineer does not know what is available or does not know how to choose among the alternatives. Perhaps the greatest deterrent to the application of coding theory today is the lack of detailed information concerning costs, complexities and interface. A potential user will implement only if he has a great deal of confidence that the solution will work and only if he knows the costs involved.

Comparative expositions on coding are required setting forth applicable factors such as:

block lengths,
redundancy ratios,
possible decoding schemes and complexities,
estimated or actual cost factors,
required decoder inputs,
methods of predicting performances,
adaptabilities to varying conditions,
required decoder delay,
adaptability to interface.

Such factors are often available for specific codes and channels. However, the various studies made usually make comparative evaluations difficult. Data are thrown out according to different rules when data gathering techniques are used; when theoretical considerations are used, the assumptions are many and varying. Ideally, the design engineer would like a cook book he could look at, follow the directions and come out with a solution which tells him what code to implement, what decoding method to use, how to interface, and the related cost. Certainly this is not fully realizable. However, more effort is required in attempting to obtain a partial solution.

4. Summary

I have described the concept of interfacing coding into a high speed tropo-multiplex environment. In doing this, I have tried to show the involved process required. A design engineer's problem is even more encompassing. Unless he is given more coding tools with which to work, coding's progress will remain slow. Do not just study codes. Assume a system and work through the details so system designers can see what is going on. This is more difficult in that it requires understanding other principles of communication. However, only by such studies will coding gain acceptance as an integral part of a system design rather than as an add-on used only when other techniques do not work.

References

G. D. Forney, Jr., 1966, *Concatenated Codes* (M.I.T. Press, Cambridge, Mass.).

A. H. Frey, Jr., 1965, Message framing and error control, *IEEE Trans. Mil. Elect.*, **Mil-9** (2), 143–147.

G. Solomon and D. J. Spencer, 1971a, Error correction/multiplex for megabit data channels, I. Analysis, TRW Report No. 7353.4-038, RADC-TR-71-176.

G. Solomon and D. J. Spencer, 1971b, Error correction/multiplex for megabit data channels, II. Implementation, TRW Report No. 7353.4-039, RADC-TR-71-176.

J. N. Srivastava et al., eds., *A Survey of Combinatorial Theory*
© North-Holland Publishing Company, 1973

CHAPTER 33

On the Problem of Construction and Uniqueness of Saturated 2_R^{K-p} Designs†

ESTHER SEIDEN ‡

Michigan State University, East Lansing, Mich. 48823, U.S.A.

1. Introduction and summary

The terminology "saturated designs in 2_R^{K-p} runs" was introduced by Box and Hunter [1961a,b], Draper and Mitchell [1967]. However, the concept is not new. In other terminology, these designs are known as orthogonal arrays of strength $t = R-1$, in two symbols 2^{K-p} columns, and K rows. The value p corresponds to the number of generators used for the construction. In the usual terminology of orthogonal arrays, p denotes the number of points beyond the points belonging to the identity matrix which are no-t-dependent in the projective space under consideration. The uniqueness and optimal size properties of orthogonal arrays discussed by Box, Hunter, Draper and Mitchell refer only to arrays constructable by geometrical methods. There is no assurance that they hold in general.

It is the purpose of this note to show that the construction of the arrays carried out by the above mentioned authors and the proof of uniqueness can be established using known geometrical methods without the help of the computer. The method used here is based on the evaluation of the function $m_4(r, 2)$, the maximum number of no-4-coplanar points in $PG(r-1, 2)$. The exact value of $m_4(r, 2)$ is obtained for $r \leqslant 8$ and an upper bound is computed for $r \geqslant 8$. The results have applications in error-correcting code problems and factorial designs.

2. The values of $m_4(r, 2)$ for $r \leqslant 7$ and a bound for $r \geqslant 8$

We may assume without loss of generality that the set of maximum number of no-4-coplanar points in $PG(r-1, 2)$ includes the rows of the identity matrix I_r. Furthermore, this implies that any additional point belonging to the set

† Lectures given at the University of North Carolina at Chapel Hill supported by the U.S. Air Force, Office of Scientific Research, under Grant No. AFOSR–68–1406.

‡ Research supported by NSF Grant GP–20537.

397

must consist of at least four ones. Thus we may conclude immediately that $m_4(4, 2) = 5$, since one can add to the points of I_4 only the point $(1, 1, 1, 1)$. This could also be determined geometrically. Take any two points belonging to the set. Through the line determined by these points pass exactly three planes. One can choose one point on each of these planes since a line joining two points of any two planes will eliminate only one point of the third plane and leaves three more points on this plane untouched. Each of these three points could be added. Now multiplying the matrix

$$\begin{pmatrix} & I_4 & & \\ 1 & 1 & 1 & 1 \end{pmatrix}$$

by the 4×2^4 matrix of all the distinct vectors of size four, we obtain the unique design 2_5^4. However, the complement of the vector space of all possible five-tuples also satisfies the condition of the design. Uniqueness in this case should be interpreted up to interchanging the names of the elements.

Next, we notice that $m_4(5, 2) = 6$ since again we may add to I_5 only one vector, either $(1, 1, 1, 1, 1)$ or $(1, 1, 1, 1, 0)$. However, the uniqueness of the design does not hold in this case. Multiplying the matrix I_5 augmented by each of the vectors mentioned with the matrix of all possible five-tuples, we obtain indeed two distinct designs 2_5^5 which cannot be obtained from each other by either interchanging the rows, columns, or renaming the elements.

Proceeding in the above manner, we see that $m_4(6, 2) = 8$. In this case, one can add to I_6 either the pair of vectors $(1, 1, 1, 1, 1, 0)$ and $(1, 1, 1, 0, 0, 1)$ or the pair $(0, 0, 1, 1, 1, 1)$ and $(1, 1, 0, 0, 1, 1)$. In this case, the design 2_5^{K-p} with $K = 8$, $p = 2$ is unique. This follows from the fact that the line determined by the points $(1, 1, 1, 1, 1, 0, 1, 0)$ and $(1, 1, 1, 0, 0, 1, 0, 1)$ is the same as the line passing through the points $(0, 0, 1, 1, 1, 1, 1, 0)$ and $(1, 1, 0, 0, 1, 1, 0, 1)$ up to interchanging the coordinates. The uniqueness of the one-dimensional subspace of the eight-dimensional vector space implies clearly the uniqueness of the complementary subspace forming the design 2_5^6.

Presently, it will be shown that $m_4(7, 2) = 11$. To prove this, let us consider all possible sets of no-4-coplanar point which can be added to I_7. It is easy to check that up to interchanging the coordinates there are exactly five such distinct sets, each consisting of four points. These are the following sets:

(I)	(II)	(III)
1 1 1 1 1 1 1	1 1 1 1 1 1 0	1 1 1 1 1 1 0
0 0 0 1 1 1 1	1 1 1 1 0 0 1	0 0 0 1 1 1 1
0 1 1 0 0 1 1	1 1 0 0 1 1 1	0 1 1 0 0 1 1
1 0 1 0 1 0 1	0 1 0 1 0 1 1	1 1 0 1 0 0 1

(IV)   (V)

```
1  1  1  1  1  0  0        1  1  1  1  1  0  0
1  1  1  0  0  1  1        1  0  0  0  1  1  1
0  1  0  1  0  1  1        0  1  0  1  1  1  0
1  1  0  0  1  0  1        0  0  1  1  1  0  1
```

This enumeration was obtained by observing the following points:

(i) If the point consisting of all ones is added, the remaining three points must have exactly four ones.

(ii) If a point having exactly one zero is added, then the remaining three points may consist of one point having five ones and two points having four ones or all three having four ones.

(iii) No more than two points can have five ones.

(iv) The maximum number of points in any subspace cannot exceed the previously established bounds.

The automorphism groups of the three-dimensional projective subspaces generated by each of the sets of the four non-coplanar points was computed. It may be interesting to notice that the automorphism group of the first set is the simple group of order 168.

It may also be worthwhile to exhibit algebraic representations of the five sets of four non-coplanar points; they are as follows:

I

$$x_4+x_5+x_6+x_7 = 0$$
$$x_2+x_3+x_4+x_5 = 0$$
$$x_1+x_2+x_4+x_7 = 0$$

II

$$x_1+x_2+x_3+x_4 = 0$$
$$x_3+x_6+x_7 = 0$$
$$x_4+x_5+x_7 = 0$$

III

$$x_2+x_3+x_4+x_5 = 0$$
$$x_1+x_6+x_7 = 0$$
$$x_2+x_5+x_7 = 0$$

IV

$$x_1+x_4+x_7 = 0$$
$$x_2+x_3+x_4+x_5 = 0$$
$$x_3+x_4+x_6 = 0$$

V

$$x_1+x_3+x_7 = 0$$
$$x_1+x_2+x_6 = 0$$
$$x_2+x_5+x_7 = 0$$

where x_i denotes the i-th coordinate of the point.

Next we shall establish the uniqueness of the design 2_3^7. This is equivalent to establishing the uniqueness of the orthogonal three-dimensional projective subspaces generated by four points obtained from each of the sets after adjoining I_4 to them and thus obtaining four eleven-dimensional points in each case. An algebraic representation of the first set can be described as follows:

$$x_4 + x_5 + x_6 + x_7 = 0$$
$$x_2 + x_3 + x_4 + x_5 = 0$$
$$x_1 + x_2 + x_4 + x_7 = 0$$
$$x_1 + x_8 + x_{11} = 0$$
$$x_2 + x_8 + x_{10} = 0$$
$$x_6 + x_7 + x_{11} = 0$$

It is easy to check that the remaining four sets can be obtained from the first set using the following mappings:

$$\text{II} \quad (1)\ (4)\ (6, 7)\ (3, 9, 2, 8, 10, 5, 11)$$
$$\text{III} \quad (2)\ (4)\ (6, 7)\ (3, 10)\ (5, 9)\ (1, 8, 11)$$
$$\text{IV} \quad (3)\ (5)\ (1, 9)\ (2, 4)\ (8, 10)\ (6, 7, 11)$$
$$\text{V} \quad (1, 7, 6, 10, 2, 4, 8, 11, 9, 3, 5)$$

Hence the uniqueness of the design is established.

We shall now make use of the enumeration of the sets belonging to PG(6, 2) to show that $m_4(8, 2) = 17$ and that

$$m_4(r, 2) \leqslant 3(2^{r-6} - 1) + 8 \quad \text{for} \quad r \geqslant 8.$$

This result follows from the following proposition:

All the enumerated sets of four points added to I_7 include a five-dimensional projective space containing the maximum number of no-4-coplanar points, i.e., 8 points. Even if any three points of each of the sets of four points is added to I_7, the same situation will prevail. If any two independent points are added to I_7, the resulting set will contain a five-dimensional projective space containing 7 no-4-coplanar points.

We are now ready to establish the bound for $m_4(r, 2)$, $r \geqslant 8$.

We may assume because of the stated proposition that PG$(r-1, 2)$ includes a five-dimensional projective subspace containing 8 points. Through this five-dimensional subspace pass $2^{r-6} - 1$ six-dimensional projective subspaces of which none can include more than 11 no-4-coplanar points. Hence

$$m_4(r, 2) \leqslant 3(2^{r-6} - 1) + 8 \quad \text{for} \quad r \geqslant 8.$$

This gives, for $r = 8$, $m_4(8, 2) = 17$.

We shall exhibit next a set of 17 no-4-coplanar points in PG(7, 2). They are:

$$I_8$$

$$
\begin{array}{cccccccc}
1 & 1 & 1 & 1 & 1 & 1 & 1 & 0 \\
1 & 1 & 1 & 1 & 1 & 0 & 0 & 1 \\
0 & 0 & 0 & 1 & 1 & 1 & 1 & 1 \\
1 & 1 & 0 & 0 & 0 & 1 & 1 & 1 \\
1 & 0 & 1 & 0 & 1 & 0 & 1 & 1 \\
0 & 1 & 1 & 1 & 0 & 1 & 1 & 1 \\
1 & 0 & 0 & 1 & 1 & 0 & 1 & 0 \\
0 & 0 & 1 & 0 & 1 & 1 & 0 & 1 \\
0 & 1 & 0 & 0 & 1 & 0 & 1 & 1 \\
\end{array}
$$

The properties of this set are under investigation in order to obtain, if possible, a characterization of the maximum number of no-4-coplanar points in PG(7, 2) and, hopefully, in higher spaces. Such a characterization would also yield a better bound for $m_4(r, 2)$, $r \geqslant 9$.

References

A. Barlotti, 1957, Una limitazione superiore per il numero di punti appartementi a una callotta $C(K, 0)$ di uno spazio lineare finito, *Boll. Un. Mat. Ital.* **12**, 67–70.

R. C. Bose, 1947, Mathematical theory of symmetrical factorial designs, *Sankhya* **8**, 107–166.

R. C. Bose, 1961, On some connections between the design of experiments and information theory, *Bull. Inst. Intern. Statist.* **38**, 257–271.

R. C. Bose and J. N. Srivastava, 1964, On a bound useful in the theory of factorial designs and error correcting codes, *Ann. Math. Statist.* **35**, 408–414.

G. E. P. Box and J. S. Hunter, 1961a, The 2^{K-P} fractional factorial designs I, *Technometrics* **3**, 311–351.

G. E. P. Box and J. S. Hunter, 1961b, The 2^{K-P} fractional factorial designs II, *Technometrics* **3**, 449–458.

N. R. Draper and T. J. Mitchell, 1967, The construction of saturated 2_R^{K-P} designs, *Ann. Math. Statist.* **38**, 1110–1126.

G. Tallini, 1956, Sulla k-callotta di uno spazio lineare finito, *Ann. Math.* **42**, 119–164.

J. N. Srivastava et al., eds., *A Survey of Combinatorial Theory*
© North-Holland Publishing Company, 1973

CHAPTER 34

Strongly Regular Graphs and Symmetric 3-Designs

S. S. SHRIKHANDE

Stanford University, Stanford, Calif. 94305, U.S.A.
University of Bombay, Bombay, India

Main results

Let b_3, v, k be nonnegative integers, $3 < k < v$. Then a 3-design $D = (b_3; v, k, 3)$ is a set V of v elements called treatments and a collection of k-subsets of V called blocks such that every 3-set of V is contained in exactly b_3 blocks. It then follows that the numbers of blocks b_i containing any i-set of V is a constant where we define b_0 as the total number of k-sets. It then follows that the following relations hold:

$$b_0\binom{k}{i} = b_i\binom{v}{i}, \qquad i = 0, 1, 2, 3,$$

$$b_i\binom{k-i}{3-i} = b_3\binom{v-i}{3-i}, \qquad i = 0, 1, 2, 3, \qquad (1)$$

$$b_i(k-i) = b_{i+1}(v-i), \qquad i = 0, 1, 2.$$

Putting $b_0 = b$, $b_1 = r$, $b_2 = \lambda$, we see that a 3-design is a balanced incomplete block design (BIBD) with parameters (v, b, r, k, λ) where b_0 is the total number of blocks, r the number of blocks containing any given treatment and λ the number of blocks containing any pair of treatments. Since the b_i's are nonnegative integers, (1) gives a set of necessary conditions for the existence of D.

By considering the blocks of D containing a given treatment x and omitting x from these blocks, we get the derived design $D_d(x)$ which is seen to be a BIBD with

$$v' = v-1, \qquad b' = r, \qquad r' = \lambda, \qquad k' = k-1, \qquad \lambda' = b_3. \qquad (2)$$

The remaining blocks then form the residual designs $D_r(x)$ which is again a BIBD with

$$v'' = v-1, \qquad b'' = b-r, \qquad r'' = r-\lambda, \qquad k'' = k, \qquad \lambda'' = \lambda-b_3. \qquad (3)$$

The design D will be said to be an extension of the BIBD $D_d(x)$. A 3-design D will be called symmetric if $D_d(x)$ for any (and hence all) x is a symmetric BIBD, i.e., if the number of treatments $v-1$ and the number of blocks b' in $D_d(x)$ are equal.

Let D be a 3-design $(b_3; v, k, 3)$ in which any two blocks intersect in either 0 or g treatments. Then any two blocks of any derived design $D_d(x)$ intersect in $g-1$ treatments. Hence $D_d(x)$ and its dual are both BIBD's. Since in a BIBD with parameters (v, b, r, k, λ), $b \geq v$, it follows that $D_d(x)$ is a symmetric BIBD and hence D is a symmetric 3-design with

$$g = 1+b_3. \tag{4}$$

Conversely, suppose D is a symmetric 3-design. Consider any 2 blocks of D having a treatment x in common. Then the subblocks obtained by omitting x from these two blocks are blocks of $D_d(x)$ which is a symmetric BIBD and hence intersect in $\lambda' = b_3$ treatments. It now follows that any 2 blocks of D are either disjoint or intersect in $1+b_3$ treatments. Thus a necessary and sufficient condition that any 2 blocks of D intersect either in 0 or g treatments is that D is a symmetric 3-design and $g = 1+b_3$. Thus necessary conditions for the existence of a symmetric 3-design are that b_i, given by (1), are nonnegative integers, and that a symmetric BIBD corresponding to (2) exists. Further necessary conditions can now be obtained by considering the design $D_r(x)$ for any x. Since any 2 blocks of $D_r(x)$ intersect in 0 or $g = 1+b_3$ treatments, let α_0 and α_g be the number of blocks in $D_r(x)$ having 0 and g intersection, respectively, with any given block of $D_r(x)$. Then

$$\alpha_0+\alpha_g = b-r-1, \qquad g\alpha_g = k(r-\lambda-1). \tag{5}$$

Hence a further necessary condition is that α_g must be a nonnegative integer.

Consider the parameters $(3; 26, 10, 3)$ of a possible symmetric 3-design D. Then (1) is satisfied with $b = 65$, $r = 25$, $\lambda = 9$. The design $D_d(x)$ has parameters

$$v' = b' = 25, \qquad r' = k' = 9, \qquad \lambda' = 3$$

and is known to exist. However, (5) gives

$$4\alpha_4 = 10(15)$$

which does not have integral solution. This proves the impossibility of a 3-design $(3; 26, 10, 3)$.

An example of an infinite family of existing 3-designs is given by the family of affine resolvable BIBD's (Sprott [1955]) with

$$v = 4t+4, \quad b = 8t+6, \quad r = 4t+3, \quad k = 2t+2, \quad \lambda = 2t+1, \quad b_3 = t$$

which can be obtained from a symmetric BIBD with

$$v' = b' = 4t+3, \qquad r' = k' = 2t+1, \qquad \lambda' = t.$$

Another possible infinite family of symmetric 3-designs corresponds to the parameters $(b_3; v, k, 3)$ of a design D where

$$v = g(g^2+3g+1), \qquad b = (g^2+2g-1)(g^2+3g+1),$$
$$r = (g+1)(g^2+2g-1), \qquad k = g(g+1), \tag{6}$$
$$\lambda = g^2+g-1, \qquad b_3 = g-1,$$

where $g \geqslant 2$ is an integer. Then $D_d(x)$ has parameters

$$v' = b' = (g+1)(g^2+2g-1), \qquad r' = k' = g^2+g-1, \qquad \lambda' = g-1 \qquad (7)$$

and $D_r(x)$ has parameters,

$$v'' = (g+1)(g^2+2g-1), \qquad b'' = (g^2+2g-1)(g^2+2g),$$
$$r'' = g^2(g+2), \qquad k'' = g(g+1), \qquad \lambda'' = g^2, \qquad (8)$$

and any 2 blocks of D and hence of $D_r(x)$ intersect in 0 or g treatments.

Consider a finite undirected graph G on v vertices having no loops or multiple edges. The graph is called regular if each vertex is adjacent to exactly n_1 other vertices and hence nonadjacent to the remaining $n_2 = v-1-n_1$ vertices. A regular graph has been called strongly regular by Bose [1963] if any two adjacent (nonadjacent) vertices are adjacent to exactly p_{11}^1 (respectively p_{11}^2) vertices, and then $(v, n_1, p_{11}^1, p_{11}^2)$ represent the parameters of such a graph. Call 2 vertices 1-associates if they are adjacent and 2-associates if they are nonadjacent. Let $p_{jk}^i(x, y)$ be the number of vertices which are simultaneously j-associates of x and k-associates of y where x and y are i-associates. Then $p_{jk}^i(x, y)$ is independent of the pair (x, y) so long as they are i-associates and we have

$$p_{11}^1+p_{12}^1+1 = p_{11}^2+p_{12}^2 = n_1,$$
$$p_{12}^2+p_{22}^2+1 = p_{12}^1+p_{22}^1 = n_2. \qquad (9)$$

The adjacency matrix of a finite undirected graph G without loops or multiple edges is a symmetric $v \times v$ matrix with 0's along the main diagonal and 1 or 0 in position (i, j), $i \neq j$ according as vertices i and j are adjacent or nonadjacent. If G is a strongly regular graph with parameters $(v, n_1, p_{11}^1, p_{11}^2)$ then each row sum of A is n_1 and further

$$A^2 = n_1 I + p_{11}^1 A + p_{11}^2 (J-I-A), \qquad (10)$$

where I is the identity matrix and J is a square matrix with all elements 1. Conversely, any $(0, 1)$ symmetric matrix with 0's along the main diagonal and satisfying (10) can be regarded as the adjacency matrix of a strongly regular graph G with parameters $(v, n_1, p_{11}^1, p_{11}^2)$.

A BIBD with $b > v$ is called quasi-symmetric if any 2 blocks intersect in either μ_1 or μ_2 treatments.

We now state a theorem whose proof is hinted in the concluding remarks by Shrikhande and Bhagwan Das [1965] (also, see Goethals and Seidel [1970]).

Theorem 1. *Let there exist a quasi-symmetric BIBD with parameters (v, b, r, k, λ) such that any 2 blocks intersect in either μ_1 or μ_2 treatments. Let G be a graph whose vertices are the blocks of this design. Define 2 blocks adjacent (nonadjacent) according as they intersect in μ_1 (respectively μ_2) treatments. Then G is strongly regular with parameters $(b, n_1, p_{11}^1, p_{11}^2)$ where*

$$n_1 = \frac{k(r-1)+\mu_2(1-b)}{\mu_1-\mu_2},$$

$$p_{11}^1 = n_1+\theta_1\theta_2+\theta_1+\theta_2,$$

$$p_{11}^2 = n_1+\theta_1\theta_2,$$

$$\theta_1 = \frac{(r-\lambda)-(k-\mu_2)}{\mu_1-\mu_2},$$

$$\theta_2 = \frac{-(k-\mu_2)}{\mu_1-\mu_2}.$$

If a symmetric 3-design D with parameters (6) exists then $D_r(x)$ with parameters (8) also exists and any 2 blocks in D intersect in $\mu_1 = 0$ or $\mu_2 = g$ treatments. The same is true of $D_r(x)$. The above theorem then implies the existence of strongly regular graphs $G(D)$ and $G(D_r)$ with associated adjacency matrices A and A_r:

$$G(D): v = (g^2+2g-1)(g^2+3g-1),$$

$$n_1 = g^2(g+2), \quad n_2 = (g+1)(g+2)(g^2+g-1),$$

$$p_{11}^1 = 0, \quad p_{12}^1 = (g+1)(g^2+g-1), \quad p_{22}^1 = (g+1)^2(g^2+2g-1),$$
$$p_{11}^2 = g^2, \quad p_{12}^2 = g^2(g+1), \quad p_{22}^2 = g^4+3g^3+3g^2-g-3. \tag{11}$$

$$G(D_r): v = g(g+2)(g^2+2g-1),$$

$$n_1 = g(g^2+g-1), \quad n_2 = (g+1)^2(g^2+g-1),$$

$$p_{11}^1 = 0, \quad p_{12}^1 = (g+1)^2(g-1), \quad p_{22}^1 = g^2(g+1)^2,$$
$$p_{11}^2 = g(g-1), \quad p_{12}^2 = g^3, \quad p_{22}^2 = (g+1)(g^3+g^2+g-2). \tag{12}$$

Mesner [1964] has defined a $NL_g(n)$ scheme as a strongly regular graph with parameters

$$v = n^2, \quad n_1 = g(n+1), \quad p_{11}^1 = -n-2+(g+1)(g+2), \quad p_{11}^2 = g(g+1).$$

If $n = g(g+3)$, we have $p_{11}^1 = 0$ and the parameters of $NL_g(g^2+3g)$ are then given by

$$v = g^2(g+3)^2, \quad n_1 = g(g^2+3g+1), \quad n_2 = (g^2+2g-1)(g^2+3g+1),$$

$$p_{11}^1 = 0, \quad p_{12}^1 = (g+1)(g^2+2g-1), \quad p_{22}^1 = g(g+2)(g^2+2g-1), \tag{13}$$
$$p_{11}^2 = g(g+1), \quad p_{12}^2 = g^2(g+2), \quad p_{22}^2 = (g+1)(g+2)(g^2+g-1).$$

Let P be the adjacency matrix of a $NL_g(g^2+3g)$ scheme. Then it can be written in the form

$$P = \begin{pmatrix} 0 & 1 & 0 \\ 1' & 0 & Q \\ 0' & Q' & R \end{pmatrix}, \tag{14}$$

where $\mathbf{1}$ is a row vector with n_1 components all 1 and $\mathbf{0}$ is a row vector with n_2 components all 0. Mesner has shown that in this case Q is the $(0, 1)$ incidence matrix of a symmetric 3-design with parameters (6), Q', the transpose of Q, is the incidence matrix of the dual configuration, and R is the adjacency matrix corresponding to the blocks of Q where two blocks are adjacent or nonadjacent according as they intersect in 0 or g treatments. The parameters of R are then given by (11). Conversely, if Q is the incidence matrix of a 3-design corresponding to (6), then P is the adjacency matrix of a $NL_g(g^2+3g)$ scheme.

The only cases known so far of 3-designs with parameters (6) are for $g = 1$ or 2 and they give rise to $NL_1(4)$ and $NL_2(10)$ schemes. Designs for $g = 2$ is given by Witt [1938] and is also contained in Mesner [1964]. It is worthwhile to consider whether such designs exist for $g \geqslant 3$.

Let A be the adjacency matrix of $G(D)$ where D is given by (6). Then without loss of generality we can write

$$A = \begin{pmatrix} 0 & B \\ B' & A_r \end{pmatrix}, \tag{15}$$

where 0 is a square matrix of order $(g+1)(g^2+2g-1)$ corresponding to the blocks of D containing a given treatment x and A_r is the adjacency matrix of order $(g^2+2g)(g^2+2g-1)$ corresponding to $G(D_r)$.

If G is a graph whose adjacency matrix can be put in the form (15) where A and A_r are the adjacency matrices of graphs with parameters (11) and (12), respectively, then we call G a pseudo-$G(D)$ graph, where D is a symmetric 3-design with parameters (6).

Theorem 2. *The existence of a pseudo-$G(D)$ graph G is equivalent to the existence of a quasi-symmetric design D_r with parameters (8), where any 2 blocks intersect in either $\mu_1 = 0$ or $\mu_2 = g$ treatments.*

Proof. Suppose a pseudo-$G(D)$ graph G exists. Then its adjacency matrix A given by (15) satisfies

$$A^2 = \begin{pmatrix} BB' & BA_r \\ A_rB' & B'B+A_r^2 \end{pmatrix}. \tag{16}$$

Also, from (10),

$$A^2 = g^2(g+2)I + g^2(J-I-A), \tag{17}$$

where I and J are square matrices of order $(g^2+3g+1)(g^2+2g-1)$. Let I_1 and J_1 be square matrices of order $(g+1)(g^2+2g-1)$ and I_2 and J_2 be square matrices of order $(g^2+2g)(g^2+2g-1)$. Then from (16) and (17) we have

$$BB' = g^2(g+2)I_1 + g^2(J_1-I_1), \tag{18}$$

and

$$B'B+A_r^2 = g^2(g+2)I_2 + g^2(J_2-I_2-A_r). \tag{19}$$

From (11) and (12) it follows that B has each row sum $r(B)$ and each column sum $c(B)$, where

$$r(B) = g^2(g+2), \qquad c(B) = g(g+1). \qquad (20)$$

Then from (18) it is obvious that B is the incidence matrix of a BIBD with parameters (8). Again from (10),

$$A_r^2 = g(g^2+g-1)I_2 + g(g-1)(J_2 - I_2 - A_r) \qquad (21)$$

and hence from (19) and (21)

$$B'B = g(g+1)I_2 + g(J_2 - I_2 - A_r). \qquad (22)$$

This implies that any two different blocks of the BIBD corresponding to B intersect in 0 or g treatments. Thus, B is the incidence matrix of a quasi-symmetric design D_r with parameters (8) and with block intersections $\mu_1 = 0$, $\mu_2 = g$ and that A_r is the adjacency matrix of $G(D_r)$.

Conversely suppose that B is the incidence matrix of a quasi-symmetric design D_r with block intersections $\mu_1 = 0$, $\mu_2 = g$ and A_r is the adjacency matrix of $G(D_r)$. Then (18), (20), (21) and (22) hold. It can then be easily proved that

$$BA_r = g^2(J_{1,2} - B)$$

where $J_{1,2}$ is a matrix with $(g+1)(g^2+2g-1)$ rows and $(g^2+2g)(g^2+2g-1)$ columns with all entries 1. Also from (21), (22) we have

$$B'B + A_r^2 = g^2(g+2)I_2 + g^2(J_2 - I_2 - A_r).$$

These relations show that if A is given by (15), then (17) is satisfied, which implies that A is the adjacency matrix of a pseudo-$G(D)$ graph.

We show that the existence of a quasi-symmetric design D_r with parameters (8) implies the existence of two other BIBD's.

The first design is obtained as follows: Consider the configuration obtained by considering only those blocks of D_r which contain a given treatment, say, x and omitting this treatment x from these blocks. The BIBD is obtained by taking the dual of this configuration and has parameters

$$(g^2(g+2), (g^2+g-1)(g+2), g^2+g-1, g^2, g-1).$$

To obtain the other design we proceed as follows: Let β be any block of D_r. Let S_1 be the set of $g(g^2+g-1)$ blocks each of which is disjoint from β. Then obviously from the value $p_{11}^1 = 0$ in $G(D_r)$ we see that any two blocks of S_1 intersect in g treatments. Let S_2 be the set of remaining $(g+1)^2(g^2+g-1)$ blocks each intersecting β in g treatments. Let y be any one of the

$$(g+1)(g^2+g-1)$$

treatments not occurring in β. Let y occur $\alpha_1(y)$ times in S_1 and $\alpha_2(y)$ times in S_2. Then obviously

$$\alpha_1(y) + \alpha_2(y) = g^2(g+2).$$

Now suppose $\alpha_1(y) > g^2$, then $\alpha_2(y) < g^2(g+1)$. Consider the $\alpha_2(y)$ block of

S_2 which contain y. Then since y must occur with each of the treatments of β exactly g^2 times and each block of S_2 contains g treatments of β, the number of pairs (y, x), x in β, which can thus be formed is $g\alpha_2(y) < g^3(g+1)$. This, however, must account for $g^3(g+1)$ such pairs in D_r as such pairs do not occur in S_1. We thus get a contradiction. Hence for any y, $\alpha_1(y) \leqslant g^2$. Since the number of treatments y is $\leqslant (g+1)(g^2+g-1)$, this implies that the total number of entries in S_1 is $\leqslant g^2(g+1)(g^2+g-1)$. However, the total number of entries in S_1 is actually $g^2(g+1)(g^2+g-1)$. Hence the number of treatments y is $(g+1)(g^2+g-1)$, and $\alpha_1(y) = g^2$ for any y. It is now easy to see that the dual of the configuration of the blocks in S_1 gives rise to a BIBD with parameters $(g(g^2+g-1), (g+1)(g^2+g-1), g(g+1), g^2, g)$.

References

R. C. Bose, 1963, Strongly regular graphs, partial geometries and partially balanced designs, *Pacif. J. Math.* **13**, 389–418.

J. M. Goethals and J. J. Seidel, 1970, Strongly regular graphs derived from combinatorial design, *Canad. J. Math.* **22**, 597–614.

D. M. Mesner, 1964, Negative Latin square designs, Institute of Statistics, University of North Carolina, Mimeo Series No. 410.

S. S. Shrikhande and Bhagwan Das, 1965, Duals of incomplete block designs, *J. Indian Statist. Ass. Bull.* **3**, 30–37.

D. A. Sprott, 1955, Balanced incomplete block designs and tactical configurations, *Ann. Math. Statist.* **26**, 752–758.

E. Witt, 1938, Über Steinersche Systeme, *Abh. Hamburg* **12**, 265–275.

J. N. Srivastava et al., eds., *A Survey of Combinatorial Theory*
© North-Holland Publishing Company, 1973

CHAPTER 35

Balanced Arrays and Orthogonal Arrays

J. N. SRIVASTAVA †

Colorado State University, Fort Collins, Colo., U.S.A.

and

D. V. CHOPRA

Wichita State University, Wichita, Kansas 67208, U.S.A.

Summary

We point out the relation between the theory of balanced arrays and various combinatorial areas of design of experiment. Recalling some combinatorial theorems from Srivastava [1971], we apply these to prove a class of new and simple but rather stringent results on the existence of balanced arrays. Applications to the special case of orthogonal arrays are also considered.

1. Introduction

Although orthogonal arrays have been known for almost a quarter of a century, the balanced arrays are much more recent. It will therefore be useful to first define the latter.

Definition 1.1. A *balanced array* (*B-array*) T with s symbols, m rows, N columns, and strength t, is an $m \times N$ matrix T with elements belonging to a set S containing s symbols, such that for every $(t \times N)$ submatrix T_0 of T, and for every vector $\mathbf{v}$ of size $(t \times 1)$ with elements from S, we have

$$\lambda(\mathbf{v}, T_0) = \lambda(P(\mathbf{v}), T_0), \quad \text{for every } P. \tag{1.1}$$

Here, $\lambda(\mathbf{v}, T_0)$ stands for the number of columns of T_0 which are identical with $\mathbf{v}$, and $P(\mathbf{v})$ is any vector obtained by permuting the elements of $\mathbf{v}$. Thus P stands for permutation.

Definition 1.2. The matrix T considered in Definition 1.1 is called an *orthogonal array* with the same parameters, if for every $(t \times N)$ submatrix T_0 of T, and for every pair of vectors $\mathbf{v}$ and $\mathbf{v}^*$ of size $(t \times 1)$ each, we have

$$\lambda(\mathbf{v}, T_0) = \lambda(\mathbf{v}^*, T_0). \tag{1.2}$$

Clearly, an orthogonal array is a special case of a balanced array. Also,

† The work of this author was partly supported by NSF Grant No. GP 30598 X.

clearly, N must be a multiple of s^t. The positive integer $\mu = Ns^{-t}$ is called the *index* of the orthogonal array.

Definition 1.3. Let one of the elements of S be denoted by zero (0), so that the remaining $s - 1$ elements of S may be called non-zero elements. Let v be any vector with elements from S. Then the *weight of* v, denoted by $w(v)$, is defined to be the number of non-zero elements in v.

Under Definition 1.1, consider the special case $s = 2$, with S having the elements 0 and 1. From (1.1), it is clear that $\lambda(v, T_0)$ will be a constant for all vectors v whose weight is the same. Thus, in this case, there exist nonnegative integers μ_k^t ($k = 0, 1, \ldots, t$) such that if $v(t \times 1)$ is of weight k, then $\lambda(v, T_0) = \mu_k^t$ for every $(t \times N)$ sub-matrix T_0 of T. The vector $(\mu_0^t, \mu_1^t, \ldots, \mu_t^t)$ is called the index set of the array.

We shall now discuss the connection of balanced arrays with other older and more well known branches of the combinatorial theory of design of experiments and coding theory.

Consider first the special case of orthogonal arrays. When $t = 2$ and $s = 2$, we get an important class of arrays which are equivalent to Hadamard matrices. On the other hand, the case $t = 2$ and $N = s^2$, corresponds to a set of $(m-2)$ mutually orthogonal Latin squares of order s.

General orthogonal arrays can be classified into three categories A, B, and C. Arrays of type A are those whose columns form the set of solutions of an equation of the form $Hx = c$, where $H(k \times m)$ is a matrix over $GF(s)$ of rank k, and $c(k \times 1)$ is a vector over $GF(s)$. Obviously, in this case, s must be a prime power and $N = s^{m-k}$. It can be shown (Rao [1946]) that the s^{m-k} solutions to the above equation will form an orthogonal array of strength t, if the matrix H has the property Q_t which states that every vector in the row space of H must be of weight greater than or equal to $t+1$. Arrays of type A may also be called "vector space" arrays, since the set of columns of such an array obviously forms a vector space over $GF(s)$.

Type A arrays are also equivalent to group codes. Indeed, the null space of the columns of H can be shown to be a d-error correcting code if $t = 2d$, and a d-error correcting and $(d+1)$-error detecting code if $t = 2d+1$. Also, type A arrays are important from the point of view of the classical theory of confounding and fractional replication as well. Thus this class of arrays forms the famous link (Bose [1961]) between the theory of factorial designs and coding theory. Both from the point of view of obtaining a code with minimum redundancy, and a fractionally replicated design having the smallest number of treatment combinations, one is interested in obtaining matrices H with property Q_t which maximize k (for given m) or maximize m (for given r, where $r = m-k$). It is well known that the latter problem is equivalent to Bose's "packing problem": Find the maximum number m

of points in $PG(r-1, s)$ such that no subset of t points lies on a hyperplane of dimension less than $t-1$. The solution of the packing problem is therefore important both from the coding and the design point of view. It may be useful to remark that the cases $t = 2$, and $t = 3$, $s = 2$, are solved (Bose [1947]). Also, the problem has been solved for many other combinations of s, t and r, with the result that almost all cases of practical interest from the design viewpoint are taken care of. However, from the coding point of view, the problem is far from solved. Here the interest lies in low values of t, particularly $t = 4$, and relatively large values of r (say between 20 to 200). Incidentally, an interesting link between the packing problem and some aspects of abstract combinatorial geometry has recently been established by Dowling [1971]. For some references to work on the packing problem, see Barlotti [1965].

We now discuss orthogonal arrays of type B. Consider the equations $Hx = c_i$ $(i = 1, \ldots, f)$, where H is $(k \times m)$, c_i is $(k \times 1)$, and rank$(H) = k < m$. We assume that H does not have the property Q_t. Let $T_i(m \times s^{m-k})$ be the array obtained by solving $Hx = c_i$. Let $T = [T_1 \, T_2 \cdots T_f]$, and let $C = [c_1, c_2, \ldots, c_f]$. Then the following (unpublished) result (Srivastava [1967]) holds: The matrix $T(m \times f \cdot s^{m-k})$ forms an orthogonal array of strength t if and only if for every nonnull vector $u'(1 \times k)$ such that weight$(u'H) \leqslant t$, the vector $(u'C)$ is an orthogonal array of (size $(1 \times f)$ and) strength 1.

The arrays of type B are obviously of a different structure than those of type A. For type B arrays, N does not have to be a power of s. Incidentally, there exist examples of type A and type B arrays with the same m and N, and m being the maximum possible for the given N.

Arrays which are not of types A or B are said to be of type C.

A useful array which is not of type A is given in Bose and Bush [1953]. The parameters are $s = 3$, $m = 7$, $N = 18$, $t = 2$, $\mu = 2$.

Coming back to balanced arrays, let us first consider the special case $s = 2$, $t = 2$. In this case it can be shown that the array is the incidence matrix of a balanced incomplete block design (BIBD) with possibly unequal block sizes, written BIBDU for short. The ordinary BIBD's form a special case of these when the block sizes are constrained to be all equal, and thus correspond to arrays with the further restriction that each column be of equal weight. This shows that certain techniques of constructing BIB designs, such as the method of difference sets, could be generalized to help in the construction of balanced arrays.

The case $s = 2$, $t = 3$ also is important. The so-called inversive planes form a sub-class of such arrays when the column weights are constant.

Balanced arrays in general are useful in the theory of factorial designs. Indeed, balanced arrays of strength 2, 3 and 4 are identical with balanced fractional factorial designs of resolutions 3, 4 and 5, respectively. We shall not discuss here the detailed properties of balanced arrays considered as fractional factorial designs. The interested reader may refer to Srivastava and Chopra [1971a,b].

From the foregoing discussion, it may appear that the theory of partially balanced designs is not related to the theory of balanced arrays. However, this is not true. As is shown in Bose and Srivastava [1964], certain important principal submatrices of the "information matrix" corresponding to a balanced fractional factorial design (i.e., a balanced array) belong to the linear associative algebras generated by certain well known partially balanced association schemes. (Incidentally, association schemes are merely a different name for what are known as strongly regular graphs.) These algebras proved very helpful in certain statistical studies by Srivastava and Chopra [1971a]. Apart from the above, there are other more direct connections between PBIB designs and B-arrays. We in fact believe that most of the association schemes can be defined using certain balanced arrays. For some work in this field, see Srivastava and Anderson [1971].

It might be important to stress here that balanced arrays not only provide a mathematically challenging field of research which unites various branches of the combinatorial theory of design of experiments, they are also urgently needed for practical problems arising in factorial experimentation.

For some literature on B-arrays in general, the reader is referred to Chakravarti [1956], Srivastava [1971] and Rafter [1971]. In some of the earlier work, B-arrays have been called "partially balanced" arrays. However, since the B-arrays are related directly to BIBD's rather than the PBIBD's, we have adopted the new terminology.

2. Preliminaries

In this section, we present without proof some results from Srivastava [1971] for later use.

Theorem 2.1. *A set of necessary and sufficient conditions for the existence of a 5-rowed B-array T, of strength 4 and index set $\mu' = (\mu_0, \mu_1, \mu_2, \mu_3, \mu_4)$ is that there exists an integer d such that*

$$d \geqslant \psi_{11}(\mu) \equiv \max (0, \mu_4 - \mu_3, \mu_4 - \mu_3 + \mu_2 - \mu_1),$$
$$d \leqslant \psi_{12}(\mu) \equiv \min (\mu_4, \mu_4 - \mu_3 + \mu_2, \mu_4 - \mu_3 + \mu_2 - \mu_1 + \mu_0). \tag{2.1}$$

Theorem 2.2. *A set of necessary and sufficient conditions for the existence of a 6-rowed array T of strength 4 and index set μ' is that there exists an integer d_0 such that*

$$d_0 \geqslant \psi_{21}(\mu) \equiv \max (0, d_{1 \cdot 2} + \theta_1, d_{1 \cdot 4} + \theta_2, d_{1 \cdot 6} + \theta_3),$$
$$d_0 \leqslant \psi_{22}(\mu) \equiv \min (d_6, d_{4 \cdot 6} - \theta_4, d_{2 \cdot 6} - \theta_5), \tag{2.2}$$

where $d_{i.j} = d_i + d_{i+1} + \cdots + d_j$, $j \geq i$; for each i, $d = d_i$ satisfies (2.1); and $d_1 \geq d_2 \geq \cdots \geq d_6$. Also, the θ's in (2.2) are given in terms of the μ's by

$$\begin{aligned}
\theta_1 &= -\mu_4, \quad \theta_2 = -\mu_2 + 2\mu_3 - 3\mu_4, \\
\theta_3 &= -\mu_0 + 2\mu_1 - 3\mu_2 + 4\mu_3 - 5\mu_4, \\
\theta_4 &= -\mu_3 + 2\mu_4, \quad \theta_5 = -\mu_1 + 2\mu_2 - 3\mu_3 + 4\mu_4.
\end{aligned} \tag{2.3}$$

Equivalently, (2.2) can be rewritten as

$$\begin{array}{ll}
\text{(a)} & d_6 \\
\text{(b)} & -d_1 - d_2 + d_6 \\
\text{(c)} & -d_1 - d_2 - d_3 - d_4 + d_6 \\
\text{(d)} & -d_1 - d_2 - d_3 - d_4 - d_5 \\
\text{(e)} & d_4 + d_5 + d_6 \\
\text{(f)} & -d_1 - d_2 + d_4 + d_5 + d_6 \\
\text{(g)} & -d_1 - d_2 - d_3 + d_5 + d_6 \\
\text{(h)} & -d_1 - d_2 - d_3 \\
\text{(i)} & d_2 + d_3 + d_4 + d_5 + d_6 \\
\text{(j)} & -d_1 + d_3 + d_4 + d_5 + d_6 \\
\text{(k)} & -d_1 + d_5 + d_6 \\
\text{(l)} & -d_1
\end{array}
\geq
\begin{array}{l}
0 \\
-\mu_4 \\
-\mu_2 + 2\mu_3 - 3\mu_4 \\
-\mu_0 + 2\mu_1 - 3\mu_2 + 4\mu_3 - 5\mu_4 \\
-\mu_3 + 2\mu_4 \\
\mu_4 - \mu_3 \\
-\mu_2 + \mu_3 - \mu_4 \\
-\mu_0 + 2\mu_1 - 3\mu_2 + 3\mu_3 - 3\mu_4 \\
-\mu_1 + 2\mu_2 - 3\mu_3 + 4\mu_4 \\
-\mu_1 + 2\mu_2 - 3\mu_3 + 3\mu_4 \\
-\mu_1 + \mu_2 - \mu_3 + \mu_4 \\
-\mu_0 + \mu_1 - \mu_2 + \mu_3 - \mu_4
\end{array} \tag{2.2}$$

Theorem 2.3. *Consider an array $T(m, N, 4; \boldsymbol{\mu}')$. Let x_i be the number of columns of T of weight i $(i = 0, 1, \ldots, m)$, so that $\sum_{i=0}^{m} x_i = N$. Then the following single diophantine equations (SDE) must hold:*

$$\sum_{i=0}^{m} \binom{i}{j}\binom{m-i}{4-j} x_i = \binom{m}{4}\binom{4}{j}\mu_j, \qquad j = 0, 1, 2, 3, 4. \tag{2.4}$$

Definition 2.1. *An array T with 2 symbols 0 and 1, and m rows is called "trim" if $x_0 = x_m = 0$.*

Theorem 2.4. *Let T be a trim array with parameters $(m, N, 4; \boldsymbol{\mu}')$, and let x_i be defined as in Theorem 2.3. Let there be g distinct solutions $(d_{r0}, d_{r1}, \ldots d_{r6})$ for the vector $(d_0, d_1, \ldots, d_6)$ under the inequalities (2.2). Note that each of these solutions would correspond to a distinct 6-rowed array; let these be denoted by T_r $(r = 1, \ldots, g)$. Let y_r denote the number of times the 6-rowed array T_r occurs as a subarray* of T, so that $\sum_{r=1}^{g} y_r = \binom{m}{6}$. Define δ_{rj} $(r = 1, \ldots, g; j = 0, 1, \ldots, 6)$ by*

$$\begin{aligned}
\delta_0 &= -\theta_3 - 6\bar{d} + d_0, \quad \delta_1 = -\theta_5 + 5\bar{d} - d_0, \\
\delta_2 &= -\theta_2 - 4\bar{d} + d_0, \quad \delta_3 = -\theta_4 + 3\bar{d} - d_0, \\
\delta_4 &= -\theta_1 - 2\bar{d} + d_0, \quad \delta_5 = -\bar{d} - d_0, \quad \delta_6 = d_0, \\
\boldsymbol{\delta}' &= (\delta_0, \delta_1, \ldots, \delta_6), \quad \bar{d} = (\tfrac{1}{6})(d_1 + \cdots + d_6),
\end{aligned} \tag{2.5}$$

* Here, by a 6-rowed subarray of T we mean an array which would be left after cutting out any set of $m-6$ rows of T. The rows of the 6-rowed array which is left are not to be permuted.

where the θ's are defined at (2.3). Let $\pi_j = \sum_{r=1}^{g} \delta_{rj} y_r$; $j = 0, 1, \ldots, 6$. Then the following triple diophantine equations (TDE) must be satisfied for the case $m = 8$:

(a) $7x_1 + x_2 = \pi_0$,   (b) $21x_1 + 12x_2 + 3x_3 = 6\pi_1$,

(c) $15x_2 + 15x_3 + 6x_4 = 15\pi_2$,   (d) $10x_3 + 16x_4 + 10x_5 = 20\pi_3$,

(e) $6x_4 + 15x_5 + 15x_6 = 15\pi_4$,   (f) $3x_5 + 12x_6 + 21x_7 = 6\pi_5$,   (2.6)

(g) $x_6 + 7x_7 = \pi_6$,   (h) $\sum_{r=1}^{g} y_r = 28$.

The equations (2.6) are termed TDE because there are three sets of variables involved, namely, the x's, the d's and the y's.

Corollary 2.1. When $g = 1$, the equations (2.6) reduce to the following double diophantine equations (DDE):

(a) $7x_1 + x_2 = 28\delta_0$,   (b) $21x_1 + 12x_2 + 3x_3 = 168\delta_1$,

(c) $15x_2 + 15x_3 + 6x_4 = 420\delta_2$,   (d) $10x_3 + 16x_4 + 10x_5 = 560\delta_3$,

(e) $6x_4 + 15x_5 + 15x_6 = 420\delta_4$,   (f) $3x_5 + 12x_6 + 21x_7 = 168\delta_5$,   (2.7)

(g) $x_6 + 7x_7 = 28\delta_6$.

In eqs. (2.7), we have dropped the subscript 1 in the $\delta_{j1}(j = 0, \ldots, 6)$.

Corollary 2.2. Suppose that for some j, we have $\delta_j = \delta_{j+2} = 0$ and $\delta_{j+1} \neq 0$. Then the DDE are contradicted.

We also recall the following from Srivastava and Chopra [1971a]:

Theorem 2.5. Let T be a B-array with $m \geqslant 8$, and $t = 4$. Then

$$\mu_1 + \mu_3 \geqslant \tfrac{6}{7}\mu_2,$$ (2.8)

$$4\mu_2^2 \leqslant \mu_2(\mu_1 + \mu_3) + 6\mu_1\mu_3.$$ (2.9)

3. Analysis of the SDE

For $m = 8$, the SDE (2.4) reduce to

$$70x_0 + 35x_1 + 15x_2 + 5x_3 + x_4 = 70\mu_0,$$ (3.1)

$$35x_1 + 40x_2 + 30x_3 + 16x_4 + 5x_5 = 280\mu_1,$$ (3.2)

$$15x_2 + 30x_3 + 36x_4 + 30x_5 + 15x_6 = 420\mu_2,$$ (3.3)

$$5x_3 + 16x_4 + 30x_5 + 40x_6 + 35x_7 = 280\mu_3,$$ (3.4)

$$x_4 + 5x_5 + 15x_6 + 35x_7 + 70x_8 = 70\mu_4.$$ (3.5)

The above equations involve nine variables. Among these, x_0 and x_8 can, in most cases, be gotten rid of by using Theorem 3.1 below.

Let $A = A(\mu_1, \mu_2, \mu_3)$ be the class of all B-arrays of strength 4 and $m = 8$ with fixed values of μ_1, μ_2, μ_3. Let $A_0 = A_0(\mu_1, \mu_2, \mu_3)$ be a subclass of A with the further restriction that if an array T belongs to A_0 then T is "trim", i.e., T does not contain any column of weight 0 or 8. Thus, if $T \in A_0$ then the value of $(x_0 + x_8)$, for T, equals zero.

Theorem 3.1. (a) *If, for given μ_1, μ_2 and μ_3, the subclass $A_0(\mu_1, \mu_2, \mu_3)$ is empty, then the whole class $A(\mu_1, \mu_2, \mu_3)$ is empty.* (b) *If A_0 is nonempty, $T \in A_0$, and T has index set $(\mu_0, \mu_1, \mu_2, \mu_3, \mu_4)$, then there exists an infinite number of arrays in the class $A \cap \bar{A}_0$ (i.e., with $x_0 + x_8 \neq 0$) with the index set of the form $(\mu_0^*, \mu_1, \mu_2, \mu_3, \mu_4^*)$, with $\mu_0^* \geqslant \mu_0$, $\mu_4^* \geqslant \mu_4$ and with inequality in at least one case.*

Proof. (a) Suppose an array $T^* \in A \cap \bar{A}_0$ exists. Then $s \neq 0$, i.e., T^* has some vectors of weight 0 and some of weight 8. Obtain an array T^{**} from T^* by omitting all the vectors of weight 0 or 8. Then $T^{**} \in A_0$, which contradicts the fact that A_0 is empty. (b) This is obvious, since an array with the index set $(\mu_0^*, \mu_1, \mu_2, \mu_3, \mu_4^*)$ can be obtained from an array with index set $(\mu_0, \mu_1, \mu_2, \mu_3, \mu_4)$ by adding $(\mu_0^* - \mu_0)$ vectors of weight 0 and $(\mu_4^* - \mu_4)$ vectors of weight 8.

Henceforth we shall assume $x_0 = x_8 = 0$ in the equations (3.1)–(3.5). In (3.1)–(3.5), we now make the following transformations:

$$x = x_4, \quad u = x_2 + x_6, \quad v = x_3 + x_5, \quad w = x_1 + x_7,$$
$$u' = x_2 - x_6, \quad v' = x_3 - x_5, \quad w' = x_1 - x_7, \tag{3.6}$$
$$\mu'' = \mu_0 + \mu_4, \quad \mu' = \mu_1 + \mu_3, \quad \mu_0'' = \mu_0 - \mu_4, \quad \mu_0' = \mu_1 - \mu_3 \overset{*}{=} 0.$$

In the above, and throughout this paper, the symbol $\overset{*}{=}$ means that the corresponding inequality holds for arrays with $\mu_1 = \mu_2 = \mu_3$.

From (3.1–3.6), we have, by simple additions and subtractions,

$$5u + 10v + 12x = 140\mu_2, \tag{3.7}$$

$$35w + 15u + 5v + 2x = 70\mu'', \tag{3.8}$$

$$35w + 40u + 35v + 32x = 280\mu' \overset{*}{=} 560\mu_2, \tag{3.9}$$

$$7w' + 3u' + v' = 14\mu_0'', \tag{3.10}$$

$$7w' + 8u' + 5v' = 56\mu_0' \overset{*}{=} 0, \tag{3.11}$$

$$5u + 6v + 6x = 14(4\mu' - \mu'') \overset{*}{=} 14(8\mu_2 - \mu''), \tag{3.12}$$

$$2v + 3x = 7(10\mu_2 - 4\mu' + \mu'') \overset{*}{=} 7(2\mu_2 + \mu''), \tag{3.13}$$

$$14w + 9u - 4x = 28(4\mu' - 7\mu_2) \overset{*}{=} 28\mu_2. \tag{3.14}$$

From (3.7), it is clear that $12x \leqslant 140\mu_2$, and $5 | x$, so that

$$x = 5k, \quad 0 \leqslant k \leqslant \tfrac{7}{3}\mu_2. \tag{3.15}$$

where k is a non-negative integer. From (3.13), we have that $7 | (2v + k)$. Let $2v + k = 7k_1$, where k_1 is a non-negative integer. Then

$$v = \tfrac{1}{2}(7k_1 - k), \tag{3.16}$$

which implies

$$k_1 \geqslant \tfrac{1}{7}k, \quad \text{and} \quad k + k_1 \text{ is even.} \tag{3.17}$$

From (3.13), we obtain

$$2k+k_1 = 10\mu_2-4\mu'+\mu'' \overset{*}{=} 2\mu_2+\mu''. \tag{3.18}$$

(3.17) and (3.18) give

$$\mu'' = 4\mu'-10\mu_2+(2k+k_1) \overset{*}{=} -2\mu_2+(2k+k_1), \tag{3.19}$$

$$\tfrac{1}{7}k \leqslant k_1 = (10\mu_2-4\mu'+\mu'')-2k \overset{*}{=} (2\mu_2+\mu'')-2k. \tag{3.20}$$

Substituting from (3.15) and (3.16) into (3.9), we have,

$$0 \leqslant u = 28\mu_2-7k_1-11k. \tag{3.21}$$

Substituting from (3.21) and (3.15) into (3.14), we get

$$0 \leqslant w = 2(4\mu'-16\mu_2)+\tfrac{1}{2}(17k+9k_1) \overset{*}{=} -16\mu_2+\tfrac{1}{2}(17k+9k_1) \tag{3.22}$$

Combining (3.21) with (3.17), we have

$$\tfrac{1}{7}k \leqslant k_1 \leqslant [4\mu_2-\tfrac{11}{7}k], \tag{3.23}$$

where $[x]$ denotes the largest integer less than or equal to x. Subtracting (3.11) from (3.10), we have

$$-5u'-4v' = 14(\mu_0''-4\mu_0') \overset{*}{=} 14\mu_0''.$$

Therefore $14|(5u'+4v')$, and u' is even. Thus we have

$$u' = 2l_2, \tag{3.24}$$

l_2 being any integer. Also, from (3.11), we have $7|(u'-2v')$. Therefore

$$u'-2v' = 7l_1, \tag{3.25}$$

l_1 is any integer. From (3.24) and (3.25), we have

$$v' = l_2-\tfrac{7}{2}l_1 = l_2-7l_3, \tag{3.26}$$

where $l_1 = 2l_3$. Hence u', v' are known as soon as we choose l_2 and l_3. Substituting from (3.26) and (3.24) into (3.11), we have

$$w' = 8\mu_0'-3l_2+5l_3 \overset{*}{=} 5l_3-3l_2. \tag{3.27}$$

Similarly, (3.27), (3.26), (3.24) and (3.10) give

$$l_2 = -\mu_0''+4\mu_0'+2l_3 \overset{*}{=} 2l_3-\mu_0''. \tag{3.28}$$

The above equations give the value of $(u, v, w, x; u', v', w')$ in terms of μ', k, k_1, l_2 and l_3. Knowing the values of the former, the equations (3.1)–(3.5) can be solved for the x_i's using (3.6). Furthermore, from (3.6), it is obvious that

$$u \geqslant |u'|, \quad v \geqslant |v'|, \quad w \geqslant |w'|, \quad \mu' \geqslant |\mu_0'|, \quad \mu'' \geqslant |\mu_0''|, \tag{3.29}$$

and

$$(u\pm u'), \; (v\pm v'), \; (w\pm w'), \; (\mu'\pm\mu_0') \text{ and } (\mu''\pm\mu_0'') \text{ are even.} \tag{3.30}$$

Now, (3.3) together with (3.16), (3.26), (3.18) and (3.28) implies that

$$\tfrac{1}{2}(k+k_1)\pm(l_2-l_3) \quad \text{and} \quad k_1\pm(l_2-2l_3) \text{ are both even.} \tag{3.31}$$

Furthermore, $v \geqslant |v'|$ and $u \geqslant |u'|$ give us respectively

$$\tfrac{1}{7}[\tfrac{1}{2}(7k_1 - k) + l_2] \geqslant l_3 \geqslant \tfrac{1}{7}[-\tfrac{1}{2}(7k_1 - k) + l_2], \tag{3.32}$$

$$\tfrac{1}{2}u \geqslant l_2 \geqslant -\tfrac{1}{2}u. \tag{3.33}$$

The x_j's, or more precisely $(x_1, \ldots, x_7)$, will be called the primary weight parameters of an array, and the set $(u, v, w; u', v', w')$ will be called the secondary weight parameters. In the above analysis, we have expressed the secondary weight parameters in terms of $(k, \mu_2, k_1, \mu'; l_2, l_3, \mu_0')$, which will be called "tertiary" weight parameters. We have not included μ'' and μ_0'' in the last set, since by (3.19) and (3.28) they get fixed when the other tertiary parameters are fixed.

Next, from these results we obtain a lower bound on the number of assemblies in a B-array of strength four with two symbols, and $m \geqslant 8$.

Theorem 3.2. *For a trim B-array with 2 symbols, $t = 4$ and $m \geqslant 8$, we have $N \geqslant \frac{3.5}{3}\mu_2$. (As will be shown later, the bound is attainable for $\mu_2 = 6$).*

Proof. Clearly it is enough to prove the theorem when $m = 8$. We first observe that

$$N = \mu_0 + 4\mu_1 + 6\mu_2 + 4\mu_3 + \mu_4 = 6\mu_2 + \mu'' + 4\mu', \tag{3.34}$$

where μ', μ'' have been defined in (3.6). Substituting from (3.19) and (3.22) in the above, we have

$$N \geqslant 28\mu_2 - \tfrac{1}{2}(13k + 7k_1) = 28\mu_2 - (x_3 + x_5 + \tfrac{7}{5}x_4). \tag{3.35}$$

Using (3.15) and (3.21) in (3.35), we obtain the result.

Corollary 3.1. *If $k = 0$, $k_1 = 2$ and $\mu_2 = 1$, then $\mu' \geqslant 3$.*

Proof. This follows from the inequality (3.22).

Theorem 3.3. *Let T be a trim B-array with*

(a) $\mu_2 = 1$ *and $x_4 = 5$, then $N \geqslant 23$;*

(b) $\mu_2 > 0$, *then μ_1 and μ_3 are not both zero, and μ_0 and μ_4 are not both zero.*

Proof. (a) From (3.15), (3.17) and (3.21) we have $k = k_1 = 1$. Then (3.22) and (3.19) give respectively $\mu' \geqslant 3$ and $\mu'' \geqslant 5$. The result follows from (3.34).

(b) That $\mu_1 = \mu_3 = 0$ is not possible follows from (3.2)–(3.4) since $\mu_2 = 1$. Similarly for μ_0 and μ_4.

Theorem 3.4. (a) *If no B-array exists corresponding to a given set $(x, u, v, w; u', v', w')$ of weight parameters and a given value of μ_2, then so is the case with the weight parameters $(x, u, v, w; -u', -v', -w')$ for that value of μ_2.*

(b) *If no array exists corresponding to a given set $(k, \mu_2, k_1, \mu'; l_2, l_3, \mu_0')$ then so is the case for the set $(k, \mu_2, k_1, \mu'; -l_2, -l_3, -\mu_0')$.*

Proof. (a) This follows by observing that if there exists a B-array T with the secondary parameters $(x, u, v, w; -u', -v', -w')$, then by interchanging 0 and 1 in T we shall get the complementary array $\overline{T}$ with parameters $(x, u, v, w; u', v', w')$.

(b) Assume, as in (a), that an array T exists corresponding to the parameters $(k, \mu_2, k_1, \mu', \mu''; -l_2, -l_3, -\mu_0', -\mu_0'')$. Then it is clear that for $\overline{T}$ the values of $k, \mu_2, k_1, \mu', \mu''$ will remain unchanged. Since the parameters $(-u', -v', -w')$ of T change to (u', v', w') in $\overline{T}$, it can be checked using successively equations (3.24), (3.26), (3.27) and (3.28) that the values $(-l_2, -l_3, -\mu_0')$ for T change to (l_2, l_3, μ_0') for $\overline{T}$. This completes the proof.

Theorem 3.5. *Let T be a B-array with parameters $(m, N, t; \boldsymbol{\mu}')$. Suppose that for some i $(0 \leqslant i \leqslant m)$ we have $x_i + x_{m-i} > 0$, where x_i is the number of columns of weight i. Then there exist two non-trim arrays T_1 and T_2 with $m-i$ and i rows, respectively, and having other parameters the same as those of T.*

Proof. There exists a column $\mathbf{v}$ in T such that $w(\mathbf{v}) = i$, say. Then T_1 (or T_2) are obtained by omitting all rows of T which have 1 (or 0) in $\mathbf{v}$. Similarly when $w(v) = m-i$.

4. Arrays with $\mu_2 = 1$

First we note a result for later use.

Theorem 4.1. (a) *There does not exist any trim B-array with $m = 7$, $\mu_2 = 1$ and $N < 28$, except when $N = 21$. The array with $N = 21$ has index set $(0, 0, 1, 3, 3)$, and is unique except for an interchange of 0 and 1.*

(b) *Let there be an array with $21 \leqslant N \leqslant 27$, $m = 7$ and $\mu_2 = 1$. Then, apart from an interchange of 0 and 1, its index set is of the form $(\mu_0, 0, 1, 3, \mu_4)$ with $\mu_0 \geqslant 0$, $\mu_4 \geqslant 3$.*

Proof. (a) The nonexistence of arrays with $21 < N < 28$ follows from Theorems 2.1 and 2.6, and Corollary 2.1 of Chopra and Srivastava [1971]. Using these same theorems, it is easily observed that $N = 21$ implies $\mu' = (0, 0, 1, 3, 3)$, apart from an interchange of 0 or 1. By using the SDE for $N = 21$, one then obtains $x_2 = 21$, from which uniqueness of the array is easily shown. (The array is obtained by taking all 7-vectors of weight 2).

(b) This follows from (a).

Theorem 4.2. *Let $T(m \times N)$ be a B-array with two symbols, $t = 4$, $m \geqslant 8$ and $\mu_2 = 1$. Then $N \geqslant 28$.*

Proof. Clearly, it is sufficient to prove the result for trim arrays T. From (3.15), we get $k = 0, 1, 2$, and hence $x = 0, 5, 10$. When $k = 2$, (3.21) gives $7k_1 \leqslant 6$, and hence $k_1 = 0$. But then (3.20) gives $k = 0$, a contradiction. Hence $x = 0$ or 5.

Now, from (2.1), we have $0 \leqslant \mu'' - \mu' + 1$, or $\mu'' \geqslant \mu' - 1$. Hence $N = \mu'' + 4\mu' + 6\mu_2 \geqslant 5(\mu' + 1)$. Then $N < 28$ implies $\mu' \leqslant 4$. Also, (2.9) gives $\mu' \geqslant 2$. Hence $2 \leqslant \mu' \leqslant 4$. From (3.9), we have $8\mu' = w + \frac{8}{7}u + v + \frac{32}{35}x = N + \frac{1}{7}u - \frac{3}{35}x$. From (3.7), we get $u + 2v + \frac{12}{5}x = 28$. Hence, since $7|(2v+k)$, the permissible values (for $N < 28$) of (x, u) are $(0, 14)$, $(0, 0)$ and $(5, 10)$. These values of (x, u) correspond respectively to $8\mu' = N+2$, N and $(N+1)$. Hence $N < 28$ gives $\mu' < 4$. When $\mu' = 2$, $(x, u) = (0, 14)$, we get $N = 14 < 5\mu' + 5$. Hence, since $7k_1 = 2v + k$ and $u + 2v + \frac{12}{5}x = 28$, and in view of Theorem 3.3(a), the possible values (for $N < 28$) of (k, k_1, μ', N) are $(0, 4, 2, 16)$, $(0, 2, 3, 22)$, $(1, 1, 3, 23)$ and $(0, 4, 3, 24)$. We consider these one by one.

Let $N = 16$. Then $\mu' = 2 = \mu''$. Hence μ' is of the form $(\mu_0, \mu_1, 1, 2 - \mu_1, 2 - \mu_0)$. If an array with this index set exists, then by interchanging 0 and 1 in this array, we can obtain an array with index set $(2 - \mu_0, 2 - \mu_1, 1, \mu_1, \mu_0)$. By adjoining the two arrays, we shall get an array with $m = 8$, $N = 32$, and index set $(2, 2, 2, 2, 2)$. Such an array is, however, known to be nonexistent (Seiden and Zemach [1966]). Hence $N \neq 16$.

When $22 \leqslant N \leqslant 24$, we get $\mu' = 3$ and $w > 0$. By Theorem 3.4, $\mu_1\mu_3 \neq 0$. But, using Theorem 3.5 with $m = 8$, $i = 1$, and also Theorem 4.1(b), it follows that we must have $\mu_1\mu_3 = 0$, a contradiction. This completes the proof.

Theorem 4.3. *There exists a unique (apart from* $(0, 1)$ *interchange) B-array of strength 4, $m = 8$, $\mu_2 = 1$, and 28 assemblies. Its index set is $(0, 0, 1, 4, 6)$, and it is obtained by taking all the distinct 8-vectors of weight 2.*

Proof. From Theorem 2.17 of Chopra and Srivastava [1971], it follows that the arrays with $m = 7$, $N = 28$ and $\mu_2 = 1$ must have index sets of the forms $(3, 1, 1, 3, 3)$ or $(\alpha, 0, 1, 4, 3+\beta)$, where $\alpha \geqslant 0, \beta \geqslant 0$. When $\mu' = (3, 1, 1, 3, 3)$ and $m = 8$, (3.18) gives $2k + k_1 = 0$, so that $k = k_1 = 0$. Then, from (3.21), $u = 28$, and hence $x = v = w = 0$. Then (3.10) and (3.11) give respectively $3u' = 0$, $8u' = 112$, a contradiction.

From Theorem 4.1, we find that $m = 8$, $\mu_2 = 1$, $N = 28$ implies that the array must be non-trim, so that in the SDE we have $x_0 = x_8 = 0$. If $\mu' = (\alpha, 0, 1, 4, 3+\beta)$, (3.11) and (3.10) give $\alpha = 0$, and hence $\beta = 3$. Again, from (3.18), $k = k_1 = 0$. Hence $u = 28$, and (3.10) gives $3u' = 84$, or $u' = 28$. This completes the proof.

5. The intermediate diophantine equations (IDE)

We now derive a new set of diophantine equations, the IDE. These will be useful for later developments. Though these equations are new, the basic idea is the same as that for deriving the TDE (Srivastava [1971]) at (2.6).

Indeed, the basic idea is simple. Suppose we are considering the existence of an $(m \times N)$ array T. Suppose further that the $(k \times N)$ subarrays T_0 of T may be of g' different types $T_{01}, \ldots, T_{0g'}$, and that for each i and j $(i = 0, 1, \ldots, k; j = 1, \ldots, g)$, we know the number of columns of T_{0j} with weight i. Now, assuming that there are x_l columns of T whose weight is l, we can compute the number of k-vectors of weight i which are contained in the columns of T. Also, we can compute this same quantity by assuming that out of the $\binom{m}{k}$ subarrays of T there are z_j arrays of type T_{0j}. Equating, we shall get $k+1$ equations, one for each value of i. The unknowns in these equations will be the x_l, the z_j, and probably some other quantities. The SDE and the TDE are obtained by developing this idea for the cases $k = t$ and $(t+2)$, respectively, where t denotes the strength of the array. The IDE correspond to $k = t+1$.

Consider a B-array T of size $(m \times N)$, strength t, and index set $(\mu_0^t, \mu_1^t, \ldots, \mu_t^t)$. Let T_0 be any $((t+1) \times N)$ subarray of T. Let d (≥ 0) be the number of columns of T_0 which equal $(1, 1, \ldots, 1)'$. Then, in equation (2.4) of Srivastava [1971], it is shown that if ϕ_i denotes the number of times any particular $(t+1)$-vector of weight i occurs as a column of T_0, then every $(t+1)$-vector of weight i occurs ϕ_i times in T_0, and

$$\phi_i = \mu_i^t - \mu_{i+1}^t + \mu_{i+2}^t - \cdots + (-1)^{t-i+2}\mu_t^t + (-1)^{t+1-i}d, \quad 0 \leq i \leq t, \quad (5.1a)$$

$$\phi_{t+1} = d. \tag{5.1b}$$

The following is helpful in the calculation of the ϕ's:

$$\phi_i = \mu_i - \phi_{i+1}, \quad i = 0, 1, \ldots, t. \tag{5.2}$$

Since each ϕ_i is nonnegative, (5.1a,b) show that d is bounded both above and below by linear functions of the μ's. Suppose d can take h different values. (Notice that when $t = 4$, $h = 1 + \psi_{12} - \psi_{11}$, where the ψ's are given by (2.1), then $g' = h$). Clearly, any $(t+1)$-rowed array is completely determined by the value of d. Thus, for simplicity, the $(t+1)$-rowed arrays will be denoted by T_{0d} (instead of T_{0j}), where d ranges over its possible set of values. Now, the number of $(t+1)$-vectors of weight i $(0 \leq i \leq t+1)$ which are contained in the columns of T (without row permutations) equals $\sum_d z_d[\phi_{id}\binom{t+1}{i}]$, where ϕ_{id} is the value of ϕ_i for the array T_{0d}. Hence we obtain the IDE:

$$\sum_{q=0}^{m} \binom{q}{i}\binom{m-q}{t+1-i}x_q = \binom{t+1}{i}[\sum_d z_d\phi_{id}], \quad 0 \leq i \leq t+1, \quad \sum_d z_d = \binom{m}{t+1}. \tag{5.3}$$

These equations contain two sets of unknowns, the x's and the z's. Since ϕ_{id} are easy to compute (relative to the δ's needed for the TDE or DDE), the IDE are (relatively) easier to use. Still, however, in many applications, one needs to go to the TDE.

For the case $m = 8$, $t = 4$, (5.3) can be written in full as

$$56x_0 + 21x_1 + 6x_2 + x_3 = \phi_0^*, \tag{5.4a}$$

$$35x_1 + 30x_2 + 15x_3 + 4x_4 = 5\phi_1^*, \qquad (5.4\text{b})$$

$$20x_2 + 30x_3 + 24x_4 + 10x_5 = 10\phi_2^*, \qquad (5.4\text{c})$$

$$10x_3 + 24x_4 + 30x_5 + 20x_6 = 10\phi_3^*, \qquad (5.4\text{d})$$

$$4x_4 + 15x_5 + 30x_6 + 35x_7 = 5\phi_4^*, \qquad (5.4\text{e})$$

$$x_5 + 6x_6 + 21x_7 + 56x_8 = \phi_5^*, \qquad (5.4\text{f})$$

where

$$\phi_i^* = \sum_d z_d \phi_{id}, \qquad i = 0, 1, \ldots, 5. \qquad (5.5)$$

6. Arrays with $\mu_2 = 2, 3$

In Srivastava and Chopra [1971b], arrays with $\mu_2 = 2$ and $N \leqslant 41$ have been studied. Other arrays with $\mu_2 = 2$ shall be studied in similar papers elsewhere. Here we consider only the case $\mu_2 = 3$.

Henceforth, if P denotes any statement (such as the IDE, DDE, etc., or any inequality such as (2.2c)), then R(P) shall mean "the rejection of P". Thus "R(TDE)" will mean "a contradiction of the TDE".

Also, let

$$\alpha = \mu_1 - \mu_0, \quad \beta = \mu_3 - \mu_4. \qquad (6.1)$$

Theorem 6.1. *Consider a B-array T with parameters $(m, N, t; \mathbf{\mu}')$, where $m \geqslant 8$, $t = 4$, $\mu_2 = 3$, and $\alpha + \beta = 0$. Then $\alpha = -\beta = 0, 1$ or -1.*

Proof. It is enough to show that if $\alpha < 0$, then $\alpha = -1$. If possible, let $-\alpha = \beta > 1$. Then (2.1) gives us $\beta = 3$ or 2.

Case (a). $\beta = 3$. Here (2.1) gives $\psi_{11} = \psi_{12} = 0$, and hence $d_1 = d_6 = 0$. Also, $\mu_0 \geqslant 3$. From (2.2c,e,h), we get $\mu_4 = 3$, and $3 \leqslant \mu_0 \leqslant 6$. Hence $\psi_{21} = \psi_{22} = 0$, and $d_0 = \bar{d} = 1$, $\delta_3 = \delta_5 = 0$, and $\delta_4 = 3$, leading to R(DDE).

Case (b). $\beta = 2$. Here $\psi_{11} = 0$, $\psi_{12} = 1$. Hence, recalling the IDE at (5.4), we get $h = 2$, and $\phi_5^* = z_1$, $\phi_3^* = 3z_1 + 2z_0$, and $\phi_2^* = z_0$. Hence, $56 = z_0 + z_1 = \phi_2^* + \phi_5^* = 2x_2 + 3x_3 + (2.4)x_4 + 2x_5 + 6x_6 + 21x_7 + 56x_8$. Now the SDE (3.4), (3.5) give $280\beta = 560 = 5x_3 + 12x_4 + 10x_5 - 20x_6 - 105x_7 - 280x_8$. Multiplying the first equation by 10 and subtracting the second, we get $x_i = 0 \, (i = 2, \ldots, 8)$. But this contradicts (5.4d). Hence the proof is complete.

Theorem 6.2. *Let T be a possible B-array with $\mu_2 = 3$, $m \geqslant 8$, $\alpha = 0$ and $\beta > 0$. Then $\beta = 1$. (Similarly, $\beta = 0$, $\alpha \geqslant 0$ implies $\alpha = 1$.)*

Proof. Let $\beta > 1$; then (2.1) shows that $\beta = 3$ or 2.

Case (a). $\beta = 3$ implies $d_1 = d_6 = 0$ and $\mathbf{\mu}' = (0, 0, 3, 6, 3)$. Here $d_0 = 0$, and this leads to R(DDE).

Case (b). $\beta = 2$. Here $0 \leqslant \psi_{11} \leqslant \psi_{12} \leqslant 1$. In the IDE, take $h = 2$. Then $\phi_{5d} = d$, $\phi_{4d} = \mu_4 - d$, $\phi_{3d} = 2 + d$, $\phi_{2d} = 1 - d$, $\phi_{1d} = \mu_0 - 1 + d$, $\phi_{0d} =$

$1-d$, and $\phi_5^* = z_1$, $\phi_4^* = 56\mu_4 - z_1$, $\phi_3^* = 112 + z_1$, $\phi_2^* = z_0$, $\phi_1^* = 56\mu_0 + z_0$, $\phi_0^* = z_0$. Hence, from (5.4), we have $0 = \phi_0^* + \phi_2^* + 3\phi_5^* - \phi_3^* = 56x_0 + 21x_1 + 8x_2 + 3x_3 + x_5 + 16x_6 + 63x_7 + 168x_8$. Hence, $x_i = 0$, for $i \neq 4$. Hence $56 = z_0 + z_1 = \phi_0^* + \phi_5^* = 0$, a contradiction. This completes the proof.

Theorem 6.3. *Let T be a B-array of strength 4 with $m \geqslant 8$, $\mu_2 = 3$. Then α and β cannot be both positive.*

Proof. Suppose $\alpha > 0$, $\beta > 0$. As before, we find $\beta \leqslant 3$ and $\alpha + \beta \leqslant 3$, so that $(\alpha, \beta) = (1, 2)$, $(2, 1)$ or $(1, 1)$.

Case (i). $(\alpha, \beta) = (1, 2)$. Here $(\psi_{11}, \psi_{12}) = (0, 0)$, $d_i = 0$ (all i), and using (2.2c,e), we find that $\boldsymbol{\mu}' = (\mu_0, \mu_0 + 1, 3, \mu_4 + 2, \mu_4)$, with $\mu_4 = 1$ or 2. Then R(DDE) follows from $\boldsymbol{\delta}' = (0, \mu_0, \mu_4 - 1, 2 - \mu_4, \mu_4, 0, 0)$. Proceed similarly for $(\alpha, \beta) = (2, 1)$.

Case (ii). $(\alpha, \beta) = (1, 1)$ with $(\psi_{11}, \psi_{12}) = (0, 0)$. We get $\boldsymbol{\mu}' = (\mu_0, \mu_0 + 1, 3, 1, 0)$. Here $d_0 = 0$, $\boldsymbol{\delta}' = (3 - \mu_0, \mu_0 - 2, 1, 1, 0, 0, 0)$, and R(DDE) follows.

Case (iii). $(\alpha, \beta) = (1, 1)$ with $(\psi_{11}, \psi_{12}) = (0, 1)$. Here $\boldsymbol{\mu}' = (\mu_0, \mu_0 + 1, 3, 1 + \mu_4, \mu_4)$ and then $0 \leqslant d_6 \leqslant d_1 \leqslant 1$. Because of (2.2f,g) all values of $\mathbf{d}'$ with $d_2 \neq d_5$ are rejected, and we consider the remaining four values one by one. Take $d_1 = d_6 = 1$. Using (2.2c,e,h,k) and (2.2 d,i), the possible values of (μ_0, μ_4) are $(0, 2)$, $(0, 3)$, $(1, 3)$ and $(1, 4)$, and $d_0 = 3 + \mu_0 - \mu_4$. Then $\boldsymbol{\delta}' = (0, 0, \mu_0, 1 - \mu_0, 1 + \mu_0, \mu_4 - \mu_0 - 2, 3 + \mu_0 - \mu_4)$, and we have R(DDE). Next, take $d_1 = d_5 = 1$, $d_0 = 0$. Then $\psi_{21} = \psi_{22} = 0$, $6\boldsymbol{\delta}' = (0, 1, 4, 3, 8, 5, 0)$, leading to R(DDE). Now, when $d_1 = d_5$, $(\mu_0, \mu_4) = (1, 3)$ corresponds to both $d_6 = 0$ or 1. Hence we use the TDE (2.6) with $g = 2$. Using (2.6a,b,c), we obtain $3x_3 = y_2$, $6x_4 = 15y_1 + 5y_2$, $x_0 = x_1 = x_2 = 0$, which contradicts (2.6d). The two cases when $d_2 = d_6 = 0$ are taken care of by the two cases already considered, by interchange of 0 and 1 in the array.

Theorem 6.4. *Let T be a B-array with $m \geqslant 8$, $\mu_2 = 3$, $t = 4$ and $\alpha + \beta > 0$. Then either $\alpha = 0$, $\beta > 0$ or $\alpha > 0$, $\beta = 0$.*

Proof. The proof follows from Theorem 6.3 if we show that $\alpha \geqslant 0$, $\beta \geqslant 0$. Suppose $\alpha < 0$. Then, as before, $\beta = 3$ or 2.

Case (a). $\beta = 3$. Here $d_i = 0$ for each i, and $\boldsymbol{\mu}' = (\mu_0, \mu_0 + \alpha, 3, 6, 3)$, $d_0 = 0$, leading to R(DDE).

Case (b). $\beta = 2$. Then $\alpha = -1$ and $0 \leqslant \psi_{11} \leqslant \psi_{12} \leqslant 1$. For the IDE, we then obtain $\phi_2^* = z_0$, $\phi_3^* = 112 + z_1$, $\phi_5^* = z_1$, $z_0 + z_1 = 56$. Then $0 = 2\phi_2^* + 3\phi_5^* - \phi_3^* = 4x_2 + 5x_3 + (2.4)x_4 + 2x_5 + 16x_6 + 63x_7 + 168x_8$. Hence $x_i = 0$ $(i = 2, \ldots, 8)$, a contradiction.

Theorem 6.5. *Consider a B-array T of strength four, with $m \geqslant 8$, $\mu_2 = 3$, $\mu_1 \leqslant 2$, and $\alpha + \beta < 0$. Then $\alpha \leqslant 0$, $\beta \leqslant 0$, with strict inequality in at least one case.*

Proof. Let $\alpha > 0$. Now α and $-\beta$ are positive. It is obvious that μ_1 and μ_4 are both nonzero. (2.1) implies $\alpha \leqslant 3$, $\alpha \leqslant \mu_1$ and $-\beta > \alpha$.

Case (i). $\alpha = 2 = \mu_1$. When $\mu_1 = 2$, we get $\psi_{11} = \psi_{12} = 1-\beta$, so that $\boldsymbol{\mu}' = (0, 2, 3, \mu_4+\beta, \mu_4)$. But this contradicts (2.2). When $\mu_1 > 2$, we have $\psi_{11} = \psi_{12} = -\beta$, and $\boldsymbol{\mu}' = (\mu_0, \mu_0+2, 3, 0, -\beta)$. From (2.2), we observe $\mu_0 = 4, 5$, so that $\boldsymbol{\mu}' = (4, 6, 3, 0, -\beta)$ and $(5, 7, 3, 0, -\beta)$. Finally (using the DDE), the nonexistence of the corresponding arrays follows from $(d_0; \bar{d}) = (-\beta; -\beta)$, $\delta_1 = \delta_3 = 0$, $\delta_2 = 3$; and $(d_0; \bar{d}) = (-\beta; -\beta)$, $\delta_0 = \delta_3 = 0$, $\delta_1 = \delta_2 = 3$.

Case (ii). Next, we take $\alpha = 3$. Then $\mu_1 \geqslant 3$, and $\psi_{11} = \psi_{21} = -\beta$, and $\boldsymbol{\mu}' = (\mu_0, \mu_0+3, 3, \mu_4+\beta, \mu_4)$. From (2.2h,j,e), we have $\mu_0 = 3$, and $-\beta \leqslant \mu_4 \leqslant -2\beta$. Hence $\boldsymbol{\mu}' = (3, 6, 3, \mu_4+\beta, \mu_4)$, $\psi_{21} = \psi_{22} = -\mu_4-2y$ and $\delta_1 = \delta_3$, leading to R(DDE).

Case (iii). $\alpha = 1$ implies $(\psi_{11}, \psi_{12}) = (2-\beta, 2-\beta)$, $(1-\beta, 2-\beta)$ or $(-\beta, 1-\beta)$ according as $\mu_1 = 1, 2$ or > 2. These are considered separately.

(a) For $\alpha = \mu_1 = 1$, we have $d_1 = d_6 = 2-\beta$, and $\boldsymbol{\mu}' = (0, 1, 3, \mu_4+\beta, \mu_4)$. From (2.2c,i), we get $3-\beta \leqslant \mu_4 \leqslant 5-\beta$, leading to $\psi_{21} = \psi_{22} = 5-2\beta-\mu_4$ and $\boldsymbol{\delta}' = (0, 0, 0, 1, 1, \mu_4-3+\beta, 5-2\beta-\mu_4)$. In the DDE, this implies $x_6 < 0$, a contradiction.

(b) When $\alpha = 1$, $\mu_1 = 2$, we get $(\psi_{11}, \psi_{12}) = (1-\beta, 2-\beta)$ and $\boldsymbol{\mu}'$ equals $(1, 2, 3, \mu_4+\beta, \mu_4)$. Because of (2.2f,g,j), we obtain (i) $d_1 = d_6 = 2-\beta$, or (ii) $d_1 = d_5 = 2-\beta$, $d_6 = 1-\beta$. In (i), using (2.2d,i), we have $4-\beta \leqslant \mu_4 \leqslant 6-2\beta$. Under (ii), we get $d_1 = d_5 = 2-\beta$, $d_6 = 1-\beta$, and (using 2.2c,e), $4-\beta \leqslant \mu_4 \leqslant 5-2\beta$. Then $\psi_{21} = \psi_{22} = 6-2\beta-\mu_4$ respectively under (i) and (ii). Therefore the values of $(d_0; \bar{d})$ and $\boldsymbol{\delta}'$ in these cases are given by $(6-2\beta-\mu_4; 2-\beta)$, $(0, 0, 1, 0, 2, \mu_4+\beta-4, 6-2\beta-\mu_4)$ and $(5-2\beta-\mu_4; \frac{11}{6}-\beta)$, $(0, \frac{1}{6}, \frac{4}{6}, \frac{3}{6}, \frac{8}{6}, \mu_4+\beta-\frac{19}{6}, \frac{11}{6}-\beta)$. Furthermore, $\mu_4 = 6-2\beta$ is impossible, since then $\delta_1 = \delta_3 = 0$, $\delta_2 = 1$. For the remaining cases, we apply the TDE with $g = 2$. We will find x_i's in terms of y_1, y_2. We observe that $x_1 = x_2 = 0$, $3x_3 = y_2$, $6x_4 = 15y_1+5y_2$ and $30x_5 = -20y_2-120y_1$. This is possible only when $y_1 = y_2 = 0$, a contradiction.

7. Arrays with $\mu_2 = 4, 5, 6$

Theorem 7.1. *Consider an* $(m \times N)$ *B-array* T *of strength* 4 *with index set* $\boldsymbol{\mu}'$, *and* $\mu_2 = 4$, $m \geqslant 8$. *Then* $N \geqslant 56$.

Proof. Using (2.9), we get $\mu' \geqslant 6$. Moreover, if $\mu' \geqslant 8$, then $N = \mu''+4\mu'+6\mu_2 \geqslant 56$. Hence $\mu' = 6, 7$. Let us consider these separately.

Case (a). If $\mu' = 6$, then (3.22), (3.15), (3.17) and (3.21) give

(a) $9k_1+17k \geqslant 160$,   (b) $0 \leqslant k \leqslant 9$,   (c) $k \leqslant 7k_1 \leqslant 28\mu_2-11k$. (7.1)

Now, using (3.17), (3.18), we get $(k, k_1) = (7, 5)$. From (3.18), (3.22), we have $\mu'' = 3$ and $w = 2$. Now (3.21) gives $u = 0$, and hence $x_2 = x_6 = 0$; and (3.15), (3.16) give $x = 35$ and $v = 14$. Therefore, the SDE reduce to

(a) $35x_1 + 5x_3 + 35 = 70\mu_0$,

(b) $35x_1 + 30x_3 + 560 + 5x_5 = 280\mu_1$,

(c) $30x_3 + 1260 + 30x_5 = 1680$,   (7.2)

(d) $5x_3 + 560 + 30x_5 + 35x_7 = 280\mu_3$,

(e) $35 + 5x_5 + 35x_7 = 70\mu_4$.

From (7.2a,c,d,e) we find that $7|x_3$, $2|(x_1 + x_3 + 1)$, $7|x_5$, $2|(x_1 + x_5)$, $2|(x_3 + x_7)$ and $2|(x_5 + x_7 + 1)$. Also, $v = 14$, $w = 2$, together with the above conditions, give us $(x_3, x_5) = (0, 14)$, $(7, 7)$ or $(14, 0)$, and $(x_1, x_7) = (0, 2)$, $(1, 1)$ or $(2, 0)$. Now $(x_1, x_7) = (0, 2)$ and $(x_3, x_5) = (0, 14)$ contradict (7.2b), and $(x_1, x_7) = (2, 0)$ and $(x_3, x_5) = (14, 0)$ contradict (6.4d). Finally, $(x_1, x_7) = (1, 1)$ and $(x_3, x_5) = (7, 7)$ is rejected by (7.2). Thus there does not exist a B-array having $x_4 = 35$, $u = 0$, $v = 14$ and $w = 2$.

Case (b). When $\mu' = 7$, we have $9k_1 + 17k \geqslant 144$. It can be easily checked that the only possible (k, k_1) in this case are $(9, 1)$, $(8, 2)$, $(7, 3)$, $(7, 5)$, $(6, 6)$, $(0, 16)$. Of these, the pair $(9, 1)$ is rejected because of (7.1c), and for the remaining possible values of (k, k_1) we find that $N \geqslant 56$. This completes the proof.

Theorem 7.2. *There does exist a B-array* $T(8, N, 4; \mu')$ *having* $N = 56$, $\mu_2 = 4$, *and index set* $(4, 6, 4, 1, 0)$. *The array is obtained by writing the 56 8-vectors of weight 3 each. This array is unique, apart from an interchange of 0 and 1. Furthermore, this is the only array with* $N = 56$.

Proof. Using the result in Section 4, check that $N = 56$ implies $(k, k_1) = (0, 16)$. Also, then, $\mu' = 7$, $\mu'' = 4$, $x = u = w = 0$, $v = 56$. Using the SDE, $5|(\mu_3 - 1)$. Hence $|v'| = 56$. This gives the result.

Theorem 7.3. *For a B-array* T *of strength* 4, *with* $m \geqslant 8$ *and* $\mu_2 = 5$, *we have* $N \geqslant 66$. (*However, it is not known whether a B-array of strength* 4 *with* $m = 8$, $\mu_2 = 5$ *and* $N = 66$ *exists.*)

Proof. We have $N = 6\mu_2 + 4\mu' + \mu'' = 30 + 4\mu' + \mu''$, which implies that μ' and μ'' have to be minimum for N to be the smallest possible. Moreover, from (3.22), μ' is such that $16\mu' \geqslant 64\mu_2 - [17k + 9k_1]$, where $[x]$ denotes the greatest integer less than or equal to x, and therefore we take that value of μ' which satisfies

$$\mu' = 4\mu_2 - \left[\frac{17k + 9k_1}{16}\right]. \qquad (7.3)$$

It is obvious from (7.3) that $17k + 9k_1$ has to be as large as possible if μ' is to be minimum. From (3.15), we obtain $0 \leqslant k \leqslant 11$. Now, using (3.17) and (3.23), we find that $k \neq 11$, and that the possible values of (k, k_1) (with k_1 a maximum) are $(0, 20)$, $(1, 17)$, $(2, 16)$, $(3, 15)$, $(4, 12)$, $(5, 11)$, $(6, 10)$, $(7, 9)$,

$(8, 6)$, $(9, 5)$ and $(10, 4)$. Furthermore, we observe that $[\frac{1}{16}(17k + 9k_1)]$ assumes the largest value 12 when $(k, k_1) = (6, 10)$, $(7, 9)$, $(9, 5)$ and $(10, 4)$. Of these four competing pairs, the minimum value of N is given by that pair for which μ'' is the least. It can be checked from (3.18) that $(k, k_1) = (6, 10)$ gives the least value of μ'', which equals 4. Hence the minimum number of assemblies required is 66.

Theorem 7.4. *Let* $T(m \times N)$ *be a B-array of strength* 4, *with* $\mu_2 = 6$, *and* $m \geqslant 8$. *Then* T *must have* $N \geqslant 70$.

Proof. From (2.9), we have $\mu' \geqslant 8$. If $\mu' > 8$, then $N \geqslant 4\mu' + 6\mu_2 \geqslant 72$. When $\mu' = 8$, the only values of μ_1 and μ_3 satisfying (2.9) are $\mu_1 = \mu_3 = 4$. For these cases, we have $N \geqslant 68$, and the non-existence of arrays with $N = 68$ or 69 will complete the proof of the theorem. Possible $\boldsymbol{\mu}'$ with $N = 68$ or 69 are $(0, 4, 6, 4, 0)$, $(1, 4, 6, 4, 0)$ and $(0, 4, 6, 4, 1)$. For $\boldsymbol{\mu}' = (0, 4, 6, 4, 0)$, from the SDE (3.1), (3.3), we have $x_5 = 224$, which contradicts $N < 70$. Similar contradictions are obtained for the other two values of $\boldsymbol{\mu}'$.

8. Orthogonal arrays

Before closing, we make a few remarks on orthogonal arrays of strength 4 and index μ. The following result throws light on how some of the preceding development can be brought to bear on orthogonal arrays.

Theorem 8.1. *An orthogonal array of strength* 4, *m rows and index* μ *exists if and only if there exists a trim balanced array with the same parameters, and index set* $(\mu_0, \mu, \mu, \mu, \mu_4)$ *with* $\mu_0 \leqslant \mu$, *and* $\mu_4 \leqslant \mu$.

It was in view of this result that in Section 3, we used the symbol "$\overset{*}{=}$" to indicate the expressions one obtains when $\mu_1 = \mu_2 = \mu_3$.

Theorems 2.1 and 2.2 can be used to get existence conditions for orthogonal arrays with 5 and 6 rows, respectively. Thus, for example, for an orthogonal array of index μ and $m = 5$, we have $\psi_{11} = 0$, $\psi_{12} = \mu$, so that there are $1 + \mu$ non-isomorphic arrays.

For six-rowed arrays, we have the following interesting result.

Theorem 8.2. *A necessary and sufficient condition that there exists a B-array with* $m = 6$, $t = 4$, *and index set* $\boldsymbol{\mu}'$, *is that there exist integers* $d^*(\geqslant 0)$ *and* i $(0 \leqslant i \leqslant 5)$ *such that* $(d_1 = d_i = d^* + 1, d_{i+1} = d_6 = d^*)$ *is a solution of* (2.2), *where for* $i = 0$ *the equality* $(d_1 = d_i = d^* + 1)$ *is to be ignored.*

Proof. The result is established by observing that the coefficient matrix (say A) of the vector $\mathbf{d}' = (d_1, \ldots, d_6)$ on the left side of (2.2) is such that each element in any column of A is larger than or equal to the corresponding element in the preceding column of A.

Notice that the above theorem greatly simplifies the inequalities (2.2), since the left hand side of (2.2) now uses only one variable d^* instead of six variables $d_1, \ldots, d_6$. For orthogonal arrays with (a general) index μ, there is further simplification since now the right hand side of (2.2) involves just μ (instead of $\mu_0, \ldots, \mu_4$). Thus it becomes a simple matter to check the existence of orthogonal arrays with $m = 6$. Indeed, using the above, we would easily find that orthogonal arrays with $m = 6$, $t = 4$ exist for all μ, except for $\mu = 1$ and 3 (Seiden and Zemach [1966]).

The considerable simplification occurring for $m = 6$ holds out promise for the study of existence of orthogonal arrays with $m \geqslant 7$.

References

A. Barlotti, 1965, Some topics in finite geometrical structures, University of North Carolina, Institute of Statistics, Mimeo Series No. 439 (Chapel Hill, N. Car.).

R. C. Bose, 1947, Mathematical theory of the symmetrical factorial design, *Sankhya* **8**, 249–256.

R. C. Bose, 1961, On some connections between the design of experiments and information theory, *Bull. Intern. Statist. Inst.* **38**, part IV.

R. C. Bose and K. A. Bush, 1952, Orthogonal arrays of strength two and three, *Ann. Math. Statist.* **23**, 508–524.

R. C. Bose and J. N. Srivastava, 1964, Multidimensional partially balanced designs and their analysis, with applications to partially balanced factorial fractions, *Sankhya Ser.* A **26**, 145–168.

I. M. Chakravarti, 1956, Fractional replication in asymmetrical factorial designs and partially balanced arrays, *Sankhya* **17**, 143–164.

D. V. Chopra and J. N. Srivastava, 1971, Optimal balanced 2^7 fractional factorial designs of resolution V, $N \leqslant 41$ (submitted for publication).

P. Dembowski, 1968, *Finite Geometries* (Springer, New York).

T. A. Dowling, 1971, Codes, packings and the critical problem, *Proc. Conf. on Combinatorial Geometry and Its Applications* (A. Barlotti, ed.; Perugia, Italy).

J. A. Rafter, 1971, Contributions to the theory and construction of partially balanced arrays, Ph.D. dissertation (under Prof. E. Seiden), Michigan State University, East Lansing, Mich.

C. R. Rao, 1947, Factorial arrangements derivable from combinatorial arrangements of arrays, *Suppl. J. Roy. Statist. Soc.* **9**, 123–139.

E. Seiden and R. Zemach, 1966, On orthogonal arrays, *Ann. Math. Statist.* **37**, 1355–1370.

J. N. Srivastava, 1967, Investigations on the basic theory of $2^m 3^n$ fractional factorial designs of resolution V and related orthogonal arrays (Abstract), *Ann. Math. Statist.* **38**, 637.

J. N. Srivastava, 1971, Some general existence conditions for balanced arrays of strength t and 2 symbols, *J. Combin. Theory*, to appear.

J. N. Srivastava and D. A. Anderson, 1971, Factorial subassembly association schemes with application to the construction of multi-dimensional partially balanced designs, *Ann. Math. Statist.* **42**, 1167–1181.

J. N. Srivastava and D. V. Chopra, 1971a, On the characteristic roots of the information matrix of balanced fractional 2^m factorial designs of resolution V, with applications, *Ann. Math. Statist.* **42**, 722–734.

J. N. Srivastava and D. V. Chopra, 1971b, Balanced optimal 2^m fractional factorial designs of resolution V, $m \leqslant 6$, *Technometrics* **13**, 257–269.

J. N. Srivastava et al., eds., *A Survey of Combinatorial Theory*
© North-Holland Publishing Company, 1973

CHAPTER 36

On Software Validation

RONA B. STILLMAN and NEIL J. STILLMAN
Rome Air Development Center, Griffiss Air Force Base, N.Y. 13440, U.S.A.

The entire computer programming industry—software houses, commercial users and government users in particular—shares the burden and suffers the consequences of unreliable software and the accompanying incomplete, incorrect and unintelligible documentation. Programs thought to be debugged, i.e. correct, will suddenly produce wrong results, no results, or behave otherwise erratically because some special condition in the data or in the environment was not accounted for in the logic of the program. Examples of this range from the annoying, such as faulty computer billing, to the expensive and dangerous, such as the false "overload" indication given by the on-board Apollo 11 computer. The possibility of errors is harrowing in programs designed to make medical diagnoses, or to handle air-traffic control.

Current practice is to design and implement a system, and then to test it for some arbitrary subset of possible input values and environmental conditions. The system is accepted when it executes the test cases correctly, or when time runs out, whichever occurs first. Clearly, no system accepted on this basis will ever be completely reliable. However, it is unreasonable to hope to test every case, since even simple programs may have an infinite input domain and an extraordinarily large number of execution paths. Consider the program below (Fig. 1), expressed as a directed graph in which each vertex represents a statement, and each edge indicates that control may flow from the statement at its base to the statement at its tip. If a maximum of twelve iterations of the loop is assumed, this eight statement program has over 2,500,000,000 possible paths of execution from START to HALT. What is required, then, is not a set of test cases, but rather a mathematical formalism within which programs can be rigorously proven to be correct. The remainder of this paper will be devoted to describing a formalism developed by Floyd, and indicating problems that must be resolved before this promising approach can yield practical results.

Clearly, the property of being "correct" is a relative one. That is, a program is said to execute correctly if it accomplishes the intentions of the programmer. Within Floyd's formalism, the verifier is provided with both the program and "a statement of the programmer's intentions", i.e., a definition of correctness. If the program does, indeed, execute as intended, it will be possible to exhibit a formal mathematical proof of that fact. Otherwise, the verifier may provide a counterexample, that is, a set of variable values and a point in the program at which these values would cause execution different from the stated intentions.

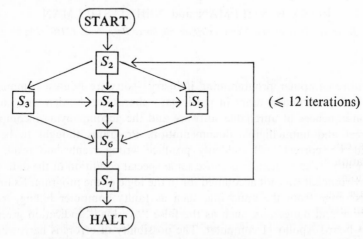

Number of execution paths from START to HALT
$$= (6 + 6^2 + \ldots + 6^{12}) \approx 6^{12} \approx 2,500,000,000 \text{ paths}$$

Fig. 1.

The language of the formalism is the first-order predicate calculus, and these are the basic notions: Consider a program (Fig. 2), in flowchart (i.e., directed-graph) form. Associate assertions over the program variables with statements in the program (i.e., vertices in the graph) with the understanding that when execution reaches statement i, then assertion $P_i(V_1, V_2, \ldots, V_n)$ is true of the program variables $(V_1, V_2, \ldots, V_n)$. In particular, a characteristic assertion called the initial assertion, $I(V_1, V_2, \ldots, V_n)$, is associated with the START statement and a characteristic assertion called the final assertion, $F(V_1, V_2, \ldots, V_n)$, is associated with the HALT statement. By means of these assertions, the programmer has defined correctness as follows: If the initial values of the program variables $(V_1, V_2, \ldots, V_n)$ satisfy condition $I(V_1, V_2, \ldots, V_n)$ and execution terminates, then the final values of $(V_1, V_2, \ldots, V_n)$ must satisfy $F(V_1, V_2, \ldots, V_n)$. The problem may now be restated more precisely: Given a program P with initial assertion I and final assertion F, devise a calculus which would enable one to verify P, i.e., to construct a rigorous proof that P is correct with respect to I and F.

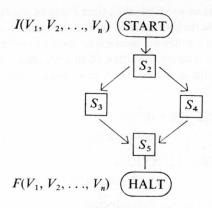

Fig. 2.

This calculus, like every system of deductive reasoning, involves the application of valid rules of inference and a set of valid axioms. In this case, the axioms are the semantic description of the statements in a programming language and have the general format $P_0\{\text{STATEMENT TYPE}\}P_1$, where P_0 is an assertion on the program variables before execution of the statement and P_1 is an assertion on the program variables after execution of the statement. The axioms say the following: If P_0 is true of the program variables before the statement is executed, and the statement is, indeed, executed, then P_1 is true of the program variables after execution. In all cases, P_1 represents the strongest verifiable consequent of the statement, given the antecedent P_0 (see Fig. 3).

GENERAL AXIOM FORMAT

$$\vdash P_0\{\text{Statement Type}\}P_1$$

AXIOM FOR ASSIGNMENT STATEMENT

$$P_0(V_1, V_2, \ldots, V_n)\{V_i \leftarrow f(V_1, V_2, \ldots, V_n)\} \exists V_i'(P_0(V_1, \ldots, V_i', \ldots, V_n) \cap$$
$$V_i = f(V_1, \ldots, V_i', \ldots, V_n))$$

EXAMPLE

$$(V_1 = 2 \cap V_2 = V_3)\{V_1 \leftarrow V_2\} \exists V_1'(V_1' = 2 \cap V_2 = V_3 \cap V_1 = V_2)$$

Fig. 3.

Consider a single statement program (Fig. 4), with initial and final assertions specified: $I(V_1, V_2, \ldots, V_n)$ STATEMENT $F(V_1, V_2, \ldots, V_n)$. This program is said to be verified iff $P_1(V_1, V_2, \ldots, V_n) \supset F(V_1, V_2, \ldots, V_n)$, where P_1 is the consequent prescribed by the statement axiom, for it is clear that if

P_1 is the strongest assertion derivable, then F can be no more restrictive than P_1 if F is to be asserted at the end of the program. $P_1(V_1, V_2, \ldots, V_n) \supset F(V_1, V_2, \ldots, V_n)$ is a potential theorem of the first-order predicate calculus, and is called the verification condition of this program. A logical proof that the verification condition is true, then, is a rigorous proof of correctness of the program.

ONE-LINE PROGRAM

$I(V_1, V_2, \ldots, V_n)\{\text{Statement}\}F(V_1, V_2, \ldots, V_n)$

APPROPRIATE AXIOM

$P_0(V_1, V_2, \ldots, V_n)\{\text{Statement}\}P_1(V_1, V_2, \ldots, V_n)$

VERIFICATION CONDITION

$P_1(V_1, V_2, \ldots, V_n) \supset F(V_1, V_2, \ldots, V_n)$

Fig. 4.

It is not difficult to extend this line of reasoning inductively to programs of any finite length. The consequent of any statement, as prescribed by the appropriate axiom, simply serves as the premise assertion for the next statement. Thus, generating a verification condition consists of repeatedly transforming the assertion at hand, each transformation being determined by the semantic definition, i.e., the axiom associated with that statement type in the programming language.

A set of very simple definitions is now required to establish a meaningful working vocabulary.

Definition. A *program* $P = (S_1, S_2, \ldots, S_n)$, $n \geqslant 1$, is a finite sequence of statements, most conveniently represented as a directed graph.

Definition. A *tagged statement* is one that has an assertion associated with it.

For instance, the START and HALT statements are always tagged statements.

Definition. A *control path* $CP = (S_{i_1}, S_{i_2}, \ldots, S_{i_m})$, $m \geqslant 1$, is a finite sequence of statements such that control can pass from statement S_{i_j} to statement $S_{i_{j+1}}$ for $1 \leqslant j \leqslant m-1$.

Definition. A *tagged path* $T = (S_{i_1}, S_{i_2}, \ldots, S_{i_m})$ is a control path such that the first statement, S_{i_1}, and the last statement, S_{i_m}, are tagged, and none between are tagged.

It should be apparent that a program P is verified iff every tagged path in P is verified. That is, a program as a whole is said to execute correctly iff every possible path in the program executes correctly. Hence, the proof of correctness of a program consists of proving the validation condition for each of the tagged paths in the program. To insure that the length of each tagged path and the total number of tagged paths remain finite, it is necessary to demand that at least one statement in every loop in the program be tagged. Thus no loop can ever be enclosed within the confines of a tagged path. Since loops are iterative by their nature, the assertions required to "cut" these loops are called iterative assertions.

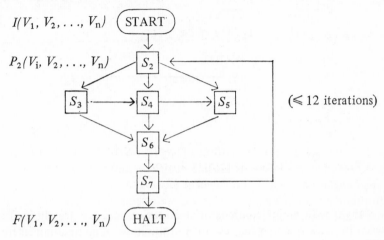

Number of tagged paths = 1 (from START to S_2) + 6 (from S_2 to HALT)
+ 6 (from S_2 to S_2) = 13
Number of control paths from START to HALT $\cong 2.5 \times 10^9$

Fig. 5.

Let us reconsider the first program we discussed, with initial, iterative and final assertions assigned as required (Fig. 5). By proving the 13 verification conditions, then, we have proven the correctness of all 2.5×10^9 control paths from START to HALT. If statement S_6 is tagged as well (Fig. 6), the number of tagged paths, and consequently the number of verification conditions, falls to 9, a minimum for this program. In general, however, the question of how many assertions to use and where to place them so as to minimize the effort required to verify a program is still open. What is known is that for a program of any reasonable size (e.g., any production program), the number of verification conditions is appallingly large, and the chance of error in proving them is depressingly high. The most appealing solution is to assign this task to the computer; that is, to have the machine automatically generate proofs for the verification conditions. It is usually thought that solving such problems requires a certain amount of ingenuity or human intelligence. In fact, however, any theorem that can be proven at all can be proven by executing some purely clerical algorithms. The problem is that these algorithms

15

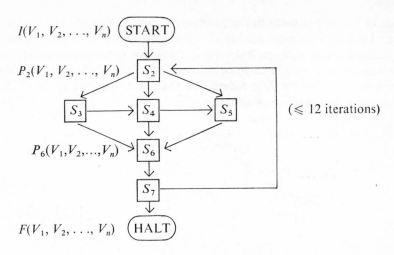

$$\text{Tagged paths} = 1 \text{ (from START to } S_2) + 6 \text{(from } S_2 \text{ to } S_6)$$
$$+ 1 \text{ (from } S_6 \text{ to } S_2) + 1 \text{ (from } S_6 \text{ to HALT)} = 9$$

Fig. 6.

are fraught with major combinatorial obstacles, and are grossly inefficient. These obstacles will now be described in the hope of stimulating interest in their solution.

Deductive algorithms are based upon the notions of unsatisfiability and refutation, rather than upon the notions of validity and proof. To prove a proposed theorem T (the verification conditions can be considered as theorems), it is sufficient to exhibit just one contradiction derived from $\sim T$. In practice, the relevant mathematical axioms, the hypothesis of T and the negation of the conclusion of T are expressed as a finite set S of sentences in the first-order predicate calculus. By using specified rules of inference, the deductive algorithm seeks to produce a logical contradiction, that is, a pair of mutually exclusive inferences, from S. Early systems employed exhaustive instantiation as their only rule of inference. They made repeated substitutions of constants for variables in the sentences of S, producing successively larger sets of instances until a complementary pair of instances were found. The substitutions were made in a systematic and exhaustive manner, guaranteeing that if a contradiction existed, it would be included, eventually, in the steadily expanding sets of instances. For problems of reasonable mathematical interest, however, this expansion is a virtual explosion. The sets of instances soon grow so large as to be absolutely beyond the limits of feasibility. The irony of the situation is that almost all of the instances generated are irrelevant to the problem, and the number which are relevant is quite small.

In 1964, J. A. Robinson formulated the resolution principle, a sound and effective rule of inference that reduced the combinatorial explosion by a factor in excess of 10^{50}. Whereas an instantiation-based system substitutes constants for variables in the sentences of S, a resolution-based system deals with the sentences of S in their more general form. Rather than blindly instantiating every sentence in S and waiting for a contradiction to surface, resolution produces a new sentence from two existing sentences only if these sentences, or instances of them, contain a pair of complementary literals (details are given in 5, but are not germaine here). Nevertheless, most of the sentences generated are irrelevant to the proof. Moreover, since a sentence once generated is never discarded, these irrelevant sentences go on to produce legions of irrelevant descendants. Hence resolution alone is still a very inefficient refutation procedure.

Various proof strategies have been devised in an attempt to make resolution-based systems more efficient. These proof strategies are logical adjuncts to resolution, and fall into three main categories:

(1) Editing strategies, which eliminate redundant sentences upon generation. These strategies are often variations of the subsumption principle, which prescribes deletion of any sentence that is an instance of any other sentence in S. Once a significant number of sentences has been retained, however, it takes a great deal of time to determine if a new sentence is redundant. Sometimes deleting a redundant sentence is not worth the time spent in discovering that it was, indeed, redundant.

(2) Search strategies, which specify the order in which sentences are to be considered for resolution, or define certain criteria which limit the candidates for resolution. For example, Wos' unit preference strategy resolves shorter sentences before longer ones, with unit sentences (i.e., sentences composed of only one literal) being resolved first. J. A. Robinson's Hyper-Resolution prohibits the resolution of two negative sentences (i.e., sentences which contain at least one complemented literal). The success of search strategies, however, is erratic and unpredictable. They perform well on some problems, and poorly on others, and it is difficult, if not impossible, to decide in advance if a given problem will run well under a given search strategy.

(3) Augmenting resolution as the rule of inference, to increase its power and efficiency. For example, many attempts have been made to devise a system that will recognize the special characteristics of the equality relation, and handle it accordingly, rather than treating it as just another predicate.

None of these approaches, alone or in combination, has been powerful enough to make automatic theorem-proving practical. Other concepts, including the application of new hardware, is currently being investigated Associative processors, which are content-addressible and offer parallel arithmetic and logic, are promising candidates. Perhaps human intelligence should intrude at key decision points in the proof process, while the bulk of the computation is performed by computer. But what are the key decisions, and can they be handled effectively by a human being? It would be useless, for instance, to expect a man to choose two sentences to be resolved from a list of two thousand, or even two hundred. Clearly, a great deal of research remains to be done if we are to overcome the combinatorial problems inherent in software verification and automatic theorem-proving.

References

R. Floyd, 1967, Assigning meanings to programs, *Proc. Symp. Appl. Math.* **19**, 19–32.

J. King, 1969, A program verifier, Ph.D. Thesis, Computer Science Dept., Carnegie-Mellon University.

J. A. Robinson, 1963, Theorem-proving on the computer, *J. Assoc. Comp. Mach.* **10**.

J. A. Robinson, 1967, Review of automatic theorem-proving, *Proc. Symp. Appl. Math.* **19**, 1–18.

J. A. Robinson, 1965a, A machine-oriented logic based on the resolution principle, *J. Assoc. Comput. Mach.* **12**, 23–41.

J. A. Robinson, 1965b, Automatic deduction with hyper-resolution, *Intern. J. Comput. Math.* **1**, 227–234.

R. B. Stillman, 1972, Computational logic: the subsumption and unification computations, Ph.D. Thesis, Systems and Information Sciences Dept., Syracuse University.

L. Wos, D. Carson and G. Robinson, The unit preference strategy in theorem proving, *Proc. AFIPS 1964 Fall Joint Computer Conf.* **26**, 616–621.

J. N. Srivastava et al., eds., *A Survey of Combinatorial Theory*
© North-Holland Publishing Company, 1973

CHAPTER 37

The Enumerative Theory of Planar Maps

W. T. TUTTE

University of Waterloo, Waterloo, Ontario, Canada

1. Planar maps

A *planar map* is the figure formed when a non-null connected graph is drawn in the plane. We allow the graph to have loops and multiple joins, and we require it to be finite. The graph separates the rest of the plane into connected disjoint open regions called *faces*. Just one face is unbounded; we call it the *outer* face.

The vertices and edges of the defining graph G are called the vertices and edges, respectively, of the resulting planar map M, and they are assigned the same incidence relations with one another in M as in G. The frontier of any face is a union of edges and vertices of G. The face is said to be incident with those edges and vertices that lie in its frontier.

The *valency* of a vertex of M is the number of incident edges, loops being counted twice. The valency of a face of M is the number of incident edges, isthmuses being counted twice.

The simplest planar map of all is the *vertex-map*. This has a single vertex and no edges. Accordingly, it has a single face. The vertex and the face each have valency zero.

Suppose G is not a vertex-map. Then each vertex is incident with at least one edge. In the neighbourhood of a vertex v, the adjacent portions of the incident edges can be regarded as short segments radiating from v. Let us call these portions the *approaches* to v. They have a natural cyclic order. We can say that this is the cyclic order in which the approaches are encountered by a moving point describing a small orbit round v in an anti-clockwise sense, and crossing each approach just once in a revolution. Replacing each approach by the corresponding edge, we obtain the *natural cyclic sequence of edges* at v. Any loop incident with v has two approaches to v, and so occurs exactly twice in this cyclic sequence of edges. But a link incident with v occurs only once.

If G has more than one vertex there is also a natural cyclic sequence of edges associated with any face F. When we mentioned an anticlockwise sense of revolution, we assumed an orientation of the plane. Accordingly, we may speak of the right and left sides of a directed edge of G. Imagine therefore a

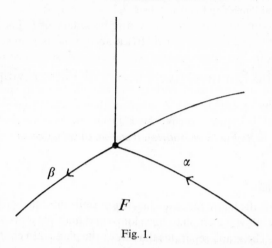

Fig. 1.

moving point traversing an edge E of G incident with F, and having F on its left with respect to its motion in E. Eventually, the moving point reaches an approach α of E to a vertex v, and moves up to v (Fig. 1). Now let β be the approach to v immediately preceding α in the natural cyclic sequence of approaches to v. We require our moving point to leave v along β. Then in its passage through v it has nothing on its immediate left except points of F. There is of course the special case in which v is monovalent. Then β is identical with α. Our moving point proceeds to v, reverses direction and travels back along α. In this case, E has the same face F on both sides. It can be seen that

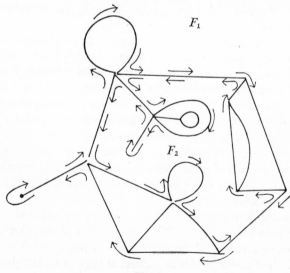

Fig. 2.

our moving point must traverse the entire boundary of F. Each edge incident with F is traversed either once or twice according as it has F on only one side or on both. An edge incident with F is traversed twice if and only if it is an isthmus of G.

We refer to the re-entrant path of the moving point as the *bounding path* of F. The edges incident with F taken in their cyclic order in the bounding path (so that isthmuses appear twice) give the *natural cyclic sequence of edges* round F. In Fig. 2 we indicate the bounding paths of two faces F_1 and F_2.

2. Rooted maps

Consider a planar map M with outer face F. If M is not a vertex-map, we select one of the directed edges R of the bounding path of F and call it the *root*. The choice of a root converts M into a *rooted map*. The negative end V of R is then the *root-vertex*, the unoriented edge E corresponding to R is the *root-edge*, and the outer face is sometimes called the *root-face*.

Two rooted maps M_1 and M_2, with graphs G_1 and G_2, respectively, are called *combinatorially equivalent* if there is a homeomorphism f of the plane onto itself that transforms M_1 into M_2, with preservation of the natural cyclic sequence of approaches to each vertex, the natural cyclic sequence of edges round each face, and the root. From now on we do not distinguish between combinatorially equivalent rooted maps.

We may now pose enumerative problems such as the following: how many (combinatorially distinct) rooted maps are there with n edges? The answer to this problem is given in (Tutte [1963]) as

$$a_n = \frac{2(2n)!3^n}{n!(n+2)!} \tag{1}$$

If we write $n = 0$ in (1) we obtain $a_0 = 1$. We may take this to refer to the vertex-map. It seems convenient to count this as a rooted map even though no root can be chosen for it. Putting $n = 1$ we find $a_1 = 2$. This result corresponds to the two rooted maps shown in Fig. 3.

Fig. 3.

Next we find that $a_2 = 9$. The nine rooted maps with two edges are shown in Fig. 4.

We shall have to consider some special kinds of rooted map. We explain some of the necessary terminology here. A *plane tree* is a planar map whose

15**

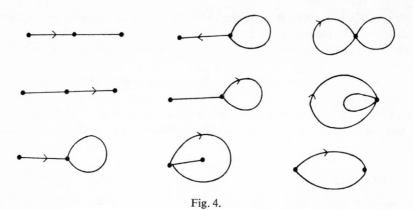

Fig. 4.

graph is a tree. Such a map has only one face, the outer face. We note that two combinatorially distinct rooted plane trees may have isomorphic graphs. They will then have different natural cyclic sequences of edges round their outer faces. A connected graph is said to be *non-separable* or 2-*connected* if it cannot be expressed as a union of two subgraphs, each with at least one edge, having only one vertex (and no edges) in common. A 2-connected graph is said to be 3-*connected* if it cannot be expressed as a union of two subgraphs, each with at least two edges, having only two vertices (and no edges) in common. A planar map is said to be 2-*connected* or 3-*connected* if its graph is 2-connected or 3-connected, respectively. A *triangulation* is a planar map in which each face has valency 3. An *inner-triangulation* is a planar map, not a vertex-map, in which each face other than the outer face has valency 3. It can be shown that a triangulation is 2-connected if and only if it has no loop, and 3-connected if and only if it has no circuit of fewer than 3 edges. A *simple* triangulation is a 3-connected triangulation in which no circuit of 3 edges separates two vertices.

In the next two Sections we give two examples of the methods used in the theory of planar enumeration.

3. Rooted trees

Suppose we wish to determine the number t_n of rooted plane trees with n edges. Considering the vertex-map, we see that $t_0 = 1$. From Figs. 3 and 4, we see that $t_1 = 1$ and $t_2 = 2$. We introduce the generating series

$$F(x) = \sum_{n=0}^{\infty} t_n x^n, \tag{2}$$

and we try to find an equation that is satisfied by $F = F(x)$. A typical plane tree T, with a root-edge, is shown in Fig. 5.

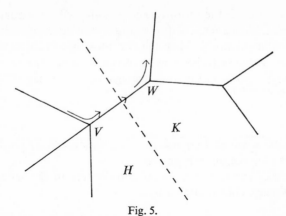

Fig. 5.

If we delete the root-edge, this tree falls apart into two smaller trees H and K, containing the negative end V and the positive end W of the root, respectively. We can regard H and K as rooted plane trees: the root of H immediately precedes that of T in the natural cyclic sequence of directed edges round the face of T, and the root of K immediately succeeds that of T in this sequence. Moreover, if arbitrary rooted plane trees are specified as H and K, we can reverse the construction to obtain a unique corresponding tree T. Since the ordered pairs $\{H, K\}$ are enumerated by the function F^2, we deduce that

$$F = 1 + xF^2. \tag{3}$$

We can now apply Lagrange's Theorem to obtain an expression for F as a power series in x. We thus find that the t_n are the Catalan numbers.

$$t_n = \frac{(2n)!}{n!(n+1)!}. \tag{4}$$

4. Rooted 2-connected inner-triangulations

Let M be a rooted 2-connected inner-triangulation. We write $m(M)$ for the valency of its outer face and $t(M)$ for the number of inner faces. We note that there is just one such map satisfying $t(M) = 0$. It is the first map of Fig. 3. Some writers would not regard the graph of this map as 2-connected, but it does satisfy the definition of 2-connection given above. For this "link-map", we have $m(M) = 2$. It is clear that $m(M) \geqslant 2$ for every rooted 2-connected inner-triangulation.

For enumerating rooted 2-connected inner-triangulations, the appropriate generating function is

$$q = q(x, z) = \sum_M x^{m(M)} z^{t(M)}, \tag{5}$$

and it is convenient to use also

$$h = \sum_N z^{t(N)}, \tag{6}$$

where N runs through the set of rooted 2-connected inner-triangulations M with $m(M) = 2$. Such a map N, if it is not the link-map, gives rise to a rooted 2-connected triangulation T when the non-root edge incident with the outer face is deleted from the graph so as to merge the outer face of N with one of the triangles of N to form the outer triangle of T. We may therefore write h also as

$$h = 1 + \sum_T z^{s(T)}, \tag{7}$$

where T denotes a rooted 2-connected triangulation and $s(T)$ is the number of its faces, counting the outer triangle.

Let us classify the rooted maps M contributing to the series q. First we note the link-map, contributing a term

$$x^2.$$

In the remaining case, the root R is incident with the outer face F and with one other face K. The face K is a triangle having as vertices the root-vertex V, the positive end W of R, and one other vertex X. We now say that M is of Type I if X is incident with the outer face, and of Type II otherwise. Figs. 6 and 7 show maps of Type I and Type II, respectively.

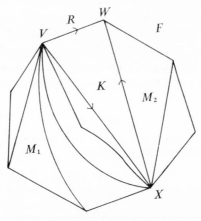

Fig. 6.

Consider a map M of Type I. When we delete the root-edge, the graph decomposes into two 2-connected subgraphs G_1 and G_2, with only the vertex X in common. We may suppose V to belong to G_1, and W to G_2. These subgraphs define 2-connected inner-triangulations M_1 and M_2, respectively. These can be regarded as rooted. The root of M_1 is the directed edge VX of the triangle K, and the root of M_2 is the directed edge XW of that triangle. It is clear that there is a unique map M giving rise in this way to any specified ordered pair $\{M_1, M_2\}$. We deduce that the terms of q corresponding to maps of Type I are the terms of the series

$$x^{-1}zq^2.$$

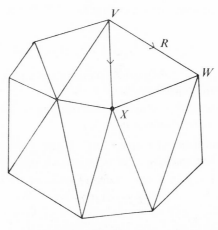

Fig. 7.

Next let us consider a map of Type II. In this case, the deletion of R gives rise to a new 2-connected inner-triangulation M'. We always have $m(M') \geqslant 3$. Subject to this condition, we can reverse the construction and obtain a unique M from a given M'. (The root of M' is taken to be the directed edge VX of the triangle K.) We deduce that the terms of q corresponding to maps of Type II are the terms of the series

$$x^{-1}z(q-x^2h).$$

From the foregoing results, we deduce the identity

$$q = x^2 + x^{-1}zq^2 + x^{-1}z(q-x^2h),$$

that is,

$$zq^2 + (z-x)q + x^3 - x^2zh = 0. \qquad (8)$$

We now consider how to solve this equation for q and h. It is not so simple as the quadratic equation that we obtained in the case of rooted plane trees. However, we can write it as

$$\{zq-(z-x)/2\}^2 = D/4, \qquad (9)$$

where

$$D = (z-x)^2 - 4z(x^3 - x^2zh). \qquad (10)$$

We may now determine h in terms of a parameter ξ initially defined by the equation

$$zq(\xi, z) + (z-\xi)/2 = 0. \qquad (11)$$

Thus ξ is a power series in z. It follows from (9) that both D and $\frac{1}{2}(\partial D/\partial x)$ take the value zero when ξ is substituted for x. We thus have the equations

$$z^2 - 2z\xi + \xi^2 - 4z\xi^3 + 4z^2\xi^2 h = 0,$$
$$-z + \xi - 6z\xi^2 + 4z^2\xi h = 0. \tag{12}$$

Eliminating h, we find that

$$z = \xi(1 - 2\xi^2). \tag{13}$$

Substituting this in either of the equations (12), we find that

$$z^2 h = \xi^2(1 - 3\xi^2). \tag{14}$$

Changing to the parameter $\theta = \xi^2$, we have

$$z^2 = \theta(1 - 2\theta)^2, \tag{15}$$
$$z^2 h = \theta(1 - 3\theta). \tag{16}$$

We can apply Lagrange's Theorem to (15) and (16) to obtain $z^2 h$ as a power series in z. The result is

$$h = \sum_{n=0}^{\infty} \frac{z^{2n} 2^n (3n)!}{(n+1)!(2n+1)!}.$$

We recall that the coefficient of z^{2n} in this series is the number of rooted 2-connected triangulations with $2n$ faces in all. (The number of faces must be even).

Using (8), (15) and (16), we can obtain an equation for q in terms of x and θ. From this, we can obtain q as a power series in x and θ. Then the various powers of θ can be expressed in terms of z by Lagrange's Theorem. It is found that

$$q = \sum_{n=0}^{\infty} \sum_{j=0}^{\infty} \frac{2^{j+1}(2n+1)!(3j+2n)! x^{n+2} z^{n+2j}}{(n!)^2 j!(2j+2n+2)!}.$$

5. The quadratic method

A procedure analogous to that of Section 4 is used in several papers on planar enumeration. It is used for example (Tutte [1962a]) to enumerate rooted 3-connected triangulations. The application to 2-connected ones was made by R. C. Mullin [1964]†. We may note also the enumeration by W. G. Brown [1963] of the rooted 2-connected planar maps with a given number of edges and a given valency for the root-face. In a sequel (Brown and Tutte [1964]) to the last-mentioned paper, W. G. Brown and W. T. Tutte use the procedure to enumerate rooted 2-connected planar maps with $i+1$ vertices and $j+1$ faces. Presumably the numbers a_n of Section 2 could be obtained by this "quadratic method", but they were originally obtained in a different way (Tutte [1963]) and this particular calculation does not seem to have been made yet.

† Our 2-connected triangulations are called "triangular maps" by Mullin.

6. Changes of connectivity

There is a procedure which has been used a number of times to deduce one enumeration from another. We illustrate it by showing how the 3-connected rooted triangulations can be enumerated by making use of the results for 2-connected ones.

It is not difficult to see that each rooted 2-connected triangulation T has an associated rooted 3-connected triangulation $S(T)$ from which it is derived by splitting some of the edges into digons and then triangulating the insides of these digons. Even the root-edge may be split in this way, but the outer triangle must be preserved. When the root is split, the new root is that "half" of the old root that is still incident with the outer triangle. Fig. 8 shows the relation between T and $S(T)$.

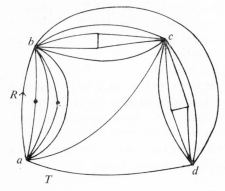

Fig. 8.

We write

$$\bar{h} = 1 + \sum_{S} z^{t(S)} = \sum_{i=0}^{\infty} \bar{h}_i z^{2i}, \tag{17}$$

where S denotes a rooted 3-connected triangulation having $t(S)$ faces in all. $\bar{h}_i$ is the number of such rooted 3-connected triangulations for which $t(S) = 2i$. For such a triangulation, the number of edges is $3i$ and the number of vertices is $i+2$. Using the relation between T and $S(T)$, we find that

$$h = 1 + \sum_{i=1}^{\infty} \bar{h}_i z^{2i} h^{3i} \tag{18}$$

$$= \bar{h}(z^2 h^3),$$

where h is as in Section 4.

Write $t = z^2 h^3$. Then by (15) and (16) we have

$$t = \frac{\theta(1-3\theta)^3}{(1-2\theta)^4}.$$

It is convenient to replace θ by the parameter

$$\lambda = \frac{(1-3\theta)}{(1-2\theta)}.$$

We then have

$$\lambda = 1 - t\lambda^{-3},$$
$$h = 3\lambda - 2\lambda^2.$$

Applying Lagrange's Theorem to these equations, we obtain h, that is $\bar{h}(t)$, as a power series in t. In this way we can show that the number of rooted 3-connected triangulations with $2n$ faces is

$$\frac{2(4n-3)!}{n!(3n-1)!}.$$

This kind of argument was used in reverse in Tutte [1962b] to go from 3-connected triangulations to 2-connected ones (in dual form). It was used in Tutte [1962a] to reach the rooted simple triangulations from the rooted 3-connected ones. In Tutte [1963], starting with the numbers a_n, the method was used to reach the rooted 2-connected planar maps with n edges, and then the rooted 3-connected planar maps with n edges. In the cases of the 3-connected planar maps and the simple triangulations, explicit formulae were obtained, but were not convenient. However, simple recursion formulae were derived. These permit the determination of the number of rooted maps for all reasonably small values of n. We reproduce below the table for $A(n)$, the number of rooted simple triangulations with $2n$ triangles.

n	$A(n)$	n	$A(n)$
1	1	12	164 796
2	1	13	897 380
3	0	14	4 970 296
4	1	15	27 930 828
5	3	16	158 935 761
6	12	17	914 325 657
7	52	18	5 310 702 819
8	241	19	31 110 146 416
9	1 173	20	183 634 501 753
10	5 929	21	1 091 371 140 915
11	30 880	22	6 526 333 259 312

R. C. Mullin and P. J. Schellenberg have used an analogous procedure in enumerating the rooted 3-connected planar maps of $i+1$ vertices and $j+1$ faces (Mullin and Schellenberg [1968]). They start from the result of Brown and Tutte [1964] on the corresponding 2-connected maps.

7. Planar maps with a partition

It seems desirable to extend the scope of enumerative map theory so as to deal with rooted planar maps with a specified number of vertices of each valency. The sequence $(q(1), q(2), q(3), \ldots)$, where $q(i)$ is the number of vertices of valency i, is called the vertex-partition of such a map. So far, this extension has been made (neglecting minor complications) only for trees and Eulerian maps.

Trees are discussed in Tutte [1964]. There, the theory is carried so far as to deal with 2-coloured trees with two vertex-partitions, one for each colour. (See also De Bruijn and Morselt [1967].) A generalization in which integral weights are attached to the edges of the trees is discussed in Tutte [1964].

Eulerian maps, that is, planar maps in which the valency of each vertex is even, are dealt with, in disguised form, in Tutte [1962c]. The "slicings" of that paper can be regarded as Eulerian maps with each vertex expanded into a "boundary curve". The main result is perhaps best stated in a later paper (Tutte [1963]), as Theorem 4.2.

Applying this result with each "face of the root-colour" contracted to a vertex, we find that the number of rooted Eulerian maps in which the root-vertex has valency $2t$ and there are just q_s other vertices of valency s is

$$\frac{(n-1)!}{(k-1)!(n-k+2)!} \frac{(2t)!}{t!(t-1)!} \prod_{s=1}^{\infty} \left\{ \frac{(2s-1)!}{s!(s-1)!} \right\}^{q_s},$$

where n is the number of edges and k is the number of vertices.

The proof has some relation to the quadratic method explained in Section 4, but is more complicated.

8. Unrooted planar maps

Sometimes we want to know how many essentially different planar maps there are of a given kind, without any choice of root. Usually, two such maps are counted as combinatorially equivalent if there is a homeomorphism of the closed plane onto itself that transforms one into the other, either with preservation of all the natural cyclic sequences of edges, or with reversal of all these sequences.

We can suppose such a map M to be drawn on a sphere. To turn it into a rooted map, we choose any face as the outer face and (allowing for inversions) any directed edge incident with this face as the root. Let $r(M)$ be the number of rooted maps corresponding to M, $e(M)$ the number of edges, and $h(M)$ the order of the automorphism group of M. It is shown in [8] that

$$r(M) = 4e(M)/h(M). \tag{19}$$

This result can sometimes be used to check catalogues of unrooted planar maps. For example, P. J. Federico [1969] discussed "the number of 9-hedra". These can be regarded as the unrooted 3-connected planar maps of 9 faces.

Having listed these figures with a given number of vertices, he was able to find the corresponding number of rooted maps by using (19). He could then check this result against the tables of Mullin and Schellenberg [1968]. In a similar way, the author has checked a catalogue of unrooted simple triangulations, using the table of $A(n)$ in Section 6.

We can try to estimate the number $B(n)$ of unrooted simple triangulations with $2n$ vertices, for large n. It is a plausible conjecture that if $C(n)$ is the number of such triangulations with $h(M) \geqslant 2$, then

$$C(n)/B(n) \to 0$$

as $n \to \infty$. However, this proposition has not yet been proved. If we accept it, we can deduce from (19) that $B(n)$ is approximately $A(n)/12n$.

Similar estimates can be made for other kinds of unrooted planar map, always on the unproved assumption that symmetrical maps are rare. The problem of unrooted plane trees is exceptional; it can be dealt with by Polya's method (Harary et al. [1964]). The problem of the unrooted triangulations and quadrangulations of the disc has also been treated rigorously (Brown [1965, 1964]).

References

W. G. Brown, 1963, Enumeration of non-separable planar maps, *Canad. J. Math.* **15**, 526–545.

W. G. Brown, 1964, Enumeration of triangulations of the disk, *Proc. London Math. Soc.* **14**, 746–768.

W. G. Brown, 1965, Enumeration of quadrangular dissections of the disk, *Canad. J. Math.* **17**, 302–317.

W. G. Brown and W. T. Tutte, 1964, On the enumeration of rooted non-separable planar maps, *Canad. J. Math.* **16**, 572–577.

N. G. de Bruijn and B. J. M. Morselt, 1967, A note on plane trees, 1967, *J. Combin. Theory* **2**, 27–34.

P. J. Federico, 1969, Enumeration of polyhedra: the number of 9-hedra, *J. Combin. Theory* **7**, 155–161.

F. Harary and W. T. Tutte, 1966, On the order of the group of a planar map, *J. Combin. Theory* **1**, 394–395.

F. Harary, G. Prins and W. T. Tutte, 1964, The number of plane trees, *Proc. Koninkl. Nederl. Akad. Wetensch.* **67**, 319–329.

R. C. Mullin, 1964, Enumeration of rooted triangular maps, *Am. Math. Monthly* **71**, 1007–1010.

R. C. Mullin and P. J. Schellenberg, 1968, The enumeration of c-nets via quadrangulations, *J. Combin. Theory* **4**, 259–276.

W. T. Tutte, 1962a, A census of planar triangulations, *Canad. J. Math.* **14**, 21–38.

W. T. Tutte, 1962b, A census of Hamiltonian polygons, *Canad. J. Math.* **14**, 402–417.

W. T. Tutte, 1962c, A census of slicings, *Canad. J. Math.* **14**, 708–722.

W. T. Tutte, 1963, A census of planar maps, *Canad. J. Math.* **15**, 249–271.

W. T. Tutte, 1964, The number of planted plane trees with a given partition, *Am. Math. Monthly* **71**, 272–277.

W. T. Tutte, 1970, On the enumeration of two-coloured, rooted and weighted plane trees, *Aequationes Math.* **4**, 143–156.

J. N. Srivastava et al., eds., *A Survey of Combinatorial Theory*
© North-Holland Publishing Company, 1973

CHAPTER 38

A Construction for Room Squares

W. D. WALLIS

University of Newcastle, Newcastle, New South Wales 2308, Australia

The Room square existence problem

A Room square $\mathcal{R}$ of side n, n odd, is an $n \times n$ array based on a set R of size $n+1$. Each cell of the array either is empty or contains an unordered pair chosen from R; all $\frac{1}{2}n(n+1)$ such pairs appear exactly once in $\mathcal{R}$, and each element of R occurs exactly once in every row and every column of $\mathcal{R}$. T. G. Room introduced the squares in Room [1955], and they have subsequently been studied by various authors. (A survey of the problem, with bibliography, will be found in Stanton and Mullin [1970].)

Room [1955] noted that squares of sides 3 and 5 are impossible; it seems likely that Room squares exist of all other odd sides. It is known (Stanton and Horton [1970], Wallis [1972]) that if there are squares of sides n_1 and n_2 then there is a square of side $n_1 n_2$. A square of side n exists when $n = 1$ and whenever n is a prime power, except for $n = 3$ or 5 and possibly $n = 257$ or 65537 [(Horton *et al.* [1971], Mullin and Nemeth [1969], Stanton and Mullin [1968]). Horton [1971] has shown that there is a square of side $5p$ for every odd prime p. We shall show

Theorem 1. *If there is a Room square of side* n (> 1) *then there is a Room square of side* $3n$.

From this theorem and the other results stated above, we immediately deduce

Theorem 2. *If there exist Room squares of sides 257 and 65537, then there are Room squares of all odd sides except 3 and 5.*

Proof of Theorem 1. We assume that $\mathcal{R}$ is a given Room square of side n $(n > 1)$ based on the set $R = \{0, 1, 2, \ldots, n\}$ and that the rows and columns of $\mathcal{R}$ have been ordered so that the cell in position (i, i) contains $\{0, i\}$.

Lemma. *There is a permutation* ϕ *of* $\{1, 2, \ldots, n\}$ *such that the* $(i, i\phi)$ *cell of* $\mathcal{R}$ *is empty for every* i.

Proof. Define an $n \times n$ matrix $A = (a_{ij})$ where $a_{ij} = 1$ when the (i, j) cell

449

of $\mathscr{R}$ is empty and $a_{ij} = 0$ otherwise. Then A is a $(0, 1)$-matrix with every row-sum and column-sum equal to $k = \frac{1}{2}(n-1)$. Since $n > 1$, k is a positive integer, so by Ryser [1963], Theorem 5.5.3,

$$A = P_1 + P_2 + \cdots + P_k,$$

where the P_i are permutation matrices. If we define $i\phi$ to be the number of the column of P_1 which has 1 in row i, we have a suitable permutation ϕ.

For $1 \leqslant a,b \leqslant 3$, we define an $n \times n$ array $\mathscr{R}_{ab}$ by deleting the diagonal entries from $\mathscr{R}$ and then replacing the unordered pair $\{x, y\}$ where it occurs in $\mathscr{R}$ by $\{x_a, y_b\}$ when $x > y$. ϕ will be a permutation satisfying the conditions of the Lemma, and $\mathscr{R}_{ab}\phi$ will denote the result of carrying out the column permutation ϕ on $\mathscr{R}_{ab}$; column j of $\mathscr{R}_{ab}$ becomes column $j\phi$ of $\mathscr{R}_{ab}\phi$. Write

$$\mathscr{S}^* = \begin{array}{|c|c|c|}
\hline
\mathscr{R}_{11} & \mathscr{R}_{22}\phi & \mathscr{R}_{33}\phi \\
\hline
\mathscr{R}_{23}\phi & \mathscr{R}_{31} & \mathscr{R}_{12} \\
\hline
\mathscr{R}_{32}\phi & \mathscr{R}_{13} & \mathscr{R}_{21} \\
\hline
\end{array}.$$

If we denote $\{0, 1_1, 2_1, \ldots, n_1, 1_2, 2_2, \ldots, n_2, 1_3, 2_3, \ldots, n_3\}$ by S, then $\mathscr{S}^*$ has every unordered pair of elements from S as an entry precisely once, except that all pairs of the form $\{0, i_a\}$ and $\{i_a, i_b\}$ are missing. To construct a Room square $\mathscr{S}$ based on S, it is sufficient to put these pairs into $\mathscr{S}^*$ in such a way that for $i = 1, 2, \ldots, n$,

   (a) $0, i_1, i_2\phi^{-1}$ and $i_3\phi^{-1}$ are added to column i,
   (b) $0, i_1, i_2\phi^{-1}$ and i_3 are added to column $n+i$,
   (c) $0, i_1, i_2$ and $i_3\phi^{-1}$ are added to column $2n+i$,
   (d) $0, i_1, i_2$ and i_3 are added to rows $i, n+i$ and $2n+i$.

We observe that the first diagonal block has empty cells in positions (i, i) and $(i, i\phi)$ for every i, and the corresponding cells are empty in the other diagonal blocks. So it is possible to carry out the following procedure for $i = 1, 2, \ldots, n$: place

$$\{0, i_1\} \text{ in cell } (i, i),$$
$$\{i_2, i_3\} \text{ in cell } (i, i\phi),$$
$$\{0, i_2\} \text{ in cell } (i+n, i\phi+n),$$
$$\{i_1, i_3\} \text{ in cell } (i+n, i+n),$$
$$\{0, i_3\} \text{ in cell } (i+2n, i\phi+2n),$$
$$\{i_1, i_2\} \text{ in cell } (i+2n, i+2n).$$

This satisfies the requirements (a), (b), (c), (d).

References

J. D. Norton, 1971, Quintuplication of Room squares, *Aequationes Math.* **7**, 243–245.

J. D. Horton, R. C. Mullin and R. G. Stanton, 1971, A recursive construction for Room designs, *Aequationes Math.* **6**, 39–45.

R. C. Mullin and E. Nemeth, 1969, An existence theorem for Room squares, *Canad. Math. Bull.* **12,** 493–497.

T. G. Room, 1955, A new type of magic square, *Math. Gazette* **39,** 307.

H. J. Ryser, 1963, *Combinatorial Mathematics*, M.A.A. Carus Monograph No. **14** (Wiley, New York).

R. G. Stanton and J. D. Horton, 1970, Composition of Room squares, *Combinatorial Theory and its Applications* (P. Erdös *et al.*, eds.; North Holland, Amsterdam), pp. 1013–1021.

R. G. Stanton and R. C. Mullin, 1968, Construction of Room squares, *Ann. Math. Statist.* **39,** 1540–1548.

R. G. Stanton and R. C. Mullin, 1970, Techniques for Room squares, *Proc. Louisiana Conf. on Combinatorics, Graph Theory and Computing* (Louisiana State University, Baton Rouge), pp. 445–464.

W. D. Wallis, 1972, Duplication of Room squares, *J. Austral. Math. Soc.* **14,** 75–81.

Author Index